Silicon Carbide—1973

Silicon Carbide—1973

Proceedings of the Third International
Conference on Silicon Carbide held
at Miami Beach, Florida, on
17–20 September 1973

under the sponsorship of
AIR FORCE CAMBRIDGE RESEARCH LABORATORIES

with the cosponsorship of
Army Materials and Mechanics Research Center
Office of Naval Research
The Carborundum Company
General Electric Company
Hughes Research Laboratories
University of South Carolina
Westinghouse Electric Corporation

in cooperation with the
INTERNATIONAL COMMITTEE
ON SILICON CARBIDE

conducted by the
UNIVERSITY OF SOUTH CAROLINA

Silicon Carbide–1973

edited by
R. C. MARSHALL
J. W. FAUST, JR.
C. E. RYAN

University of South Carolina Press
Columbia, South Carolina

Published by the University of South Carolina Press, Columbia, S.C., 1974

Manufactured in the United States of America

Library of Congress Cataloging in Publication Data

International Conference on Silicon Carbide, 3d, Miami
  Beach, Fla., 1973.
  Silicon carbide—1973; proceedings.

  Under sponsorship of the Air Force Cambridge Research
Laboratories; conducted by the University of South Carolina.
  1. Silicon carbide—Congresses. I. Marshall, Robert C., ed. II. Faust, John William,
1922–   ed. III. Ryan, Charles Edward, 1938–   ed. IV. United States. Air Force.
Cambridge Research Laboratories. V. South Carolina. University. VI. Title.
QD181.S6I57 1973   546'.683'2   74–2394
ISBN 0–87249–315–6

Preface

This volume is the Proceedings of the Third International Conference on Silicon Carbide. It follows in the tradition of the first conference, which was held in Boston on 2–3 April 1959, and the second conference, which was held at Pennsylvania State University on 20–23 October 1968.

A detailed account of the history of silicon carbide research as chronicled in these Conference Proceedings is given in the lead paper, "Perspectives on Silicon Carbide," by R. C. Marshall.

The Conference was conducted by the University of South Carolina under Air Force Cambridge Research Laboratories Contract Number F 19628–72–C0033. Professor J. W. Faust, Jr., was Principal Investigator under this contract and served as Conference Director. He was assisted by day organizers for each day of the Conference. R. C. Marshall and R. W. Brander served as organizers for the Crystal Growth sessions on the first day; J. W. Faust, Jr., and W. J. Choyke for the Electrical and Physical Properties discussed on the second day; A. R. Kieffer and H. D. Batha for the Non-Electronic Applications presented on the third day, and R. B. Campbell and C. E. Ryan for the Device Techniques and Devices session on the final day of the Conference.

The papers for this conference have been received in the form of photo-ready copy. In the interests of rapid publication, the changes in the editorial process have been limited to obvious technical errors rather than concern with the niceties of sentence structure.

The editors are indebted to Roseanna Hutchins and Y. Tung for their invaluable assistance in preparing this volume.

Executive Committee

R. C. Marshall, *Air Force Cambridge Research Laboratories*, Chairman
C. E. Ryan, *Air Force Cambridge Research Laboratories*
J. W. Faust, Jr., *University of South Carolina*, Conference Director

Program Committee

R. B. Campbell, *Westinghouse Astronuclear Laboratory*, Chairman
H. D. Batha, *The Carborundum Company*
J. W. Faust, Jr., *University of South Carolina*
A. R. Kieffer, *Technical University*, Vienna, Austria
R. C. Marshall, *Air Force Cambridge Research Laboratories*

International Organizing Committee

C. E. Ryan, *Air Force Cambridge Research Laboratories*, Chairman
R. W. Brander, *Post Office Research Department*, London, England
W. J. Choyke, *Westinghouse Research Laboratories*
A. R. Kieffer, *Technical University*, Vienna, Austria
W. F. Knippenberg, *Philips Research Laboratories*, The Netherlands
P. Krishna, *Banaras Hindu University*, India
P. Margotin, *Centre de Recherches de la C.G.E.*, Marcoussis, France
W. von Muench, *Technische Universitaet*, Hanover, West Germany
H. Tanaka, *National Institute for Research in Inorganic Materials*, Ibaraki, Japan
Yu. M. Tairov, *Electrical Engineering Institute*, Leningrad, USSR

Contents

PART II. POLYTYPISM

PART III. PHYSICAL PHENOMENA

PART IV. NON-ELECTRONIC APPLICATIONS

PART V. DEVICES AND DEVICE TECHNIQUES

PART VI. APPENDIXES

PART VII. INDEXES

PART I
CRYSTAL GROWTH

Perspectives on Silicon Carbide

Robert C. Marshall

The first International Conference on SiC was held in Boston, Massachusetts, on 2-3 April 1959. This Conference was initiated by C. E. Ryan of the Air Force Cambridge Research Laboratories during the summer of 1958. The need for such a conference was amply demonstrated by the attendance of more than 500 scientists representing ten nations of the world. Forty six papers were presented covering the areas of single crystal growth, structure, imperfections, surfaces, electrical and optical characteristics, radiation effects and device techniques and devices. The Proceedings of the Conference[1] were published by Pergamon Press in 1960 as a 520 page volume. Over 1500 copies have been printed.

For almost 10 years there were no formal get togethers by scientists of the world in the area of SiC. AFCRL conducted a minimal program of In-House and contracted research to maintain an awareness of the state-of-the-art and acted as an informal information center for the exchange of information on all aspects of silicon carbide as an electronic material. In 1967, Professors Roy and Henisch of Pennsylvania State suggested a small "Working Party" of active workers in the SiC field spend a few days informally discussing problems of mutual interest. Since it was highly desirable to invite a few scientists from other countries, AFCRL was solicited to contribute financial as well as technical support. It soon became necessary to arrange a more formal Conference with invited and contributed papers, schedules, publications, etc. On October 20-23, 1968 the 2nd International Conference on SiC was held at Pennsylvania State University under the co-sponsorship of AFCRL and the Carborundum Company. Thirty four papers were presented with contributions by 58 authors. The attendance at this Conference was approximately 125 from seven nations. The Proceedings of the Conference were

Feitknecht	Brown Boveri	Margotin	GE, France
Kamath	Norton Company	Marsh	Hughes Research Labs.
Kieffer	Univ. of Vienna		

A list of meeting places, special tours and special materials discussed is given here:

New York City	Tour/Bell Labs Murray Hill	Silicon,GaP
Vienna, Austria	Tour/Prof. Kieffer's Lab	other carbides
Miami Beach, Florida	AF Materials Symposium	
Neuremburg, Germany	Seimans Labs.	III-V
Phoenix, Arizona	Motorola Laboratories	Silicon
Grenoble, France	CRNS	II-VI
Columbia, S. C.	Univ. of South Carolina	GaN
Amsterdam, the Netherlands	Phillips Laboratories	Liquid crystals

Financial support of the committee was provided jointly by the Air Force and each members organization. The Air Force provided its support through a contract originally with Pennsylvania State University, subsequently then the University of South Carolina[3]. This contract authorized administrative support for arranging and conducting the meetings with Prof. J. W. Faust, Jr. as Chief Investigator. The individual member companies paid all members expenses with the exception of travel which was paid through the contract.

The format of the meetings included formal and informal presentations and discussions of mostly unpublished information and data of research efforts on crystal growth, measurement and analysis techniques, junction formation, device techniques, electrical evaluation, and any other information that would assist the research and development in this area.

Since January 1969, the committee has met nine times with major in-depth discussions of one technical area such as crystal growth, and measurement techniques. These meetings were extremely helpful in assisting ongoing programs and a place where technical problems could be discussed with world experts and hopefully, solutions found or new avenues of approach suggested.

The results of these meetings were compiled in a series of "Minutes" which were provided each member of the Committee. These volumes are over 1500 pages and do not include correspondence and data between individual members or administrative matters. We at AFCRL feel that these meetings are extremely rewarding and informative and hope that members from other organizations have a similar feeling.

At the Amsterdam meeting in September 1972 the Committee decided that a third International Conference should be scheduled. The intent of the Conference would be to provide a timely and accurate assessment of the current status of silicon carbide as a material of technological value because of its unique properties. Many advances have been made since the last Conference in 1968. This Conference we feel will provide the basis for scientific, technical and managerial evaluations and decisions on the directions of future research and the most propitious application for development.

For the Air Force, I wish to thank all the members of the International Committee on Silicon Carbide and the guest scientist at the various meetings for their enthusiastic and cooperative attitude at these meetings. Without the full and free exchange of information and data on their ideas and research efforts the Committee would not have been such a success.

In closing I think it is important to this Conference for me to spend a few minutes listing some of the new and hopeful trends that have happened at AFCRL since the last Conference. As I previously stated, most of the research in this area is subthreshold - not only at AFCRL but in most of the organizations working on silicon carbide. We feel however that the progress in this material has been positive the last few years and I would like to indicate some of the results we have had in the past or will report at this Conference.

Since the 1968 Conference, scientists at AFCRL have published 25 papers on silicon carbide and will present eight more papers at this Conference. Early fundamental research in our laboratories with 2H SiC needles provided useful information on polytype conversion without a change in crystal morphology. This new information led to the investigation of thin films deposited by CVD and cathodic sputtering on both silicon carbide and selected foreign substrates. Films deposited by these processes could be converted from amorphous and poly- crystalline to single crystal by a controlled thermal annealing technique. In our investigations, beta SiC thin films were stable at least up to 2000°C for 8 hours which we have attributed to better structure and chemical perfection of the films. The trend in CVD from methyltrichlosilane to silane enabled growth of SiC thin films at substrate temperatures of 1000°C or less. This opens up the possibility of a wide variety of substrate materials for thin film deposition.

published as a special issue of the Materials Research Bulletin[2] in 1969. Over 800 copies have been printed.

During the Conference, discussion with attending scientists revealed the fact that many felt that a small working group meeting semiannually for technical review and exchange of data and specimens would be highly desirable, especially since much of the work on SiC was being accomplished at a sub-threshold level. It was felt that discussing each others problems and exchanging data might advance the SiC state-of-the-art at a more rapid rate.

AFCRL, because of its interest, history and program in SiC, was the logical organization to initiate such a group.

Professor J. W. Faust, Jr., of Pennsylvania State University, who had successfully organized and run the Second International Conference was contacted by Messrs. Ryan and Marshall/AFCRL and several discussions were held on the subject. It was decided to contact all the major companies in the United States and Europe who had an ongoing program in silicon carbide to determine if there was a real interest in such a committee and if the organizations would approve and support one of their scientists to be a member of the committee.

The results of these visits and discussions indicated complete approval of such a committee and in January 1969 the International Committee on SiC was formed with C. E. Ryan and J. W. Faust, Jr., as co-chairmen, and R. C. Marshall as Executive Secretary.

The organizational meeting of ICSiC was held at the Commodore Hotel, New York City, March 1969 prior to the Electrochemical Society Symposium on Silicon Technology. During the meeting the purpose and objectives of the committee were established, memberships confirmed and financial arrangements discussed. Although the main concern of the committee was SiC it was decided that at each meeting a short time would be spent on another major wide band semiconductor material and if possible and appropriate, a guest speaker would be invited to discuss the subject. As far as possible each meeting was held in conjunction with another major society meeting in the United States and Europe.

A list of the members of International Committee on Silicon Carbide is as follows:

Ryan	AFCRL	Schaffer	The Carborundum Company
Faust	Univ. of S. Carolina	Brander	GEC, England
Blank	General Elec Co., USA	Knippenburg	Phillips, Netherlands
Campbell	Westinghouse Elec Co.	Marshall	AFCRL

One of the more important of these materials is silicon, and we have reported on
the successful deposition of beta SiC (single crystal) films on silicon substrates.

Preliminary investigations are also being conducted on liquid epitaxial
techniques. Reports will be given by AFCRL scientists in this area both today
and on Thursday. Liquid epitaxy holds a potential for making reproducable
devices in controlled environments. In the crystal growth area, the use of high
purity elemental silicon and carbon in place of commercial SiC grit could well
lead the way to silicon carbide single crystals with controlled impurity levels.
The systematic exploration of crystal growth in the high temperature regions
of 2800°C plus at pressure up to 50 atmospheres and the high temperature
annealing of dense polycrystalline structures is just beginning to bring interesting results. Some of these experiments will be reported during the Conference.

We see SiC as a material with tremendous potential[4] for electronic and
optical applications on the basis of its dielectric constant, thermal conductibility, saturation velocity and mechanical characteristics. The material
problems are indeed formidable but new ideas, some of which I have mentioned are
being generated and are being investigated. I am sure as this Conference
progresses many of these and other ideas will instill enthusiasm in all attendees.
At its close, we will all be in a better position to evaluate the true merits
of this material and to determine the proper direction for future research.

REFERENCES

1.   J. R. O'Connor, J. Smiltens, ed., Silicon Carbide, A High Temperature
        Semiconductor, Pergamon Press (1960).

2.   H. K. Henisch, R. Roy, ed., Mat. Res. bul., Special Issue (1969).

3.   Air Force Contract F19628-71-C-0033, University of South Carolina

4.   C. E. Ryan, Mat. Res. Bull. 4, 1-12, (1969).

Epitaxial Growth of SiC Layers

R. W. Brander

The epitaxial growth of silicon carbide is reviewed and the improvements in
control which are obtainable compared with single crystal growth methods are
indicated. The merits of vapour phase and solution epitaxy are discussed in
terms of the limitations imposed by the system design and by the substrate
material.

Silicon carbide offers the device manufacturer many useful features which are
not available from other semiconductor materials; in particular, its refractory
nature, inertness and high energy gap allow high power dissipation and superior
reliability. In order to utilise these features to the full the material
quality must at least approach that available with other semiconductors. These
advantageous properties,however, introduce difficulties in fabrication which
result in the quality of single crystal material being extremely difficult to
control. Epitaxial techniques, by reducing the growth temperature and providing
a more controlled growth ambient, are capable of producing material of superior
quality and reproducibility.

VAPOUR PHASE DEPOSITION ON SiC SUBSTRATES

The deposition of SiC by thermal dissociation or by hydrogen reduction of
vapour species is a logical extension of the well characterised and commercially
viable epitaxial growth of silicon. Unfortunately SiC, being a much more inert
material and exhibiting considerably stronger binding forces, is a much less
mobile species than Si and therefore requires higher growth temperatures.
Although these temperatures are not as high for epitaxial growth as they are for
single crystal growth, they still give rise to the major problems associated
with the preparation of good quality layers.

8

In basic structure SiC epitaxial reactors differ little from those used in silicon epitaxy and consist of a purified gas supply, sources of silicon and carbon containing vapours, metering valves, and a reaction chamber containing a heated substrate support. Fig 1 shows a typical schematic and indicates the type of sources used, both liquid and gaseous, each having its own particular advantage. Initially only liquid sources were available with the purity necessary for crystal growth but recently gaseous sources of sufficient purity for most purposes have become commercially available and are becoming increasingly popular since they simplify system construction and avoid the introduction of chlorine into the system thus reducing the etching of the substrate and susceptor. Hydrogen, purified by diffusion through palladium-silver cells, is generally used as the carrier gas as this is necessary for the reduction of many of the sources, however, some sources, such as silane (SiH_4) and dimethyldichlorosilane $[(CH_3)_2 SiCl_2]$, can be thermally decomposed and can therefore be employed with argon carrier gas [1]. The elimination of hydrogen from the system results in an additional reduction in the etch rate of the substrate and susceptor. Typical of the combinations employed are CCl_4 and $SiCl_4$ [2, 3], $HSiCl_3$ and C_6H_{14} [4], SiH_4 and C_3H_8 [5, 6], and $SiCl_4$ and C_6H_{14} [7], all in hydrogen. In addition the sublimation of SiC in an Al containing atmosphere has been investigated [8]. In the above cases α-SiC layers were grown on α-SiC substrates. The growth of β-SiC layers on β-SiC substrates has been studied, using SiH_4 and C_3H_8 or $CH_3 SiCl_3$ in hydrogen [9].

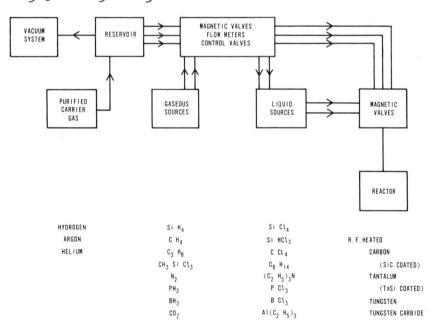

HYDROGEN	$Si\,H_4$	$Si\,Cl_4$	
ARGON	$C\,H_4$	$Si\,HCl_3$	R.F.HEATED
HELIUM	$C_3\,H_8$	$C\,Cl_4$	CARBON
	$CH_3\,Si\,Cl_3$	$C_6\,H_{14}$	(SiC COATED)
	N_2	$(C_2\,H_5)_3N$	TANTALUM
	PH_3	$P\,Cl_3$	(TaSi COATED)
	BH_3	$B\,Cl_3$	TUNGSTEN
	CO_2	$Al(C_2\,H_5)_3$	TUNGSTEN CARBIDE

Fig 1   Schematic diagram of vapour epitaxial growth systems

The reaction in the vicinity of the heated substrate will vary according to the source species and ambient gas. Since SiC is inherently stoichiometric a one to one ratio will be desired and any significant excess of silicon or carbon will be contained as inclusions. The mole ratios in the gas phase will differ from this unity ratio due to the different reaction kinetics of the two species and the relative vapour pressures of the silicon and carbon species including any hydrides and chlorides which become established in any particular system. The source concentrations, therefore, are generally determined empirically since theoretical predictions are difficult to make. Harris et al [10] have, however, analysed the equilibrium partial pressures in the system hydrogen-silicon-carbon and have concluded that the efficiency of SiC formation decreases with increasing temperature and increasing silicon to carbon ratio but that a carbon excess can be readily tolerated due to the buffering action of the many carbon species capable of existing in the vapour phase at significant partial pressures. These results cannot readily be extrapolated to other systems since they do not take into account the diffusion of the different species through the boundary layer or the preferential deposition of the different species in other parts of the reactor. The addition of other constituents to the reactor, such as chlorine, will also considerably influence the equilibrium, as will different susceptor materials. In practice silicon to carbon mole ratios of 1:3 to 5:1 have been successfully used in different systems.

It has been demonstrated experimentally that at high total mole ratios in hydrogen the balance of the two species becomes more critical and closer to unity if good single crystal growth is to be obtained. This is particularly so above 10^{-2} mole ratio in hydrogen which results in high growth rates and polycrystalline deposits. Below about 10^{-3} mole ratio in hydrogen a relatively broad tolerance in silicon to carbon ratio exists and single crystal growth is more easily obtained. The net growth rate will be the difference between the deposition rate and the etch rate and at these low concentrations is generally in the range 0.2 - 1.0 μm min^{-1} depending on the growth temperature which is typically 1500 - 1800°C. Above 1800°C hydrogen etching is the dominant process. Smoother layers are generally obtained at the lower growth rates.

The quality and rate of growth has been shown to vary according to which (0001) surface is used as the substrate but not with any consistency. Under suitable conditions layers of good crystalline quality can be grown on either face but the relative rates may differ by up to a factor of two in either direction depending on the growth system employed. The net effect is obviously complicated and will be a balance between the preferential etching rate of hydrogen on the carbon face and any preferential deposition rate which may exist due to the relative activities of the two surfaces. Since the relative growth rates will be affected by the preferential adsorption of trace impurities and possibly even by the ratio of silicon to carbon at the growth interface, it is not unreasonable to expect the differences in the results reported by the various authors.

At the temperatures employed for growth, the susceptor material plays an important part in the reaction kinetics and can be either a major source of growth constituent, an etchant vapour or a contaminant. Graphite susceptors are an obvious choice for the growth of SiC but they do contribute significantly to the partial pressure of carbon in the system, particularly in the presence of hydrogen. In a steady state situation an allowance for this additional pressure could be made and the silicon to carbon ratio adjusted accordingly; however, the close proximity of the 'source' and substrate will encourage uneven growth and the build up of polycrystalline material at the edge of the crystal. In addition since SiC is being continuously deposited on the susceptor the reactivity of the surface is continuously changing with a resultant fluctuation in carbon vapour pressure due to the repeated cracking and re-sealing of the polycrystalline SiC layer. The effect seems to be most pronounced in vertical reactor systems where the substrate is supported horizontally on top of the susceptor but can be minimized in systems where the gaseous ambient flows over the substrate and susceptor as in vertical systems with steeply sloping crystal supports [7] or in horizontal systems. Many alternative susceptor materials and covering materials for graphite susceptors have been investigated. Covering materials introduce considerable temperature drops and require the use of higher susceptor temperatures. In addition most refractory oxide materials are attacked by hydrogen at elevated temperature and cannot be used. Refractory metals and their associated carbides and silicides offer considerable possibilities and most can be readily coupled to induction heating systems. Typical of the materials used are tungsten, tungsten carbide [4], tantalum [5, 6], and tantalum silicide coated tantalum [3]. The maximum temperature of operation of the metal susceptor is limited by eutectic formation with the deposited Si and C and their useful life is very limited when used at above $1700^{\circ}C$. An advantage of TaSi is its inertness to hydrogen. Tungsten carbide can be used at temperatures, up to

2000°C but is significantly attacked by hydrogen at this temperature and even contributes a noticeable quantity of carbon to the growing ambient at 1600°C.

A disadvantage of all susceptor materials is their tendency to introduce contamination into the growing layers. Graphite and metallic materials can be obtained in relatively pure form but bonded refractories tend to be impure and many contain binding agents; self bonded materials must be used. It has also been shown that the metallic materials are incorporated in the layers to a small extent, about 10^{16} atoms cm^{-3}, and that these can significantly affect the luminescent properties of the materials [4]. One technique which has been proposed to eliminate the effect of the susceptor is the direct radiant heating of the SiC substrate [11]. Some support for the crystal still has to be provided but this can be of considerably smaller mass than that required for a susceptor and can be small SiC polycrystalline rods of high purity. Additionally these supports will be at a lower temperature than the growing surface whereas in the RF heated system the converse applies. The difficulty with radiant systems is the high losses encountered in heating a significant crystal area to above 1600°C and the difficulty of achieving uniform temperature over the entire crystal surface.

The achievement of good crystal quality in the epitaxial layer depends as critically on the preparation of the substrate as it does on the growth environment. SiC substrates are generally prepared by the Lely process and frequently have flat surfaces which could be used for epitaxy. These surfaces, however, have experienced highly supersaturated environments during cooling and as a result contain small spurious nuclei which give rise to polycrystalline inclusions during subsequent growth. These inclusions must be removed from the surface prior to growth and either mechanical polishing or etching techniques are employed. Mechanical polishing alone is not good enough as residual minor scratches and strain result in non-uniform growth. Chemical etching techniques are better in this respect since no damage is produced. However, it is difficult to ensure a perfectly flat surface without the presence of some etch pits; in addition only one crystallographic face can be employed due to the asymmetry of most etches [12]. Whichever technique or combination of techniques is employed the substrate is always given a gaseous etch immediately prior to growth. Since hydrogen is employed as the carrier gas in most systems this is conveniently used as the etchant at 1600°C or above to remove up to $10\,\mu$m of material. The actual etch rate is influenced by the system design and by the susceptor

material [13]. Growth is initiated immediately subsequent to etching by introducing the reactants into the gas stream; the silicon source being activated
first in order to prevent carbon deposition on the crystal surface.

The quality of the silicon carbide layers is best checked by X-ray topography
which gives a picture of the crystal at each Bragg reflection and indicates
defects, twinning and polytype changes [14]. In conjunction with optical
microscopy this can be used to characterise many of the defects which occur in
epitaxial growth. Good quality layers frequently exhibit a lower surface defect
density, as observed by etch pit counts, than the original substrate and have
smooth almost featureless surfaces of a single polytype over the majority of the
substrate area. If the substrate surface was originally of a single polytype
after etching then the epitaxial layer will usually be of the same polytype.
Spurious random nucleations can occur at the edge of the crystal and can spread
inwards, this can be very prevalent on carbon susceptors and can result in
polycrystalline growths. Changes in polytype can occur, a typical example being
when β-SiC is nucleated and spreads over the entire crystal. In general this
material is severely twinned but almost completely single crystal layers
(plate 1) have been grown; the small spots of opposite contrast to the main
crystal reflection represent small twinned inclusions. The β-SiC regions have
been shown to nucleate at points and to spread over the surface of the crystal
as growth proceeds. Plate 2 shows some poorly developed β-SiC regions nucleated
in a 6H layer. In this case the β-regions grew through a p-n junction formed in
the epitaxial layer and their shape has been revealed by observing the forward
and reverse biased electroluminescence. Plate 2c shows the shape of the inclusion. The propagating mechanism for the growth of this inclusion has not been
clearly established since the direction of growth could not be linked consistently with either substrate orientation or temperature gradient. It may be
surmised, however, that the initiation of these β growths is due to particle
formation at the growth interface. The growths are particularly prevalent when
the silicon to carbon ratio deviates from the ideal value and may be related to
the nucleation of silicon or carbon at a defect or irregularity on the surface.
Impurities also have a significant effect on the growth quality and it is
essential to make sure these are eliminated from the system. Oxygen is particularly disadvantageous in this respect and most systems are evacuated and carefully leak tested in order to minimize its introduction. Dust and other foreign
particles have a much more severe effect and introduce inclusion of β-SiC of

an extremely coarse polycrystalline nature. When growth conditions deviate substantially from the optimum values, but before polycrystalline growth sets in, surfaces with noticeable boundaries are obtained. This is a strong indication of the presence of multiple polytypes and it can be shown that the visible boundaries correspond to the edges of the different polytypes (Plate 3).

In an attempt to obtain greater control over the growth of β-SiC layers on α-SiC substrates thin layers of metal have been introduced between the interface and the vapour species [15]. This is a combination of solution growth with a vapour source and is frequently termed the vapour-liquid-solid technique. An important feature of this method is the substantial reduction in the temperature at which well ordered growth will occur. Single crystal β-layers with (111) twinning were obtained reproducibly at 1250°C with nickel and gold intermediates whereas in the absence of the metal only polycrystalline material could be grown at that temperature. The effect of annealing on the quality of these layers is discussed in a following paper.

VAPOUR DEPOSITION ON FOREIGN SUBSTRATES

Foreign substrates offer the best hope of overcoming the size limitations imposed by bulk SiC crystals and may also enable β-SiC to be grown more easily by removing the influence of the α-SiC substrate. The most obvious choice of substrate from the viewpoints of chemical compatibility, availability and size is silicon. This choice, however, introduces problems of expansion coefficient and lattice constant miss-matches, and imposes a limitation on the maximum temperature for growth nucleation of about 1400°C resulting in a low surface mobility for SiC nuclei. SiC has been grown on silicon slices by numerous techniques: reaction of hydrocarbons with single crystal surfaces [1, 16, 17, 18]; reactive sputtering or evaporation of silicon in a hydrocarbon environment [16, 19]; reduction of silicon hydrocarbons or silicon and carbon compounds [1, 18, 20] and by direct evaporation [21].

When single crystal silicon is heated to temperatures in excess of 900°C in an hydrocarbon environment, the hydrocarbon is reacted at the silicon surface to form SiC. Either direct reaction or pyrolysis could take place depending on the relative supplies of silicon and carbon species at the reacting surface. In some experiments [16] no trace of free carbon could be observed on the surface indicating that pyrolysis was not occurring, whereas in other reactions [1] excess carbon, which gives rise to random orientation, has been observed. The

SiC nucleates initially at dislocations [1] and is generally of the β-polytype. The film grows to provide continuous coverage of the substrate but is never completely single crystal although grain sizes up to a fraction of a millimeter have been obtained. The material is, however, epitaxial with varying degrees of misorientation and twinning depending on the substrate orientation. The growth mechanism has been investigated in detail and it has been shown conclusively by ^{14}C experiments [17] that the predominant mechanism is silicon diffusion through defects in the growing SiC layer to the outer reacting surface. The rate of supply of silicon will vary with the quality and grain size of the SiC and this latter will be controlled by the relative amounts of Si and hydrocarbon at the surface, low hydrocarbon pressures will allow more time for Si surface diffusion before SiC formation and will result in the growth of larger grains [18]. Although layers of many microns thickness can be grown, the diffusion process limits the thickness of good quality material to a fraction of a micron. The mechanism is discussed further by Mogab and Leamy in following papers.

To obtain thicker films it is necessary to provide an additional source of Si and this can be achieved either by evaporation or sputtering of silicon in low partial pressures of hydrocarbons or by supplying silicon in the vapour phase at atmospheric pressure as previously discussed. Layers nucleated directly in Si and C containing environments have shown similar properties to those produced by carbonization, namely epitaxial growth with a preferred orientation and considerable twinning. Rai-Choudhury [1] observed that dimethyldichlorosilane $(CH_3)_2 SiCl_2$ gives the best preferred orientation on (100) Si with multiple twinning occurring when (110) and (111) silicon substrates were used. This latter compound deposits excess silicon if used in hydrogen at 1200°C and argon was therefore used as the carrier gas. Jacobson [20] did not encounter this problem at 1365°C using the same susceptor material, SiC coated graphite, although he did find that he required very pure materials in order to obtain well oriented growth. In order to ensure that the prepared silicon surface did not have an effect on the nucleation he initially grew an epitaxial silicon layer from $SiCl_4$ before introducing the organic compound. Oriented growths were obtained from 5×10^{-3} - 2×10^{-2} mole ratio of $(CH_3)_2 SiCl_2$ in hydrogen at the relatively high growth rates of $4 \mu m \, min^{-1}$ compared with those used for single crystal growth on SiC substrates at higher temperatures. The degree of preferred orientation increased as the film thickness increased to reach an optimum at $12 \mu m$, further increase in thickness caused deterioration and no oriented growth was evident at film thicknesses greater than 86 μm.

Kuroiwa and Sugano [18] used $SiCl_4$ and CH_4 at 10^{-3} mole ratio in hydrogen at 1300-1360°C to thicken up β-SiC layers nucleated by carbonization. They grew layers up to 40μm thick and in contrast to the above found no improvement in quality with increasing thickness and a significant deterioration in quality if the mole fraction was increased into the range used by Jacobson.

Reactive sputtering or evaporation of Si in methane and acetylene have been used to grow films in the temperature range 500-1200°C [16, 19]. Impinging rate ratios of greater than 30:1 C_2H_2 to Si were necessary to leave no unreacted Si on the surface at 1100°C. The resultant films although grown at a much slower rate are substantially similar to those grown by the other techniques except that below 950°C Matsumoto et al [19] succeeded in depositing 2H material with some stacking disorder. Below 900°C the films exhibited random stacking.

In all cases the layers grown on silicon substrates were crystalline and epitaxial and exhibited varying degrees of preferred orientation.

Plane of Si Substrate	Plane of SiC Layer	Orientation		Ref.
(001)	(001)β	(100) β ||	(100) Si	
	(110)β	(001) β || (1$\bar{1}$0) β ||	($\bar{1}$10) Si ($\bar{1}$10) Si	(20)
(110)	(110)β	(1$\bar{1}$1) β ||	(1$\bar{1}$1) Si	(16)
(111)	(111)β	(1$\bar{1}$0) β || ($\bar{1}$10) β ||	(1$\bar{1}$0) Si (1$\bar{1}$0) Si	(16)
(111)	(0001) 2H	(11$\bar{2}$0) 2H |	(1$\bar{1}$0) Si	(19)

Table 1 Epitaxial orientations of SiC deposited on Si substrates.

The orientations observed in the best quality films are summarised in table 1. Slight angular misorientation of the grains of up to one or two degrees is frequently observed and occasional grains may be completely misorientated. The tendency for twinning to occur generally results in an uneven surface appearance although if the grain size is sufficiently small, smooth almost featureless layers result.

The deposition of layers on substrates other than silicon is likewise predominated by randomised nucleation with a tendency towards some preferred orientation as the temperature is raised above 900°C. Onuma [21] investigated the electron beam evaporation of α-SiC onto mica and obtained α-SiC deposits. Learn and Hag [16] grew β-SiC layers on molybdenum by evaporating and d.c. sputtering silicon in an acetylene ambient. Deposition onto the alumina substrates has been investigated by Rai-Choudhury [1] using various reactions. Etching of the Al_2O_3 substrate during deposition proved to be a major problem and only when using $(CH_3)_2 SiCl_2$ in argon were deposits obtained. The material was very poor quality polycrystalline β-SiC.

SOLUTION GROWTH

Surprisingly little work has been undertaken on the epitaxial growth of silicon carbide from solution despite the ready availability of silicon and many metals as solvents and of graphite as a container. This is particularly so in view of the considerable effort which was devoted to the growth of freely nucleated SiC from solution and the commercial success of solution growth for the III-V compounds. The reluctance to use silicon no doubt stems from the difficulties imposed by the very low solubility of SiC in silicon, less than 10^{-2} atomic %, at temperatures which would be regarded as reasonable, say, below 2000°C. Metals with their much higher solubilities for SiC have been resorted to in a few instances but in general the technique has not been developed to its full potential. The prime advantage of solution growth is the likely gettering effect of a solution on many of the impurities which can not be eliminated from other techniques. Additionally it is possible to dope melts with large percentages of elements which are difficult to incorporate in crystals grown from the vapour phase or are just difficult to obtain in a convenient volatile form. A general disadvantage of the technique is the possible inclusion of the solvent into the layer thus limiting the choice to materials with low solid solubilities in silicon carbide. The substrates used for solution growth have usually been α-SiC crystals prepared by similar techniques to those discussed. The final etching stage is usually carried out in the solution by reversing the temperature gradient so that the solution round the crystal is unsaturated and dissolves a few microns of material before growth commences.

Silicon carbide epitaxial layers have been grown from bulk solutions [22, 23] and from thin solvent zones [24, 25, 26]. Knippenberg and Varspui [22] grew epitaxial layers by floating their substrates on a chromium melt contained in a silicon carbide crucible, Fig 2a. The melt was held at a constant temperature and the helium ambient pressure reduced to allow chromium evaporation and super-saturation of the melt. Epitaxial layers of the same polytype as the α-SiC substrates were grown although polytype changes, in one example from 15R to 4H, did occasionally occur during growth. The growth of β-SiC on substrates with cubic overgrowths could only be obtained when silicon was included in the melt. Chromium was found to be incorporated in the layers as an acceptor with an activation energy of 0.25eV at concentrations up to 10^{19} cm^{-3}.

In order to avoid the potential contamination of a third constituent Brander and Sutton [23] employed silicon as the solvent and overcame the problem of low solubility by setting up a temperature gradient in the melt such that carbon was dissolved from the crucible walls which were maintained at a higher temperature than the seed crystals ($T_1 > T_2$) (Fig 2a). At a temperature of 1650°C growth rates of a few microns per minute could be obtained although the best quality layers were grown at 0.5μm min^{-1} and were of the same polytype as the substrate surface. The versatility of the technique for investigating the effect of different dopants was also demonstrated, numerous materials being added to the

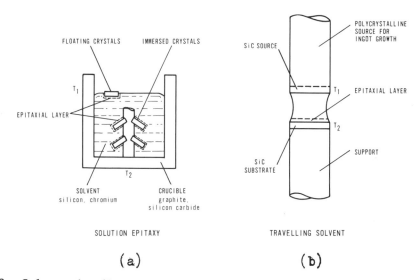

Fig 2   Schematic diagrams of solution epitaxy techniques   a) the immersion and floating techniques   b) the thin zone technique.

melt and their effect on luminescence spectra evaluated. Apart from the standard dopants, iron, copper, titanium, beryllium and oxygen were investigated. In all cases a substantial quantity of material had to be added to the melt for its presence in the layer to be significant, indicating segregation coefficients considerably less than unity and confirming the gettering action of the melt. In the case of nitrogen, however, the segregation coefficient was greater than unity and very small traces in the melt gave rise to quite high background n-type conductivities.

In the thin zone technique the solvent is contained between two SiC crystals, one acting as the substrate and the other as the source, Fig 2b. The driving force for growth is a temperature gradient in the zone between the two crystals $(T_1 \neq T_2)$, the source crystal being maintained at a higher temperature than the substrate and thus undergoing dissolution. Diffusion of the solute takes place down the concentration gradient and the liquid at the cooler crystal becomes supersaturated giving rise to deposition. This diffusion of the solute is believed to be the rate controlling step in the growth process. The quantity of solvent employed is much smaller than in the total immersion technique and any significant evaporation cannot therefore be tolerated. This additionally restricts the choice of zone material to one with a low vapour pressure at its melting point and effectively rules out the use of silicon unless some technique to prevent its transport to the cool reactor surfaces, as will be discussed in a later paper by von Munch, can be provided. Chromium has been the most widely used solvent in this technique [24, 25, 26] although rare earth solvents [27] have been employed. To obtain satisfactory uniform growth it is necessary to maintain a uniform temperature across the substrate and to have a zone which is sufficiently thin to avoid convection currents and r.f. stirring spoiling the temperature gradient. Most workers have placed their sandwich on heated supports, frequently RF susceptors, and have relied on radiation to cool the upper crystal so that this latter acts as the substrate $(T_1 < T_2)$. Although layers can be grown by this technique, generally termed the travelling solvent method, the resultant temperature gradient across the zone tends to cause solidification at the edges with the resultant inclusion of voids in the regrown layer. Greater success has been obtained using edge heating and in addition this allows the lower crystal to be cooled and used as the substrate, so minimising the effect of convection currents $(T_1 > T_2)$. Certain configurations of this technique are referred to as the travelling heater method.

The major difficulty which has been encountered in putting the technique into practice has been the achievement of reproducible wetting of the SiC interfaces. With chromium zones it is generally necessary to deposit chromium by evaporation onto crystals heated to $1300^\circ C$ so that alloying takes place prior to building the zone structure. Zones > 40μm thick are necessary to dissolve residual contamination before any SiC can be transported [26]. Thin zones have a tendency to break-up and thick zones tend to be susceptible to convection currents; hence an optimum thickness must be determined. For chromium zones thicknesses in the range 100-200 μm are generally employed. The wetting of the zone material to the SiC can be improved by adding a more reactive material to the system either as a gas e.g. hydrogen or as a constituent in the zone e.g. Si and Ta [25]. The material grown by this technique appears to suffer considerably from the incorporation of defects particularly at the nucleating interface. This may be due to the incorporation of solvent into the layer in quantities suffic- ient to affect the expansion coefficients but is more likely to be due to the relatively fast growth rates employed, up to 10μm min^{-1}. The maintenance of single crystal growth over the whole surface has also been a problem with this technique and although it has the potential for the growth of large single crystal ingots, twinning and polytype changes generally occur after a few microns have been grown.

An adaptation of the solution growth techniques has been used by Kamath (private communication) to grow thin layers of α-SiC on α-SiC substrates. In this method the substrate is supported on a graphite pedestal located in the centre of a well containing silicon. The crystal is heated to 2000°C and the silicon migrates up the graphite pedestal and forms a layer between the crystal and the carbon. SiC is transported across this layer and deposited on the crystal which is at a lower temperature due to heat lost by radiation. The method has many similarities to the thin zone technique except that the solution is continuously being renewed.

CONCLUSIONS

Epitaxial growth techniques allow silicon carbide layers to be produced at much lower temperatures than can be utilised for single crystal growth. As a result the material is more perfect, suffers rarely from fluctuations in poly- type and can be purer and more controllably doped. Since growth can be carried out substantially below the diffusion temperature for electrically active dopants, abrupt p-n junctions can be readily fabricated.

A number of problems, however, remain and further detailed investigation is required. In particular the ideal susceptor for vapour growth is still elusive metal susceptors while giving the best control over growth conditions and the

best quality material contaminate the crystals sufficiently to interfere with
the luminescent properties; coated graphite susceptors with their inherent
problem of carbon vapour species control still offer the best hope of obtaining
pure material in conventional systems. The possibility of using a direct
substrate heating technique, such as radiant heating, although technically more
difficult, holds out some promise of achieving both well controlled growth and
uncontaminated material. The choice of source material for vapour epitaxy seems
to be one of individual preference, although the results indicate that more
control over growth can be exercised if separate silicon and carbon sources are
employed. Irrespective of the source, however, there is an extreme lack of
information regarding the control of the reaction and the effect of mole ratios
and temperature on growth. More complete understanding of the reaction
processes is required before the technology can be significantly advanced beyond
the present empirical level which has resulted in many apparent contradictions
being reported. Solution growth is a technique which is undoubtedly worthwhile
developing in view of the much purer material which can be produced. In the
techniques used so far the control has been less precise than in the vapour phase
reaction and some new or modified technologies are desirable. Silicon in
graphite or silicon carbide containers is the obvious choice and the only one
which is likely to lead to contamination free growth. Since the solubility of
carbon in silicon is low this implies the use of a temperature gradient
technique.

The supply of suitable high quality substrates for epitaxial growth is the
major problem to be faced if production of epitaxially grown devices is
envisaged. Crystal platelets grown by sublimation are small and far from
perfect. The possibility of growing thick epitaxial layers and ultimately ingots
is hampered by polytype fluctuations which occur as growth times are prolonged.
Only a thorough investigation of the causes of these changes, whether it be
temperature fluctuations or contamination, will allow them to be understood and
controlled. The use of foreign substrates for silicon carbide epitaxy has not
so far helped overcome any of the problems and the material is still of inferior
quality to that produced on α-SiC substrates. A major problem on both substrates
is the very ready twinning of β-SiC, the feasibility of growing near untwinned
material, however, has been demonstrated and further work should result in its
control at least at the higher temperatures which can be attained with SiC
substrates.

REFERENCES

1   P Rai-Choudhury, N P Formigoni; J Electrochem Soc 116 p 1440-3 (1969)

2   V J Jennings, A Sommer, H C Chang; J Electrochem Soc 113 p 728-31 (1966)

3   R B Campbell, T C Chu; J Electrochem Soc 113 p 825-28 (1966)

4   A Todkill, R W Brander; Mats Res Bull. 4 p S293-302 (1969)

5   S Minagawa, H C Gatos; Jap J Appl Phys 10 p 1680-90 (1971)

6   J M Harris, H C Gatos, A F Witt; J Electrochem Soc 118 p 335-8 (1971)

7   G Gramberg, M Koniger; Solid-state Electronics 15 p 285-92 (1972)

8   S Yamada, M Kumagawa; J Cryst Growth 9 p 309-13 (1971)

9   R W Bartlett, R A Mueller; Mat Res Bull. 4 S341-53 (1969)

10 J M Harris, H C Gatos, A F Witt; J Electrochem Soc 118 p 338-340 (1971)

11 M P Callaghan, R W Brander; J Crystal Growth 13/14 p 397-401 (1972)

12 R W Brander, A L Boughey; Brit J Appl Phys 18 p 905-912 (1967)

13 J M Harris, H C Gatos, A F Witt; J Electrochem Soc 116 p 380-383 (1969)

14 B J Isherwood, C A Wallace; J Appl Cryst 1 p 145-153 (1968)

15 I Berman, J J Comer; Mat Res Bull 4 S107-118 (1969)

16 A J Learn, K E Hag; J Appl Phys 40 p 430-1 (1969), Appl Phys Letters 17
        p 26-29 (1970)
    A J Learn, I H Khan; Thin Solid Films 5 p 145-155 (1970)

17 J Graul, E Wagner; Appl Phys Letters 21 p 67-9 (1972)

18 K Kuroiwa, T Sugano; J Electrochem Soc 120 p 138-40 (1973)

19 S Matsumoto, H Suzuki, R Ueda; Jap J Appl Phys 11 p 607-8 (1972)

20 K A Jacobson; J Electrochem Soc 118 p 1001-6 (1971)

21 Y Onuma; Jap J Appl Phys 8 p 401 (1969)

22 W F Knippenberg, G Verspui; Philips Res Repts 21 p 113-21 (1966)

23 R W Brander, R P Sutton; J Phys D, 2, p 309-18 (1969)

24 L B Griffiths, A I Mlavsky; J Electrochem Soc 111 p 805-10 (1964)

25 M A Wright; J Electrochem Soc 112 p 1114-6 (1965)

26 M Kumagawa et al; Jap J Appl Phys 9 p 1422-23 (1970)

27 V I Pavlichenko et al; Sov Phys - Solid State 10 p 2205-6 (1969)

ACKNOWLEDGEMENT

The author is indebted to his former colleagues at GEC, Hirst Research Centre, particularly R V Bellau, M P Callaghan and B J Isherwood, for their past help and for providing the photographs for this review.

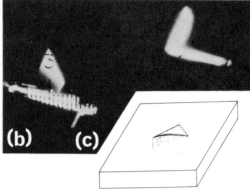

Plate 1. X-ray topograph of a β-SiC
layer grown on an α-SiC substrate by
vapour epitaxy.

Plate 2. a) Forward and b) reverse
electroluminescence of an α-SiC layer
containing a β-SiC inclusion c) a
schematic of the inclusion

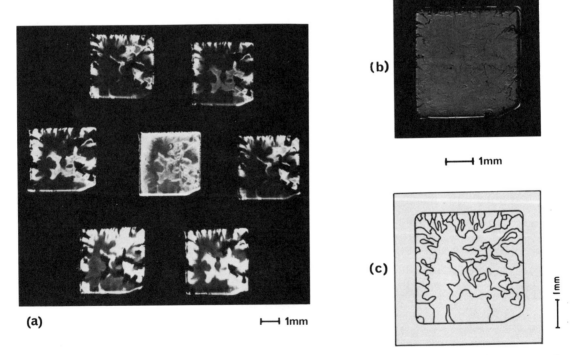

Plate 3. Multiple polytype formation in vapour epitaxial SiC showing correlation
between surface markings and individual polytypes; a) X-ray topograph, b) optical
micrograph, c) polytype boundaries. (after Isherwood and Wallace)

Epitaxial Growth of Silicon Carbide by Chemical Vapor Deposition

B. Wessels, H. C. Gatos and A. F. Witt

Alpha-silicon carbide (hexagonal-6H polytype) was grown on α-SiC substrates from the silane-propane-hydrogen system. Optimum results in terms of crystalline perfection and electrical characteristics were obtained by growing on the Si(0001) substrate surfaces at 1600°C employing a Si/C ratio greater than one. The epitaxial growth of SiC was found to be an activated process best described by adsorption-desorption kinetics, with adsorption of Si being the rate-determining step. As grown epitaxial SiC was n-type with a carrier concentration of $2 \times 10^{17}/cm^3$ and an electron mobility of 240 cm^2/V-sec at room temperature. p^+-n and n^+-p junctions were prepared by growing nitrogen-doped layers on aluminum-doped substrates and boron-doped layers on nitrogen-doped substrates, respectively. The p^+-n junctions exhibited blue electroluminescence with maxima at 2.62, 2.72, and 2.89 eV, whereas the n^+-p junctions exhibited yellow electroluminescence with a maximum of 2.11 eV.

EXPERIMENTAL PROCEDURE

The apparatus used in the chemical vapor deposition of epitaxial SiC was similar to that reported elsewhere.[1]

Since nitrogen is an electrically active impurity in silicon carbide, precautions were taken to minimize its presence in the system. The reactants were silane and propane, with hydrogen as the carrier gas (99.999% pure hydrogen diffused through a palladium-silver alloy cell). The reactants, silane and propane, were diluted with pure hydrogen to one per cent mixtures. The propane, only available in instrument grade purity (99.5%), was the main source of impurities.

The substrates used in this study were grown by a modified Lely technique. As received, the net donor concentration in the n-type substrates was approximately $5 \times 10^{18} cm^{-3}$ and in the p-type the net acceptor concentration was

$2 \times 10^{19} \text{cm}^{-3}$. The p-type samples were heavily compensated with nitrogen.

Prior to use, the substrates were polished with diamond paste. For identification of the surface orientation, a preferential etchant consisting of a boiling, saturated solution of $NaOH-K_3Fe(CN)_6$ was used.[2] Since the etchant did not yield a smooth surface, the silicon carbide substrates were used as polished. Before growth the polished substrates were cleaned in boiling trichloroethylene. Subsequently they were etched in HF to remove any SiO_2 film; finally they were washed in de-ionized water.

The substrates prepared as above were placed on a tantalum susceptor coated with silicon carbide and positioned in the quartz reaction chamber; the system was subsequently evacuated to ten microns and purged with H_2.

Typical growth conditions included a deposition temperature of 1600°C and a hydrogen flow rate of 0.5 liter per minute at one atmosphere pressure. The silane-hydrogen and propane-hydrogen mole ratios were 8×10^{-4} and 5×10^{-5}, respectively. The total growth period was usually two hours. After the growth cycle, the system was allowed to cool to 1000°C and then purged with argon for ten minutes to prevent hydrogen embrittlement of the susceptor. The thickness of the grown layer was obtained by two methods: either by beveling the epitaxially grown junction and scanning the surface with a thermoelectric probe[3] or by electrolytically staining the grown junction in 3N HF.[4]

RESULTS

Dependence of Growth Rate on Silicon Supersaturation

It was found that the growth rate of epitaxial SiC at 1650°C was linearly dependent on the silane pressure for a constant propane input as shown in Fig. 1. This result suggests that silicon controls the rate of SiC growth. The pressure dependence of the growth rate was sensitive to surface orientation; for the Si(0001) orientation a silane pressure of 7.5×10^{-4} atm. (extrapolated value) is required before growth occurs, whereas the extrapolated silane pressure necessary for growth on the $C(000\bar{1})$ orientation is at least an order of magnitude less. This dependence of the growth rate on orientation is consistent with thermal etching rate data[5] on beta SiC whereby the $C(\bar{1}\bar{1}\bar{1})$ surfaces exhibit a slower etching rate than the Si(111) surfaces.

Temperature Dependence of Growth Rate

The growth rate on the $C(000\bar{1})$ surfaces was found to be an exponential function of temperature. For this surface orientation a plot of the log of the growth rate versus the reciprocal temperature yielded an activation energy of 22 kcal/mole, which is in excellent agreement with the value of 20 kcal/mole reported by Jennings et al.[6] for the activation energy of epitaxial growth of SiC from chlorides. For the Si(0001) orientation, however, a complex temperature dependence of the growth rate was found as shown in Fig. 2.

The results of the present study are consistent with a growth model in which the rate-limiting step for epitaxial deposition is the adsorption-desorption of reactants at the growth interface.[7] The growth rate, J, in this case is given by:

$$ J = \frac{(p_{Si} - p^*_{Si})}{(2\pi m_{Si} kT)^{1/2}} \, k_{Si} \, \exp(-Q_A/RT) \qquad (1) $$

where p_{Si} is the pressure of silicon, p^*_{Si} is the equilibrium partial pressure of silicon over silicon carbide, m_{Si} is the mass of the adatom, k_{Si} is the temperature independent term of the sticking coefficient, and Q_A is the activation energy of adsorption.

The equilibrium partial pressure p^*_{Si} is exponentially dependent on temperature as follows:

$$ \frac{p^*_{Si} \, a_C}{a_{SiC}} = \exp\left(\frac{-H_s}{RT} + \frac{S_s}{RT}\right) \qquad (2) $$

where a_C and a_{SiC} are the activities of carbon and silicon carbide, respectively, H_s and S_s are the molar enthalpy and entropy of sublimation, respectively.

Thus the simple exponential temperature dependence of the growth rate for the $C(000\bar{1})$ orientation is in agreement with Eq. 1 for the case where the equilibrium partial pressure of silicon, p^*_{Si}, is negligible. For the Si(0001) orientation, p^*_{Si} is appreciable, resulting in a complex temperature dependence of the growth rate.

Considering Eq. 1, the results presented in Fig. 2 for the Si(0001) surfaces can be explained as follows: At high temperatures, i.e., above 1650°C, where decomposition predominates, there is a sharp drop in growth rate. When the growth temperature is decreased to below 1650°C, the growth rate becomes

essentially temperature independent due to the changeover from decomposition-
controlled kinetics to adsorption-controlled kinetics. Finally, at temperatures
below 1500°C, adsorption is rate-limiting and a simple exponential temperature
dependence similar to that for the $C(000\bar{1})$ surface is observed. The curve in
Fig. 2 was fitted to the experimental data with Eq. 1, using two variables,
the sublimation enthalpy H_s and the sticking coefficient k_S. The activation
energy of adsorption, Q_A, was assumed to be the same as for the $C(000\bar{1})$ surface,
i.e., 22 kcal/mole. The value of the sublimation enthalpy H_s, which gave the
best fit, was 80 kcal/mole; this value is in agreement with that of 80 ± 5 kcal/
mole,[8] required to evaporate liquid silicon, suggesting that desorption of
silicon controls the decomposition rate for the Si(0001) surface.

Dependence of Surface Morphology on Supersaturation and Si/C Ratio

In view of the fact that the Si surfaces have a larger equilibrium silicon
vapor pressure, p^*, than the C surfaces, the vapor phase supersaturation
$(p - p^*)/p^*$ is smaller for Si surfaces than C surfaces at the same ambient
silicon pressure p. Since layers grown near equilibrium, i.e., at low supersaturation, usually have a greater degree of perfection than layers grown at
high supersaturation, it is expected that layers grown on the Si(0001) surfaces
should have a higher degree of crystalline perfection than those on the $C(000\bar{1})$
surfaces. This dependence of crystalline perfection on surface orientation was
indeed experimentally observed, as shown in plate 1 and 2.   It can be seen
that the C surface exhibits a high defect concentration; the dominant defect
in this case is a domain structure, consisting of several different polytypes
(identified by x-ray analysis) intergrown within the epitaxial layer; the Si
surface on the other hand does not exhibit an identifiable domain structure,
but only isolated hillocks.

In addition to the silicon carbide substrate orientation, it was found
that the Si/C ratio in the ambient also affects the morphology of the grown
layers. For instance, in a carbon-rich ambient (Si/C = 0.5) at a silane pressure of 1×10^{-3} atm and a propane pressure of 6.6×10^{-4}, at 1645°C, the
epitaxial layer exhibited a fine-grained micro-structure, which was opaque to
light, as shown in the photomicrograph 3. This morphology is interpreted as
the result of concurrent deposition of graphite and silicon carbide. Minagawa
and Gatos[9] predicted on the basis of thermodynamic calculations for the

silane-propane-hydrogen system that single-phase silicon carbide should be deposited under the above carbon-rich conditions, since any carbon in excess of single phase SiC is retained in the vapor phase as hydrocarbons. In contrast with above thermodynamic predictions, it was found that when layers were grown at Si/C ratios of 1/1 (silane pressure 2 x 10^{-3} atm) simultaneous deposition of graphite and silicon carbide took place. Evidently the single-phase growth region is more restricted than predicted from pure thermodynamic arguments and controlled by deposition kinetics.

Deviation from the predictions of Minagawa and Gatos were also observed for epitaxial growth under silicon excess conditions. It was found that up to Si/C ratios of 5/1, well into the predicted two-phase region, no evidence of Si deposition was encountered. In fact, the best epitaxial layers, as the one shown in **plate 4**, resulted from growth at Si/C ratios of 5/1.

The apparent discrepancy between experimental findings and thermodynamic calculations can be attributed to the uncertainty in the value of the free energies used in the thermodynamic calculations. This uncertainty is especially large for silicon, since the vapor pressure of silicon calculated from the JANAF tables[10] is two orders of magnitude lower than that reported by Nesmaynov.[8] The observation that no silicon phase did form under silicon excess conditions may be explained by the fact that at the growth interface the silicon vapor pressure is greater than that associated with the silane input pressure. Thus, any silicon in excess of that required for single-phase formation of SiC would vaporize. Silicon vapor would have a "buffering effect" similar to that of acetylene for the growth of SiC under carbon-rich conditions, as described by Harris et al.[11] However, it appears that excess silicon is more effectively retained in the gas phase than excess carbon.

Electrical Properties of Epitaxial Layers

Carrier concentration and electron mobility for the epitaxial layers were obtained by Hall effect measurements using the Van der Pauw configuration. For the as grown 6H layers on the Si(0001) surface, room temperature electron concentrations as low as 9.8 x 10^{16} cm^{-3} and electron mobilities of 235 cm^2/V sec were observed at 20°C, which is comparable to that observed in the bulk material prepared by the modified Lely technique.[12]

Diodes were prepared by growing doped epitaxial layers on substrates of opposite conductivity type; the electroluminescence spectra of the diodes depended on the specific dopants used and the substrates. At room temperature, the luminescence of forward biased diodes, prepared from undoped epitaxial layers grown on aluminum-doped substrates, (p^+-n), gave a broad band with a peak of 2.26 eV (5450Å) for injection levels at 80 mA/mm^2. As the injection level was increased, a peak emerged at 2.72 eV (4550Å), in the blue region of the spectrum. At current levels of 100 mA/mm^2 the blue band became dominant. Also, at high injection levels two other bands were observed with maxima at 2.62 eV (4750Å) and 2.89 eV (4300Å). These epitaxial layers did not exhibit any photoluminescence at room temperature.

For epitaxially prepared diodes doped with boron, the electroluminescence exhibited a broad band with a maximum at 2.11 eV (5900Å). Yellow luminescence was observed for epitaxially grown junctions prepared by depositing two epitaxial layers of different conductivity type on aluminum-doped p-type substrates (p^+-p-n structure) and by depositing a boron-doped epitaxial layer on nitrogen-doped substrates (n^+-p structure). Both diode structures exhibited strong room temperature photoluminescence with the same characteristic spectrum as in electroluminescence. For the p^+-p-n structure the yellow electroluminescence was relatively bright with an intensity much greater than that of the blue, violet and green luminescence of the **blue** (p^+-n) diodes for comparable current densities.

Acknowledgements

The authors are indebted to the North American Philips Corporation for sponsoring this work.

REFERENCES

1. S. Minagawa and H. C. Gatos, Japan J. Appl. Phys. 10, 1680 (1971).

2. J. M. Harris, H. C. Gatos and A. F. Witt, J. Electrochem. Soc. 116, 672, (1969).

3. L. J. Kroko and A. G. Milnes, Solid-State Electronics 8, 829 (1965).

4. R. W. Brander and A. L. Boughey Brit. J. Appl. Phys. 18, 905 (1967).

5. R. W. Bartlett and R. A. Mueller Silicon Carbide-1968 (Pergamon Press, New York, 1969) p. 341.

6. V. J. Jennings, A. Sommer and H. C. Chang, J. Electrochem. Soc. <u>113</u>, 728 (1966).

7. M. M. Faktor and I. Garrett, J. Cryst. Growth <u>9</u>, 12 (1971).

8. A. N. Nesmaynov, <u>Vapor Pressure of the Chemical Elements</u> (Elsevier, New York, 1963).

9. S. Minagawa and H. C. Gatos, Japan J. Appl. Phys. <u>10</u>, 844 (1971).

10. JANAF Thermochemical Tables, Dow Chemical Co., Midland, Michigan.

11. J. M. Harris, H. C. Gatos and A. F. Witt, J. Electrochem. Soc. <u>118</u>, 339 (1971).

12. S. H. Hagen and C. J. Kapteyns, Philips Res. Repts. <u>25</u>, 1 (1970).

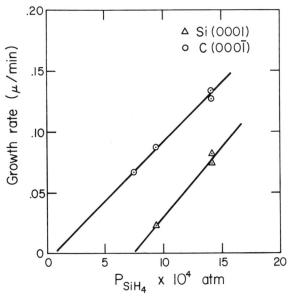

Figure 1

Linear growth velocity vs. SiH_4 partial pressure for SiC layers grown on Si(0001) and C(000$\bar{1}$) surfaces at 1650°C $P_{C_3H_8}$ = 3 x 10^{-5} atm

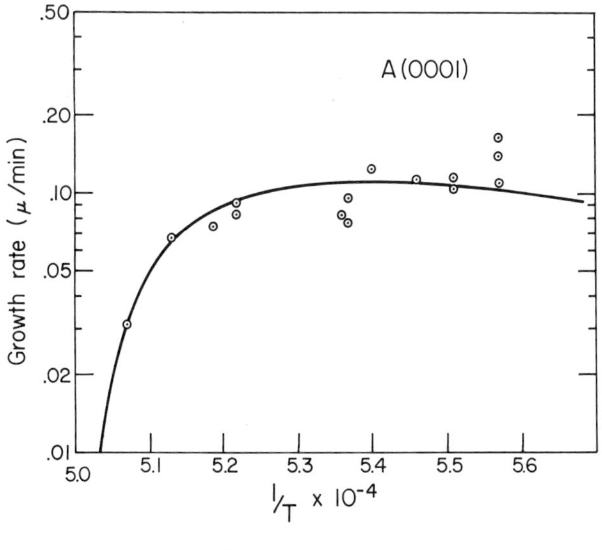

Figure 2

Linear growth velocity vs. reciprocal temperature for SiC layers grown on Si(0001) surface, SiH_4 = 1.4 x 10^{-3} atm, Si/C = 5.

Plate 1.

a. Silicon carbide layer grown on
Si(0001) surface of α-SiC substrate
(150X).

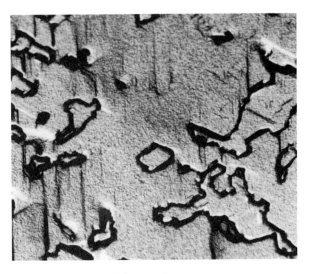

Plate 2.

b. Silicon carbide layer grown on
C(000$\bar{1}$) surface of α-SiC substrate
(400X).

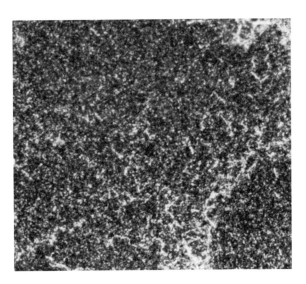

Plate 3.

c. Silicon carbide layer grown under
carbon-rich conditions (500X).

Plate 4.

d. Silicon carbide layer grown under
silicon-rich conditions (150X).

CVD-SiC on Carbon Fibres

E. Fitzer, D. Kehr and M. Sahebkar

This paper is concerned with the vapour deposition of SiC from methylchlorosilane on endless carbon fibres as substrates.

The aim of these studies was on one hand to prepare high strength silicon carbide fibres by vapour deposition of relatively thick carbide coatings on carbon monofilaments as indicated in the left side of figure 1.

On the other hand, thin uniform SiC coatings on each monofilament within a carbon fibre yarn were prepared as shown in the right side of figure 1 in order to improve the surface qualities of carbon fibres.

All experiments were performed under steady state conditions in a tubular reactor with a hot wall arrangement shown in figure 2.

The apparatus consisted of a graphite tube 40 cm long, heated by induction, and of the equipment for fibre transport, placed at both ends of the graphite tube. The whole apparatus is surrounded by a quartz glass shell which permits the controlled supply and removal of the reaction gases.

For the preparation of SiC fibres, carbon monofilaments of 30μ thickness have been used as substrates. For the coating experiments of carbon fibre yarn, various THORNELL type yarns have been used. Deposition temperatures between 1100 and 1400°C were applied. The rate of the SiC deposition is strongly dependant on the nucleation conditions. The diagram in figure 3 exhibit the diameters of the obtained SiC monofilaments as function of the silane feed with a residence time of 17.5 min., in the left side at 1100°C and in the right side at 1400°C.

Although the deposition rate increases with rising silane supply, one can recognize an inhibiting effect at low silane supplies. Also a decrease of SiC fibre growth after a maximum rate is observed, in some cases of a high silane feed,

33

namely at high silane partial pressures. In these cases the gas phase nucleation
is promoted and SiC-soot is formed. The complex influence of the various deposition parameters on the deposition rate is caused by the ratio between substrate
and gasphase nucleation.

A similar influence can be observed on the structure and the mechanical properties of the deposites. The electron scanning micrographs in figure 4 show the
fracture surfaces of SiC fibres with their carbon cores. The left sample was
deposited with smooth surfaces and glassy fracture are obtained, if such low
deposition temperatures have been used. These fibres exhibit a tensile strength
of 10,000 MPa and Youngs' moduli of 500,000 MPa, corresponding to 1.4 mill. psi
strength and 70 mill. psi Young modulus respectively. Higher deposition temperatures yielded coarse-grained deposites with decreased mechanical properties. The
SiC fibre shown in the right micrograph was prepared at $1400^{o}C$.

The complex influence of the deposition parameters on the nucleation and
crystal growth is reflected also in the strength properties. The diagram in
figure 5 shows the measured room temperature strength values of the total SiC
fibres including the carbon substrate, whose own strength is not better than 500 MPa
(70,000 psi). Best strength values are obtained at the lowest deposition temperatures and low silane supplies, causing substrate nucleation and low rate of crystal
growth. The high deposition temperatures such as $1400^{o}C$ cause coarse-grained
deposites with strengths at about 3000-4000 MPa corresponding to 420-560,000 psi.
The strength minimum, found with the fine-grained samples deposited at temperatures
between $1200^{o}C$ and $1300^{o}C$, is probably caused by a co-deposition of elemental silicon as found by x-ray diffraction. The complex chemistry of CVD from various
methylchlorosilanes and the competition between SiC, Si and C deposition has been
discussed previously (1-5).

Although the outstanding strength values of some of these endless SiC fibres
seem very promising, the problems arising from brittleness of this material are
not yet solved, i.e. the winding difficulties. Nevertheless, silicon carbide
fibres on carbon fibre substrates seem to be good candidates for reinforcement
components for composite materials.

Contrary to the strength problems, studied with thick SiC coatings on single
carbon filaments, other problems arise during the coating of carbon yarns with
thin SiC layers, because the coating procedure of all monofilaments within the
yarn has to be performed simultaneously. Carbon fibre yarns consist of several

thousand monofilaments and are geometrically complicated porous bodies. The deposition rate must be controlled by the chemical reaction rate and not by the diffusion rate. Otherwise, not only the outer monofilaments of the yarns would be coated preferably, but also SiC bridges would be formed, as indicated in the schematic figure at the left hand of figure 6.

In order to study the influence of the deposition conditions on the uniformity of the coating, the inpore-deposition of SiC has been studied previously with a porous graphite substrate, whose pore spectrum is given in the diagram at the right - hand of figure 6. One can recognize that the pore sizes as well as the pore distribution are in the same order of magnitude of at about 1-10 microns as the expected slits between the monofilaments.

Figure 7 shows the apparatus used for the study of the inpore deposition of SiC. The porous graphite is used as succeptor for induction heating. The reaction gases flow towards the surface of the graphite tube under atmospheric pressure. The deposition temperature was varied between 900 and 1400°C and the silane partial pressure between 1 and 100 mbar.

At the deposition temperature of 900°C, the whole substrate was impregnated as can be seen by the micrograph (figure 8). At 1000°C the substrate was impregnated only to a depth of 2.5mm, whereas the depth of penetration at 1100°C was only 0.5mm corresponding to the radius of a carbon fibre yarn. Above 1200°C SiC was deposited only on the substrate surface. One can conclude that the reaction rate controls the total deposition rate at temperatures below 1200°C.

The kinetic problem of the inpore deposition is similar to the so-called "effectiveness factor" of porous catalyst grains in heterogeneous gas catalytic reactions, and the same criterion can be used. As known the penetration depth can be calculated for a given effectiveness factor from the reaction rate coefficient. In figure 9 the calculated penetration depths for different pore radii are plotted against the temperature as dashed and dotted lines, the experimental values are shown as full lines. The experimental values agree well with the precalculated ones, if one takes into account that the pore diameter is diminished continuously during the CVD treatment. As a result of these experiments one can say that the coating of a carbon fibre yarn should be possible up to temperatures of 1200°C at least. The geometry of the slits within the yarn will improve the penetration of the deposition gas in comparison with the pores in a bulk graphite.

Figure 10 shows the cross section of a coated carbon yarn. Uniform coatings of about 0.5μ thickness were prepared on all filaments of the bundle without formation of strong SiC bridges up to deposition temperatures of 1400^oC. The deposition rate at relatively high silane concentrations was 15μ/hr. at 1000^oC, 52μ/hr. at 1100^oC, and 73μ/hr. at 1200^oC. A kinetic evaluation of the experimental results with the aid of the material balance of the ideal tubular flow reactor yielded values of the reaction rate constants of 1.25; 4.83; and 6.6 cm/sec. for deposition temperatures of 1000, 1100 and 1200^oC respectively. The activation energy was 47 kcal/mol. SiC layers of about 1 micron thickness can be obtained at 1200^oC with a residence time of 49 sec., at 1100^oC in 69 sec., and at 1000^oC in 4 minutes.

The strength of the carbon fibre is not influenced by such thin coatings if the rupture strain of the substrate is comparable with that of the CVD-SiC. (Table 1)

Fibre type	Layer thickness [microns]	Tensile strength [kp/mm^2]	Young-Modulus [kp/mm^2]	Elongation at break [%]
Thornel 25	uncoated	124	14880	0,84
"	0,25	120	13050	0,92
Thornel 50	uncoated	150	28400	0,53
"	0,25	146	35500	0,41
WYB	uncoated	64	3700	1,62
"	0,03	61	4100	1,49
"	0,1	59	3640	1,62
"	0,17	55	3750	1,4
"	0,23	50	4040	1,25
"	0,32	48,5	3900	1,24
"	0,52	39	4200	1,13

This is true in the case of stretch graphitized THORNELL type graphite fibres. However, fibres with a higher strain as WYB type will become more sensitive to notches and will break at a lower stress, depending on the SiC-layer thickness. Therefore, the ultimate strength seems to be slightly reduced by the thin coatings in these cases.

In any case the wetting behaviour of the fibres by polymeric and oxydic matrixes is improved by SiC coatings and better interlaminar shear strength properties of fibre reinforced composites are obtained.

SUMMARY

High strength high modulus silicon carbide fibres can be prepared by CVD of SiC from methylchlorosilane on carbon monofilaments at deposition temperatures around 1100°C. At 1400°C the strength was lower but always as high as 4000 MPa. In the second part, studies concerning the uniform coating of all monofilaments within a carbon fibre yarn are described. Experiments on inpore CVD of SiC in a porous polycrystalline graphite explain the reaction rate control of the total deposition rate.

(1) E. Fitzer, D. Kehr and M. Sahebkar, Chemie-Ingenieur-Technik, 45 (1973) Nr. 19.

(2) M. Bonnke, E. Fitzer, Ber. dtsch. Keram. Ges., 43 (1966), 180/87.

(3) L. Aggour and W. Fritz, Chemie-Ingenieur-Technik, 43 (1971), 472.

(4) C. F. Powell, J. H. Oxley and J. M. Blocher, Chemical Vapour Deposition, John Wiley and Sons Inc., New York 1966.

(5) J. M. Blocher and J. C. Withers, Chemical Vapour Deposition, Second Intern. Conf, The Electrochem. Soc., Inc., New York 1970.

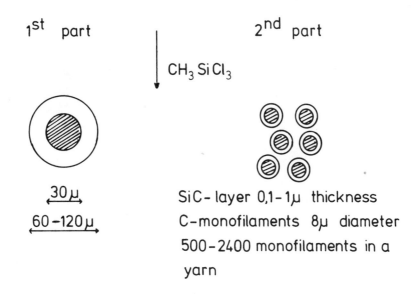

Fig. 1 CVD-SiC as SiC fibre (left) and SiC coating on carbon yarns.

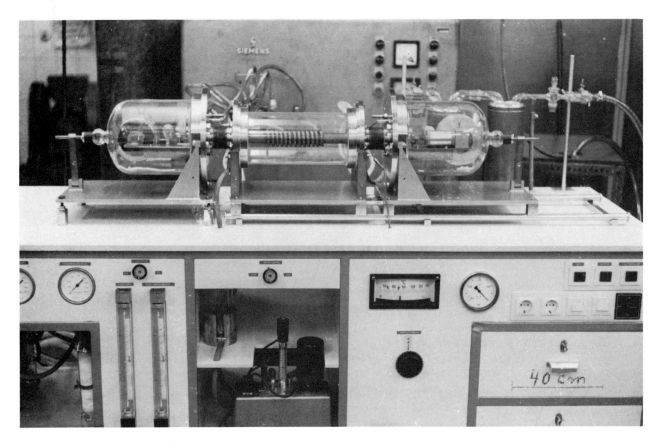

Fig. 2   Apparatus for CVD of SiC on carbon fibres.

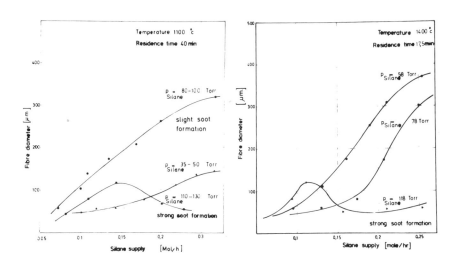

Fig. 3   SiC fibre-thickness, obtained by CVD from
CH_3SiCl_3 in H_2.

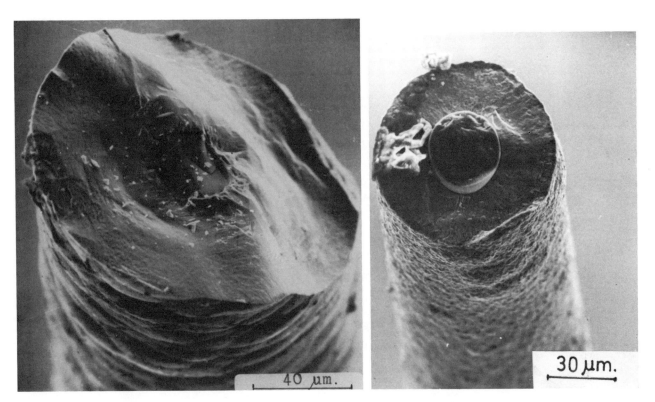

Fig. 4   Electron scanning micrographs of SiC fibres.

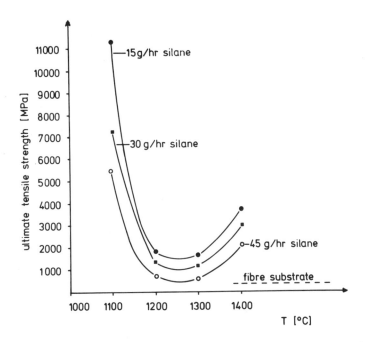

Fig. 5   Ultimate tensile strengths of CVD–SiC fibres
as function of the deposition temperature.

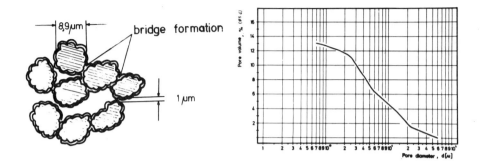

Fig. 6   Slits in a carbon yarn compared with the
         pore size distribution of a porous poly-
         cristalline graphite.

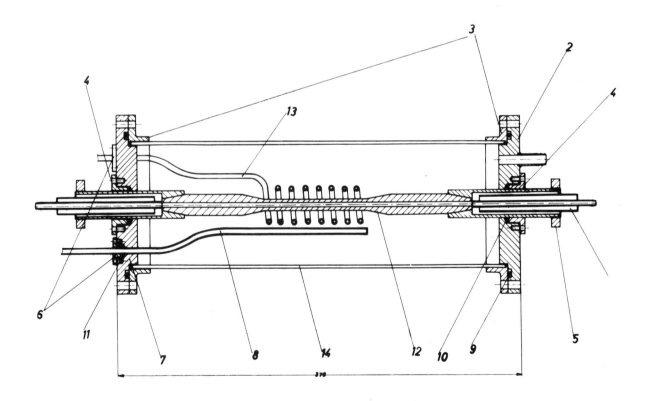

Fig. 7   Apparatus for the inpore deposition of SiC
         in graphite.

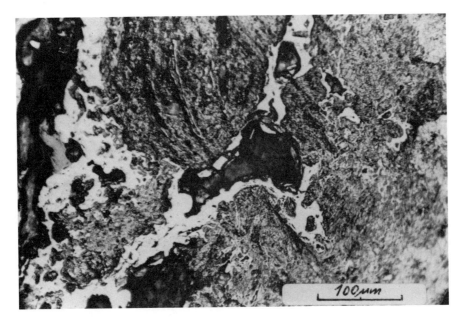

Fig. 8 Graphite impregnated with CVD-SiC.

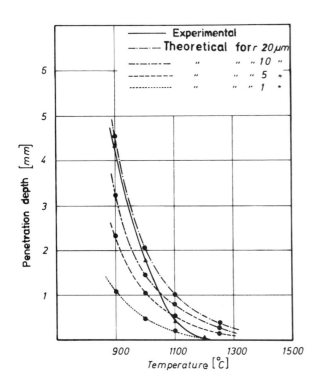

Fig. 9 Precalculated and experimental
(full line) SiC-impregnation
depth.

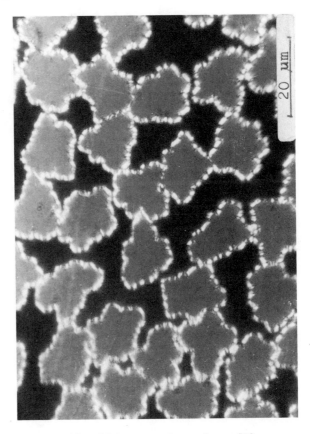

Fig. 10 SiC-coated carbon fibre.

Annealing of Sputtered β SiC

I. Berman, R. C. Marshall and C. E. Ryan

Abstract

Using an RF Sputter and Etching Module, silicon carbide was deposited upon the surface of (100) and (111) silicon, (111)(110) (100) tungsten, & (0001) alpha silicon carbide. Almost all deposits were initially amorphous. After thermal annealing the thin films on the (100) silicon and (0001) alpha became single crystal. Silicides of tungsten were formed on the tungsten substrates after annealing.

Introduction

Single crystals of silicon carbide have been grown by sublimation, epitaxial growth and by growth from solution. To these growth techniques, the method of recrystallization has been added.

Phase transformations in silicon carbide through recrystallization have been the subject of a number of investigations. The differences in reported results appear to be dependent upon the techniques and conditions under which the original material was grown[1].

In a detailed study of the phase transformation from 2H through 3C to 6H, Krishna, Marshall, and Ryan[2] have reported that a solid state transformation occurred. Their conclusions were based upon the annealing results of 2H **needles** (or thick base whiskers).

Bootsman et al[3], in a published paper "Phase Transformation, Habit Change and Crystal Growth in SiC" also observed a transformation. Their conclusions were that the transformation was of crystal habit change and was rate controlled. Bootsman's results were based on annealing thin base whiskers.

Berman et al[4], annealed thin films of chemical vapor deposited polycrystalline beta and found a solid state transformation to single crystal beta. This polytype remained beta even when held at $2000^{\circ}C$ for 8 hours. The stability of the beta was attributed to the purity of the deposited thin films. These disclosures, together with the results of others,[5,6] show the feasibility of tailoring silicon carbide by controlling growth conditions and heat treatment. Under properly controlled parameters, it should be possible to deposit an amorphous film layer of silicon carbide and cause a transformation to the required single crystal polytype.

The above results were reported on chemical vapor deposits at elevated substrate temperatures. It seemed a logical sequence to investigate the phase changes in silicon carbide formed at the lowest temperature possible. Then through a series of heat treatment cycles crystallize and/or recrystallize the silicon carbide to any possible polytype or crystal perfection. In this report thin films of amorphous-like SiC, deposited at about room temperature, were annealed and examined for any changes in crystallization or phase. Sputter deposit and annealing was the technique used.

Procedure

At the 1968 Silicon Carbide Conference at Penn. State, Khan[7] reported that single crystal silicon carbide had been deposited by DC sputter techniques. The temperature of the substrate was between $700^{\circ}C$ and $800^{\circ}C$. When the silicon substrate temperature was less than $700^{\circ}C$, single crystal deposits could not be attained.

RF sputtering deposits have been prepared at AFCRL. The sputter module has etching capabilities. The substrate holder, which also serves as the etching station, was cooled with recirculating chilled water. During deposits the substrate was cooled.
Plate 1 shows the RF sputter module. It has two target set up with a movable water cooled substrate holder.

The substrates used in this investigation were:

1. Alpha SiC commercially grown (0001)
2. Silicon (111) and (100)
3. Single crystal tungsten (111) (110) (100)

These substrates were chosen because of lattice match, availability, and because they possess the potential for good device fabrication. The gas ambient was

5 nines pure argon as stated by the manufacturer. For added purity control, the
argon was passed through a carbon dioxide cooling trap and finally over heated
titanium. The operating pressure of the system was eight microns. The target
material was 12.7 cm in diameter and composed of 99+% pure alpha. The distance
between the target and the substrate was 5 cms. Peak R.F. voltage was 1.2 KV.

Alpha substrates were prepared by polishing the rougher side $(000\bar{1})$ to 1/4
micron diamond finish. This was done to insure an intimate contact with the
water cooled substrate holder. Sputter deposits were made on the natural grown
smooth surface (0001). The other substrates in this investigation were polished
to 1/4 micron, on one side, and chemically etched. Deposits were put upon the
polished face.

All substrates were ultrasonically cleaned in trichloroethylene and etched
in HF to remove oxides. On the first silicon samples, a thin film of nickel was
initially deposited[8]. The substrates were placed in the module, sputter
etched, and then sputter deposited.

Data:

Plate 2 shows the etch and deposit rates on the different surfaces. The
sputter deposit rate, as measured with a Dektak, was 166 Å per minute on all of
the surfaces. The etch rates which were also measured with a Dektak, showed
(111) silicon to have etched faster than (100) silicon. The comparison of etch
rates for the tungsten faces gave different results. The (110) face etched fas-
test followed by the (100). The (111) face had the slowest etch rate.

With alpha silicon carbide the $(000\bar{1})$ face etched faster than the (0001).

An electron diffraction pattern of the sputtered SiC after annealing showed
that the tungsten had reacted to give compounds with silicon. On the (111)
silicon the beta film was not completely single crystalline. Good single crystal
beta was obtained on (100) silicon and on (0001) silicon carbide.

Plate 3 is an electron diffraction pattern of a before and after annealing
of SiC sputter deposited on (111) silicon surface. The layer was still fibrous
crystalline beta after a total of 2 hours of annealing at 1250°C.

Because silicon substrates are readily available with high crystalline perfection, the rest of the data reported is of sputtered SiC on (100) silicon.

In previous CVD thin film work, the use of a thin film nickel intermediate[8]
on silicon gave single crystal layers consistently, so for that reason our first
sputter deposits employed that technique.

Plate 4 shows the results of annealing the sputter deposited layers of different thicknesses on (100) silicon. Film thickness, as controlled by deposit time, was the variable. The deposit rate was 166 Å/min. Before annealing all deposits were amorphous. After annealing, the control (or no nickel) sample appeared to give the best results. Thirty minutes of deposit, or a 5000 Å thick layer, could not be annealed into single crystals in 30 min. Layers of 500 Å to 2500 Å did become single crystal in the same annealing procedure.

The electron diffraction patterns of some of these samples are shown in the next four plates:

Plate 5 is the electron diffraction patterns of a 900Å deposited SiC layer before and after annealing. The annealed layer shows a good beta pattern with twinning. The before annealing E.D.P. shows the layer to be amorphous.

Plate 6 is the electron diffraction patterns of a 1700Å deposited SiC layer before and after annealing. The before annealing E.D.P. shows amorphous structure while after annealing it shows single beta.

Plate 7 is the electron diffraction patterns of a 5000Å SiC deposit before and after annealing. The before annealing shows amorphous material. The surface after annealing shows a polycrystalline beta structure.
In all above annealed diffraction patterns there is evidence of some polycrystalline β.

The results obtained by omitting the nickel intermediate is shown in Plate 8. These electron diffraction patterns are of a 1700Å deposit of SiC on a (100) silicon surface before and after annealing. The diffraction pattern shows good single crystal beta after the annealing had occurred.

A 1700Å SiC deposit was made upon a single slice of silicon. The slice was was then scribed and broken into 1/2 cm squares. These squares were annealed at different temperatures to determine the temperature of transformation. The time of each anneal was thirty minutes. Electron diffraction patterns were taken before and after annealing. It appeared that temperatures less than 1200°C would not yield a single crystal surface in 30 minutes. The data is shown in Plate 9.

Although the electron diffraction pattern can identify the polytype and the crystallinity, the morphology of the surface is not defined. In an effort to resolve the nature of the beta film, one annealed film was removed from a silicon surface and examined in transmission with the electron microscope.

One of the SiC surfaces with pre-sputtered nickel showed imperfections after annealing when viewed under low magnification. Because of the structure of these imperfections, the study was made on this surface. An electron diffraction pattern in transmission of this particular silicon carbide deposit showed it to be beta, mostly single crystal, but with a few rings showing some polycrystalline structure.

The sputtered layer was approximately 2000Å thick.

Plate 10 shows a 500X optical microscope view of the annealed surface before the thin film was removed.

Plate 11 shows a 10K magnification of an imperfection on the same surface as seen by a scanning electron microscope. It shows a square configuration. At this point the thin film was removed and mounted in the electron microscope.

Plate 12 is a transmission electron microscopic examination of the area surrounding the imperfection and shows a continuous surface. The magnification is 40, 000X.

Plate 13 is an electron diffraction pattern taken in transmission through one of the imperfections. It shows a single crystal pattern dominating. A few rings denote some polycrystalline material present.

Conclusions:

An RF sputter module was used to deposit silicon carbide upon the surface of (100) and (111) silicon, upon (111), (110) and (100) tungsten, and upon (0001) silicon carbide. At the substrate temperature of the experiment, the deposits were amorphous - like. After thermal annealing at 1250°C for 30 minutes, the material crystallized to single beta. No 2H or other polytype was detected. Probably because the crystal growth rate exceeded that required by the kinetics of their growth. Under the same conditions of temperature cycling, the SiC films on the tungsten surfaces chemically reacted to give silicides.

A SiC sputter deposit rate of 166Å per minute gave a good control of the deposit thickness. In order to effect a conversion to single crystal beta, the deposit thickness had to be limited to 2500Å or less.

Further studies will be made to see if chemical vapor depositions made upon these recrystallized substrates will enable thick layers of single beta to be deposited.

The authors wish to thank Mr. J. Comer for performing the electron diffraction examinations and Mr. J. Littler for annealing of the substrates.

REFERENCES

1.   W. F. Knippenburg and G. Verspui, Mat Res Bull., Vol. 4 (1969) S33-S44,
         Pergammon Press.

2.   P. Krishna, R. C. Marshall and C. Ryan, Journal of Crystal Growth <u>2</u> (1971)
         129-31, North Holland Pub. Co.

3.   G. A. Bootsman, W. F. Knippenberg and G. Verspui, Journal of Crystal Growth
         <u>8</u>, (1971), 341-353, North Holland Pub. Co.

4.   I. Berman and C. Ryan, R. C. Marshall, and J. Littler, AFCRL-72-0737,
         19 Dec 72, Phys. Science Rsch Paper No. 522.

5.   J. A. Powell and H. A. Will, J. of Applied Phys. <u>8</u>, Vol 43 (1972),
         1400-1408.

6.   L. Patrick, D. R. Hamilton, W. J. Choyke, Physical Review, Vol. 143, No. <u>2</u>,
         (1966), 526-536.

7.   I. H. Khan, Materials Res. Bull., Vol. <u>4</u>, (1966), S285-S292, Pergammon
         Press, Inc.

8.   I. Berman and J. J. Comer, Materials Res. Bull. Vol. <u>4</u>, (1969), S107-S118,
         Pergammon Press, Inc.

SUBSTRATE	ETCH RATE	GROWTH RATE AT 1.2 KV
Si		
100	278 Å/MIN	
111	300 Å/MIN	166 Å/MIN
W		
100	200 Å/MIN	
111	147 Å/MIN	166 Å/MIN
110	277 Å/MIN	
SiC		
0001	166 Å/MIN	
000$\bar{1}$	222 Å/MIN	166 Å/MIN

Plate 2.   Etch and Deposit Rates.

Plate 1.   Sputter Module 8620.   Plate 3.   E.D.P. Before and After (111).

SiC SPUTTERED ON Si WITH A NICKEL INTERMEDIATE

NO.	SPUTTER TIME (MIN)	ELECTRON DIFFRACTION ANALYSIS	
		BEFORE ANNEAL	ANNEAL 1250°C - 30 MIN
40A	1	AMORPHOUS	SOME POLY SINGLE β SiC, TWINNED
40B	3	COMPOUNDS OF Ni AND Si	β SiC FIBER ORIENTED
40C	5	AMORPHOUS	SINGLE β SiC TWINNED
40D	15	AMORPHOUS	SOME POLY SINGLE β SiC TWINNED
40E	30	AMORPHOUS	POLY β SiC
CONTROL	15	AMORPHOUS	SINGLE β SiC NO TWINNING

Plate 4.   Thickness of Deposit vs. Crystallinity.

Plate 5.   E.D.P. Before and After
           Anneal 900 Å

Plate 6.   E.D.P. Before and After A
           Anneal 1700 Å.

Plate 7.   E.D.P. Before and After
           Anneal 5000 Å.

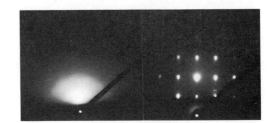

Plate 8.   E.D.P. Before and After
           Anneal 1700 Å (no nickel).

EFFECT OF ANNEAL TEMPERATURE		SPUTTER EXPERIMENT
	β SiC ON SI	
No.	ANNEAL TEMP. °C	ELECTRON DIFFRACTION ANALYSIS
40E-1	350°	AMORPHOUS
40E-2	400°	POLY CRYSTAL
40E-3	500°	POLY CRYSTAL
40E-4	800°	POLY CRYSTAL
40E-5	1000°	POLY CRYSTAL
40E-6	1200°	SINGLE CRYSTAL β SIC

Plate 9.   Data Anneal Temp. vs. Crystallinity.

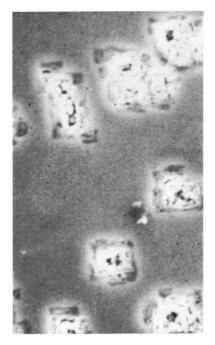

Plate 10.   Surface in E. D. Transmission

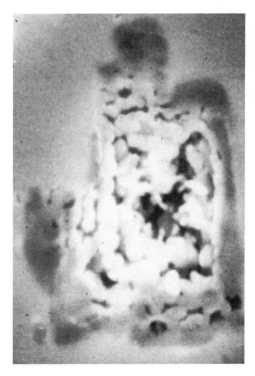

Plate 11.   Geometry of
            Imperfection

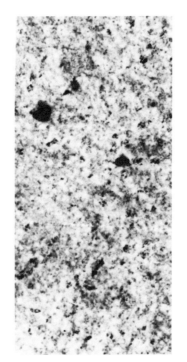

Plate 12.   50K Transmission
            of Surface.

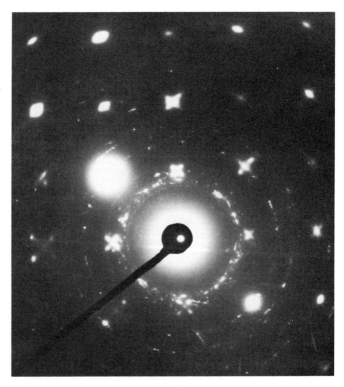

Plate 13.   E.D.P. in Center of
            Imperfection.

Liquid-Phase Epitaxy of Silicon Carbide by the Travelling Heater Method

W. von Muench and K. Gillessen

Single crystal SiC-layers have been produced by the travelling heater method at 1800 °C using silicon as a solvent. The silicon evaporation has been suppressed by a cold-to-hot vapor transport mechanism with iodine. Layers of the 6 H and 15 R polytypes were grown on various seed crystals. N-type doping was achieved by adding nitrogen; cubic SiC-layers were obtained at high levels of nitrogen doping.

INTRODUCTION

By analogy to the technology of III-V-compounds it may be expected that solution growth will yield more perfect SiC-layers (especially for opto-electronic devices) than vapor growth processes. Due to the low solubility of carbon in silicon (0.1 atom % at 1800 °C) and the high vapor pressure of molten silicon, most experiments of liquid phase epitaxy (LPE) of silicon carbide in the past were performed with a metal solvent (e.g. chromium). Such a solvent, however, will inevitably introduce undesired impurities at a level determined by the solid solubility in silicon carbide.

There are three methods for the growth of silicon carbide from a flux of molten silicon: 1. crystal growth in a crucible, 2. LPE by the travelling solvent method (TSM), and 3. LPE by the travelling heater method (THM). Nelson et al. (1) have investigated the stability of the crystal/solution interface in the case of silicon carbide and a silicon melt containing carbon. Applying Nelson's stability criterion to different growth methods, it has been shown that TSM and THM give a better chance to achieve stable growth than crystallization from a saturated silicon melt in a crucible (2). Only THM, however, combines stable growth with

51

constant growth conditions and a rather high growth velocity. In addition, the THM arrangement features a temperature maximum within the liquid zone; this is a necessary condition for the compensation of silicon evaporation by vapor transport as described in the following section.

SUPPRESSION OF SILICON EVAPORATION IN THE THM

Liquid silicon has a rather high vapor pressure (0.6 Torr at 2000 $^{\circ}$C), which renders the evaporation of silicon a major problem in the growth of silicon carbide from a silicon melt. Usually, the silicon evaporation is retarded by a surrounding high pressure atmosphere of argon. There is no complete suppression of the evaporation by this method, however, and the high pressure causes serious technical problems. Furthermore, any nitrogen content of the gas is limiting the purity of the crystals.

Using the travelling heater method, the evaporation of silicon can be completely compensated by a halogen transport reaction. Under certain conditions, the reactions

$$SiI_2 \rightarrow Si + 2I \quad , \quad SiI_4 \rightarrow Si + 4I$$

are dominant in a closed silicon/iodine system, resulting in a silicon transport from the cold regions to the hot silicon melt. In practice, an amount of 10^{-2} g iodine per cm^3 of volume in the gas phase is adequate to prevent silicon evaporation at 1800 $^{\circ}$C (2). The whole system, of course, has to be kept at elevated temperatures to avoid condensation of iodine.

SILICON CARBIDE GROWTH BY THE THM

Solution growth of silicon carbide requires temperatures around 1800 $^{\circ}$C. Since there is no crucible material which is resistant to molten silicon at these temperatures, a float-zone arrangement according to figure 1 was chosen for the travelling heater growth. The lower silicon carbide rod serves as source; the single-crystal seed is fixed to the upper rod. A certain amount of liquid silicon between seed and source is heated by induction coils. The addition of iodine prevents silicon evaporation during the growth experiment of several hours. To establish a temperature maximum within the melt, as a condition for THM and correct function of the gas transport mechanism, it is essential to have a

perfect r.f. power concentration. A slow upward motion of the ampoule then causes
a slightly asymmetrical temperature distribution within the liquid zone. Thus,
transport of silicon carbide from the lower rod to the seed is initiated.

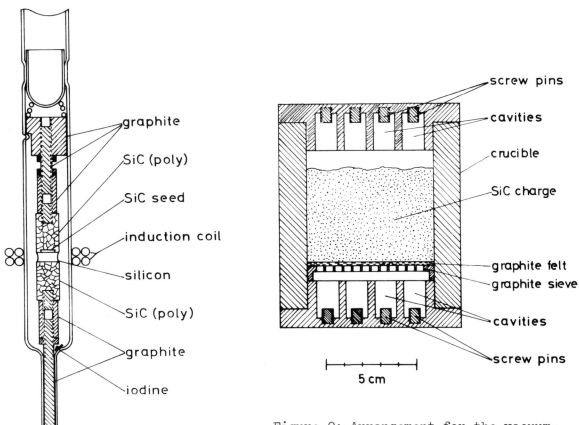

Figure 1: Sealed tube arrangement
          for the travelling heater
          growth of SiC

Figure 2: Arrangement for the vacuum-
          sublimation of SiC rods

For the above described travelling heater arrangement, it is of great importance
to use high-density polycrystalline silicon carbide rods. In addition, the
thermal conduction properties (especially anisotropy) are most critical to the
success of a growth experiment. The best results so far were achieved with polycrystalline material produced by vacuum-sublimation at 2400 °C in an arrangement
according to figure 2. A screw pin of graphite inserted into each growth cavity
serves to produce a thread at the end of each silicon carbide rod which facilita-
tes the mechanical connection with the other (graphite) parts of the travelling

heater system. Plate 1 shows an axial cut through a silicon carbide rod with
thread grown by vacuum-sublimation. The oxidation of the surface reveals large
crystal grains (mostly 6 H type) with different orientations; generally, the
c-axis of the crystallites is aligned to the axis of the rod.

The complete growth system consists of the quartz ampoule (fig. 1), the r.f.
generator, two resistance furnaces, an automatic optical pyrometer, and a
variable speed precision drive. The r.f. generator is turned on after the coolest
part of the ampoule has reached a temperature of 400 °C (During growth this temperature is about 700 °C). When a temperature of 1800 °C is established at the
upper end of the molten silicon, crystal growth can be started by moving the
ampoule in upward direction. The duration of a growth experiment ranges between
0.5 and 5 hours, the (average) growth rate varies between 4 and 20 µm/min.

RESULTS AND DISCUSSION

During growth experiments of several hours, phase boundaries shaped according to
fig. 3 were established. This shape can be understood by considering the electro-
magnetic and gravity forces, the temperature gradients, and the convection within
the liquid silicon. In several cases, however, the shape of the upper and lower

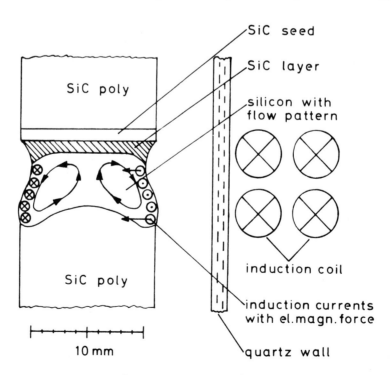

Fig. 3: Phase boundaries between silicon and silicon carbide

phase boundaries was inverted, probably due to poor wetting of the lower SiC rod. All growth experiments were performed on the (0001)-faces of 6 H seed crystals. In one case, however, the seed crystal was 15 R, probably (or some other rhombohedral polytype). Although the ampoule was moved in one direction at a constant velocity, the actual growth rate may vary considerably during a growth run. Even a negative growth rate (i.e. dissolution of the seed) has been observed initially. This is due to an asymmetry of the thermal conduction along the axis. Thus, a typical growth run may include three different periods: an initial etching stage, a slow growth period, and, finally, a period with a growth velocity close to, or exceeding the average velocity.

After the growth experiments, thin samples were cut along the axis of the rods, parallel to the $(11\bar{2}0)$-faces of the seed crystals. These samples were polished and oxidized (6 hours at 1070 °C in wet oxygen). Under these conditions, the oxide layers are exhibiting the following interference colors:

    first order: light blue on 6 H n-type crystals
                   royal blue on 6 H p-type crystals
                   light yellow on 15 R n-type crystals

    second order: blue-green to violet on 3 C n-type crystals.

According to the small difference of the band-gap of 6 H and 15 R silicon carbide the oxidation rates of these polytypes are only slightly different. Since the interference color of the oxide is also dependent on the doping level, the orientation, and the perfection of the SiC crystal, it is sometimes difficult to distinguish between 6 H and 15 R regions simply by microscopic inspection of the oxidized samples. X-ray Laue transmission diagrams have been taken in these cases.

Several 6 H silicon carbide layers on different 6 H seeds were obtained by the travelling heater method. Plate 2 shows an oxidized sample with the following regions (from top to bottom): end of upper rod (polycrystalline), seed (6 H, n-type), SiC layer grown by the THM (6 H, n-type). The latter region apparently was grown under varying growth conditions: there is, at first, a perfect layer (probably grown at a low growth rate), then a layer with rather gross imperfections, and, finally, a polycrystalline area (probably due to a high growth rate). It may be concluded from several THM experiments at temperatures between

1800 and 1850 °C that perfect SiC layers can be grown at a rate less then 8 μm/min, approximately. A growth rate between 8 μm/min and 20 μm/min will end up with major imperfections, especially with respect to silicon inclusions which cause internal cracks during the cooling cycle. A growth rate exceeding 20 μm/min inevitably results in polycrystalline growth.

The silicon carbide layer grown on the 15 R seed at an average velocity of 6.7 μm/min was also 6 H. As can be seen from plate 3, the major part of the epitaxial layer is quite perfect. At the bottom, however, there are several inclusions of silicon causing fine cracks.

A few SiC layers grown by the travelling heater method were 15 R. Plate 4 is a sample taken from a 15 R layer grown on a heavily doped (p-type) 6 H seed.

N-type doping was achieved by filling the ampoule with nitrogen (0.5 and 20 Torr). Cubic SiC layers were obtained at the highest level of nitrogen doping. Experiments with boron doping failed, so far, i.e. the layers were n-type (like those without intentional doping). It is, therefore, concluded that boron evaporates from the melt and/or is transported by iodine from the hot zone to the walls of the ampoule (3).

REFERENCES:

(1) W.E. Nelson et al., AFCRL - 66 - 579 Report No. 3 (1966).

(2) K. Gillessen and W. v. Muench, J. Crystal Growth 19 (1973) 263.

(3) H. Schaefer, J. Crystal Growth 9 (1971) 17.

Plate 1: Polycrystalline silicon
          carbide rod grown by vacuum-
          sublimation

Plate 2: 6 H silicon carbide layer on
          6 H seed

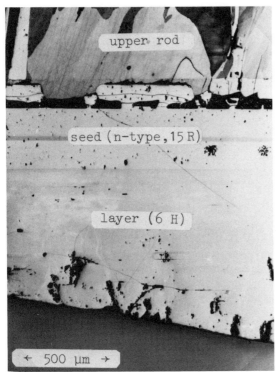

Plate 3: 6 H silicon carbide layer on
          15 R seed

Plate 4: 15 R silicon carbide layer
          on heavily doped (p-type)
          6 H seed

Epitaxial Growth of β-SiC on Si: Kinetics and Growth Mechanism

C. J. Mogab and H. J. Leamy

INTRODUCTION

Previous studies on the formation of thin β-SiC films by reaction of single crystal Si with gaseous hydrocarbons can be separated into two groups according to the growth technique employed: (1) reaction in a flowing gas mixture under conditions of viscous flow and (2) reaction in high (HV) or ultrahigh (UHV) vacuum chambers at hydrocarbon pressures corresponding to free molecular flow. The former process requires temperatures in excess of 1100°C and, in the case of reaction with CH_4, the rate of SiC formation is controlled by the diffusion of Si through the product layer.[1,2] In contrast, the vacuum growth technique yields highly oriented or epitaxial β-SiC layers at temperatures as low as 800°C.[3,4] However, the kinetics and growth mechanism for this reaction have not been ascertained.

EXPERIMENTAL

The experimental details of film growth have been given in the preceding paper. The extent of reaction was measured by determining the thickness of the SiC layer formed under a given set of reaction conditions. The film thickness was measured by use of the fundamental lattice absorption band of SiC at 12.6 μm. The details of this measurement technique are described elsewhere.[5]

RESULTS

Initial results indicated substantially smaller growth rates than reported by Learn and Khan,[4] particularly for HV growth. In addition, we noted a lack of reproducibility in the growth rate from run to run when nominally identical growth conditions were employed. For example, two wafers from the same ingot, reacted under identical conditions but in different vacuum cycles might have layer thicknesses differing by as much as 50%. This scatter precluded a thorough kinetic analysis. Consequently the results given below have been selected to illustrate typical observations rather than to provide comprehensive kinetic data.

Effect of Surface Preparation and Vacuum Conditions

We have observed no significant difference in growth rate between the HV and UHV systems for crystals reacted under otherwise identical conditions. Similarly, initial surface preparation or majority carrier type had no substantial effect on growth rate. Preannealing above 1150°C in UHV to obtain a clean Si surface[6] also had no effect on growth kinetics. Heating of a wafer in HV at temperatures near 1000°C for up to 2 hrs., with no intentional hydrocarbon introduction yielded β-SiC films approximately 100 Å in thickness, while for UHV the average film thickness obtained was $\lesssim$50 Å.

Dependence of SiC Film Thickness on Growth Conditions

The dependence of SiC film thickness on reaction time for the temperature range 950-1100°C is conveniently separated into two regimes. For C_2H_2 pressures below approximately 10^{-5} Torr, a linear dependence of SiC layer thickness on time is invariably found for crystals reacted in the same run, that is, the growth rate is constant. Figures 1 and 2 show typical data obtained at a temperature of 950°C. Figure 3 illustrates the effect of temperature on layer thickness for a C_2H_2 pressure of 3×10^{-6} Torr and a reaction time of 1 hr. It is generally observed that increasing temperature increases the growth rate; however, due to the scatter in the data, the temperature dependence of the linear rate constant has not been determined.

At C_2H_2 pressures above approximately 10^{-5} Torr the SiC layer thickness exhibits considerable scatter even for crystals reacted in the same run as shown in Fig. 4. It appears that after some relatively short initial period there is little or no further growth. As shown in the previous paper, SiC layers formed at C_2H_2 pressures above 10^{-5} Torr differ morphologically from those grown at lower pressures.

DISCUSSION

The linear growth kinetics (constant growth rate) observed at C_2H_2 pressures below 10^{-5} Torr indicate that the reaction is not controlled by diffusion in this pressure range. Moreover, the empirical diffusion coefficients which can be derived from the results of Graul and Wagner[2] and Nakashima, et al.[1] indicate that at $P_{C_2H_2} \lesssim 10^{-5}$ Torr, 900°C $\lesssim$ T $\lesssim$ 1100°C silicon transport through the defect free epitaxial regions of the film plays no significant part in film growth. Instead, the porous defect regions of the film provide a rapid "short circuit"

path through which Si is transported, by surface diffusion, to the film-vacuum
interface were reaction takes place. This conclusion is based on several characteristics of the observed defect morphology.[7]

   The linear growth kinetics imply that, for $P_{C_2H_2} \lesssim 10^{-5}$ Torr, surface transport is faster than some other step in the overall reaction which fixes the rate.
No definite choice from amongst the alternative rate controlling steps is possible on the basis of the limited kinetic data available. However, the general
observations of a rather weak temperature dependence for the growth rate (Fig. 3)
and a roughly linear pressure dependence (Fig. 2) suggest that adsorption of C_2H_2
or desorption of H_2 may be rate controlling.

   The growth kinetics and film morphology for C_2H_2 pressures exceeding $\approx 10^{-5}$
Torr differ markedly from those for growth below this pressure, implying a change
in the growth mechanism. In particular, the kinetic data suggest that growth is
almost completely suppressed following the formation of an initial thin layer of
SiC (Fig. 4). This would be expected if diffusion through the growing layer
became rate controlling once a continuous film had formed. The transition to a
bulk or grain boundary diffusion mechanism requires, of course, that the defects
be inoperative with respect to the faster surface diffusion process which occurs
at lower pressures. Since the defect density increases with pressure, it follows
that the defect channels must be "sealed-off" early in the growth for
$P_{C_2H_2} > 10^{-5}$ Torr.

   On the basis of the kinetic and microstructural data given here the following growth model is proposed. Initial reaction results in the formation of isolated SiC nuclei at the Si surface. In the low pressure regime (below $\approx 10^{-5}$ Torr
C_2H_2) these nuclei are mainly in preferred orientation, i.e., they are epitaxial.
The nuclei grow more rapidly laterally than vertically with the Si reactant being
supplied from adjacent unreacted surface regions. At the onset of reaction
impinging C_2H_2 molecules adsorb mainly on unreacted Si. The Si reactant is
removed uniformly over the surface as growth proceeds. Continued lateral growth
of the epitaxial SiC nuclei leads to eventual impingement and coalescence of
adjacent nuclei. During this growth process the amount of SiC surface increases
at the expense of unreacted Si surface. Consequently, adsorption of hydrocarbons
occurs to an increasingly greater extent on SiC rather than on unreacted Si.
Since diffusion of Si through the epitaxial carbide regions is quite slow, the
main source of silicon for further growth becomes the occluded regions of
unreacted Si, which experience an increased demand for Si, since they must now

supply a much larger area than during the initial stages of growth. Compliance with this demand leads to pitting of the substrate, which exposes other than (111) crystallographic faces and results in the formation of polycrystalline material. At the lower C_2H_2 pressures the rate at which Si can be supplied to the gas-solid interface, by surface diffusion from the defects, exceeds the rate at which C is supplied by adsorption and decomposition of C_2H_2. Consequently, film growth is uniform (constant growth rate) and the defects remain porous although they contain polycrystalline product material.

Clearly, in this scheme, the size and number density of initial nuclei will determine the size and number density of the resulting defects. According to classical nucleation theory the size and number density of nuclei would be expected to vary with temperature and acetylene pressure as observed here. That is, higher temperatures and lower C_2H_2 pressures result in a smaller number of larger nuclei being formed at the onset of reaction. The nuclei distribution is reflected in the defect distribution which is produced upon coalescence of the individual nuclei.

At higher C_2H_2 pressures the initial nuclei are mainly randomly oriented (non-epitaxial) and are present at a much higher number density, in accord with nucleation theory. Thus, at the point when coalescence begins, the occluded, unreacted regions are much smaller in size. In addition, at the higher pressures the rate at which C is supplied by adsorption and decomposition of C_2H_2 exceeds the rate at which Si can be supplied by surface diffusion. Growth is then no longer uniform but tends to occur preferentially around the defect (the Si source) leading to hillock formation. The pile-up of product material around the defects together with their smaller size and separation and the random orientation of the matrix leads to a sealing off of the defect channels at an early point in the growth sequence. Subsequent growth must then proceed by diffusion of Si through a dense polycrystalline layer, a process which is virtually precluded in the temperature range of interest.

CONCLUSIONS

The reaction of single crystalline Si with gaseous acetylene at elevated temperatures in a clean vacuum environment produces thin β-SiC overgrowths. At low C_2H_2 pressures ($\lesssim 10^{-5}$ Torr) the product growth rate is constant and the reaction proceeds by diffusion of Si through defects incorporated in the overgrowth. These defects are formed in the initial stage of reaction due to depletion of the Si reactant in localized regions. They consist of shallow (~ 1000 Å) pits in the

Si crystal, over which the growing SiC assumes a porous polycrystalline morphology. The formation of these defects is an intrinsic feature of the growth process resulting from the low growth rates and impermeability of the defect-free portion of the product phase to the diffusion of the reactants.

Higher C_2H_2 pressures ($\gtrsim 10^{-5}$ Torr) result in higher initial growth rates and a sealing-off of the defect channels at an early stage in the growth process. Subsequent reaction is nearly completely suppressed (at temperatures $\lesssim 1100°C$) due to the impermeability of the SiC product layer.

REFERENCES

1. H. Nakashima, T. Sugano and H. Yanai, Japan. J. Appl. Phys., 5, 874 (1966).

2. J. Graul and E. Wagner, Appl. Phys. Lett., 21, 67 (1972).

3. I. H. Khan and R. N. Summergrad, Appl. Phys. Lett., 11, 12 (1967).

4. A. J. Learn and I. H. Khan, Thin Solid Films, 5, 145 (1970).

5. C. J. Mogab, J. Electrochem. Soc., 120, 932 (1973).

6. R. C. Henderson, R. B. Marcus and W. J. Polito, J. Appl. Phys., 42, 1208 (1971).

7. C. J. Mogab and H. J. Leamy, to be published.

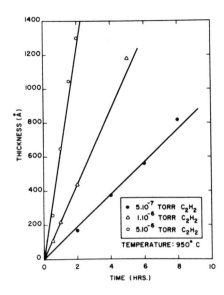

Figure 1. The dependence of β-SiC layer thickness on reaction time for a reaction temperature of 950°C.

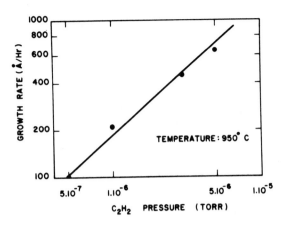

Figure 2. The dependence of growth rate on C_2H_2 pressure as determined from the data shown in Fig. 1.

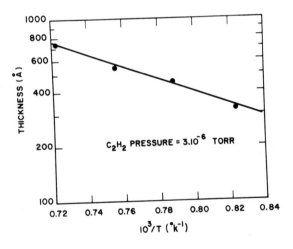

Figure 3. Typical temperature dependence of β-SiC layer thickness after one hour of reaction at 3×10^{-6} Torr C_2H_2.

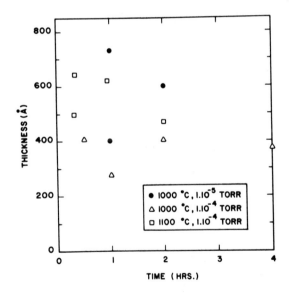

Figure 4. Variation of β-SiC layer thickness with time for C_2H_2 pressure $\geq 10^{-5}$ Torr.

Epitaxial Growth of β-SiC on Si: Growth Morphology and Perfection

H. J. Leamy and C. J. Mogab

The formation of SiC films by the reaction of single crystal Si with gaseous hydrocarbons at elevated temperatures has been studied by a number of investigators (1-8). Epitaxial films grown in both HV and UHV vacuum systems, as opposed to viscous flow systems, exhibit characteristic "pit" and "hillock" type defects (5-8), in addition to the usual crystallographic defects (stacking faults, twins, etc.) commonly found in thin epitaxial films. These pits and hillocks seriously compromise any conceivable electrical device applications for the films, but their origin is uncertain and it is not known whether they are inherent to vacuum growth or could be eliminated by some alteration of the growth process.

In an attempt to obtain a better understanding of the vacuum growth process particularly as regards the rate controlling mechanism, the effect of vacuum environment, and the origin of the characteristic defects, we have studied the growth of βSiC films by reaction of C_2H_2 with Si single crystals in HV and UHV systems. In this presentation we report on a detailed study of the formation and structure of the growth defects. The following paper contains an analysis of the growth kinetics and mechanism.

Both p and n type single crystal Si wafers were used in this study. The {111} oriented wafers were nearly dislocation free and of low oxygen content. They were reacted with C_2H_2 in both a HV and a UHV chamber. The HV system was pumped with a 6" diffusion pump containing DC 704 oil. The UHV system was an all metal, bakeable system consisting of a 400 liter/sec ion pump, a titanium sublimation pump, and a 200 liter/sec turbomolecular pump. A base vacuum of 3×10^{-10} Torr was routinely achieved in this system. Reactions were carried out over the temperature range 800-1150°C and for C_2H_2 pressures ranging from 1×10^{-7} Torr to

5×10^{-4} Torr. In all, more than 180 wafers were reacted under carefully controlled conditions. A detailed description of the experimental procedure appears elsewhere (9). The reacted wafers were examined with optical and scanning electron microscopy, reflection electron diffraction, and transmission electron microscopy.

Light optical microscopy of films formed at all but high ($P_{C_2H_2} \gtrsim 1\times10^{-5}$) pressures revealed "growth defects" similar to those observed in earlier work. For example, the micrographs of Plate I were obtained from specimens which had been reacted for various times at constant C_2H_2 pressure and temperature. It is evident that defect growth is concurrent with SiC film growth. Confirmation of this apparent trend is shown in Fig. 1, which indicates that the areal density increases linearly with film thickness, <u>regardless</u> of reaction conditions within the range: $1\times10^{-7} < P_{C_2H_2} < 1\times10^{-5}$ Torr, $800°C \lesssim T \lesssim 1050°C$. Films formed at pressures above $\sim 1\times10^{-5}$ Torr appeared to be relatively smooth and defect free by optical microscopy. Optical microscopy also revealed that surface defects, particularly scratches, act as sites for preferential formation of growth defects. In general, surface scratches were decorated with defects and a zone of defect free material was observed around every decorated scratch.

The structure of a typical, thin SiC layer formed at low C_2H_2 pressure (5×10^{-7} Torr) is shown in the scanning electron micrographs which comprise Plate II. The growth defects are revealed as roughly triangular regions of high relative secondary electron emissivity, which are bounded by irregular patches of darker, low emissivity material (Plate

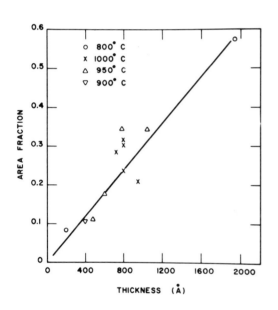

Fig. 1

IIa). The regions of high emissivity shown at higher magnification in Plate IIc are seen to consist of 2000-3000 Å particles which form a porous, connected structure. The irregular patches of low secondary electron emission are clearly delineated in Plate IIb which was produced by collecting high energy, back-

scattered electrons, which easily penetrate the thin SiC layer. The dark patches
are regions of low relative backscatter coefficient which extend beneath the high
emissivity, porous regions. The nature of these dark regions may be deduced by
noting that the backscatter coefficient increases linearly with atomic number (10).
Dark regions therefore appear when the electron probe samples a region of lower
relative density. As will be shown subsequently, these dark regions are in fact,
shallow, flat bottomed pits in the underlying silicon.

It is apparent that, since β SiC is nearly transparent in the visible
range, the defects observed by optical microscopy are these shallow pits. The
growth defects thus consist of a porous aggregate of small crystallites situated
over a shallow, flat bottomed pit in the silicon substrate. Direct confirmation
of this morphology is given in Plate III, which is a cross section view obtained
by imaging a cleaved wafer. Plate III clearly shows that the SiC/vacuum inter-
face is smooth, while the underlying silicon is pitted to a depth of $\sim$ 2000 $\overset{\circ}{A}$ on
a lateral scale equivalent to that of the pit distribution which was observed by
imaging the SiC/vacuum surface.

The effect of pressure on growth defect morphology is shown in Plate IV.
It is immediately apparent that growth defects are formed at number densities
which increase with pressure up to $P_{C_2H_2} \gtrsim 10^{-5}$ Torr, where the entire film is
covered with 200-500 $\overset{\circ}{A}$ diameter volcano like hillocks at a density of $\sim 8\times10^8 cm^{-2}$.
Quantitative determination of the defect number density revealed that it increases
linearly with $P_{C_2H_2}$ up to 1×10^{-4} Torr, where the hillock density closely matches
the expected pit density.

It is therefore established that the growth defects consist of porous
aggregates of individual crystallites which occur in conjunction with pits in the
underlying silicon. Their density increases linearly with increasing hydrocarbon
pressure and their area (and volume if a constant pit depth is assumed) is proportional to the extent to which the reaction has occurred; i.e. to the film
thickness.

Plate Va is a secondary electron image of a SiC film which has been
removed from the Si substrate. As might be expected, no evidence for below background scattered electron contrast could be obtained. The single crystallinity
of the defect free region of this specimen is confirmed by the electron transmission results shown in Plate Vb. Vb is a dark field scanning transmission image
of the area shown in Va, which was produced by collecting the electrons diffracted
by the $(2\bar{2}0)$ planes. Diffracting regions appear light, while regions bent away

from the Bragg diffraction orientation appear dark. In addition to the broad
bend contour visible in Vb, triangular areas of below background contrast corresponding to the locations of growth defects are visible. For example, the growth
defect labeled "A" in Va appears as a dark triangle in Vb. It is therefore evident that the dark triangular portion of this defect, delineated in Vb is not
epitaxially oriented.

Confirmation of the polycrystalline nature of the defects in given in
Plate VI. VIa is a conventional selected area diffraction pattern of a defect
free region of the film shown in Plate V and VIb is a diffraction pattern of
the triangular porous region of a typical defect. VIb is typical of films produced at $P_{C_2H_2} \gtrsim 1 \times 10^{-4}$ Torr, and shows that the porous region is composed of
randomly oriented crystallites. VIa is representative of the interdefect regions
of all films grown at low pressures, and contains evidence for the presence of
crystallographic defects (stacking faults, etc.) within this nominally single
crystalline, epitaxial film.

Our observations may be summarized as follows: (1) at a given C_2H_2
pressure the size and number density of the defects increases and decreases,
respectively, with increasing growth temperature. (2) At a given temperature, the
defect size decreases with increasing C_2H_2 pressure while the number density of
defects increases. (3) At a given temperature and C_2H_2 pressure, the defect
density is independent of time, but the defect size increases with film thickness
and thus with time. (4) Scratches in the substrate generally show a higher density of smaller defects than scratch-free regions, and are bounded by a defect-
free zone. In the following paper a mechanism for SiC layer growth consistent
with these observations is described.

References

1. W. G. Spitzer, D. A. Kleinman and C. J. Frosch, Phys. Rev. 113, 133 (1959).

2. H. Nakashima, T. Sugano and H. Yanai, Japan J. Appl. Phys. 5, 874 (1966).

3. P. Rai-Choudhury and N. P. Formigoni, J. Electrochem. Soc. 116, 1440 (1969).

4. J. Graul and E. Wagner, Appl. Phys. Letters 21, 67 (1972).

5. I. H. Khan and R. N. Summergrad, Appl. Phys. Letters 11, 12 (1967).

6. K. E. Haq and I. H. Khan, J. Vacuum Sci. Technol. 7, 490 (1970).

7. A. J. Learn and I. H. Khan, Thin Solid Films 5, 145 (1970).

8. A. S. Brown and B. E. Watts, J. Appl. Cryst. <u>3</u>, 172 (1970).

9. C. J. Mogab and H. J. Leamy, to be published.

10. P. R. Thornton, <u>Scanning Electron Microscopy</u>, Chapman and Hill Ltd., London, 1967.

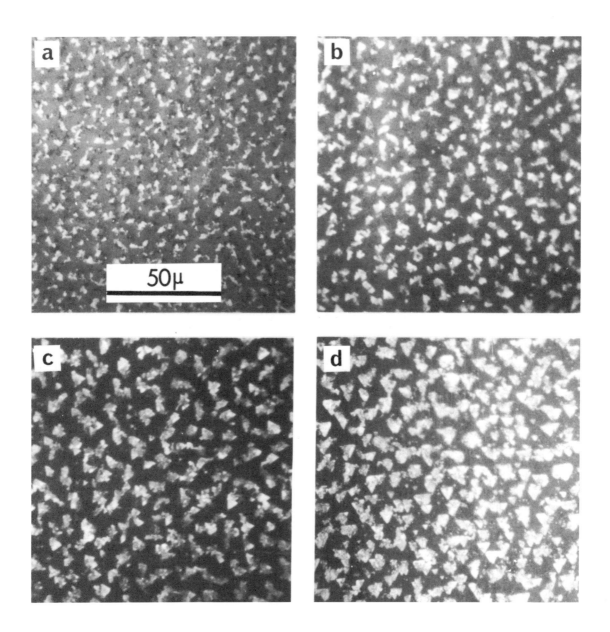

PLATE I: Light optical micrographs of SiC films formed by reaction at $P_{C_2H_2} = 5\times10^{-7}$ Torr, T = 955°C and: a) 2 hours, b) 4 hours, c) 6 hours, d) 8 hours.

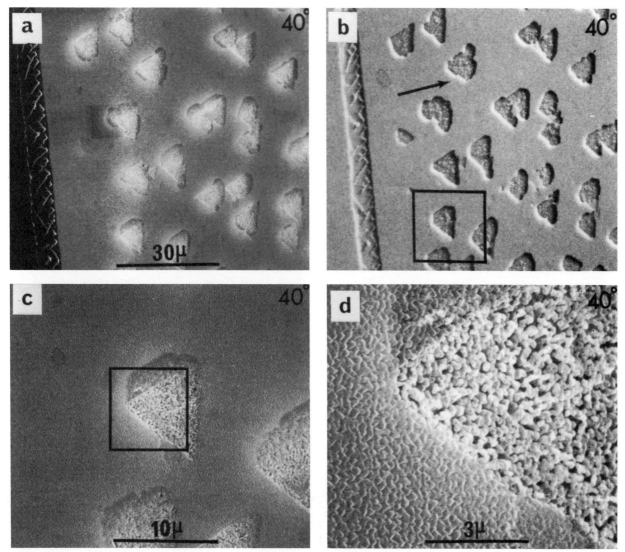

PLATE II: Scanning electron micrographs of an approximately 800 Å SiC film formed at T = 1050°C and $P_{C_2H_2}$ = 5x10^{-7} Torr. Figures IIa and IIb are secondary and backscattered electron images of the same area of the specimen. The arrow in Fig. IIb points toward the backscattered electron detector. Figure IIc is a higher magnification image of the area outlined in IIb, and Figure IId shows the area outlined in Figure IIc.

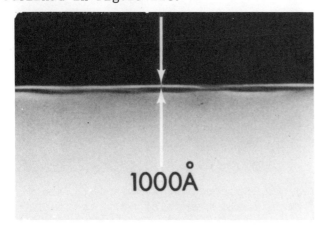

PLATE III: Edge on view of a 1000 Å SiC film and Si substrate. Note the shallow pits in the substrate.

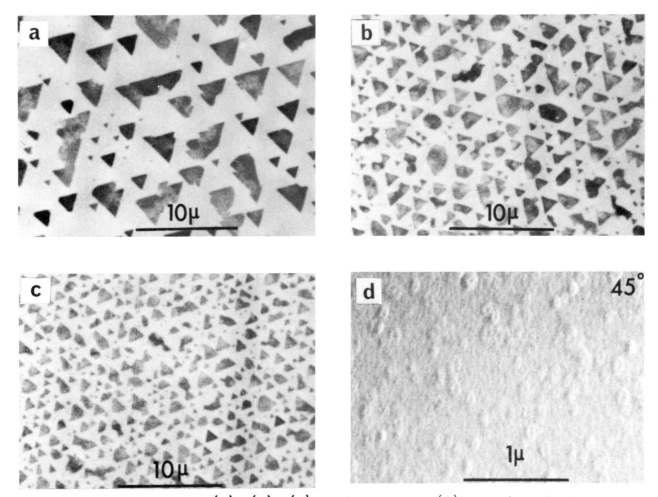

PLATE IV: Backscattered (a), (b), (c), and secondary (d) scanning electron micrographs of SiC films produced by reaction at 950°C for: (a) 4 hrs, $P_{C_2H_2}$ = 1×10⁻⁶ Torr, (b) 30 min, $P_{C_2H_2}$ = 5×10⁻⁶ Torr, (c) 30 min, $P_{C_2H_2}$ = 1×10⁻⁵ Torr, and (d) 3 min, $P_{C_2H_2}$ = 1×10⁻⁴ Torr.

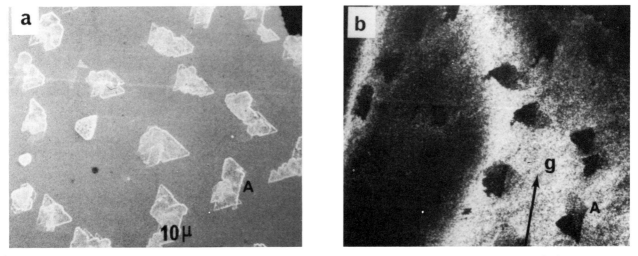

PLATE V: Scanning electron micrographs of an unsupported SiC film. (a) is a secondary electron image and (b) is a $(2\bar{2}0)$ dark field transmission micrograph of the area shown in (a).

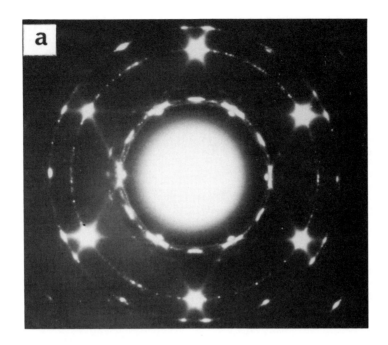

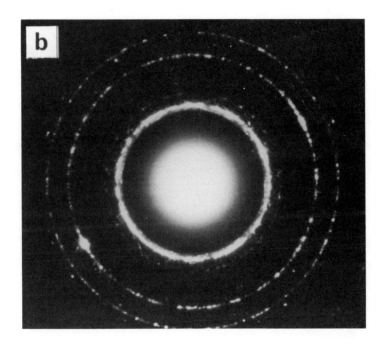

PLATE VI: Selected area electron diffraction patterns of: (a) a growth defect free region of the film shown in PLATE V, (b) the nonepitaxially oriented, porous region associated with a typical growth defect.

Epitaxial Growth of SiC on AlN Substrates

R. F. Rutz and J. J. Cuomo

INTRODUCTION

The match of the lattice constants for SiC and AlN are quite close and the growth of crystalline AlN on SiC has been observed.[1] The opposite process of growing single crystal SiC on AlN has not been widely explored, probably because of the unavailability of suitable large area AlN crystals and also because at the high temperatures usually used for SiC growth ($\sim$ 1800°C) the AlN tends to decompose. Another serious problem is that the coefficients of expansion of AlN and SiC, while close, are sufficiently different over the wide temperature ranges involved that stress is introduced, leading to crazing and cracking of one material or the other, but most frequently the AlN. This paper describes growing large area flat and smooth single crystals of tungsten and sputtering AlN on these wafers at $\sim$ 1000°C on the order of one μm thick. The AlN with the tungsten still adhering is used as a substrate to grow single crystal SiC layers, using a close-spaced pyrolytic vapor deposition technique.[2] These layers are normally doped n-type, but p-type layers have been grown by adding elemental aluminum in the growth chamber. The quality of the SiC is sufficiently good to make p-n junction light emitting diodes.

PREPARATION OF THE TUNGSTEN AND SPUTTERED AlN SUBSTRATES

Using the hydrogen reduction of tungsten hexafluoride chemical vapor deposition process, layers of (111) single crystal tungsten can be epitaxially grown on (0001) sapphire wafers.[3,4] It has been found that, with carefully prepared sapphire substrates[5] processed at a temperature of about 550°C, tungsten layers more than 10 μm thick can be produced on wafers one inch in diameter. When the sample is cooled to room temperature, the tungsten will

72

separate cleanly from the sapphire at the interface due to the thermal expansion difference between the two materials and the weak interfacial bonding. The mechanical attachment at the edge of the wafer is readily removed by grinding.[6] When separated from the substrate, the tungsten is a replica of the sapphire surface; it is smooth, essentially strain free and self supporting with the [111] orientation normal to the surface. X-ray examination often reveals twinning.

These smooth tungsten wafers are used as substrates onto which AlN is sputtered. The sputtering process which has been previously described[7] consists of using a 99.999% pure aluminum target and, after suitable outgassing procedures, reactively sputtering in pure nitrogen with an rf driven bias of -50v dc applied to the substrate. For good epitaxial growth in the system used the tungsten wafers are heated to 1100°C with a power density of 8w/cm^2 at the target. The rate of deposition is typically 2Å/sec. Under these conditions, the deposited AlN can be single crystalline, even on tungsten wafers in which x-ray examination reveals twinning, and serve as suitable substrates for the production of epitaxial single crystal SiC. Although the AlN has a higher average coefficient of linear expansion than tungsten in the temperature range involved, if the sputtered layer is kept thin ($\sim$ one μm) with respect to the tungsten thickness, the warping of the wafers is not appreciable. The AlN coated tungsten is cleaved into squares approximately .250" on a side in order to fit into the crucible used for SiC growth.

GROWTH OF SiC

For the growth of SiC, a tungsten heater strip crucible holder made from 1/8" thick tungsten shown in Fig. 1(a) is used. The crucible holder depression is 0.070 inches deep and .375 inches in diameter. A pyrolytic carbon crucible with the high thermal conduction direction vertical fits the depression. It is .150" high with a .025" wall thickness. In it a presynthesized SiC crystal source is placed with the AlN surface of the tungsten-AlN seed resting on the SiC as shown in the cross-sectional view of Fig. 1(b). The heater strip is heated by 60 $\sim$ AC current controlled by a variac. A current of approximately 800 amperes is required to achieve a typical growth temperature of 1860°C as measured by an optical pyrometer focused on a slot .005 inches wide and .015 inches deep cut into the side of the heater strip. The heater strip is clamped into massive water cooled copper electrodes and enclosed in a 1.5 inch diameter molybdenum light shield slotted for pyrometer viewing and insulated from the electrodes by quartz rings and into which Ar forming gas (15% H$_2$) is introduced through a molybdenum pipe. The entire assembly is covered with a nine inch diameter hemisphere of pyrex, with the forming gas escaping from the hemisphere at the rim which rests loosely on an aluminum base.

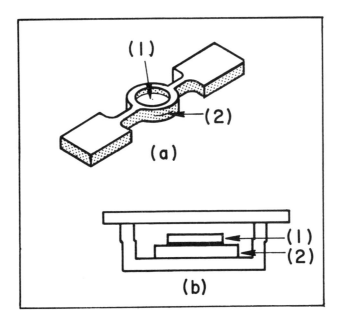

Figure 1

Tungsten heater strip and crucible. (a) Heater strip (1) crucible depression (2) Temp. measuring slot. (b) Crucible (1) W-AlN seed (2) SiC source crystal.

It has been found that a tungsten heater temperature of approximately 1860°C provides conditions within the carbon crucible to promote transfer of SiC from the SiC source crystal to the AlN at a growth rate of 0.1 μm/min. The temperature of the SiC is not known, but is somewhat cooler than this and the AlN coated tungsten wafer is possibly 50°C cooler than the SiC since there is a vertical temperature gradient.

It is desirable to have the SiC source wafer polished to a smooth finish since the transfer of SiC to the substrate is very dependent upon spacing and the surface of the transferred layer will be a negative replica of the SiC source crystal surface if depressions or ridges are more than a few micrometers in depth and height.[2] Except for the effects of the aluminum and nitrogen from decomposition of the AlN, the doping of the layer will be determined by that of the SiC source.

At the growth temperature the AlN decomposes, but since this is reached in a short time (typically five minutes), only a small amount escapes before a layer of SiC is formed inhibiting further decomposition of AlN except at the periphery of the sample. An example of the SiC growth is shown in Plate 1. In this sample, the thickness of the SiC layer is $\sim$ one μm and the SiC near the edge overhangs the region from which the AlN has disappeared. Another example is shown in crosssection in Plate 2. The surface of the grown SiC layer generally shows hexagonal or triangular patterns, or sometimes a terraced appearance. However, some samples show a corrugated surface. Several samples have been measured by x-ray analysis by J.Angilello of the IBM Research Center who has found the layers to be strained, hexagonal single crystalline, and generally oriented (0001). The (0001) plane of the SiC in this case is parallel to the (111) of the tungsten and, in the plane, the [11$\bar{2}$0] direction in the SiC corresponds to an equivalent [211] direction in the tungsten. In the corrugated samples, the (11$\bar{2}$, 12) orientation tends to occur.

If the growth is continued long enough, all of the AlN will disappear, while the SiC layer continues to grow with the tungsten resting on the growing layer. The grown layer of SiC only lightly adheres to the tungsten and the source crystal and is readily removed from them. Plate 3 shows a crystal grown in this manner. The darker spots on the surface are spurious nucleations formed during the cooling period which can be quite abrupt (a minute or two). Plate 4 shows a top and cross-sectional view of another sample grown this way at a higher magnification. The grown SiC typically is doped n-type, presumably due to the nitrogen from the decomposing AlN, even when moderatly doped p-type crystal sources are used. However, if a few milligrams of elemental Al is added to the crucible, heavy p-type layers are grown. The crystals are blue, but more so on the edges than the center of the larger samples, showing that the Al doping is non-uniform.

DEVICE APPLICATIONS

Some thick samples of SiC heavily doped with Al with the AlN and tungsten removed were made into p-n junction diodes by alloying one side to a tungsten tab for ohmic contact and alloying silicon fragments 10 mils in diameter to the other side in a nitrogen atmosphere to form an N region. Biased in the forward direction, these emitted light of a pronounced bluish color with a reasonably high quantum efficiency of $1.1 \times 10^{-3}\%$ at 30 ma bias. A spectrum of one diode

measured by R. Tsu of the IBM Research Laboratory is shown in Fig. 2. In some
layers, a p-n junction with similar light emitting qualities occurred during
growth with the N type evidently being due to the nitrogen from source crystal
and the decomposing AlN in the initial states of growth and the p-type due to

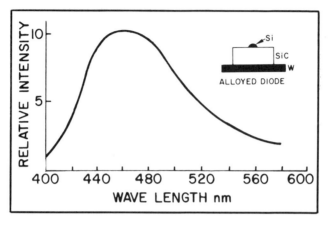

Figure 2

R.T. spectral response of SiC p-n junction diode 30 ma. forward bias.

the added Al charge in later stages. In some of the alloyed diodes, tunnel
diode characteristics were observed indicating that the Al doping in these
diodes is on the order of 10^{20} parts/cm^3.[8]

Another type of device that has been fabricated is a bistable switch with
memory. This is made from samples in which shorter growth times were used so
that the SiC, AlN and W portions are still integral. Sections 5 mils square
were cut, and contacts applied to the SiC and tungsten which act as terminals
to the AlN, the active material of the device. These operate with characteristics similar to Si-AlN-W switches[7] in that they may be cycled at room
temperature from the high to low and back to high impedance states by application of an alternating voltage at 10 mc/sec. rates and retain either the high
or low impedance state for many months at zero bias. They have the added
feature of operating although at lower cycling rates at temperatures as high as
500°C. Encapsulation will be required for long periods of operation since
tungsten oxidizes. Plate 5 shows the characteristics of a unit being switched
at 500°C at 100 Hz.

SUMMARY AND DISCUSSION

In summary, a novel method of preparing SiC has been demonstrated which
offers the promise of obtaining relatively large area single crystal layers

of SiC thick enough and of a quality suitable for the production of useful
electronic and electro-optical semiconductor devices. The furance geometry
and method of growth have limited the size of present crystal layers to approximately 6.5 mm, but the extension to much larger areas limited only by the size
of available single crystal sapphire wafers appears to be straightforward. AlN
crystals have also been grown at $\sim$ 1780°C with the same apparatus on the
tungsten-AlN substrates using a sintered AlN source, but cracking almost always
occurs if thicknesses greater than a few microns are grown. Other methods of
SiC growth that are compatible with AlN and tungsten such as sputtering and
some chemical vapor deposition techniques, should be applicable at a lower
temperature than that required by the method used here and present fewer difficulties with respect to decomposition of the AlN and stress due to the different
rates of thermal expansion of the materials. If growth occurs below 1700°C so
that alloying of SiC with tungsten does not occur, the AlN layer may not be
necessary.

The authors wish to thank E. W. Harden and W. W. Molzen for their contributions to the heater strip design and sputtering techniques.

REFERENCES

(1)   W. F. Knippenberg and G. Verspui, Silicon Carbide-1968, Proc. Inter. Conf.
SiC, Univ. Park, Penna., Oct. 20-23, 1968, p. 51.

(2)   R. F. Rutz, U.S. Patent No. 3,577,285 (1971).

(3)   A. Brenner and W. E. Reid, U.S. Patent No. 3,072,983 (1973).

(4)   A. F. Mayadas, J. J. Cuomo and R. Rosenberg, Jour. Electrochem. Soc., 18,
1742 (1969).

(5)   A. Reisman, A. M. Berkenblitt, J. J. Cuomo, S. A. Chan, Journ. Electrochem.
Soc., 118, 1653 (1971).

(6)   J. J. Cuomo, IBM Technical Disclosure Bulletin, 15, 564 (1972).

(7)   R. F. Rutz, E. P. Harris, and J. J. Cuomo, IBM Jour. Res. and Develop.,
17, 61 (1973).

(8)   R. F. Rutz, IBM Jour. Res. and Develop., 8, 539 (1964).

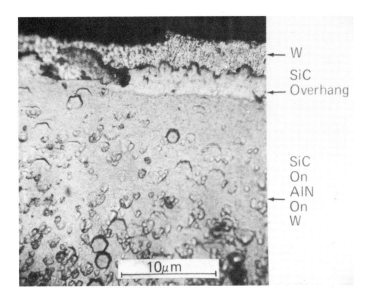

Plate 1

Top surface near the edge of a SiC layer
grown on an AlN layer sputtered on a
tungsten crystal.

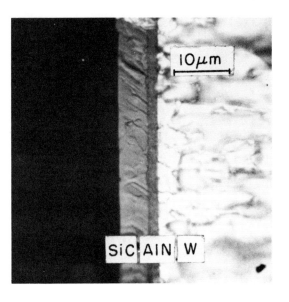

Plate 2

Cleaved edge of a SiC layer grown
on an AlN layer sputtered on a
tungsten crystal.

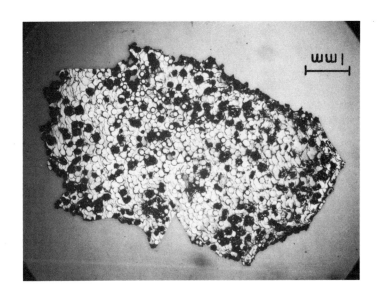

Plate 3

SiC crystal approximately 30 μm thick separated
from the tungsten during growth by total sublimation of the AlN layer.

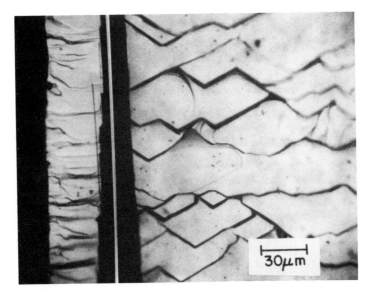

Plate 4

Cleaved edge and top view of a SiC crystal
separated from the tungsten during growth
shown at same magnification. The right edge
of the cleaved section is the top of the crystal.

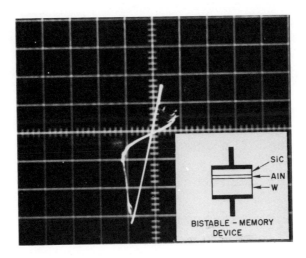

Plate 5

I-V characteristics of a SiC-AlN-W bistable-
memory device being cycled at 100 Hz at 500°C.
0.5 v/div. horizontal, 0.4 ma/div. vertical.

Chemically Vapor Deposited SiC for High Temperature and Structural Applications

J. R. Weiss and R. J. Diefendorf

Chemical vapor deposition has been investigated as a means of producing high purity silicon carbide of sufficient size that studies of its potential as a high temperature structural material could be made. A wide range of deposition conditions provided the limits within which pure silicon carbide could be deposited from methytrichlorosilane and hydrogen. The relationship between structure, properties and deposition conditions has been analyzed with special emphasis on the determination of residual stresses in the deposits.

Chemical vapor deposition has produced high purity ceramic and refractory materials, but usage of these materials in bulk form often has been limited by poor strength, high residual stresses, and high cost. Past performance of CVD SiC coatings on nuclear fuel particles has prompted the present investigation of this process for the production of monolithic parts [1-5]. A very wide range of deposition conditions was studied to prescribe the limits for deposition of pure silicon carbide. Specimens made within the region where pure SiC deposited were characterized for crystal structure, preferred orientation, microstructure, and surface morphology. Transverse bend strength and residual stress were determined on selected samples.

80

Experimental Procedure

Silicon carbide was deposited from methyltrichlorosilane/ hydrogen mixtures inside eight inch long graphite tubes of 1/2 and 1 inch bore. (Several runs which produced 2½" x 7 3/4" plates of CVD SiC were also made.) These deposition tubes were maintained at constant temperature within a resistance heated graphite vacuum furnace (Fig. 1), and deposition tube temperature was monitored by

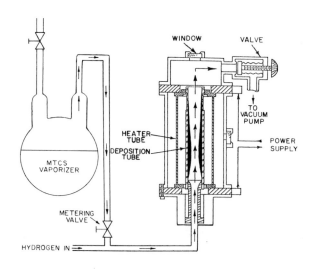

Figure 1

Schematic of Deposition Furnace Apparatus

optical pyrometry. A rotary pump was used to provide reduced pressures. The methyltrichlorosilane concentration was controlled by maintaining the liquid methyltrichlorosilane at a constant, but elevated temperature, and leaking the vapor through a micrometer needle valve into the furnace. The duration of each run was adjusted to permit sufficient deposit buildup to reduce substrate effects but not so thick as to change flow conditions. The deposition tubes were sectioned, the deposition rate profiles measured, and the morphology examined by optical and scanning electron microscopy. X-ray diffraction was used to determine crystal structure, preferred orientation, and the presence of excess carbon and silicon. Infrared spectrometry also was used to detect impurities. Density, bend strength and residual stress were found for selected runs. Strength was measured by four point bending with the major span of 1 3/8 inches.

Residual stresses were measured by using a variation of the Sach's technique. The thin tubular deposits provided an easy method of determining residual stress. After noting the diameter of the sample (with the graphite removed), the cylinder was sliced axially, and the change in circumference noted. The residual stress within the ring can then be calculated from:

$$\sigma_{R.S.} = \frac{Et \ \Delta S}{4\pi r^2 - 2\pi \Delta S}$$

where   E = Young's modulus
        t = deposit thickness
        r = radius of cylinder
        ΔS = change in circumference.

Correlations between deposition conditions and residual stress were made by multiple regression analysis.

Results and Discussion

Traditionally, methyltrichlorosilane has been used since it contains one silicon and one carbon atom, and hence, stoichiometric silicon carbon would supposedly deposit   .

(1-8)

However, the need for and importance of certain thermally produced molecular species in the deposition is apparent from the
position of the maximum deposition rate along the inside of the
deposition tube (Fig. 2).

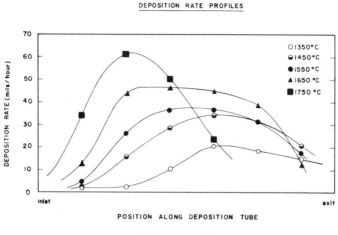

DEPOSITION RATE PROFILES

Figure 2

The effect of temperature on the deposition of SiC as a
function of distance. The deposition geometry is that
shown in Fig. 1. The SiC was deposited from hydrogen:
methyltrichlorosilane with a 2:1 ratio, at a pressure of
5 torr; a flow of 5 cfh and a tube diameter of 0.5 inch.

If the distance along the tube were purely that required to heat

the feed gas molecules to the deposition temperature, the position

of this maximum should be independent of temperature. Actually the

distance from the cold-hot junction to the maximum in rate is found

to decrease sharply as the deposition temperature is raised.

Methyltrichlorosilane either fragments to more reactive free

radicals or decomposes to other more reactive molecules. It is

from these or subsequently formed free radicals or molecules that

silicon carbide is formed. The exact nature of the species is

somewhat speculative, but primary fragmentation to CH_3 and $SiCl_3$ is

most likely. It is the reaction of these species with the surface
that may deposit silicon carbide, free silicon, or free carbon. At
temperatures exceeding $\approx$ 1600°C, and at 20 torr or less, pyrolytic
graphite or soot is deposited along with the silicon carbide. At
lower temperature with pure methyltrichlorosilane, patches of
pyrolytic graphite nucleate on the substrate along with the silicon
carbide. The deposition of carbon can be inhibited by the addition
of hydrogen, and experimentally, it has always been found necessary
to add hydrogen to the methyltrichlorosilane to eliminate carbon
deposition and produce high quality deposits. However, these
hydrogen additions can cause deposition of excess silicon at
temperatures below approximately 1300°C.

The relation of microstructure to the deposition process is
most easily comprehended in terms of nucleation theory. Over the
range of temperature that is commercially attractive for SiC deposition, the deposition rate increases relatively slowly. Hence, an
increase in temperature, which will increase surface mobility, will
increase grain size. Similarly, a decrease in concentration of
carbon and silicon in the gas-phase will decrease deposition rate
and increase grain size at constant temperature. At the lower
deposition temperatures, the crystallite size is sufficiently small
that the surface morphology and microstructure are determined by
surface inhomogeneities and the appearance is similar to pyrolytic
graphite (Plate 1). As the temperature is increased, the crystal-

lite size will attain dimensions similar to those of the growth
cones caused by surface inhomogeneities. At this temperature or
any higher temperature, the microstructure and surface morphology
depend on the crystallite size. Plate 1 shows the effect of increasing the surface mobility and decreasing the deposition rate.
There is also a change in crystal habit from cubic, to rhombohedral
to hexagonal as the temperature is increased. At pressures greater
than 20 torr, accentuated morphological details such as dendrites,
whiskers, and nodules with deep crevices between them   form. All
are probably caused by the high concentration gradient which exists
under the diffusion controlled deposition at these pressures. The
relation between surface morphology (which can be related to microstructure) and deposition conditions is shown in Fig. 3.

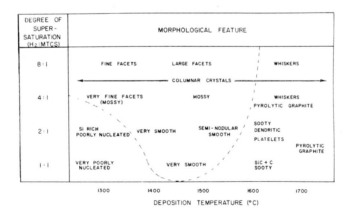

Figure 3

The morphology of SiC deposits as a function of deposition conditions. Material was taken from ring IV.
Tube diameter is 0.5 inch; flow = 5 cfh.

Within the ranges of conditions producing pure silicon carbide,

the material was translucent, yellow in color, and essentially

theoretically dense as determined by immersion and X-ray techniques.
Infrared transmission and specular reflectance measurements of 1-2mm
thick plates correspond to previously reported results on single
crystal β-silicon carbide, although the absolute transmission was
proportionately higher[9]. The crystal structure is β-SiC with
occasional diffraction lines corresponding to some α inclusions or
caused most likely by stacking faults as reported[10]. Material of
predominantly cubic SiC had a strong preferred orientation ($\pm$ 6°) of
of the (111) planes perpendicular to the growth direction. Other
preferred orientations were observed to occur with different
crystal structures.

Room temperature strength values of between 50 and 80 KPSI
were common (four point bending - courtesy of J. Palm, General
Electric Co., Research and Development Center, Schenectady, N.Y.).
However, when the same material was tested at 1500°C the strength
had increased to an average of 180 KPSI. Subsequent examination
shows that specimen preparation appears to affect the strength at
room temperature. Diamond grinding and polishing leaves remnants
of microcracks and chips.

Though the majority of the SiC produced was relatively strong,
residual stresses sometimes exceeding the actual strength of the
part was observed. Obviously, when constrained shapes such as
cylindrical tubes are produced, the residual stress decreases the
useful strength of the part by its magnitude. Though the smoothest

finest grained material is most desirable, from a strength stand-

point, it is this material that contains the highest residual

stress. If deposits are produced at a set of conditions yielding a

very crystalline material the residual stress is apt to be low

(< 5000 PSI). When the material is very fine grain, the residual

stress will usually be greater than 10,000 PSI (see Plate 1). An

explanation of this lies in the relative rates with which individual

atoms or layers of SiC are being added. At low deposition rates an

individual atom is permitted more time to migrate to a growing front

of an individual crystallite. As deposition rates increase, re-

arrangement time is shortened and a less ordered structure is

"frozen in" (see Fig. 4). The rates can be appreciated by calcu-

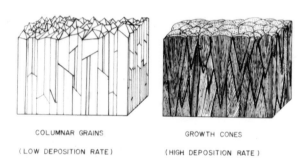

COLUMNAR GRAINS   GROWTH CONES

(LOW DEPOSITION RATE)   (HIGH DEPOSITION RATE)

Figure 4

Typical cross-section through the last deposited surface
of two pieces of CVD SiC. The low deposition rate model
shows a highly crystalline structure that resulted from
high surface mobility due to either high temperature or
low reactant concentration. At high deposition rates,
the structure is "frozen-in" and a last deposited
surface, typical of pyrolytic graphite, results.

lating the number of atomic layers per unit time being added. The

distance between (111) planes is $\approx$ 2.5Å. Therefore, at 15 mils/hr deposition rate, layers are being added at the rate of ≈ 450 per

second. A plot of residual stress versus deposition rate for 1 inch

bore deposition tubes confirms the theory (see Fig. 5). Note that

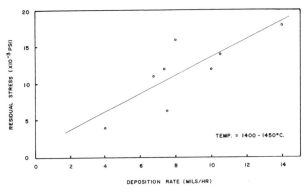

Figure 5

Residual stress as a function of deposition rate. Each
point corresponds to the measured residual stress for a
particular run.

though the run to run scatter is high, a trend does exist. Further

evidence of a correlation is seen from the relationship between

residual stress and deposition rate for a particular deposition

tube as shown in Fig. 6.

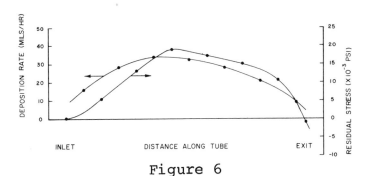

Figure 6

Deposition rate and residual stress versus distance along
a particular deposition tube showing the strong correlation within a run.

One consequence of residual stress is that although SiC can be

deposited in the final shape of a desired part, attention must be

paid to whether the deposit is made internally or externally to make use of the sign of the residual stress in minimizing applied stresses. Superimposed on this is the necessity of using substrates with expansion coefficients that match that of the CVD SiC ($\approx$ 4.5 x 10^{-6}/°C) otherwise stresses from the mismatch can augment those of the residual stress and cause the material to crack up during cool down.

Conclusions

High strength CVD SiC can be produced but flaws can be introduced by simple machining and polishing that reduce the strength of the sample four fold. By proper control of deposition conditions, residual stress can be minimized but generally at an added expense of increased deposition time. Finally, when constrained shapes are to be made, attention should be paid to mandrel design, since the proper residual stress may reduce the stresses applied in use.

Acknowledgment

The authors are indeed grateful to the General Electric Corporate Research and Development Center, Schenectady, N.Y., for partial support of this research and in particular to Mr. J. Palm of that organization for his help in obtaining the strength data. For many helpful discussions about organosilicon compounds and for providing us with the methyltrichlorosilane, thanks are due Mr. T. Selin of the General Electric Co., Silicone Division. Partial support of the project was also provided by both the General Electric Co. Reentry Systems Division and the Eastman Kodak Co. and is gratefully acknowledged.

Special thanks are due Mr. T. Mints, Jr. president of Sargent Welch Scientific Co. for providing the vacuum equipment repairs and to David Chua of RPI for excellent scanning electron photomicrographs of the samples. The research was conducted at the NASA Materials Research Center at RPI in Troy, N.Y.

References

1.　Gulden, T. D., J. Amer. Cer. Soc. <u>51</u> (8),(1968).

2.　Price, R. J., Amer. Cer. Soc. Bull. <u>48</u> (9), (1969).

3.　Gulden, T. D., J. Amer. Cer. Soc. <u>52</u> (11), (1969).

4.　Kaae, T. L., J. Amer. Cer. Soc. <u>54</u> (12), (1971).

5.　Ivanova, L. M., Pletyushkin, A. A., in "Silicon Carbide", I. N. Frantsevich, Plenum Pub. Corp., 1970.

6.　Kawashima, C., Setaka, N., and Nakagawa, J., J. Cer. Asso. Japan <u>75</u> (2), (1967).

7.　Phillips, E. and Lukas, K., AECL-3674, Feb. 1971.

8.　Popper, P. and Mohyuddin, I., in "Special Ceramics", (1964), Ed. by P. Popper, Academic Press, N.Y. 1965.

9.　Lipson, H. G., in "Silicon Carbide" 1960, Ed. by J.R.O'Connor and J. Smiltens, Pergammon Press, N.Y. 1960.

10.　Gulden, T. D., J. Amer. Cer. Soc. <u>54</u> (10), (1971).

Plate 1

Scanning electron photomicrographs of last deposited durface
morphologies of CVD SiC deposited at stated condition variations.

a) Low temperature, high reactant
concentration. Deposition rate
was 21 mils/hr. Low surface
mobility resulted in a relatively high residual stress.

b) Higher temperature than (a);
all other conditions maintained.
Higher surface mobility results
in the disappearance of growth
cones. (Rate 34 mils/hr.)

c) Decrease in reactant concentration with all other conditions
held constant, results in a lower
deposition rate and lower
residual stress. (Rate=19 mils/hr.)

d) At very low concentrations the
SiC becomes very crystalline and
well defined crystallites for the
last deposited surface. (Rate =
11 mils/hr.)

Growth Mechanisms of Silicon Carbide in Vapour Deposition

W. F. Knippenberg, G. Verspui and A. W. C. von Kemenade

1. Introduction

A study of the growth mechanisms of silicon carbide is a matter of practical importance and of scientific relevance.

It is a matter of practical importance for the obvious reason of the control of morphology. It provides us with the methods of fulfilling the varying needs of research and industry for materials in the form of single crystals differing in shape. The growth mechanism to a large extent also determines the nature of the defects and the distribution of impurities, both of prime importance in mechanical, electrical and most probably in all other conceivable applications. This applies not only to particulate crystals but also to materials in the form of epitaxial layers and compact oriented overgrowths.

It is a matter of scientific interest, although SiC will not be unique in the sense that a case study of many compounds will soon reveal the fragmentary and qualitative nature of our understanding of the processes of nucleation and growth. This is especially true in Chemical Vapour Deposition (C.V.D.) of compounds depending on the formation- and decomposition reaction of definite molecular species.

The layered growth of a faceted crystal from surface- or dislocation nuclei, in an adsorption-desorption equilibrium of gaseous species, is generally understood by the terrace-ledge-kink transport mechanism[1]. In the past the growth phenomena in SiC

around screw dislocations have been given close attention. Actually,
however, they play a minor role in the growth of SiC crystals;
growth is governed by surface nucleation and ledge movement by
face-specific reactions. The equilibrium shape of β-and α SiC,
predicted[2] on the basis of estimated values of the Gibbs surface
function for low index planes, is that of a tabular crystal with a
ratio of diameter to thickness of less than 2. Such a shape is
only reached in the temper reaction of small particles of SiC
because the measure in which the faces are present in prolonged
growth will depend on the ratio of their growth velocities and
respective orientation. The basic growth shape of β-and α SiC is
lamellar with the plane of the lamella perpendicular to cubic [111]
or hexagonal [0001]. This growth shape can be found exaggerated in
the semi-dendritic growth of lamellae. The ratio of diameter to
thickness of the lamellae can be as large as 10^3. The pronounced
non-equilibrium shape of dendrites is the result of a direction-
ally enforced anisotropy in the growth velocities: stimulation of
growth in the length direction and/or retardation or blocking of
the advancement of the side planes; aspect ratios of 10^5 can be
obtained. This enforced growth can be due to localization of the
nucleation by the localized presence of a foreign substance or by
the favouring of supply to a tip as the place of highest supersaturation, e.g. in a diffusion-controlled growth system. These
effects are counteracted by the forces promoting morphological
stability as far as surface flow allows. There are two more reasons
why the growth of β SiC, having the sphalerite structure, may deviate from the idealized isometric form. Since β SiC belongs to a
cubic space group which lacks a centre of symmetry ($F\bar{4}3m$), the
trigonal axes are polar and {111}, {221}, {211} and {321} give a
positive and negative form of a hemieder with different properties
as to etching and growth. By the lack of a centre of symmetry in
the hexagonal crystals there are also two forms for the basal
and pyramidal planes. Traditionally the positive form {0001} has
been taken as being bounded by Si atoms, the negative {000$\bar{1}$} being
bounded by C-atoms. It was found by comparison with absolute
configuration determinations that the face with the Si atoms shows

hexagonal etch pits while the face with the C-atoms is homogeneously attacked. By analogy with the naming of the hexagonal forms we
will, for cubic SiC, accept the positive form $\{111\}$ as being bound
by Si atoms; the negative form $\{\overline{1}\overline{1}\overline{1}\}$ as being bound by C atoms.[3]
The second additional reason for deviations of the idealized form
is the low "stacking fault energy" of a stacking fault in $[111]$.
Such a stacking fault extending through the crystal can be looked
upon as the boundary of two parts of the crystal in rotation twin
position[4]. It has been shown that for a number of AB compounds of
the sphalerite structure a succession of such boundaries offers a
non-vanishing re-entrant edge for easy nucleation and extended
growth[5]. Nuclei containing stacking faults will be favoured and,
in consequence, the simplest form in which a cubic crystal occurs
is that of a trigonal plane-parallel platelet with a $\{111\}$ habit
face elongated in $[1\overline{1}0]$. The platelet has twin lamellae parallel
to $\{111\}$. The growth differs in thickness on either side of the
lamellae.
By the formation of parallel and intersecting twin lamellae (Plate
1) an almost isotropic growth mode is still possible (nodular
growth). The growth on the twin lamellae proceeds by the formation of surface nuclei and layered growth, columnar dendrites in
the $[111]$ direction, or in the case of intersecting lamellae, also
by a kind of hopper growth from the edges.
It is the response of these competitive mechanisms on the flux
and nature of the condensing gaseous species which at a certain
temperature governs the morphology of the deposit.

2. Growth in C.V.D.

In a C.V.D. process, a chemically reactive gas mixture,with
molecular species containing elements in addition to Si and C, is
thermally activated to deposit material by a heterogeneous reaction on a substrate held in a reaction vessel. The method is
employed because it allows growth at much lower temperatures and
at much greater "supersaturation" than the Physical Vapour Deposition process,which uses the condensation from molecular species
obtained by heating SiC, containing Si- and C atoms solely. The

latter is, strictly spoken, also a C.V.D. process as, in the
incongruent evaporation of SiC, definite molecular species are
formed on which disproportionation growth depends.

The steady state growth process after primary nucleation is
a series of consecutive reactions which in extreme cases can each
be rate limiting. The following steps can be distinguished: the
generation of the reactive gas mixtures; the transport of the
gaseous species to the crystal/gas interface; the interface
processes; the transport of reaction products away from the
interface.

2.1. Reactive gas mixtures

The gas mixtures most frequently used are mixtures of
methylchlorosilanes in hydrogen at 1 atm pressure. With appropriate conditions SiC can be deposited roughly between 1000°C and
2000°C, e.g. according to the gross reaction $SiCl_3CH_3 \rightarrow SiC+3HCl$.
Besides SiC, the condensed phases which can arise in this system
are solid carbon and solid or liquid silicon; the Cl/H ratio is
not affected by the deposition process. In the 4-component system
with the variables T, P and Cl/H fixed only three phases, one
gaseous and two condensed, are allowed. Silicon and carbon are
known to occur besides SiC in the deposits for certain combinations of the system parameters[4].

To investigate whether this co-deposition is equilibrium
controlled, the composition of the gas phase in equilibrium with
the condensed phases SiC and silicon and SiC and carbon was calculated as a function of the Cl/H ratio, the temperature and the
pressure (Fig. 1).[6]
Methylchlorosilane concentrations up to tens of percents can be
used without homogeneous gas reactions preponderating so that a
large range of "supersaturations" can be applied in principle.
The methylchlorosilane molecules will dissociate in the high
temperature field around the substrate and, by fast radical
reactions, transform into the species having greatest stability
at these temperatures. As can be seen from Fig. 1 it is to be
expected that the silicon and carbon bearing gaseous species which

reach the gas/solid interface are predominantly $SiCl_2$ and CH_4, respectively. It is only at temperatures below 1000°C that the kinetics of the dissociation reaction, characterized by a relatively large activation energy comes into play.

2.2. Transport to the interface

The methylchlorosilanes are introduced into the reactor by the aid of a (hydrogen) gas stream.
The material transport near the gas/solid interface is described by the formal introduction of a diffusion layer of a definite thickness: the Nernst diffusion layer. The thickness of this layer is correlated with the hydrodynamic conditions of the gas in the reactor as determined by the geometry of the system, kinematic viscosity of the gas and its mean velocity by natural and forced convection[7]. In the event of a rapid interface reaction the concentration near the interface will have the equilibrium value.

It stands to reason that over an extended substrate it is only by special precautions that a locally and temporarily equal and constant flux of material can be assured. Only for very low efficiency and ideally mixing reactors can depletion effects be neglected; the input Si-C concentration is effectively lowered .

Compared with the chemical reactions, the diffusion transport step is characterized by a relatively small temperature dependence, and will be rate limiting at high temperatures and low concentrations, for reactions which are not equilibrium controlled.

2.3. Growth

The deposits grow from isolated nuclei on the substrate. Their growth rate, composition and morphology are a function of temperature, flux of the reacting species and Si/C ratio in the gas phase. The two general modes of deposition viz. compact layers and particulate crystals will be discussed on the basis of experiments in a vertical hot rod-cold wall reactor; diameter 10cm, height 50cm[4]. Details regarding the substrate dimensions and imposed gas flow will be given with the illustrated examples.

2.3.1. Compact layers

Compact layers are characteristically obtained for high fluxes of methylchlorosilanes. For monomethyltrichlorosilane (Si/C=1) the dependence of the growth velocity on temperature is given in Plate 2. In the high temperature region of the curves 1 and 2, carbon is found in pseudomorphosis of silicon carbide crystals[8]. This deviates from the carbon inclusions in the form of graphitic lamellae which were found by Fitzer et al[9] for Si/C < 1. The thermodynamic calculations show, that for Cl/H ratios between 10^{-2} and 1, above 1500°K, from a feed gas Si/C=1 only SiC+C can be deposited in thermodynamic equilibrium. The fact that SiC is deposited and carbon occurs only in an etch process points to kinetic difficulties in the carbon deposition.

The region of the co-deposition of silicon and silicon-carbide is also indicated in Plate 2. For the reason stated above they cannot occur here in thermodynamic equilibrium. The occurrence of silicon must be due to the fact that under these conditions the reduction of $SiCl_2$ by hydrogen outweighs the CH_4 pyrolysis. The silicon was found in the form of interrupted lamellae.

The growth rate varies linearly with the input concentration and by about the square root of the input velocity of the gas. For the higher fluxes the effective activation energy tends to a value of more than 30 kcal. Combined with the linear dependence on the input concentration mentioned this suggests kinetic control by the interface processes. The nature of the interface reactions is a matter of concern. Generally one can expect that the chemical reactivity of the chemisorbed species will depend on their adsorption state and that the unreacted or partly reacted species will have the greatest mobility. In the growth stage the final dechlorination of the Si-species and dehydration of the C-species will take place when they are chemisorbed at the kinks of the growth ledges. In the formation of surface nuclei, at low enough temperature, the reaction will take place preferentially on the most favourable position of the molecules on the terraces. The layers were investigated by X-ray and metallographic techniques

and by investigating the growth pyramids at the surface with the
aid of a scanning electron microscope. Three typical growth modes
can be distinguished from the surface appearance of the layers:
denticulated, nodular and smooth. The denticulated layers show
multi-star twin tips, with many re-entrant corners, and the more
faceted endings of columnar crystals with only a few twin plane
lamellae; they have a columnar structure with the columns grown
perpendicular to the surface and extending through the layer. The
layers of Plate 3 (a and b) were obtained at a temperature of
1350°C, a 7% concentration of the gas and an input flow of
6ltr/min. This growth mode is often encountered as the first
growth stage after nucleation, which then changes into a
structurally related but otherwise independent growth of smaller
crystals, giving rise to a strong nodular appearance of the surface.
The layers of Plate 3 (c and d) were obtained at a temperature of
1420°C, a 5% concentration of the gas and an input flow of 6ltr/min.
These deposits are rich in pores. The structure of the so-called
smooth layers (micro spherulites) obtained around 1200°C is
comparable to that of the latter growth albeit in much smaller
dimensions. The layers of Plate 3 (e and f) were obtained at a
temperature of 1250°C, a 15% concentration and an input flow of
2 ltr/min.

    To obtain a certain morphology for the deposit a combination
of system parameters is needed: a change in one of them can often
be "corrected" by the adjustment of the others.

2.3.2. Particulate crystals

        Particulate crystals are characteristically obtained for
low fluxes of reactive species. The specimen reported up to now
were dominated in their growth by the twin plane re-entrant edge
mechanism and were more or less complicated twin structures
extended in the $[1\bar{1}0]$ direction: tabular, columnar, skeletal.

    We have found some examples of another growth mode Plate 4
shows β SiC crystals with the sphalerite structure grown in a
$[111]$ direction. They were found to grow in an extended temperature range (from 1300° to 1500°) for the case shown with a 5%

concentration of the input gas and an input flow of 1 ltr/min.
They grow by a layered growth from the tip on spherulites of
the sphalerite structure. Plate 5 shows crystals of α-SiC with
the wurtzite structure grown in a similar way on the spherulites
of the sphalerite modification. They were only found at temperatures below 1400°C; for the case shown from a feed gas containing 0,7% methyltrichlorosilane and an input flow of
1 ltr/min.

There is considerable confusion in the literature concerning
the growth mechanism in C.V.D. of crystals with the rare and metastable wurtzite structure. The crystals are always found to grow
with a columnar or whisker like habit elongated in a $[0001]$
direction. Layered growth by a vapour-solid reaction nucleated at
(screw) dislocation sources in the top face, and vapour-liquid-
solid growth via a liquid silicon layer on top, have both been
discarded and invoked repeatedly [10,11].

We have investigated the top of these crystals by electron
microscopy and found that the growth proceeds by layers which
nucleate at the tip. The layers spread over the top of the crystal,
bunch to increasingly thicker layers on the low index surfaces of
the dihexagonal pyramid $\{21\bar{3}\,1\}$, hexagonal pyramid of the second
kind $\{11\bar{2}\,1\}$, hexagonal pyramid of the first kind $\{10\bar{1}\,1\}$ and the
hexagonal prism of the first kind $\{10\bar{1}0\}$ respectively. If there
is no longer nucleation at the tip the end form of a crystal
appears to be a hexagonal column bounded by $\{10\bar{1}0\}$ side faces
and $\{0001\}$ basal faces (Plate 6). Steps in the form of a spiral
ramp were not observed. We have etched these crystals and found
that the sides exhibit etch pits in the form of isosceles triangles (Plate 7) with the apex pointing in the direction of growth.
By comparison with lamellar crystals whose absolute configuration
was established it could be determined that the crystals grow in
the direction of the negative c-axis, thus towards the basal
face covered with C-atoms.
The tip, but also the complete crystal can be overgrown at the
later stages of growth by nodular or more perfectly crystallized
material of the sphalerite structure. On the facets of this

nodular material a renewed growth of columns of the wurtzite modification is possible (Plate 8 and 9).

From the experiments it must be concluded that the crystals grow by a diffusion controlled, tip-growth mechanism in a vapour-solid reaction, by layered growth from surface nuclei, on $\{\overline{1}\overline{1}\overline{1}\}$ facets of the sphalerite modification. The growth of the wurtzite structure appears to be an extreme example of the growth of a metastable structure by an epitaxial on facet growth, reported for thin films of CdTe on the S-face of CdS, up to a definite temperature[12]. The growth of the wurtzite structure could be due here to the presence of $SiCl_2$ as active species. On absorption of these species on the $\{\overline{1}\overline{1}\overline{1}\}$ faces, the most reactive position of the absorbed $SiCl_2$, with the Si atom bound to a C atom exposed by the C-face, would be the configuration in which the chlorine atoms are in an eclipsed position above the Si atoms of the lattice to which the C atom is bound in its tetrahedral bonding. This position would favour the attachment of C-atoms in an eclipsed position, thus the formation of the wurtzite structure[4].

SiC crystals which show a simultaneous growth in the [111] and [$\overline{1}\overline{1}\overline{1}$] direction are shown in Plate 10. They were obtained from a feed gas containing 5% methyltrichlorosilane and an input gas flow of about 6 ltr./min at temperatures between 1300° and 1400°C.

Under conditions in which co-deposition of SiC and silicon occurs, Si crystals grow in the form of lamellae with a $\{111\}$ habit plane extended in the [1$\overline{1}$0] direction. The platelets are characteristically twinned, sometimes forming hollow pipes on the β-SiC spherulites. The crystals of Plate 11a were obtained at a temperature of 1400°C a 0,1% concentration of the gas and an input flow of 1 ltr/min. Si-crystals formed from intersecting twin lamellae, to be compared with columnar β-SiC crystals, are shown in Plate (11b and c). The crystals of Plate 11b were obtained at a temperature of 1400°C, a 0,8% concentration of the gas and an input flow of 0,6 ltr/min; the crystals of 11c at a temperature of 1400°C, a 1% concentration of the gas and an input flow of 1 ltr/min.

It stands to reason that the hot rod-cold wall reactors are not intended and not apt to grow large particulate crystals due to the lack of long term thermal, hydrodynamic and chemical stability of the system; the prerequisites for smooth interface growth as discussed by Tiller for growth from solution [13].These conditions can be better met in the process in which the material supply to the system is in effect by diffusional transport alone: the Lely method (See Growth Mechanisms of Silicon Carbide in Vapour Deposition II)

References

1) N. Cabrera and R.V. Coleman, The Art and Science of Growing Crystals, 3, Ed. J.J. Gilman, John Wiley and Sons, Inc. New York - London (1963).

2) G.A. Bootsma, W.F. Knippenberg, G. Verspui, Proc IV All-Union Conf. of U.S.S.R. Tzakhkazor, Sept. (1972).

3) R.W. Bartlett and R.A. Mueller, Mat. Res. Bull.4, S341(1969).

4) W.F. Knippenberg, Philips Res. Repts. 18, 3 (1963).

5) J.W. Faust, Jr and H.F. John, J. Phys. Chem. Solids 25, 1407 (1964).

6) A.W.C. van Kemenade and W.F. Knippenberg, to be published.

7) W. Vielstich, Z. Electrochem. 57, 8, 647 (1953).

8) A.W.C. van Kemenade and C.F. Stemfoort, J. Crystal Growth, 12, 13 (1972).

9) E. Fitzer, Chem. Ing. Techn. 41, 5 and 6, 331 (1969).

10) I. Berman and C.E. Ryan, J. Cryst. Growth 9, 314 (1971).

11) N. Setaka and K. Ejiri, J. Am. Ceram. Soc., 52, 400 (1969).

12) M. Weinstein and G.A. Wolf, Crystal Growth Proceedings of an International Conference on Crystal Growth, Boston, 20-24 June 1966 (Ed. H.S. Peiser) Pergamon Press, 536.

13) W.A. Tiller, J. Crystal Growth 2, 69, (1968).

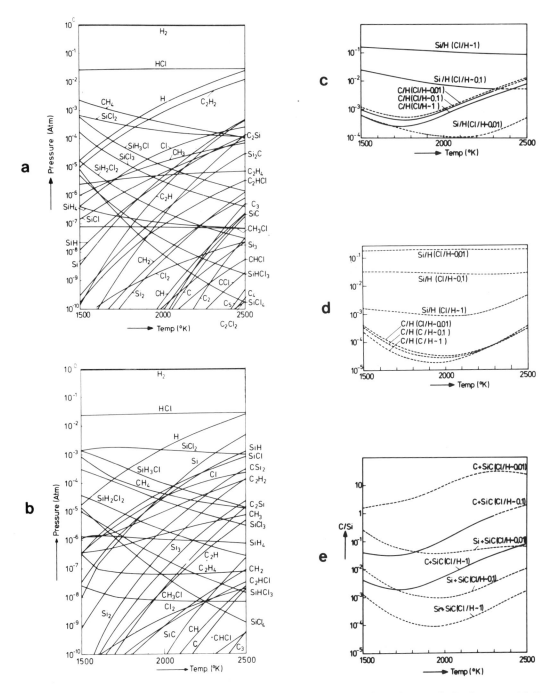

Fig. 1 a) Composition of the gas phase in the system Si-C-Cl-H in equilibrium
with the condensed phases SiC + C, for a Cl/H ratio ≈10^{-2} and a total pressure
of 1 atm. b) Idem (a) for the condensed phases SiC + Si. c) Si/H and C/H
ratios for different Cl/H ratios and a total pressure of 1 atm with the condensed
phases SiC + C. c) Idem (c) for the condensed phases SiC + Si. e) C/Si ratios
calculated from c) and d). Interrupted lines refer to equilibrium systems which
cannot be produced from Ch_3SiCl_3-H_2 mixtures.

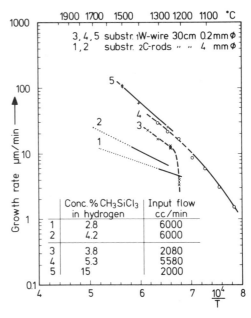

Plate 1.   Cross sections of β SiC crystals perpendicular to the growth direction, viewed in polarized light with crossed nicols, indicating the position of the twin lamellae.

Plate 2.   Growth rate vs temperature of compact layers of β SiC from CH_3SiCl_3 in hydrogen of 1 atm pressure: Solid line SiC, interrupted line co-deposition of SiC and Si, dotted line co-deposition of SiC and C.   The curves are numbered in increasing order of effective concentration.

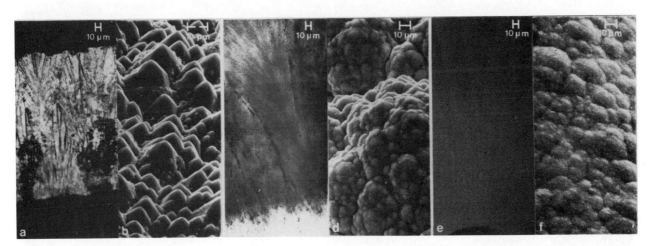

Plate 3.   Compact layers of β SiC a) Transverse section of a denticulated layer after etching. Substrate: inside carbon tube; diameter 7 cm; height 7 cm.   Input gas directed through tube; b) top surface of a); c) transverse section of a nodular layer viewed in transmitted light. Substrate: carbon rod 3mm diameter, length 20 cm; d) top surface of c); e) transverse section of smooth layer viewed in polarized light.   Substrate: tungsten wire 0,2 mm diameter, length 20 cm; f) topsurface of e).

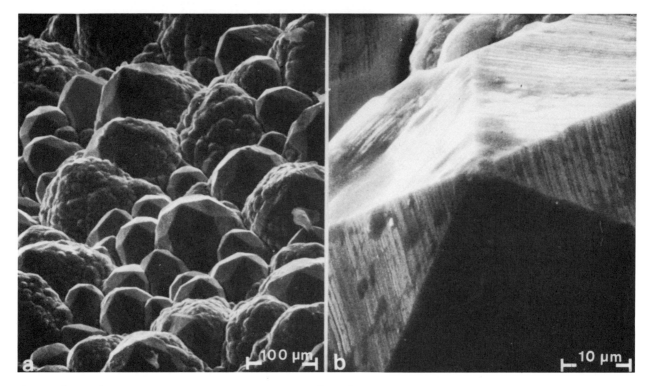

Plate 4. a) β SiC crystals grown in a [111] direction on β SiC spherulites.
Substrate: carbon rod; diameter 3mm, length 20 cm; b) Top of crystals of
a) indicating the layered growth from the tip.

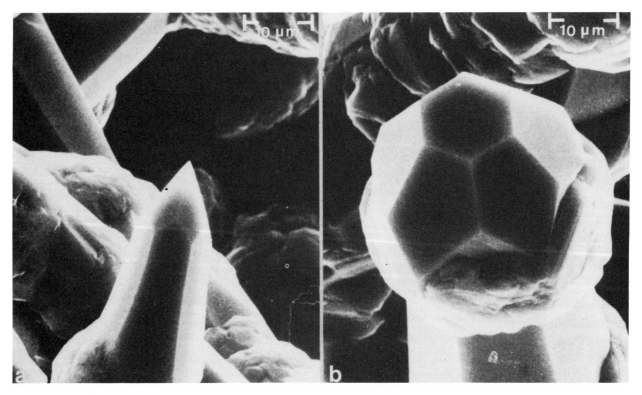

Plate 5. (a and b) α SiC crystals grown in a [0001] direction. Substrate: carbon
bar; diameter 6 cm, height 7 cm.

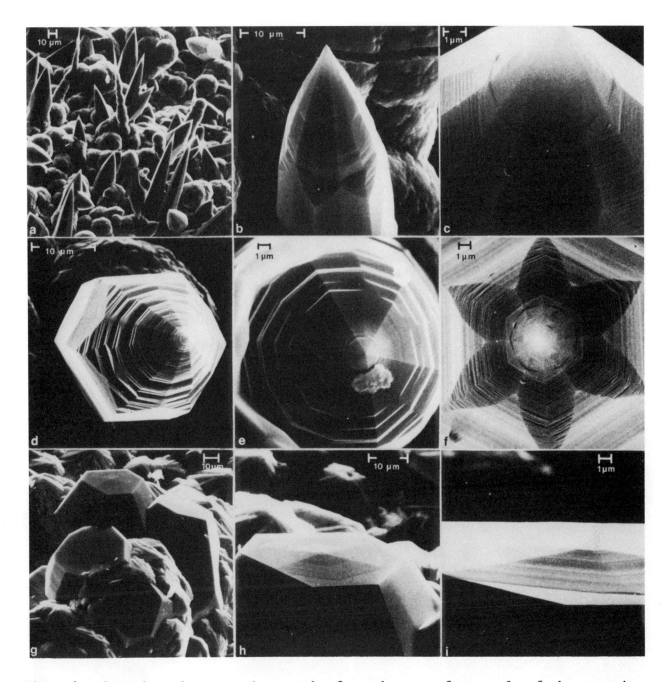

Plate 6. Scanning electron micrographs from the top of crystals of the wurtzite modification evincing the layered growth from the tip and the gradual development of face combinations.

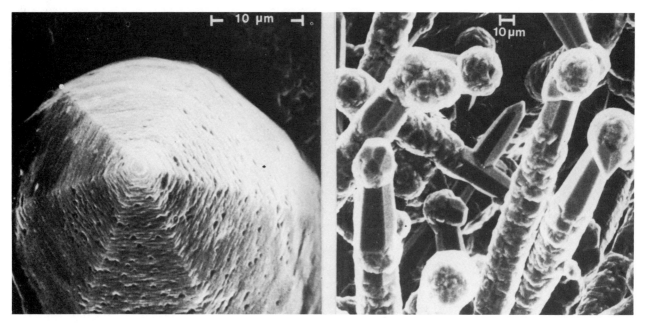

Plate 7.   The characteristic etch pattern
of wurtzite crystals by molten Na_2CO_3:
isosceles triangles with the apex pointing to the top.

Plate 8. α-SiC crystals of the wurtzite
modification intergrown with nodular
β-SiC.

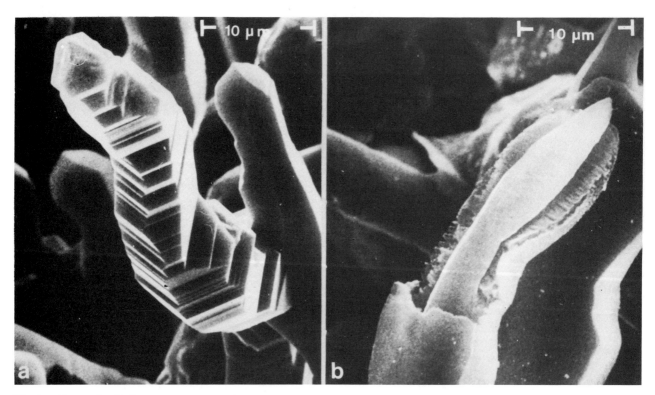

Plate 9.   a) Columnar crystals overgrown with perfectly crystalized β-SiC;
b) Columnar crystals overgrown with β-SiC after etching. The difference in etch
rate of overgrowth and core is clearly visible.

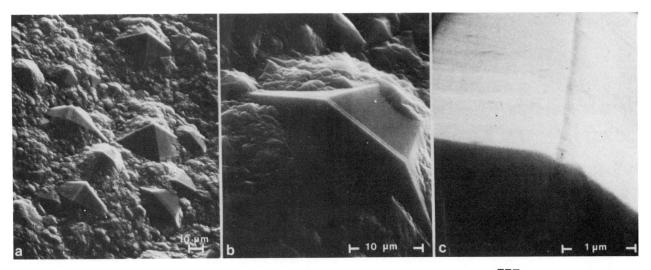

Plate 10. a) β-SiC crystals showing growth in the [111] and [$\bar{1}\bar{1}\bar{1}$] direction.
b) and c) Enlargements showing details of the twin lamellae. Substrate: inside
carbon tube; diameter 7 cm, height 7 cm (Input gas directed through tube).

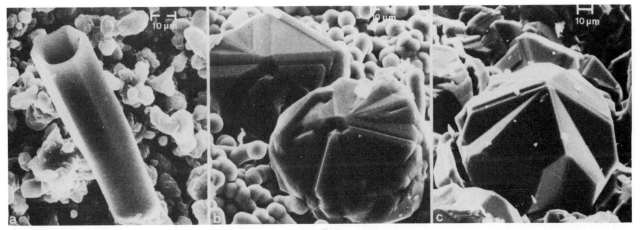

Plate 11. a) Si-crystal grown in the [$1\bar{1}0$] direction by a dendritic growth mode
of lamellae with a (111) habit plane. Substrate: carbon rod 3mm diameter,
height 15 cm b) Si-crystals grown from intersecting twin lamellae to be compared
with b). Substrate: carbon rod; diameter 3mm, height 20 cm.

Growth Mechanisms of Silicon Carbide in Vapour Deposition II

W. F. Knippenberg and G. Verspui

1. <u>Growth in the Lely process</u>

In the Lely process[1] a porous, thick walled cylinder of SiC
is heated in a vertical position enforcing a flux of gaseous
dissociation products to the interior. In practice this is
achieved by placing the cylinder in a closed carbon crucible.
When heating is applied from the side, the radiation losses to
the ends will effectuate a lower temperature at the inside than
at the outside wall. The temperature of the ends is depending on
the degree of thermal insulation. The cylinder progressively
carbonizes from the outside to the inside. In an inert atmosphere
at 1 atm pressure, preferentially at about 2550°C, particulate
αSiC crystals grow on the inside wall of the cylinder,
compact layers being obtained on the closing lids. It was found
empirically that the largest and most perfect crystals were
obtained when the temperature differences in the cavity were made
very small, in order of a few degrees, and the inert gas
pressure was not below 1 atm.[2] Improving the insulation at first
diminishes the number of crystals and leads to carbonization of
the inner wall. A further increase of insulation leads only to
carbonization.

2. <u>Reactive gas mixtures</u>

The composition and pressure of the gas obtained by heating
silicon carbide was obtained from evaporation studies[3] and mass
spectroscopic investigations[4]. From the latter it follows that
the reactive gas mixtures obtained by heating solid SiC in an
inert gas consist, besides this gas, predominantly of the
species Si, Si_2C and SiC_2. For the system solid SiC, carbon and

108

gas, the Si/C ratio in the gas at 2500°C is about unity; it
increases towards lower temperature. (Plate 1) In the two
component system Si-C with only the temperature fixed, three
phases, one gaseous and two condensed are allowed in thermodynamic equilibrium. Co-deposition of silicon carbide
and silicon is found when a stoking is cooled relatively rapidly;
possibly the only way, in view of its high vapour pressure, to
recover the silicon. Co-deposition of silicon carbide and carbon
is found for stokings towards the high temperature side of the
optimal growth temperature; the carbon being perceptible as a
slight grayish tint in pure, colourless crystals. The equilibrium
condensation ratios of SiC and silicon or SiC and carbon obtained
by cooling the gas can be estimated from the thermodynamic calculations for the system silane -propane-inert gas by Minagawa
and Gatos [5]. At 1900°K the deposition of SiC was found to be
limited to a narrow range of almost equal silicon and carbon
concentrations. At higher temperatures, due to the higher silicon
pressure, the boundary between the pure SiC deposition and the
SiC and liquid silicon co-deposition will move into the Si-excess
side in the gas; the increase in the carbon pressure will be much
smaller and the corresponding shift of the boundary between the
pure SiC deposition and the SiC and solid carbon co-deposition to
the carbon excess side will be smaller too.

In any case at higher temperature it can be expected that pure SiC
can be deposited from a more extended range of Si/C ratios.
The effect of additional elements can be judged from their
influence on the carbon or silicon potential of the gas by the
formation of stable molecular species, or on the evaporation/con-
densation coefficient (see Sec.3). The addition of hydrogen will
shift the boundary of the co-deposition of silicon carbide and
carbon to the C- excess region by the increased carbon solubility
in the gas by hydro-carbon formation.[6]. This explains the fact
that the turbidity of the crystals decreases if they are grown
in a hydrogen atmosphere instead of in a noble gas.
The increase in growth rate found in the presence of aluminium
will be partly due to the increase of the carbon potential of the

gas by the formation of Al-C gaseous species.[7] [8] In the
Norton version of the Lely process[9] in which Si is added, the
temperature range of pure SiC condensation will be smaller than
in the Lely process and the co-deposition of SiC and (liquid)
Si quite likely.

3. Transport to the gas/solid interface

In growth cavities of the usual sizes, even in the presence
of a quite small temperature gradient, there will be a circula-
tory gas flow by natural convection. This gas flow will homoge-
nize the temperature and pressure in the interior. The material
transport near the gas-solid interface can be formally described
as taking place by diffusion over a boundary layer. In fact the
wall and the deposits are effectively short-circuited by the gas
flow. In the absence of transport constraints, the evaporation
(or condensation) flux of the molecules (F) is obtained from the
Hertz-Knudsen equation:

$$F = (\alpha_v p_e - \alpha_c p) / (2\Pi \bar{m} kT)^{\frac{1}{2}}$$

In which p_e = equilibrium pressure;

p = actual vapour pressure;

$\alpha_{v,c}$ = constants accounting for evaporation/condensation
constraints;

$\bar{m}$ = average molecular mass;

and k is Boltzmann's constant and T the absolute
temperature

The actual vapour pressure in the homogenized flow will be
determined by the equilibrium pressure of the crystals on the
wall namely by their lowest temperature.

By the coupling of the respective temperatures of wall, crystals
and closing lids, it is clear that for particulate crystal growth
there will be optimal combinations of cylinder dimensions and the
temperature difference between wall and closing lids, as experimentally elaborated.

In view of the favourable position of the equilibrium (sublimation
energy ≈ 140 kcal/mole) and the high transport- and interface
condensation reaction rates at the high temperature involved,

proximity effects will dominate: The net transport to the
crystal has the characteristics of a near surface transport
through the boundary layer. This depletion of the wall flux
by the crystals grown on the wall will of course also be
present, although less effective, in the case of a pure diffusional regime in the cavity.
The gross deposition flux for particulate crystals is limited
because it results from the difference of the (limited) evaporation flux of the wall and the almost constant condensation flux
to the closing lids. The crystals will grow to a final size (total
surface area) at which their gross condensation flux equals
their net evaporation flux. From the vacuum evaporation experiments the evaporation coefficient is estimated to be 0.1-0.01.

4. Growth

The deposits grow from isolated nuclei. The material mainly
has the hexagonal 6H structure. The two general modes of deposition viz. compact layers and particulate crystals will be
discussed; the respective total amounts varying about a factor
of four.

4.1. Compact layers

Compact layers are formed on the closing lids of the
cylinder. The crystallites are laminar with a {0001} habit face
and are mainly oriented parallel with the lids. Especially for
growth at the highest temperature large single crystal regions
of great perfection can be found in the volume. Inclusions of the
metastable β SiC modification have been found, which due to a lack of a
reconstructive, near surface mechanism or dislocation movement
did not transform into 6H SiC[10] (Plate 2).
Smiltens tried to apply the Bridgman-Stockbarger method for the
growth of one large single crystal to this system by moving
the crucible vertically in a temperature gradient, with reduced
pressure promoting this deposit[11]. The lack of a growth
tendency in {0001} will, however, not favour seed selection by
the directional effect and the perfection of the crystallites will

limit boundary movements. So it will be difficult to attain this growth of one monocrystal as it is crossed by the nucleation and growth mode and not facilitated by annealing.

4.2. Particulate crystals

The individual growth rates of the crystals as well as of the individual faces are characerically influenced by their temperature conditions as determined dominantly by radiative heat transfer.[12]

In the first stage of the growth all crystals are tabular, they develop in the further stages of growth into crystals which are columnar, lamellar-tabular, or ribbon like in the presence of definite impurities. Columnar crystals tend to disappear in prolonged growth.

Columnar crystals

Typical columnar crystals are shown in Plate 3. They are extended along the c-axis; Their morphology has been described before.[13] In all investigated crystals one or more hollow channels were found parallel with the length axis. These channels, evidently the hollow core of a bundle dislocations (diameter $\approx 1\mu$) widened towards the top face (diameter $\approx 10\mu$); at the intersection with the basal face they did not give rise to a pattern of growth spirals in the greater part of the crystals investigated.[12] (Plate 4) X-ray investigations showed that they were strictly of one polytype (if not macroscopically intergrown) and showed no disorder.[14]

Lamellar-Tabular crystals

Typical plane parallel hexagon shaped lamellar crystals are shown in Plate 5. They are extended perpendicularly to the c-axis. They grow by addition of material to the side faces, not necessarily showing growth in thickness. They may have channels parallel to the c-axis, these being the hollow cores of screw dislocations leading to a system of growth spirals on the top face.[15] In many crystals investigated, however, they are absent, demonstrating their incidental character.[16] The thinnest lamella ever found

had a thickness of 7μm; it measured 8x6mm^2 on the sides; it was
curves as can be expected according to Frank[17] for a thin
crystal bounded by the two habit surfaces {0001} of different
surface energy.

The tabular crystals consist of a succession of lamellae.
These can be perfectly intergrown; if this is not the case a
boundary occurs, giving rise to disordered intergrowth or to the
growth of a crystal with a deviating structure. The side faces
are plane and smooth, growth steps or growth spirals were never
observed.

The lamellar crystals are thermally favoured; they outgrow the
columnar crystals in such a way that the columnar crystals,
remaining close to the wall, start to evaporate. Most investigated columnar crystals were therefore found after having been
thermally etched: possibly the trivial reason for the absence of
a recognizable growth pattern. From the comparison of columns
and lamellae it can be concluded that with the experimental conditions the basal face only grows   from dislocation nuclei while the
side faces develop from a semi-dendritic growth of lamellae. At
the high temperatures involved the surface diffusion of the side
faces is evidently strong enough to maintain morphological
stability and smooth interface growth.

In the Norton version of the process, where Si is added, there is
an increase in the relative amount of columnar and tabular
crystals. The tabular crystals show an appreciable growth in
thickness (//c-axis) with no difference in growth velocity of that
part of the basal face immersed in liquid silicon and the part
extending in to the "gas phase". This suggests that condensation
on the basal face in this system is facilitated by the presence
of silicon (Plate 6.) No dislocation structure is needed here
to effectuate growth on the basal face; the appearance of hollow
channels //c-axis in columnar crystals is no longer the rule.

The thermal etching of these columnar and tabular crystals,
indicating their degree of perfection, is demonstrated in Plate 7.
An influence of the starting material on the perfection of the
crystals was observed: starting from fine grained 6H SiC the

crystals are predominantly plane parallel, starting from twinned
cubic SiC many platelets are not plane but twinned over{ho.l} h>l
(Plate 5). The initial growth rate depends exponentially on the
temperature, with an activation energy corresponding with the
activation energy of evaporation (≈ 140 kcal) 3) 12) 18) 19).

Ribbon like crystals

The morphological stability is no longer preserved in the
presence of impurities e.g. such as lanthanum.[19] Its presence
leads to the growth of ribbonlike dendrites with a {0001}
habit face. (Plate 8) These dendrites result from a parallel
growth of lamellae; the thinner lamellae having a thickness of
the order of 0.5 μm, width 10μm and a length of up to 5cm.
It has been concluded that lanthanum is adsorbed on the faces of
SiC and hampers the surface diffusion.[19]
In conditions of growth this leads to the development of
protrusions and favoured tip growth as in systems showing
constitutional supercooling. The growth in the side directions,
{1$\bar{1}$00} faces, is decreased by a factor of ten whereas the growth
of the tip faces is increased by the same factor, as compared
to the growth rate of the {1$\bar{1}$00} faces when no lanthanum is
present   the activation energy is not significantly
different. (Plate 9).
The direction of growth is along the hexagonal a-axis.
The tip is rounded or sometimes ends in a plane {11$\bar{2}$0} face.
(Plate 8b) There is no such face in the crystals grown in the
absence of lanthanum, indicative of its higher growth velocity.
Etching experiments have shown that lanthanum also interacts with
the {11$\bar{2}$0} faces, so that it is not unexpected that their growth
velocity is also diminished by adsorption of lanthanum. This must
have been the case to an even greater degree than for the {1$\bar{1}$00}
faces, because from the criterium of maintaining both faces
in the growth form, the ratio of the growth velocities of the
{1$\bar{1}$00} and {11$\bar{2}$0} faces must be equal to or larger than $\frac{1}{2}\sqrt{3}$.

It has been shown by Schlipf [20] that many dendrites grow in the direction of the second to the slowest growing face.

Polytypes

The crystals are predominantly of hexagonal 6H structure with a small admixture of 15R. Doping, for instance with aluminium increases the amount of 4H for growth temperatures up to 2350°C, and with nitrogen doping an increased amount of 15R was found up to 2450°C.

Higher polytypes are found as overgrowths or intergrowths of the simple types mentioned. To obtain larger crystals of more complicated polytypes a seeding procedure was developed. Seed. crystals were grown at about 1200°C by a V.L.S. method, using Fe globules [21] of about 20μm diameter on a substrate consisting of SiC of a definite polytype and a gas atmosphere obtained by heating SiO_2 and carbon in hydrogen. SiC is deposited epitaxially on the substrate conserving the structure of the seed.

 In this way needles of a definite polytype can be grown. These needles were positioned in the wall of an SiC cylinder which was further treated as for growth in the Lely procedure.

By an epitaxial growth process the needles thickened with conservation of the polytype.

Cubic needles transform into 6H platelets so that by this method we could not produce a large thick single crystal of cubic SiC. The rate at which other types transform into the 6H structure by near surface transport depends on the probability of formation of coherent nuclei of 6H at surface faults.[10] [18] (Plate 10).

In these communications on the growth of SiC-crystals from the vapour phase the results of a number of exploratory investigations have been reported. It must be looked upon an an examination of older work and work in progress, with special attention to the growth mechanism; several connected problems such as phase stability and phase transformations could not be given due attention. It will be clear that many interesting detailed investigations remain to be performed before a fully quantitative theory of the growth of SiC crystals from the vapour can be developed in all its aspects.

Acknowledgements

The help of the analytical departments has been greatly appreciated. The authors are indebted to Mr. J.L.C. Daams for the scanning electronmicrographs and to Mr. P. Vries for the X-ray analyses. They would also like to express their gratitude to Mr. J.G. Stockmans for his assistance in the experiments.

References

1. J.A. Lely, Ber. Deut. Keram. Ges. 32, 229 (1955)

2. W.F. Knippenberg, Philips Res. Repts 18, 3, 244 (1963)

3. Idem, page 238.

4. J. Drowart and G. de Maria, in: Silicon Carbide, A High Temperature Semiconductor, Eds. J.R. O'Connor and J. Smiltens, Permagon, London, 1960 p. 16.

5. S. Minagawa and H.C. Gatos, Jap. J.Appl.Phys. 19, 7, 88 (1971).

6. B. Lersmacher, H.Lydtin, W.F. Knippenberg and A.W. Moore, Carbon 5, 205 (1967).

7. W.F. Knippenberg and G. Verspui, Mat. Res. Bull. 4, S45(1969).

8. S. Yamada and M. Kumagawa, J. Crystal Growth 9, 309 (1971).

9. G.S. Kamath, Mat. Res. Bull, 4 S57 (1969).

10. W.F. Knippenberg, G. Verspui and G.A. Bootsma, Colloques internationaux C.N.R.S. no. 205, Odeillo, 27-3 sept. 1971.

11. J. Smiltens, Mat. Res. Bull. 4, S85 (1969).

12. L.J. Kroko, J. Electrochem. Soc. 113, 801 (1966).

13. W.F. Knippenberg, Philips Res. Repts 18, 3, 261 (1963).

14. W.F. Knippenberg and A.H. Gomes de Mesquita, Z. Kristallographie 121, 67 (1965).

15. J.P. Golightly, Z. Kristallographie 130, 310 (1969).

16. F. Arrese, Acta Cryst. 18. 279 (1965).

17. F.C. Frank, Growth and Perfection of Crystals, 3, Proceedings
    of and International Conference on Crystal Growth (1958)
    Eds. R.H. Doremus B.W. Roberts.

18. G.A. Bootsma, W.F. Knippenberg and G. Verspui, J. Crystal
    Growth 8, 342 (1971)

19. G. Verspui, W.F. Knippenberg and G.A. Bootsma, J. Crystal
    Growth 12, 97 (1972).

20. J. Schlipf, Z. Kristallographie 107, 35 (1956).

21. G.A. Bootsma, W.F. Knippenberg and G. Verspui, J. Crystal
    Growth 11, 297 (1971).

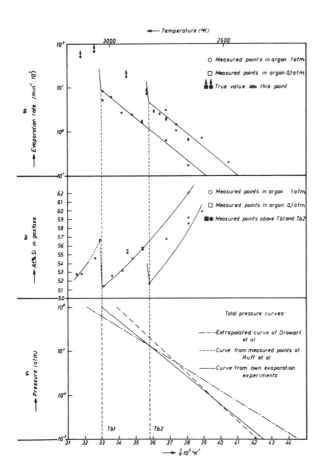

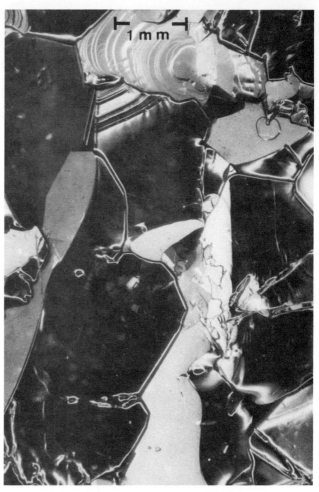

Plate 1.   a) Evaporation rate of SiC
(See ref. 3).   Relative fraction of
the SiC originally present escaping
per minute versus 1/T.   b) Total Si
content in vapour above dissociating
SiC in at. %.   The pressures of the
argon ambient are respectively 1 and
0.1 atm.   c) Total pressure curves
of the system SiC + C.

Plate 2.   Compact layer of sublimated
6H SiC viewed perpendicularly to the
substrate showing the dominant parallel
orientation of the lamellar crystals
with the substrate.

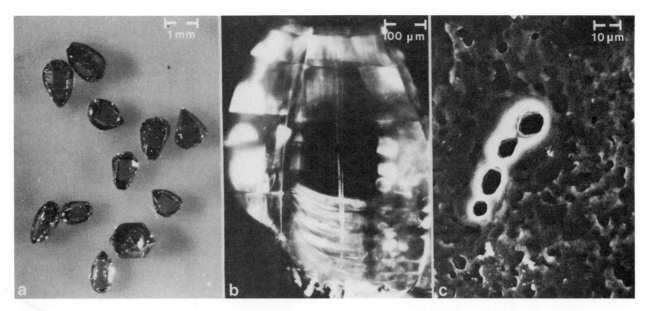

Plate 3.   a) Typical columnar crystals of α SiC.   b) Hollow channels parallel to the c-axis in a columnar crystal.   c) Cut through a crystal perpendicular to the channels after etching in molten soda to remove sawing dust out of the channels.

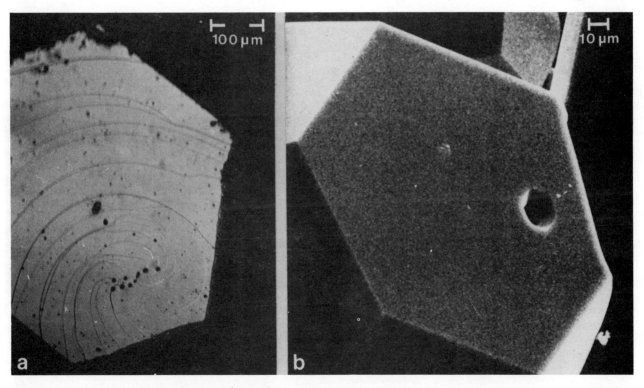

Plate 4.   a) Growth spirals on top of columnar crystal centred on hollow channels (unusual).   b) Top face of columnar crystal showing funnel-type ending of channel; no growth pattern visible on top face (usual).

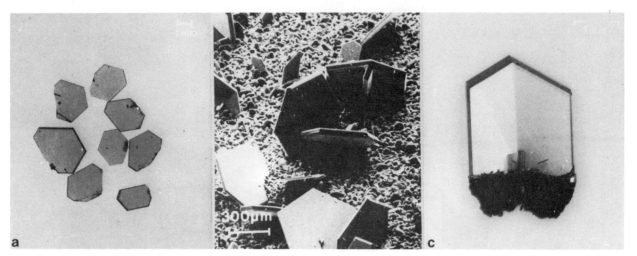

Plate 5. a) Typical plane parallel lamellar crystals. b) Orientation of the crystals on inside wall. c) Twinned α-SiC crystal grown from twinned β-SiC.

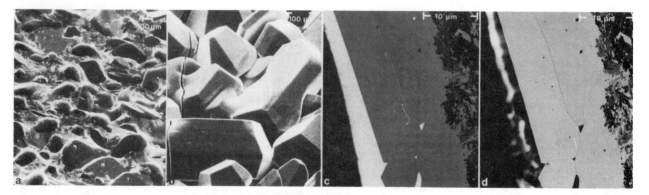

Plate 6. a) Crystals at the start of the growth in the Norton version of the Lely process; solidified silicon between the crystals. b) Deposit from a) after removal of the silicon in acid. c) Cross section of a); white parts silicon, gray parts silicon carbide. d) Idem c) after removal of the silicon.

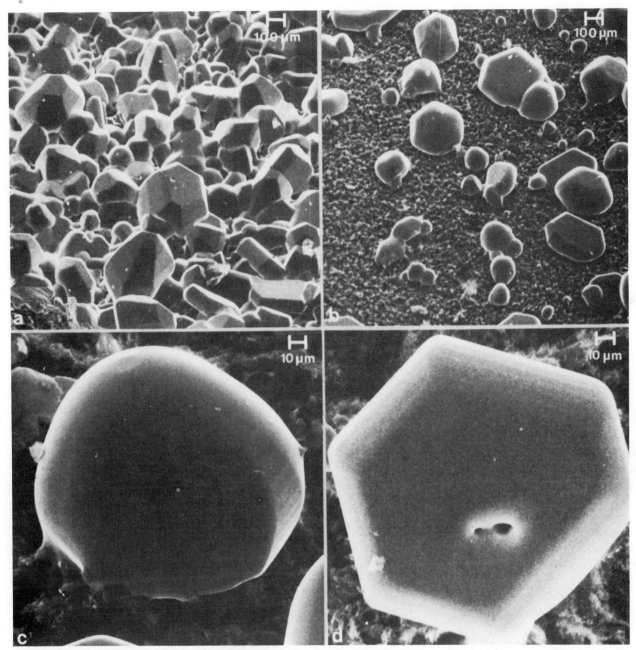

Plate 7.   a) α SiC crystals grown in the Norton process (initial growth). b)
crystals of a) after thermal etching. c) columnar crystal of b) without hollow
channels. d) lamellar crystal of b) with hollow channels (incidental).

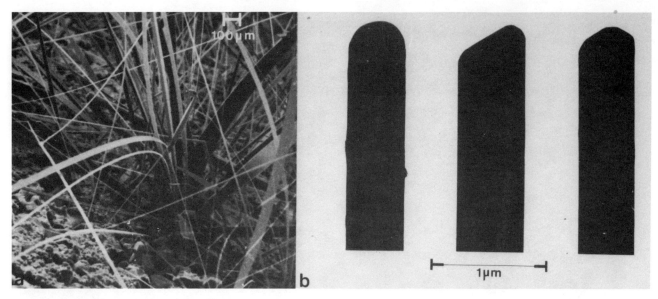

Plate 8. a) Ribbon like dendrites of α SiC. b) Electron-photomicrographs of tips.

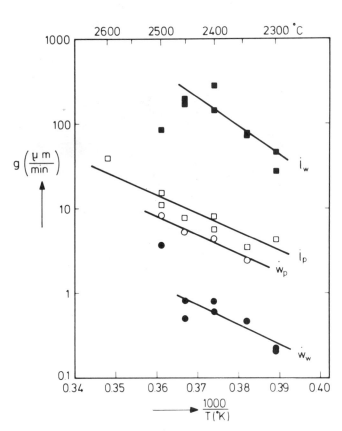

Plate 9. Growth rates in length and
width direction of platelets ($\dot{l}_p$ and $\dot{w}_p$)
and whiskers ($\dot{l}_w$ and $\dot{w}_w$).

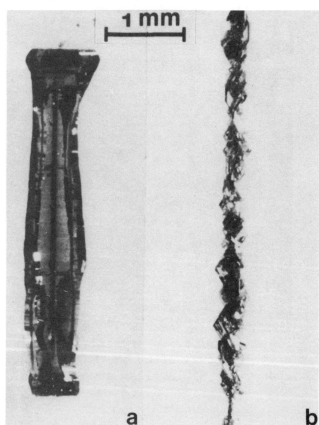

Plate 10. Epitaxial growth of hexagonal
polytypes; dendrite serving as seed crystal is visible because of difference in
doping; b) oriented growth of 6H platelets out of a β–SiC whisker.

New Growth Experiments in SiC

R. C. Marshall, C. E. Ryan, J. Littler and I. Berman

During the past 30 years there have been perhaps a dozen significant technical papers on the growth of silicon carbide. The large scale manufacture of SiC by the Acheson process using a mixture of silicon and carbon with a few percent of sawdust and common salt is well known. Lely[1], who wanted purer crystals in order to investigate their semiconducting properties, essentially brought the Acheson process into the laboratory using a hollow cylinder fabricated from lumps of technical grade SiC. This process has been almost the exclusive method of growing alpha silicon carbide platelets with several variations[2,3,4,5,6] developed over the years by other researchers.

A new technique developed by Norton[7,8] uses elemental silicon in a graphite crucible rather than a hollow cylinder of commercial silicon carbide. The advantages are quite obvious especially from the standpoint of purity, costs, and convenience.

The literature on crystal growth readily shows that most experts differ on what is happening inside the crucible during the growth process. The classical approach has been to consider Lely growth as a sublimation process. But a pure sublimation process calls for conditions which we do not believe exist completely in the growth cavity. Because of the many variables that can exist such as time, temperature, temperature gradient, ambient gas, ambient pressure, impurities, and starting materials, it is almost impossible to reproduce exact results from one laboratory to another or, in most cases, even from one experiment to another.

We feel the normal Lely process has reached its limit of technological development. This necessitates the exploration of new and/or novel methods to meet the scientific need.

Our experiments have been directed toward exploring some of these methods, to investigate their feasibility and to determine the effect of growth variables such as pressure, temperature and charge composition on the crystal growth mechanism.

For the prupose of performing the above experiments at high temperature and pressures, a furnace system was designed to allow experiments over the temperature and pressure range desired.

The design objectives of this furnace system required a relatively large volume, high-temperature, high-purity crystal growth zone in which the crucible could be heated at closely controllable rates to temperatures approaching $3000^{o}C$ under controller environments ranging from a vacuum up to 10^{-6} torr to 50 atmospheres of inert gas pressure. Unique characteristics of this system include precise control of furnace temperature and growth zone profile; high-purity internal graphite components; good axial and radial visibility into growth zone; ability to cycle directly from high vacuum to high pressure at elevated furnace temperatures; rapid removal of insulation and heater packages; and quick-opening, full access to furnace interior.

Plate 1 is an over-all view of the furnace system[9] showing the main features mentioned. The furnace is shown on a steel platform 150 cm above the floor. Suspended below the furnace are the high-pressure valve, high-vacuum valve, and oil diffusion pump. The furnace length is 80 cm with a 60 cm diameter. The radiation detector is 25 cm long. Plate 2 shows an interior view with the pressure shell, copper heat sink, lower and main heater, insulation package with the lower hearth assemblies and Grafoil insulation package installed. Heating of the work area is provided by passing large currents thru three independently powered graphite heating elements (Plate 3). The main cylindrical heater is powered by a three-phase, 150 KVA step-down transformer. The top and bottom flat trim heaters are each powered by a 25 KVA single-phase step-down transformer. The bottom heater is assembled within the insulating hearth which supports the crucible. The area available for crystal growth is 15 cm in length.

In order to maintain high purity conditions within the furnace, all vessel components are stainless steel, nickel-plated copper or nickel-plated carbon steel. All graphite parts within the furnace have a purity of 99.5 percent prior to vacuum outgassing.

Temperature control is provided by three radiation detectors, two looking axially at the top and bottom of the crucible and one radially at the center of the crucible. Temperature can also be controlled by using power transducers which monitor voltage and current of the heater circuits. The temperature control system (Plate 4) incorporates three individual SCR power controllers with automatic temperature control units. Selection of work zone temperature profiles is obtained by matching networks and switching devices to permit either independent temperature control for each heater, or a selection of a master-slave mode whereby any two slave heaters can be controlled by the master. A controllable temperature differential can be maintained. A master programmer is used for programmed automatic operation.

The vacuum capability of the furnace system is 10^{-6} torr at 1800°C. Isolation of the high vacuum pumping system is made with a pneumatically controlled high-pressure stainless steel ball valve. The semiautomatic high vacuum pumping system consists of a pneumatically controlled stainless steel high vacuum valve, 1500-liter-per-second oil diffusion pump, two-stage mechanical pump, and interlocked remote vacuum component controls. The pressure vessel is constructed of 5A105II steel with tinsel strength of 70,000 PSI. Maximum operating pressure is 800 PSI at a skin temperature of 250°F. The complete system was tested at a pressure of 1200 PSI.

Procedure

Initial application of the furnace system was directed toward the growth of single crystals of alpha silicon carbide at temperatures up to 2850°C and over-pressures ranging from a few PSI up to 750 PSI.

The exact program can vary with each experiment but the general program is as follows: The system is evacuated and heated to 1200°C until a vacuum of 10^{-5} torr is achieved. If high pressures are scheduled the system is slowly back-filled until the desired overnight pressure is achieved. This is normally 20% less than the scheduled growth temperature to allow for expansion of the ambient gas to avoid crucible failure.

The original insulation material used in this furnace system was composed of multiple layers of highly reflective non-gas absorbing graphite tape. Because most heat losses above 2000°C are thru radiation, it was felt that this material would be ideal for temperatures above this level. However, in our system, the thermal loss at a temperature of 2800°C was extremely large requiring current levels up to 1000 amperes on the trim heaters and up to 2000 amperes on the main heater. The operating life of these heaters, feedthru's and electrodes

was very short. In addition, the aluminum content of the tape exceeded acceptable levels for semiconductor materials. The insulation was replaced with a
special high fired (2650°C) high pruity graphite felt[10] using multiple winding
of 3.2 mm thick material. The current levels on the main heater decreased 350
amperes at 2800°C. The thermal conductivity of graphite tape at 2000°F in
BTU/ft/hr/ft^2/$^\circ$F is about ten times higher than graphite felt. Another important
factor to consider when conducting crystal growth experiments in large volume
furnaces at extremely high temperatures is the thermal conductivity of the
ambient gas. The thermal conductivity coefficient of noble gases can be derived
by kinetic theory as follows

$$\lambda = k\eta \frac{C_v}{M}$$

where

λ = thermal conductivity coefficient $C_v = \frac{3}{2}$ R heat capacity at constant
                                                              volume
k = proportionality constant = 2.5

η = viscosity   M = molecular weight

                                                   R = universal gas constant

Data for helium and argon gas[11] are plotted in Plate 5 for a pressure of
one atmosphere. In the first approximation it can be said that the thermal
conductivity of gas is inversely proportional to the molecular weight.

Experiments:

Crystal growth experiments were conducted over a temperature range from
2400°C to 2880°C and a pressure range from 10 PSI helium to 550 PSI helium or
argon. Crucible charge included silicon alone or silicon with carbon powder in
a ratio 10 to 1. An effort was made to maintain as small a vertical temperature
gradient as possible. As pressure levels increased, it was increasingly difficult for the top and bottom trim heaters to maintain the desired gradient mainly
due to the thermal cyclone effect. Time cycles were normally standard for all
experiments with five hours at temperature and three hours for down-cycling to
2300°C. Data indicating different periods resulted from equipment malfunction.

Results:

Mass spectrographic analyses indicated a dramatic decrease in the amount of
aluminum in the crystals after installation of the high purity graphite felt

insulators as shown in Plate 6. Aluminum concentration decreased from 30 ppm to
1.4 ppm. Markedly decreased levels were also found for the elements Boron,
nitrogen, sodium and potassium. We feel the major contributors to the impurities
still present include the graphite powder used in the charge material and the
high temperature graphite cement used in sealing the crucibles. Improvements
in impurity levels could be achieved by improved crucible purification techniques,
higher purity silicon and carbon powder, higher purity inert gas, and increased
pressure-vacuum cycling prior to initiating the growth program cycle.

The reaction at standard growth temperature and pressures (2550°C and 10 PSI)
resulted in the growth of crystals on the inner wall of the crucible similar to
the Lely process with a large portion of the walls covered with crystals.

Plate 7 is a photograph of several alpha crystals grown using a silicon and
carbon powder charge. When a large excess of silicon was available, a high percentage of the crystals had silicon inclusions. A surprisingly large number of
the inclusions had definite geometric shapes. Plate 8 shows such an inclusion
with a typical silicon deposit on the upper right corner.

The solubility[12] of carbon in silicon is 8 atom percent at 2550°C and 19
atom percent at 2830°C. This high solubility indicates that growth from solution
is feasible. The vapor pressure of silicon over silicon[13] and over silicon
carbide at this temperature is in the order of one atmosphere and rising rapidly.
The need for a pressurized inert atmosphere to maintain stable conditions for
any length of time is apparent. A number of experiments were conducted at
various temperatures and pressures with the results indicated in Plate 9. The
experiments at 10 PSI produced an increase in size and number of crystals as the
temperature increased. At 100 PSI, a 100°C increase in temperature for the same
period produced thicker crystals. At 250 PSI, only one crystal larger than 2 mm
grew. When the temperature was increased to 2700°C the number of platelets increased and a large number of Columnar cyrstals grew from the graphite wall
extended in the direction of the C axis. The dimensions of these crystals are of
the order of two mm with the largest about four mm. Plate 10 shows several crystals with numerous silicon droplets. Note one of these crystals has a platelet
growing in a direction parallel to the crucible wall. Plate 11 shows
additional crystals photographed with transmitted light. The pyramidal planes
have developed very well. The fuzz at the end of some of the crystals is the
remains of the crucible wall after removing the graphite crucible.

At 550 PSI, no crystal platelets grew after 6 hours at 2550°C. At 2650°C, some small thin platelets and columnar crystals grew. X-ray analyses of some of the platelets gave a clear 6H pattern. For experiments above 2650°C, at this pressure, a right circular graphite cone was inserted in the crucible such that the apex of the cone was at the lowest temperature. Experiments in the 2800°C regime were normally limited to three hours or less because of furnace component temperature stress. The maximum temperature achieved during these experiments was 2880°C. No crystal platelets grew on the walls of the crucible at temperatures up to 2880°C in the times indicated. A picture of one of the boules is shown in Plate 12. The half boule on the left has been etched for several hours with HF-HNO$_3$ solution. Notice the intertwining platelets. Very small (1 mm) yellow crystal platelets had grown between the major structures. A smaller number of brown and reddish crystals could be observed. X-ray analyses of some of the yellow platelets indicated 15R polytype with large slacking faults.

Conclusions:

Crystals grown using a silicon carbon charge are similar to those grown by the normal Lely technique. The use of silicon and carbon however, allows independent control of the purity of the constituents and eliminates the necessity of using commercial silicon carbide grit. The high temperature, high pressure experiments show that silicon evaporation can be managed and indicate the feasibility of solution growth of silicon carbide under achievable levels of pressure and temperature. The techniques are now at hand to open the way to a totally new approach to solving growth problems. Exploration by Czochralski[14], Kyropoulos and Freeze-gradient techniques could provide new answers to the growth problems. Further investigation into ambient conditions favoring columnar type growth could provide further possibilities in this area.

Acknowledgements:

The authors wish to express their thanks to R. M. Barrett, Director of the Solid State Sciences Laboratory for his interest and support of this program, to Maynard Hunt for the mass spectrographic analyses of the crystals, to J. Hawley for assisting in many of the experiments and to W. Jackson and staff for design and fabrication of many of the graphite crucibles, heaters and special parts. Our thanks also to J. Rohan for the stimulating discussions and suggestions on various aspects of this program, and to J. W. Faust, Jr. for the x-ray data.

REFERENCES

1.   J. A. Lely, Ber. dtsch., Keram Ges, 32, 229-231 (1955).

2.   D. R. Hamilton, J. Electrochem. Soc., 105, 735 (1958).

3.   W. F. Knippenberg, Phillips Res. Rpts. 18, 161 (1963).

4.   L. J. Kroko, J. Electrochem Soc., 113, 801, (1966).

5.   Y. Inamato et al, J. Ceramic Soc. Japan, 79(8); 259-263
         [12-6(Emm 72-275)].

6.   Y. Inamato, Z Inoue, M. Ota; J. Ceram Soc., Japan, 81(1): 11-15 (1973).

7.   E. C. Lowe, U. S. Patent 3,343,920 (1967).

8.   G. S. Kamath, Mat. Res. Bul. 4, 57-66 (1969).

9.   Astro Industries, Santa Barbara, California.

10.   Fiber Materials, Inc., Graniteville, Mass.

11.   Y. S. Touloukian and P. E. Viley, Plenum Pub. Corp., Thermophysical
         Properties of Matter, Vol. 3, p(5&6)(33&34) 1970.

12.   R. I. Scace, G. A. Slack, J. Chem. Phy. 30, 1551 (1959).

13.   R. E. Honig, J. Chem. Phys. 22, 1610 (1954).

14.   R. C. Marshall, Mat. Res. Bull. 4, 73-84, 1969.

Plate 1.   High Pressure Furnace.

Plate 3.   Furnace Control System.

Plate 2.   Interior View.

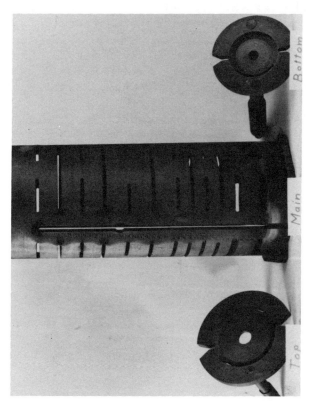

Plate 4.   Furnace Heaters.

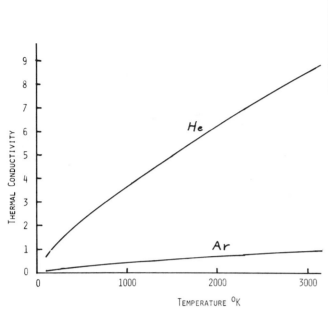

Plate 5.   Thermal Conductivity versus
           Temperature.

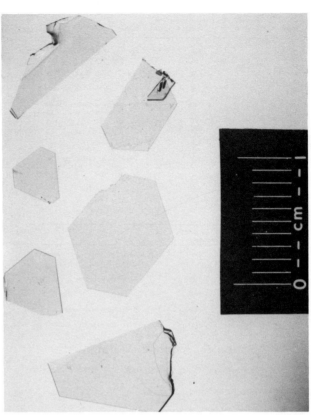

Plate 7.   Alpha Crystals.

Element	Detected Values in PPM		Element	Detected Values in PPM	
	Old Insulation	New Insulation		Old Insulation	New Insulation
Cu	0.033	ND			
B	1.6	0.34	Ni	ND	ND
N	1.4	0.13	Cr	ND	ND
O	2.4	3.9	Ca	0.58	ND
Na	1.8	0.086	Al	30	1.4
Cl	0.11	ND	Ti	0.32	0.16
Mg	ND	3.5	F	0.37	1.2
P	ND	ND	K	0.68	0.14
Fe	ND	ND	W	0.77	ND

Plate 6.   Mass Spectrographic of SiC
           Single Crystals Analysis.

Plate 8.   Alpha Crystal with
           Inclusion

Number	Pressure PSI Helium	Temperature in Degrees C	Time Hours	Results
68	10	2400	5-3	Few small crystals 2-3 mm average largest dimension
69	10	2450	5-3	Slightly larger crystals than No. 68
70	10	2500	5-3	Double quantity of Run No. 68
78	100	2560	5-3	Large number crystals - thickness of platelets 216μ 6H & 15R mixture
81	100	2650	5-3	Good crystal growth - some quite thick
65	250	2580	8	Only one crystal-small size
74	250	2700	7	Very few small crystal platelets Large number of columnar xtals Thickness 100μ - clear 6H pattern
66	550	2550	6	No crystal platelets

Number	Pressure PSI Helium	Temperature in Degrees C	Time Hours	Results
73	550	2650	5-3	Small # platelet - some columnar xtals Clear 6H pattern - thickness 75μ
84	550	2880	1	No crystal platelets - Cone
87	550	2660	3	No crystal platelets - Cone
88	550	2850	3	No crystal platelets - Cone
89	550	2850	1-2	No crystal platelets - Cone

Plate 9.   Growth Chart.

Plate 10.   Columnar Crystals.

Plate 11.   Columnar Cyrstals.

Plate 12.   SiC-Si Boule.

Growth of SiC Single Crystals from Silicon Vapor and Carbon

Y. Inomata, Z. Inoue, M. Ota and H. Tanaka

Inclusion free 5~20 mm alpha-SiC single crystals were grown by transporting silicon vapor from molten silicon into a graphaite growth cavity at 2500°C for 10~15 hours. The crystals obtained are similar to those grown by ordinary sublimation method with regard to growth state, shape and structure. The advantages and disadvantages of this new method were discussed and further improvement of this method were proposed.

INTRODUCTION

Lely's sublimation growth of SiC is the most successful method presently known for the preparation of large single crystals. But the crystals obtained by this method often include fine and dispersed carbon particles, because the composition of vapor in the growth cavity is shifted more to the carbon rich side than to that of equillibrium in the SiC–C system (1).

In a previous report (2), one of authors has proposed a new method for preparing of alpha-SiC single crystals which don't include fine carbon particles. The present report is a review of succeeding work on this new method.

APPARATUS AND PROCEDURE

The experiment were performed in a graphite resistance furnace with a single heating element as shown in Fig. 1. The reaction system used for crystal growth is shown in Fig. 2.

The main parts consist of the growth cavity, the connecting pipe with radial shield plates and the crucible which holds the molten silicon. They were all made of purified dense graphite (bulk density, greater than 1.80 ; ash content, less than 20 ppm ; without halogen treatment).

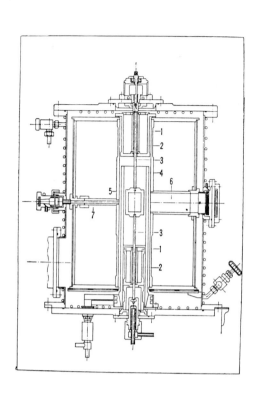

Fig. 1.   Graphite furnace

(1) graphite cylinder, (2) insulating block,
(3) graphite rod, (4) heater, (5) crucible,
(6) sight pass for pyrometer, (7) sight pass
of radiation pyrometer for temperature
control

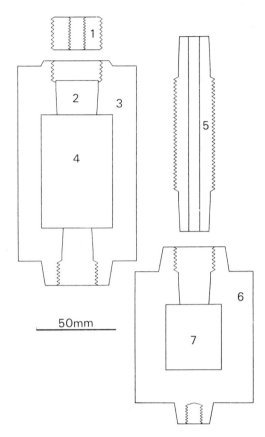

Fig. 2.   Schematic drawing of the
reaction system

(1) screw plug, (2) tapered plug, (3) crucible,
(4) growth cavity, (5) connecting pipe,
(6) crucible, (7) space to hold molten silicon

These parts were connected each other with screw and tapered plugs to avoid the leakage of the silicon vapor.   The assembled reaction system shown in Plate 1 was suspended in a graphite heater (inner diameter, 100 mm ; effective length, 300 mm) by a graphite rod of 10 mm in diameter.

The furnace was evacuated to the order of $10^{-5} \sim 10^{-6}$ Torr and then filled with highly pure argon (nitrogen content, less than 5 ppm) of 1 atm.

At first, the reaction system was preheated rotating it at the rate of 2 rpm under the condition of uniform temperature distribution to avoid breaking of the crucible containing silicon.   After being held at 1700°C for 30 minutes, the electric power was turned off and the system allowed to cool to room temperature.

Then the reaction system was moved downward in the furnace to get a favourable temperature difference between the growth cavity and the crucible holding molten silicon by utilizing the temperature gradient of the furnace.

Good results were obtained when the temperature of molten silicon was about 2200°C and that of growth cavity was 2500°C. The best position of the reaction system in the furnace was sought by trial and error process.

After choosing such condition, the second heating was then started. The temperature was raised up to 2500°C being measured on the outside wall of the growth cavity and kept 10~15 hours at this temperature.

RESULTS

Grown Crystals

Plate 2 gives pictures of the growth cavity after cutting in half parallel to the axis of the cavity. Crystals grown on the wall of the cavity were same as those grown by ordinary sublimation method. Plate-like, pale green crystals, 5~10 mm in diameter were grown with their c-axis coinciding with the axis of the cavity.

A dark filed photomicrograph of a typical crystal obtained by this new method is shown in Plate 3. No inclusion of fine and dispersed carbon particles were observed in the crystal. The bright wavy lines in the photograph correspond to the growth pattern on the basal surface.

Plate 4 shows an example of relatively large crystals obtained by this new method. Generally, basal surface of large crystals is not smooth. This tendency is seen in the crystal shown in Plate 4.

In the case of present new method, it is very important that the amount of silicon vapor being supplied to the growth cavity has to correspond with the growth rate of the growing crystals. If the supply of the silicon vapor to the growth cavity is dificient, the crystals tend to include fine carbon particles or receive thermal etching and drop from the inner wall of the growth cavity. On the other hand, if the supply of the silicon vapor is excess, the crystals tend to include the thin layers of silicon.

Plate 5 is an example of such a bad crystal. The round dark spot seen on the crystal surface are the silicon condensed from vapor. Other irregular shaped dark parts are thin silicon layers included in the crystal.

Plate 6 shows an example of a bad crystal obtained by the ordinary sublimation method. The crystal includes numerous fine and disperesed carbon particles.

The rate of supply of silicon vapor was controlled in this new method by adjusting the inner radius or length of the connecting pipe and also the temperature difference between the growth cavity and the molten silicon.

The mechanism of crystal growth in this method is not clear. But, it was assumed that the silicon carbide layers which formed on the inner wall of the growth cavity by the reaction between silicon vapor and graphite wall act as a source of subliming materials. There are, however, several puzzling aspects of the growth feature that are difficult to understand on a simple vapor growth approach. Liquid solution phase with carbon dissolved in silicon also may take part in the crystal growth (3).

Structure of Crystals

The structure of several crystals obtained by this new method were shown in Table 1. The kind of polytypes and the probability of their occurrence was the same as those obtained by ordinary sublimation method. Syntactic coalescence of 6H and 15R was consistently observed. In the table, an open circle 0 represents stacking disorder in the direction of the c-axis as determined by the presence of diffused streaks in the vibration x-ray photographs.

Table 1. Structure of crystals obtained by present
method at 2500°C for 10 hours.

No.	Structure	Diameter (mm)	Thickness (mm)	Thick. / Dia.	Stacking disorder
1.	6H + 15R	7.8	0.60	0.08	0
2.	6H + 15R	7.7	0.44	0.06	0
3.	15R	7.5	1.12	0.15	0
4.	6H + 15R	7.1	1.10	0.15	0
5.	6H	6.8	0.76	0.11	
6.	6H	6.2	0.84	0.14	
7.	6H	6.0	0.74	0.12	
8.	6H + 4H + 15R	5.9	0.50	0.08	
9.	6H	5.7	0.62	0.11	0
10.	6H + 15R	5.7	0.62	0.11	
11.	15R + 6H	5.3	0.58	0.11	0

Electron Mobility

The Hall mobility of crystals obtained are shown in Table 2. Data for SiC crystals obtained by ordinary sublimation method and for commercial crystals were also shown for comparison. The measurement were done at room temperature by Van der Pauw method. No large differences were observed among them except commercial crystals.

Kamath (4) reported that a mobility of over 250 cm²/V.sec was readily obtained for alpha SiC crystals with 10^{17} at. cm^{-3} carriers. Knippenberg (5) showed that a mibility of 300 cm²/V.sec and a over 500 cm²/V.sec was obtained for 6H and 15R crystals with 10^{16} at.cm^{-3} carriers, respectively.

The present values in Table 2 are slightly small than those in above mentioned literatures. This is probably due to the lack of purification of dense graphite used for reaction system.

Table 2. Resistivity and Hall measurement
for a-SiC at 20°C

Method		Resistivity (ohm·cm)	Mobility (cm²/V.sec.)	Carrier Concentration (10^{17} at.cm^{-3})
Present method	1*	0.091	200	3.5
	2*	0.053	120	6.4
Ordinary sublimation method	1	0.17	210	1.7
	2**	0.046	520	2.6
	3	0.21	120	2.5
	4	0.085	200	3.7
	5	0.074	190	4.6
Commercial	1	0.36	57	3.1
	2	1.4	55	0.83
	3	0.66	81	1.2

* each crystal consists of 6H and 15R.

** 15R is predominant.
all other crystals are 6H.

DISCUSSION

The followings are the merits of this new method.

1) It is possible to get crystals which do not include fine and dispersed carbon particles.

2) This process eliminates the need for fabricating a mould from pre-synthesized silicon carbide used in the ordinary sublimation method.

There are also several disadvantages.

1) Since the reaction system is tightly closed, it is difficult to evacuate inner atmosphere of growth cavity.   But it will be possible to evacuate by drilling a small hole in the top plug or suspending rod.   This hole may be sealed with SiC deposite at high temperature.

2) The crucible containing molten silicon tends to break near the melting point of silicon when melting occurs under a large temperature gradient.   However, if the silicon is melted quickly by heating without thermal gradient, crucible does not be easily broken and resist to the second heating even in a thermal gradient condition. This is probably because the SiC layer formed on the inner wall of the crucible by the preliminary heating will stops the diffusion of silicon to the graphite wall during the second heating.

3) The necessity of requiring double heating for a growth run is very tedious.   But, this can be improved by using a furnace having two separately controllable heaters.   By doing this, it will be easier to control the temperature gradient as well as to heat the reaction system homogeneously.

REFERENCES

1.   Y. INOMATA and H. TANAKA. Jour. Ceram. Soc. Japan, 78, 323 (1970).

2.   Y. INOMATA. Jour. Crystal Growth, 12, 57 (1971).

3.   G. S. KAMATH. Mat. Res. Bull. 4, S-57 (1969).   Special Issue for Silicon Carbide—
        1968.

4.   ibid.

5.   C. J. KAPTEYNS and W. F. KNIPPENBERG. Jour. Crystal Growth, 7, 20 (1970).

(a)

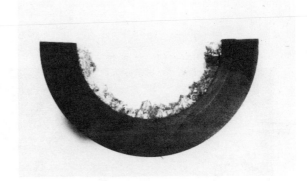

(b)

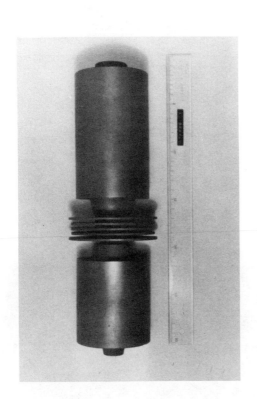

Plate 1.

The assembled reaction system.

Plate 2.

Features of the growth cavity after cutting in half.
A photograph (a) is taken from the perpendicular direction to the axis of cavity, and (b) is that taken along the axis.

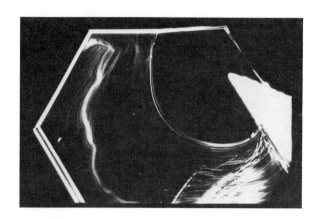

Plate 3.

Dark-field photomicrograph of plate-like 6H crystal obtained by present new method.   No carbon inclusions are observed.

Plate 4.

An example of large crystal.

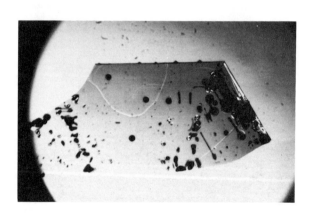

Plate 5.

A crystal grown under the condition in which silicon vapor was present in excess.

Plate 6.

Dark-field photomicrograph of a crystal obtained by ordinary sublimation method. Numerous fine inclusions are observed.

Inclusions Found in Solution-Grown β-SiC

T. Tomita and M. Ishiwata

ABSTRACT

Many hollow crystals are found in the solution-grown crystals of β-silicon carbide. Inside the hollows of some crystals, inclusions are found and determined to be a kind of silicon-rich solid solution by X-ray analysis. The mechanism of crystal growth by the re-entrant edge is difficult to explain the formation of these hollow crystals. Another mechanism is discussed.

INTRODUCTION

The crystals of β-silicon carbide so far obtained have been mostly small and dendritic. Nelson et al. (1) developed fluid-dynamically a solution method in which pure silicon is melted in a carbon or graphite crucible. They gave the optimum temperature gradient and circulation of the melt under which many crystals of β-silicon carbide grow in needle-like platelets of good habit.

The crystals thus grown are twinned through a (111) plane parallel to the largest crystal surfaces. These crystals have supposedly grown at the re-entrant edge formed between two crystal matrices twinned to each other, as proven for germanium by Wagner (1, 2, 3).

Among many crystals of β-silicon carbide obtained from one crucible, large hollow crystals of thin plates or of rhombic sections are found as well as the others of good habit, dendritic or more complex forms. If these crystal hollows were maintained or filled with some material other than silicon carbide while the crystals were growing, the concept of crystal growth by re-entrant angle may fail to explain the growth of these crystals.

X-ray diffraction topographs obtained for respective crystal matrices twinned with each other show poor correlation in the distribution of dislocations and stacking faults even for the crystals for which we can hardly recognize the crystal hollow as shown in our previous paper (4). This fact would suggest that the growth of both crystal matrices does not originate in the twin boundary.

Here, we shall show a variety of hollow crystals and the results of X-ray analysis of the inclusions occasionally found inside the crystal hollows.

VARIETY OF CRYSTAL FORMS AND THE STRUCTURE OF INCLUSIONS

Envelope-type Crystal

As shown in the previous paper (4) and in Figure 1, some hollow crystals have two wide (111) plates attached together by narrow sides, the section being trapezoidal and rarely hexagonal. The two (111) plates twin with each other, but each plate is never twinned by itself.

Some crystals show inclusions extending inside the crystal hollows as shown in Figure 2, and sometimes the inclusion remains locally in a wide hollow as shown in Figure 3. In many cases, these inclusions give X-ray diffuse reflections as shown in Figure 4a. Extinction of X-ray reflections and the lattice spacing obtained by X-ray oscillation photographs (Figure 4b) show that the inclusions are nearly silicon. However, when the inclusion shows sharp X-ray reflections, we can recognize the discrepancy between the reflections of the inclusion and that of pure silicon as shown in Figure 5. In the case shown in Figure 5, the inclusion has the rhombohedral lattice represented by hexagonal a_o of 3.8605 ± 0.0005Å and c_o of 9.408 ± 0.001Å. The amount of inclusion is enough to give strong coherency in X-ray reflections. Then, the inclusion should be some silicon-rich solid solution. Such a solid solution usually melts near the melting point of pure silicon and would be liquid when the hollow crystal is growing. Observed diffuseness of X-ray reflection suggests that the inclusions are much distorted to be a substrate for the growth of envelope-type crystals.

Crystal of Rhombic-Section

The hollow crystals other than the envelope-type may be categorized in this way where complex development of crystal wings is observed. A typical example of this kind is shown in Figure 6a and 6b, respectively showing its figure along the long side and at the top of the crystal. It is difficult to see the rhombic section on these figures but it may be seen in Figure 7 by the X-ray reflection spot arrowed. As shown in Figure 7, some two wings of rhombic circumference show oriented twin (the reflections of B and C). The inclusion is hardly recognized in this crystal using the optical microscope, but it gives the reflections indicated by D in Figure 7.

DISCUSSION

It is now obvious that the hollow crystals contained some melt of silicon or silicon-rich solid solution when it was growing. The widest crystal sides of the envelope-type crystal are always twinned with each other, but every one of them is not twinning by itself. Then, we may conclude that surface tension of viscous melt attracts adjacent crystal wings just grown and makes them stabilize especially when the crystal wings are in a twinned relationship.

As for the stability of the twinned crystals, the following consideration is one of the available.

We shall consider a thin crystal lamella built of an ideal sequence of the hexagonal close-packed layers as shown in Figure 8. When this lamella is twisted, the atoms of upper and lower layers would slip out of the sites relative to the middle layer. The directions of easy slip are directed in three lowest potential barriers and are opposite for upper and lower layers. Then, when the uppermost layer is easily dilated relative to the central layer, the lowermost layer is easily deflated. On the other hand, if the lamella is twinned, all the directions of easy twist in one side are opposite to those in the other side. Accordingly, the twinned crystal lamella would stand against twist more strongly than the single crystal lamella of the same thickness, as far as these crystal lamella were free from crystal defects.

The authors are much indebted to Dr. R. W. Bartlett, Dr. R. C. Marshall, Dr. W. E. Nelson and Dr. C. E. Ryan for their kind arrangement in giving the chance to study on many crystals.

---

(1) W. E. Nelson, A. Rosengreen, R. W. Bartlett, F. A. Halden and L. E. Marsh, AFCRL-67-0218 (1967).

(2) R. W. Bartlett and G. W. Martin, J. Appl. Phys. 39, (1968), 2324.

(3) R. S. Wagner, Acta Met. 8, (1960), 57.

(4) T. Tomita and T. Yuasa, Proc. VI Int. Conf. X-ray Optics and Microanalysis, (1972), 685.

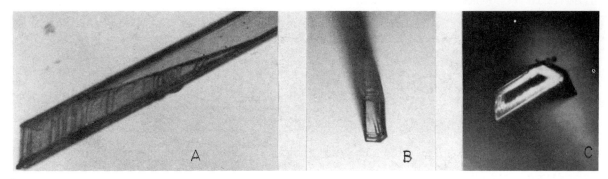

Figure 1. Microscopic views of the hollow crystal, (A) along the long side, (B) at the top and (C) sectional of another crystal.

Figure 2. A micrograph of an evelope-type crystal showing dark inclusion extended.

Figure 3. A micrograph of an envelope-type crystal showing local inclusion.

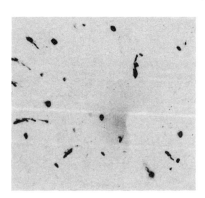

Figure 4a. X-ray Laue photograph of the crystal shown in Figure 2.

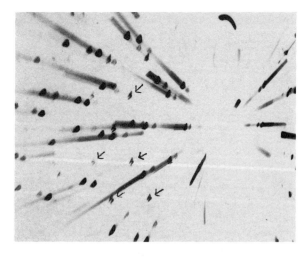

Figure 4b. X-ray oscillation photograph of the crystal shown in Figure 2. MoK$_\alpha$ 45KV. Arrowed are reflections of inclusion.

Figure 5. X-ray oxcillation photograph obtained for
inclusion is shown superposing with that of pure
silicon. CuK$_\alpha$, 25KV.

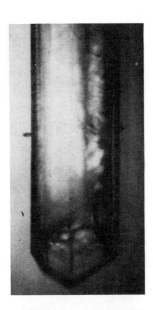

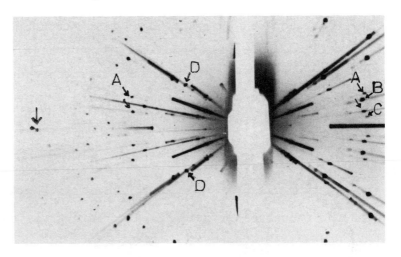

Figure 7. X-ray oscillation photograph obtained for
a crystal of rhombic section, (A) normal twin spots,
(B) and (C) misoriented twin spots, (D) the reflections of inclusion, MoK, 45KV.

Figure 6. Microscopic
views of a crystal of
rhombic section.

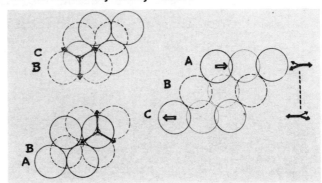

Figure 8. Schematic view of a sequence of the hexagonal close-packed layers in f.c.c. structure. The
directions of easy slip by twist are shown for upper
and lower layers.

Investigation of Silicon Carbide Crystal Growth from Vapour Phase

Yu. M. Tairov and V. F. Tsvetkov

ABSTRACT

The control of the process of growing structurally perfect single crystals with controlled doping is impossible without studying the mechanism and kinetics of crystal growth. Growing SiC crystals and producing SiC crystals by sublimation, in particular, represents the superposition of several processes, namely the evaporation of the starting material, nucleation, diffusion or convection of the vapour to the growing crystals, and specific heat-exchange mechanisms. Experimental observations indicate that all these processes are interrelated. The purpose of this paper is to consider these processes and their connections. The investigation of the processes of growing SiC crystals was carried out with high-temperature devices described in detail in[1,2].

THERMAL ETCHING (EVAPORATION) OF STARTING SiC

The investigation of SiC evaporation was carried out both on polycrystalline silicon carbide and on separate single crystals in the presence of a temperature gradient. In all the cases studied, we observed the evaporation of seeds and single crystals of SiC to take place with the formation of whiskers up to 20-50μ in diameter and up to 2 mm long (Figure 1,a). The mechanism of crystals etching with the formation of whiskers was investigated for the 2200 - 2600°C temperature range. According to our estimations, the temperature gradient in the crucible was 2 - 5 deg/cm. The following results were obtained:

1. The type of etching did not depend on the polytype of single crystals (polytypes 3C, 6H, 15R, 21R, and 4H were studied).

146

2. The etching rate and the type of etching (whiskers formation) did not depend on the crystallographic orientation of single crystals (crystals with (0001), (000$\bar{1}$), (1$\bar{1}$00) and (11$\bar{2}$0) orientation were investigated).

3. The whisker axis was always directed in the direction of heat flow.

4. When etching single crystals of 3C-SiC which have a lamella protruding at a certain angle to the surface, it was found that at the site of the lamella a canal is formed which reproduces its shape and location in the crystal. According to [3], the lamella is defective area in the crystal, since it is a set of disordered crystalline planes.

5. In the course of etching a negligible amount of amorphous carbon in the form of a web is formed between the whiskers.

6. The surface of etching at the base of the whiskers has a stepped appearance, with the steps passing on to the whiskers body (Figure 1b).

7. Whiskers have different habit (round, hexagonal, etc). The investigation of luminescence, on the whisker body and surfaces showed that the different facets appear as a result of whisker growth in thickness.

The results mentioned enable us to represent the mechanism of whiskers formation during etching in the following way. Under negative supersaturation over the crystal surface, the flat front of evaporation is distorted at the places on the surface where the chemical potential is higher (the appearance of screw dislocation, lamellae, stresses in crystals left after grinding, etc.).

The pits formed begin to deepen rapidly since the pit bottom temperature is higher than the crystal surface temperature due to the temperature gradient through the thickness of the crystal. The evaporation products are brought out of the pits by diffusion in the vapour phase between the whiskers and over the whisker surface, a part of the etching products settle down on the whisker body, with the result that they acquire crystallographic facets. Because of their growth in thickness the whiskers may intergrow, forming "a leg" out of several whiskers, the intergrowth of whiskers beginning at their tops and resulting in the formation of "a roof" (self-sealing of the evaporation cavity).

SiC SINGLE-CRYSTAL FORMATION

Silicon carbide crystal growth, when growing in the Si-C system (the sublimation method, direct synthesis out of elements, Bridgman - Stockberger method), passes through a number of intermediate steps, at which the dissociation and condensation of SiC phases take place in the presence of temperature gradients, which help the transfer of dissociation products.

The common feature of growth and evaporation in this system is the stability of the flat front in the direction opposite to that of the heat flow and the distortion of stability (dendrite growth, evaporation whisker formation) in the direction of the heat flow. The same regularity exists in the canals formed with the deposition of SiC cake on the graphite crystallization tube (G.C.T.) with holes[4]. The nucleation of secondary crystals and their subsequent development is observed either on the polycrystalline layer of condensing silicon carbide, or more often at the top of evaporation and growth whiskers. According to the results of x-ray analysis the crystallographic orientation of the whisker determined the angle of the intergrowth of the whisker and the plate and, consequently, the orientation of the plate in the growth cavity. Figure 2 illustrates the dynamics of the formation and development of SiC platelet at the whisker top (the photographs of three whiskers are given, which were formed near each other). Up to now in the theory of crystal growth from the vapour phase, according to Frank, the transformation of the whisker into a platelet is one of the most vulnerable points. Of particular interest in Figure 2 is the formation at the whisker top of a "spherical" nucleus of SiC with pits out of which faces of the $(1\bar{1}00)$ are subsequently formed. We did not succeed in revealing screw dislocations on the evaporation whisker axis, and therefore, we may conclude that in the case of silicon carbide single crystal growth from the vapour phase at 2300 - 2600°C the transformation of a whisker into a platelet occurs due to the fact that this process is energetically profitable, since under the conditions of heat dissipation carried out mainly by radiation (see below), the platelet is a more stable crystalline form. The place of platelet formation was more often the "leg" formed as a result of several whiskers intergrown together. In the majority of cases in the platelet - "leg" intergrowth, a screw dislocation was found in the re-entrant angle with the dendrite (whiskers) intergrowth as a result of intergrowth fault, if one is to judge by the distribution of growth stages[5].

The specific features of silicon carbide platelet formation in the growth cavity when the graphite crystallization tube (G.C.T.) is used, have been considered by us in[4]. We should like to stress once more that all the regularities mentioned above also act in the canals formed in SiC cake at G.C.T. holes. Whisker formation occurs here in accordance with the heat field in the canal too.

During its growth the whisker "curves" and intergrows into the growth zone, where at its top a platelet crystal is formed. Since in the growth zone the whisker always transforms into a platelet, it once more confirms that the heat conditions play a determining role in the mechanism of whisker transformation into a platelet. It is obvious that the crystal orientation in the growth zone is determined by the crystallographic orientation of the intergrown whiskers.

THERMAL CONDITIONS DURING SiC SINGLE CRYSTAL GROWTH

It is evident that the crystals which were nucleated in the growth cavity and which have a favorable orientation for the dissipation of the heat of crystallization and for feeding with crystalline material, develop faster than crystals with other orientations. In[6,7,8] it was observed that the crystals oriented horizontally in the growth canal are in a better condition for heat dissipation. By computation, it is proved that at $2600^{\circ}C$ the heat dissipation process is principally one of radiation.

This conclusion was proved experimentally by studying the growth rates of crystals placed in G.C.T. and oriented at different angles and by investigating the influence of shielding effects. SiC crystals shield a part of G.C.T. which results in a considerable evaporation of the SiC cake adjoining the G.C.T.

When the crystal was placed horizontally in a thin-wall graphite tube, positioned inside the growth cavity, the tube surface was coated with a polycrystalline deposit of SiC. Under the crystal the graphite tube was partly sublimated, and the SiC deposit was absent (Figure 3).

The distribution of temperature throughout the growing crystal is of great importance in the process of crystal growth. The present level of development does not permit experimental determination of temperature distribution throughout the crystal with pyrometers; therefore, they were obtained by computation.

The computations were done for a crystal 15 mm long and 10 mm wide, which was placed in the middle of a growth cavity of 30 mm in diameter and 60 mm high. The crystal was subdivided into three sections, 5 mm long. The computations were performed by the method used in[9]. Neglecting the heat conductivity of the crystal, we obtained an estimate of the upper value of the temperature gradient along the crystal. The latent heat of crystallization can be neglected, owing to its insignificant value as compared to the large value of radiant flow[6,9,10]. The dependence of the computed crystal temperature as a function of the distance from the G.C.T. wall is shown in Figure 4. The computation demonstrates that as the crystal grows its temperature rises owing to changes in the conditions of heat exchange with the surrounding cavity. At the specified temperature gradients along the axis and the radius of the crucible ($\sim 4^{\circ}$K/cm), a temperature gradient of $0.50 \pm 0.05^{\circ}$K/cm is formed along the crystal. The calculation performed makes it possible to draw the following important conclusion. The temperature difference between the G.C.T. wall and the crystal is about $\sim 8^{\circ}$K. This temperature difference drops on the whisker or on the "leg", on which the crystal originated. In our case the "leg" length was of the same order of magnitude as the G.C.T. thickness, i.e. 0.3 - 0.4 mm. Hence, a temperature gradient of 200°K/cm is created along the "leg" axis. According to Billig[11] and Indenbom[12] a dislocation density of 3×10^{4} cm^{-2} must arise in the crystal "leg" as a result of the temperature gradient. When studying the production of dislocation free silicon crystals[13,14,15] by the Czochralski method, it was established that dislocations existing in the seeds are inherited by the growing crystals. Experimental investigations of dislocation distribution in the grown crystals of silicon carbide[16] indicated that a maximum density of dislocations of $\sim 10^{4}$ cm^{-2} was observed at the places of "crystal - leg" intergrowth. The results of our calculations and the data mentioned are in good agreement and enable us to draw a conclusion that the formation of dislocations in the crystal root is connected with a large temperature gradient in the crystal "leg". The results cited make it possible to conclude that in order to reduce the dislocation density arising from thermal stresses in the growing SiC crystals, it is necessary to reduce the radial and mainly the axial temperature gradients in the crucible. These conclusions are confirmed in the experiments carried out in our laboratory.

PECULIAR FEATURES OF SiC SINGLE CRYSTAL GROWTH WITH THE DIFFUSIVE MECHANISM OF VAPOUR DELIVERY

When investigating the process and mechanism of single crystal growth one must know the distribution of crystal-feeding material. In the growth cavity various habits can be observed (flatly-parallel platelets, "melted" and stepped crystals with a single naturally smooth face). By introducing nitrogen "rings" we investigated the crystal growth rate along the C axis. It was established that the ratio of the growth rate of the natural face (0001) to the stepped-face of the crystals growing in the middle part of the crucible amounts to 1, whereas for the crystals growing at the lids of the crucible this ratio is reduced to 10^{-4}, i.e. at the crucible lids the predominant growth occurs towards higher temperatures; a phenomenon which up till now was not satisfactorily explained.

For calculating the concentration distribution of SiC vapour in the crystal growth zone one must know the mechanism of vapour propagation from the growth cavity walls towards the lids (whether it is by diffusion or by convection). In order to determine the heat transfer mechanism in the closed cavity we used the experimental dependence of convection factor ε_k on the product of the similarity criteria of Grashof (Gr.) and Prandtle (Pr.). For the construction of the crucible which is used in our laboratory we received Gr.Pr. 67. From the relationship $\varepsilon_k + f$ (Gr.Pr.) one may conclude that the heat transfer in the growth cavity of the design considered is obtained through heat conduction and, hence, the SiC vapour transfer will be by diffusion in argon (convection takes place when Gr.Pr. > 1000). The calculated results were verified by experiment. To investigate the vapour propagation in the growth cavity, a thin vertical graphite partition 0.4 mm thick was inserted along the diameter. After a prolonged heating of the crucible at $2600^{\circ}C$ a deposit of polycrystalline SiC was formed on the partition; the shape of the deposit boundaries reflected the shapes of different concentration lines as will be shown further.

For the stationary case corresponding to the conditions of crystals growth, the distribution of vapour concentration in the growth zone was obtained by solving the Laplace equation in cylindrical coordinates. Equilibrium concentrations of vapour on the cavity walls specified the boundary conditions. The Laplace equation was solved by the method of nets on the central computer for the radial gradient of 4 deg/cm and for the axial ones of 0, 1.6, 4 deg/cm. The computation accuracy was 2%. As an illustration, the received distribution of vapour concentration (n), equilibrium vapour concentration (n_s) and supersaturation ($\alpha + \frac{n - n_s}{n_s} \cdot 100\%$) are shown in Figure 5a. The data permitted the concentration distribution of SiC vapour in the growth cavity, depending on the axial temperature gradient to be obtained (Figure 5b). In the figure it is seen that the SiC vapour concentration is highest with zero axial temperature gradient and that with this gradient one can achieve a maximum uniformity of SiC vapour distribution along the r and z axes. These results agree fairly well with Knippenberg's experimental data$^{(17)}$.

The computation of vapour concentration distribution (n) and supersaturation (α) in the growth cavity with SiC crystals growing in it was carried out as well. Two rows of crystals, 6 mm wide, were inserted at the height z = ± 9 mm. The temperature distribution throughout the crystal was specified taking into account the above computations. In Figure 6a, it is seen that a considerable redistribution of vapour concentration and vapour supersaturation has taken place in the growth cavity. Crystals in the growth zone resulted in the formation of SiC vapour flow towards the faces of growing crystals. The results of computations of the flow towards the upper crystal face to the flow towards its lower face are given in Figure 6b. It can be seen that evan at 9 mm distance from the middle part of the crucible the lower face growth rate will exceed the upper face growth rate > 2.5 times. The computation data are in good agreement with experimental results for the measurement of growth rates with nitrogen "rings" as well as with Kroko's$^{(9)}$ and Knippenberg's$^{(17)}$ results.

KINETICS OF EVAPORATION AND GROWTH OF
SiC SINGLE CRYSTALS

One of the main problems in considering the kinetics of a process is to find the slowest step which determines the growth rate of crystals. The process of dissociative evaporation of the starting silicon carbide was studied. When dealing with the experimental relationship the decomposition rate of starting SiC was estimated according to three parameters:

1. The rate of the decrease of the starting silicon carbide diameter ($V_d = \Delta d/\Delta t$. This parameter characterizes the decomposition of the SiC cylinder as a whole.

2. The rate of the decrease of the total crucible weight ($V_{m_T} = \Delta m/\Delta t$). This parameter characterizes the vapour flow throughout the outside wall of the crucible.

3. The rate of SiC crystals growth and deposit of SiC on the crucible lids ($V_{m_k} = \Delta m_k/\Delta t$). This parameter characterizes the vapour flow into the SiC crystal growth cavity.

The results obtained are plotted in Figure 7. One may see that all the relationships obtained are linear with a sufficient accuracy.

The time of the whole SiC cylinder decomposition can be easily calculated on the basis of the relationships obtained.

The investigation of evaporation and crystallization in the range of 2420 - 2750°C was carried out simultaneously. The rate of decrease of the SiC cylinder diameter V_d and the rate of decrease of the crucible mass served as characteristics of the evaporation process. Characteristics of the crystallization process were given by the average linear rate of single crystal growth in the [1$\bar{1}$00] direction V_M, estimated by the nitrogen "rings" and by polycrystalline deposit growth rate on the crucible lids V_{m_k}. The experimental results are given in Figure 8. The activation energy of the evaporation process, obtained from the graphs, was found to be $E_{evap} = 120 \pm 30$ kcal/mol, and the activation energy of the crystallization process was $E_{cryst} = 95 \pm 30$ kcal/mol. The values of E_{evap} and E_{cryst} are given with instrument error (principally because of the inaccuracy of temperature measurement with optical pyrometers), since it is higher than the rms deflections of the straight lines drawn through the experimental points. The results obtained are in good agreement with the data of [9,17], within the range of the experimental error. Since E_{evap} and E_{cryst} correspond to each other within the range of measurement error, we may conclude that the rate of SiC single crystal growth is determined by the process of dissociative evaporation of starting SiC cylinder. The possibility of accelerating the evaporation of the starting SiC can be achieved by inserting elements lowering the activation energy of evaporation on account of chemical interaction. Thus, in [18], for example, it is shown that in hydrogen the activation energy of SiC decomposition is diminished to 59 kcal/mol owing to the chemical interaction of hydrogen with SiC. Our experimental data indicate that insertion of aluminum into the growth zone brings about the acceleration of evaporation and of SiC crystal growth, respectively. The evaporation of starting SiC can be accelerated by growing crystals in rarefied atmospheres. We have performed experiments of growing SiC single crystals at argon pressure of 300 torr. The experiments demonstrate that in this case the crystal growth rate increased $\sim$2.5 times (Figure 9).

In the course of kinetic investigations it was found that the linear rate of crystal growth tends to diminish with time. This observation can be explained on the basis of non-uniform distribution of SiC vapour concentration in the growth zone in accordance with the computation (Figure 5) as well as on the basis of the fact that even with a constant rate of SiC vapour supply into a growth zone and with a respectively constant weight velocity of platelet crystal growth, the linear growth rate will decrease inversely as time squared. In Figure 10 one

can see the change in the linear growth rate as a function of time for four
crystals separately grown at 2500°C.

Thus, the set of investigations carried out indicates that growing of silicon
carbide crystals, and in particular, by sublimation, is a complicated process
including many factors, and the understanding of the role and importance of each
factor allows one to choose correct technological conditions for growing crystals.
The investigations performed and the developed techniques of studying crystal
growth conditions are used in our laboratory at present for producing SiC crystals
by Bridgman - Stockbarger[19] method and by direct synthesis out of silicon and
carbon[20].

(1)   J. G. Pichugin, Yu. M. Tairov, D. A. Yas'kov, Pribory Tekh, Eksper, (USSR)
         4, 176, (1963).

(2)   J. G. Pichugin, Yu. M. Tairov, D. A. Yas'kov, Zavodskaya Laboratoija, 30,
         10, (1964).

(3)   S. N. Gorin, A. A. Pletushkin, Rost Kristallov, USSR, Moscow, 6, 210, (1965).

(4)   Yu. M. Tairov, V. F. Tsvetkov, IV Vses. Sovesh, po Rostu Kristallov,
         Tsakhcadzor, part 3, (1972).

(5)   G. G. Lemmlain, E. D. Dukova, Kristallografija, 1, 351, (1958).

(6)   D. Hamilton, J. Electrochem. Soc., 105, 735 (1958).

(7)   D. Hamilton, Silicon Carbide, Pergamon Press, New York, 43 (1960).

(8)   Yu. M. Tairov, Silicon Carbide, "Naukova dumka", Kiev, 189 (1966).

(9)   Yu. M. Tairov, Rost Kristallov, USSR, Moscow, 6, 199 (1965).

(10)   L. Kroko, J. Electrochem Soc., 113, 301 (1966).

(11)   E. Billig, Proc. Roy. Soc., A235 37 (1956).

(12)   Y. L. Indenbom, Kristallografija, 2, 594 (1957).

(13)   W. Dash, J. Appl. Phys., 29, 736 (1958).

(14)   E. Yu. Kokorish, N. N. Sheftal, Rost Kristallov, USSR, Moscow, 3, 338 (1961).

(15)   A. D. Beliaev, V. N. Vasilevskaya, E. G. Miseluk, Rost Kristallov, USSR,
         Moscow, 3, 380 (1961).

(16)   A. S. Tregubova, J. L. Shul'pina Fiz. Tverd Tela, 14, 2670 (1972).

(17)   W. Knippenberg, Philips Res. Rept., 18, 161 (1963).

(18)   M. Kumagawa, H. Kuvabara, S. Yamada, Jap. J. Appl. Phys., _8_, 421, (1969).

(19)   J. Smiltens, Mat. Res. Bul, _4_, S85, (1969).

(20)   G. Kamath, Mat. Res. Bul, _4_, S57 (1969).

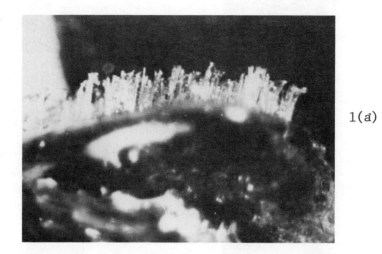

1(a)

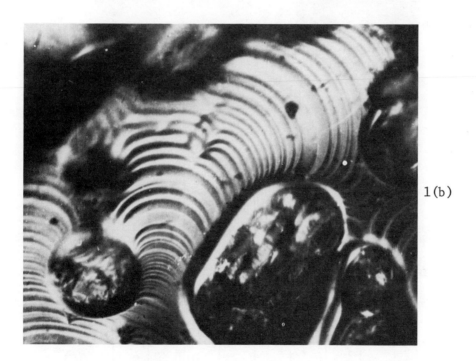

1(b)

Figure 1.   a) Formation of etching whiskers; b) Stepped appearance of the surface of etching at the base of the whiskers.

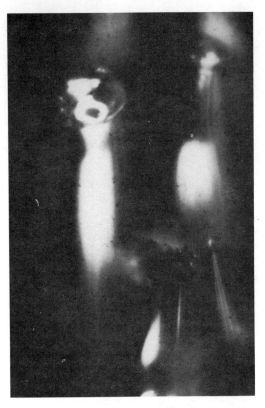

Figure 2.   Formation and development of   Figure 3.   Shielding effect of SiC crystal.
SiC platelet at the whisker top.

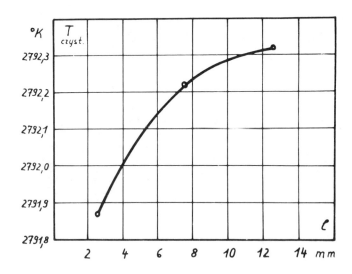

Figure 4.   Dependence of a computed crystal temperature as a function of the distance from GCT wall.

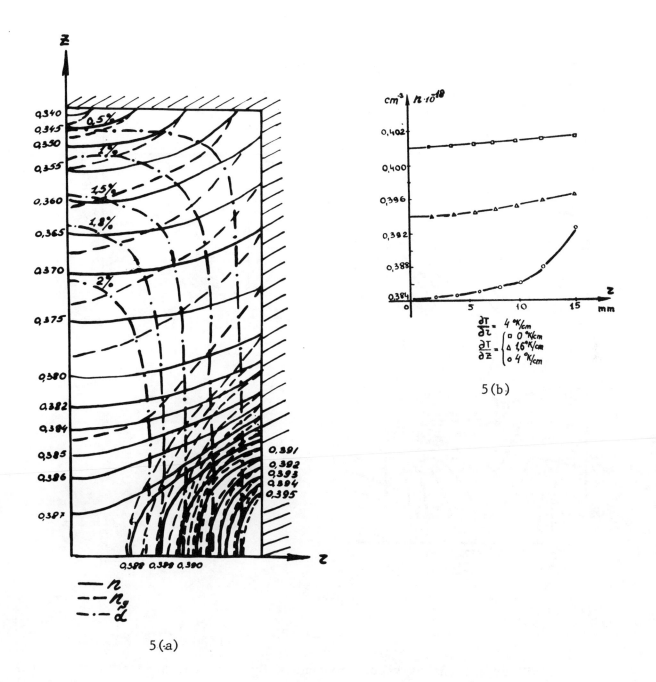

5(a)

5(b)

Figure 5. a) Distribution of vapour concentration (n), equilibrium vapour concentration (n$_s$) and supersaturation (α) in the growth covity. Both horizontal and
vertical temperature gradients are 4 deg/cm.   (b) The concentration distributions
of SiC vapour in the growth cavity, depending on the axial temperature gradient.

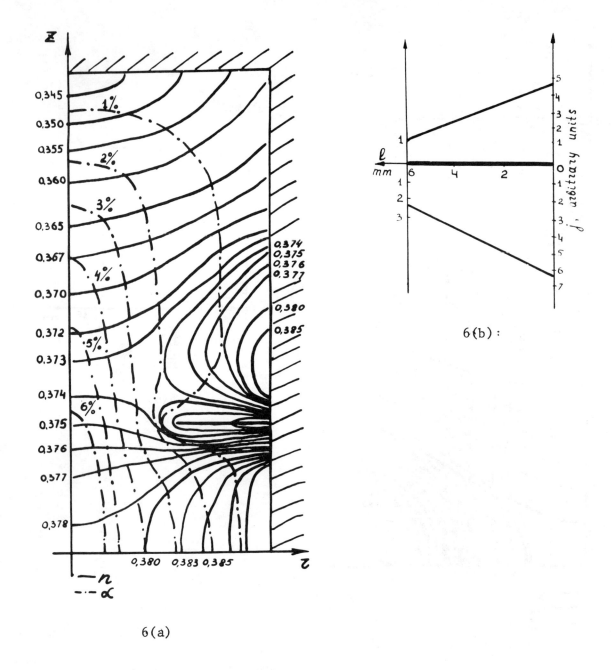

6(a)

Figure 6.   a) Distributions of vapour concentrations (n) and supersaturations (α) in the growth cavity with two rows of crystals.   Both the horizontal and vertical temperature gradients are 4 deg/cm.   (b) Distributions of the flows towards the upper and lower crystal faces.

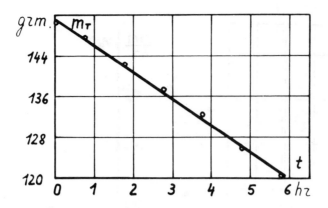

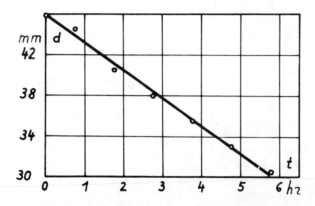

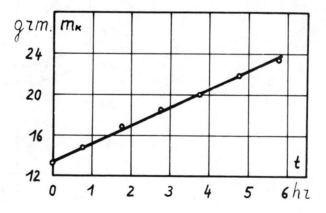

Figure 7. Kinetics of evaporation and crystallization of the starting SiC.

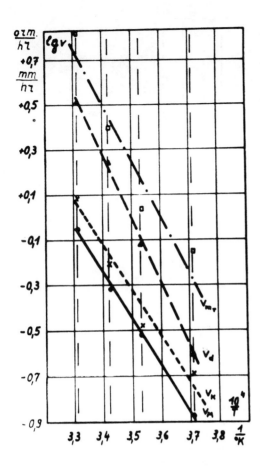

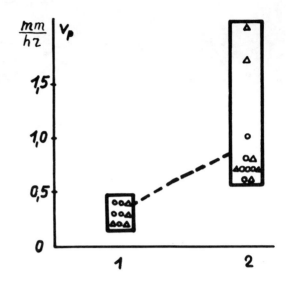

Figure 9.   Growing SiC single crystals at argon pressures of 760 torr (1) and 300 torr (2).

Figure 8.   The investigation of evaporation and crystallization in the range of 2420 - 2750°C.

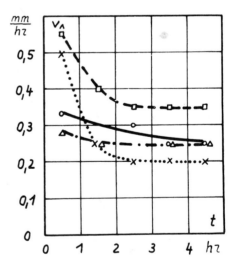

Figure 10.   The change in the linear growth rate as a function of time for four crystals separately grown at 2500°C.

LPE Growth of SiC Using Transition Metal–Silicon Solvents

P. W. Pellegrini and J. M. Feldman

ABSTRACT

Seeded solution growth of SiC layers has been investigated in an inert atmosphere at 300 psi. Ti-Si alloys melted in high-density, high-purity graphite crucibles were used for the flux material. The Si-rich portion of the $TiSi_2$-Si-SiC region in the Ti-Si-C ternary phase diagram was utilized because of the following solvent properties: (1) High SiC solubility is indicated, (2) the growth temperature is lowered from 2300°C for vapor recrystallization techniques to the 1600°C to 1700°C range, (3) melt evaporation problems encountered with other transition metal-silicon solvents are alleviated because of the low vapor pressure of titanium. Optical examination of the regrown material has shown that layers up to 100μm have been precipitated on the silicon face, while non-uniform islands were grown on the carbon face. The color of the layers on the silicon face varied from clear to light green. Oscillation x-ray techniques have been used to verify that highly oriented regions of 6H material were precipitated on 6H seeds; however, the regrown material was not truly epitaxial in the sense that a single crystal layer was not deposited on the substrate.

INTRODUCTION

Success in preparation of SiC for semiconductor devices has been marginal to date. The two main reasons for this failure are (1) the high impurity levels encountered, and (2) the inability to control polytypism. Since SiC does not form a stable liquid phase at pressures below 500 psi, most investigations of its growth properties have been attempted using vapor recrystallization methods with temperatures in the 2350°C to 2600°C range.[1] A few investigators have also used various solution growth schemes. Notable contributions in solution growth have been made by Brander,[2] Bartlett,[3] and Marshall.[4] The primary advantage which liquid phase epitaxy has over the vapor recrystallization method is that it permits lowering the growth temperature to a point where vapor pressure and technological limitations do not control the growth process.

SiC growth by spontaneous nucleation in a carbon crucible containing molten

161

Si has been discussed by Halden,[5] and has been effectively accomplished by
Bartlett.[3] Brander[2] has reported growth of epitaxial layers on prepared SiC
substrates from this system. Other than this work, there has been little success
reported on the growth of SiC layers from solution. This paper reports the
deposition of 6H SiC layers on 6H substrates from a molten transition metal-
silicon solution contained in a carbon crucible.

THE GROWTH SYSTEM

Material Selection:

Choosing a solvent is an important problem in solution growth. Once an
adequate solvent material has been found, growth studies can usually proceed in
a straightforward manner. In the present work, alloys of Si with all of the
transition metals of the first transition metal group from Ti to Ni inclusive,
and Y, Nb, and Zr of the second transition metal group were studied for their
suitability as a solvent for SiC. Previous work has shown that unalloyed
transition metals are not suitable solvents for SiC solution growth.[6]

The suitability of these transition metal-silicon alloys was determined by
placing a small amount of alloy (10 mg) on a SiC platelet in a closed carbon
susceptor and inductively heating the combination to a temperature of 1800°C.
The alloys were prepared in an electric arc furnace from starting materials
which were all more than 99.9% pure. Alloy composition in these solubility
experiments was approximately 67 atomic percent silicon. Optical examination
was used to determine which materials were the best solvents for SiC at 1800°C.
From the initial group of 10 transition metals, only silicon alloys of titanium,
zirconium and chromium demonstrated a substantial solubility for SiC.

The most outstanding material in these preliminary experiments was the Ti-
Si alloy, $TiSi_2$ with excess Si. It dissolved much more SiC from the platelet
than did any other alloy. Its vapor pressure at 1700°C is two orders of
magnitude lower than that of Cr and should present fewer technological difficulties than Cr-Si alloys.

Theory of Operation:

Growth experiments were carried out in the furnace shown in Plate 1, which
was fabricated from high-density, high-purity graphite. Graphite crucibles were
used because previous work by Brander[2] and Halden[5] has demonstrated successful
growth. SiC crucibles were not used because they were not available in high
enough purity to allow high-purity SiC growth. Also, attempts at using refractory metal oxide crucibles as melt containers has led to technological failures

such as crucible fracturing and reaction with the melt.

The graphite crucible was filled with about 40 gm. of the Ti-Si alloy of composition 40 w/o Ti and 60 w/o Si. This composition was chosen for the following reasons. First, it is in a region of the Ti-Si binary phase diagram[7] where there is an excess of Si. When this Si reacts with the walls of the carbon crucible, it forms the SiC nutrient for layer growth. Second, this mixture has a specific gravity less than that for SiC; hence any spontaneously generated crystals should sink to the bottom of the melt, thereby reducing con- tamination of the growing seed by these crystals. Finally, it is assumed that SiC is the substance of primary crystallization in the region of the Ti-Si-C diagram enclosed by tie lines joining $TiSi_2$-SiC-Si.

A carbon cap was placed on the top of the crucible to reduce radiative heat losses from the top of the melt. A small hole was made in the top for the introduction of the seed, and a narrow slot was likewise placed in the cap to allow pyrometer measurements and optical observation of the liquid-seed interface.

Vapor recrystallized SiC seed crystals of two different purities were obtained for substrate materials; dark green to black seeds with 99.5% purity, and light green seeds with less than 100 ppm impurity. These seeds were attached to a carbon rod for introduction into the melt. Two different seed orientations were utilized, the orientation being determined by which crystal- lographic axis was parallel to the axis of rotation of the seed holder.

EXPERIMENTAL

The loaded crucible was outgassed under vacuum for one hour at 1200°C. High purity helium was introduced and allowed to reach a pressure of 300 psig. This pressure insured that minimal melt evaporation occurred. The temperature was increased to 1450°C and the seed was introduced into the melt and rotated for the remainder of the experiment. As the temperature of this combination was raised to the operating temperature some of the surface of the seed was etched back, thereby creating a clean growth surface.

Material transport in this system is caused by an imposed temperature gradient. The gradient on the crucible is established by proper placement of the crucible in the rf coil. The temperature at the melt-seed interface is maintained constant by controlling the amount of rf power. The liquid becomes saturated with SiC after a few minutes at the operating temperature. The temperature gradient causes a concentration difference of SiC to exist in the liquid. This concentration difference in addition to the natural convection

forces SiC to go from the warmer bottom to the cooler top. The SiC in the top

zone is supersaturated. This non-equilibirum situation is alleviated by preci-

pitation on the seed crystal and by spontaneous nucleation in the bulk.

   The temperature at the melt-seed interface and the gradient along the

crucible walls was measured optically. Because of optical variations, losses

in the various pieces of glass, and errors in reading and calibration the

absolute temperatures quoted here are accurate to $\pm$ 20°C. The crucible gradients

are subject only to reading errors, and their accuracy is $\pm$ 2°C.

   Growth runs lasted from two to eight hours with experimental temperatures

ranging from 1450°C to 1700°C. The gradients imposed on the crucible varied

from -30°C (top hotter than bottom) to + 70°C (top cooler than bottom). Seed

shaft rotations were maintained constant in the range of 10 rpm to 60 rpm. At

the termination of an experiment the seed was slowly extracted from the liquid

to insure that a minimal amount of the regrown material would be fractured by

large thermal gradients. The seed was then etched in a 1:1 mixture of HF + HNO$_3$

to remove all solvent.

RESULTS AND DISCUSSION

   These experiments have defined a proper range of system parameters to

insure liquid phase epitaxial growth of SiC layers on SiC substrates using the

Ti-Si-C ternary system. It has been determined that temperatures in excess of

1600°C are necessary to obtain good layer growth. Growth on the silicon face

below this temperature ranged from negligible at 1450-1500°C, to sporadic growth

at the seed edges from 1500-1550°C, to small islands in the middle of the silicon

face from 1550-1600°C. Gradients from + 10°C to + 30°C (crucible top cooler)

were necessary to effect stable layer precipitation on the entire surface with

the seed oriented such that the (0001) direction was perpendicular to the axis

of shaft rotation. Gradients less than + 10°C caused seed etching, while

gradients larger than + 30°C caused growth at the top of the seed and etching

at the bottom. Seeds oriented such that the (0001) direction was parallel to

the axis of rotation demonstrated good growth over a somewhat larger range of

gradients.

   Examination of the seeds after etching showed light green to nearly clear

layers, indicative of α-SiC, had been precipitated on both the silicon and car-

bon faces. Microscopic examination showed a stepped and layered hexagonal

growth characteristic. (See Plate 2) Layer thickness was dependent on the length of time that the seed was immersed in the melt. Two hour runs had 30-40 micrometer layers, while eight hour runs had layers 120-150 micrometers thick.

No noticeable difference in growth habit or quality was observed in comparing growth on seeds of different purities. However, crystal growth was affected by seed orientation. In the case where the (0001) direction was parallel to the axis of shaft rotation, lateral growth on the silicon face was very rapid, and growth beyond the boundary of the seed crystal occurred (Plate 3). These overgrowths increased the surface area of the substrate by as much as 50%. Since the thickness of these layers was only 100 micrometers, they were easily fractured. When the dark, impure seeds were used, the outline of the dark substrate was visible to the unaided eye beneath the layer of regrown material.

When the (0001) direction was perpendicular to the axis of rotation, growth was less pronounced in the lateral direction but some overgrowth was noted (10% area increase) on the silicon face. The growth on this face was essentially flat with a few steps on the surface. The carbon face, which was also exposed in this orientation was covered with small green islands of various height and lateral extent as shown in Plate 4. These islands were similar to those noted on the silicon face when the growth temperature was below $1600^{o}C$. Also, the substrate edges showed some outward growth, but it was considerably less than that observed on the flat faces.

X-ray oscillation techniques were used to determine the polytypes of both the starting material and the regrown layers. Material deposited on the silicon surface was examined using the 15 degree a-axis oscillation technique. The regrown material on the substrate edges and the edges of the overgrown regions was examined with a 15 degree c-axis oscillation. The 6H polytype was found on both the regrown material and the substrates in all crystals. Crystalline perfection in the regrown material, as determined by streaking in the oscillation photographs, was always inferior to that of the substrate material. It was also noted that there was less streaking in material grown with the (0001) direction perpendicular to the axis of shaft rotation.

Since all the substrate seeds were of the 6H polytype, no conclusions can be made about the ability of this growth scheme to grow truly epitaxial layers. Only further investigation with different polytype seeds can determine this. The statement can be made, however, that 6H substrates always yielded regrown layers of 6H material.

Finally, it is suggested that other transition metal disilicides dissolved in a graphite crucible might be used as SiC solvent materials. $ZrSi_2$, whose solubility for SiC has been demonstrated, should also be adequate for growth experiments similar to those described here. Even some of the transition metal disilicide materials which did not demonstrate as much SiC solubility as Ti and Zr at 1600-1700°C in preliminary experiments, might be useful at, say, the eutectic composition between the disilicide and Si. In point of fact these other materials might be energetically more favorable solvents in the 1600-1700°C temperature range and allow better single crystal layers to be deposited on the SiC substrates than have been reported here.

ACKNOWLEDGEMENTS

The authors would like to thank C. S. Sahagian, C. E. Ryan and R. C. Marshall of AFCRL for their aid and helpful discussions about these experiments.

REFERENCES

1. G. S. Kamath, in Proceedings of the International Conference on Silicon Carbide, H.K. Henisch and R. Roy eds., Pergamon Press, New York, (1969), p. 61.

2. R. W. Brander, and R. P. Sutton, Brit. J. Appl. Phys. (J. Phys. D), 2, (2), 1969, p. 309.

3. R. W. Bartlett, W. E. Nelson, and F. A. Halden, J. Electrochem. Soc., 114 (11), 1967, p. 1149.

4. R. C. Marshall, in Proceedings of the International Conference on Silicon Carbide, H.K. Henisch and R. Roy eds., Pergamon Press, New York, (1969), p. 73.

5. F. A. Halden, in Proceedings of the International Conference on Silicon Carbide, J. R. O'Connor and J. Smiltens, eds., Pergamon Press, New York, (1960), pp 115-122.

6. P. W. Pellegrini, B. C. Giessen, and J. M. Feldman, J. Electrochem. Soc., 119 (4) April 1972, pp 535-537.

7. M. Hansen and K. Anderko, Constitution of Binary Alloys, second edition, McGraw-Hill, New York, 1958, p. 1197.

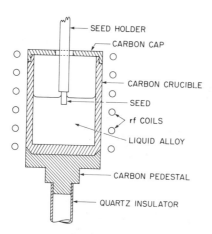

Plate 1. Diagram of graphite furnace system showing crucible and seed holder.

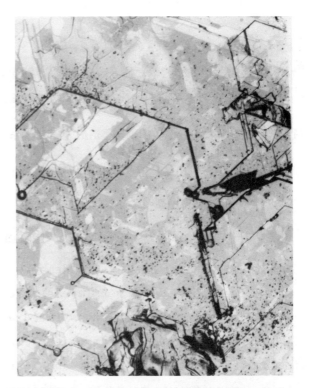

Plate 2. Regrown SiC layer showing growth steps on the silicon surface. (100X)

Plate 3. A regrown layer on the silicon face. The light region, whose thickness is 40 micrometers, shows the overgrown a area, while the dark region indicates the extent of the substrate. Reflections can also be seen which show the stepped, layered nature of the growth. (75X)

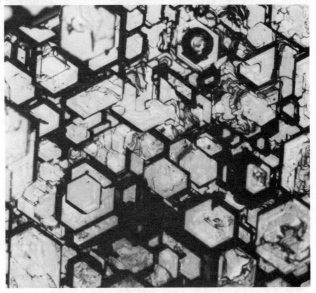

Plate 4. Regrown islands on the carbon face. Seed orientation was such that the (0001) faces were perpendicular to the axis of seed shaft rotation. (20X)

High Strength, High Modulus Silicon Carbide Filament via Chemical Vapor Deposition

H. E. DeBolt, V. J. Krukonis and F. E. Wawner, Jr.

INTRODUCTION

During the last five years demands on filamentary composite systems have reached increasingly higher levels. For example, in 1964 boron filament/epoxy matrix composite systems demonstrated a 350° F capability. In 1967 boron filament/aluminum matrix composites exhibited a 600° F capability. For advanced aircraft structural and engine applications silicon carbide filament in a titanium matrix composite was found to exhibit superior properties at temperatures of 1000° F. Several concerns have developed and produced silicon carbide filament (SiC) prior to the work at these laboratories; for example, United Aircraft Research Labs, General Technologies Corporation, and Service des Poudres (in France) have produced the filament in small amounts.

In April, 1972, a program sponsored by the Air Force Materials Laboratory, WPAFB, was initiated at Avco to improve the manufacturing process for making silicon carbide filament. The objectives of the program were to develop a SiC filament with tensile strength over 400 Ksi, which would retain over 85% of its room temperature strength up to 1000° F, and which would cost less than $200/lb at a production level of 2000 lb/yr.

The optimized production process is described below in the first section. The dominant species synthesized in the decomposition reactor are discussed in the next section. The electrical conduction phenomena of the semi-conductive SiC sheath are reviewed in the final section.

SiC DEPOSITION PROCESS

The SiC used as a reinforcing filament is vapor deposited onto a tungsten or carbon substrate. The reactive gases in the vapor phase include hydrogen and a mixture of alkyl silanes. The tungsten or carbon substrate is heated electrically to a deposition temperature of about 1300° C.

Figures 1 and 2 are schematics of the reactor and silanes recovery systems used to manufacture high strength SiC filament on a tungsten substrate. A gas mixture containing 70% H_2 and 30% silanes is introduced at the top end of the reactor tube where the tungsten substrate 0.5 mil in diameter is introduced. The reactor tube normally used is made of glass, about 0.5 inch ID by about 6 ft long. The substrate is drawn into the reactor through a saphire jewel hole 14 mils in diameter sealed by mercury, supported on its own surface tension. The mercury also serves as an electrode contact with the filament. The substrate is heated by a combination of direct current of about 250 milliamperes and VHF at a frequency of about 60 MHz, the VHF current varying from about 10 ma at the top to about 100 ma at the middle and bottom, in order to achieve the optimum temperature profile. After SiC deposition, requiring a residence time of about 20 seconds to make 4 mil filament, the filament is withdrawn through a second mercury-sealed jewel hole electrode and taken up on a plastic spool. The gas mixture exhausted at the bottom, containing about 95% of the original mixture, plus HCl, synthesized and decomposition products, is passed through a -80° C condenser to recover the unused silanes. The reclaimed silanes are reinjected back into the supply pressure vessel via a fractional distillation unit which extracts and rejects the high boiling point synthesized products.

ANALYSIS OF THE ORGANOSILANE SPECIES SYNTHESIZED

Figure 2 shows the silanes recovery system used to reclaim and purify the silanes from the reactor exhaust. Two liquid streams were analyzed, one the reclaimed and recycled liquid contained in the main silane supply pressure vessel, the other the rejected high boilers from the still pot of the fractional distillation unit. The analyses were made with an F&M Model 720 dual column temperature programmed gas chromatograph equipped with thermal conductivity detectors. The detector was held at 265° C and the injection port at 185° C for the analyses. Helium was used as the carrier gas at a flow rate giving the lowest HETP (Height Equivalent to a Theoretical Plate) for the column used. The instrumental conditions chosen are similar to those described by Burson and Kenner in their separations of chlorosilanes, methylchlorosilanes, and siloxanes[1]. The column used was a 12 ft long by 1/4 inch diameter stainless steel tube packed with 10% SE30 silicone gum rubber on acid washed Chromosorb W. The column was temperature programmed at 10° C per minute from 30° C to a maximum of 250° C. The SE30 column separates compounds according to increasing boiling point. The elution temperature versus boiling point characteristic of this column was determined by measuring the elution temperature of known hydrocarbons, primarily the linear hydrocarbons up to C15.

Figure 3 shows the chromatogram of the liquid in the main silane supply pressure vessel. Two of the compounds are those added to the system continually as makeup: dimethyldichlorosilane (Me_2SiCl_2) and methyldichlorosilane ($MeHSiCl_2$). These are added to the system in the proportions 3:1, and are present in the pressure vessel liquid in the proportions about 14:1. The other species were identified at Dow Corning Research Center, Midland, Michigan using a mass spectrometer coupled with a gas chromatograph. The other species in this mixture include dichlorosilane (H_2SiCl_2), trichlorosilane ($HSiCl_3$), and silicon tetrachloride ($SiCl_4$), the decomposition products from the injected silanes, and vinyltrichlorosilane ($ViSiCl_3$), and ethymethyltrichlorosilane ($EtMeSiCl_3$) as synthesis products. Some methyltrichlorosilane ($MeSiCl_3$) was detected, coeluting with the $MeSiCl_2$).

Figure 4 shows the chromatogram of the liquid rejected from the still pot of the fractional distillation unit. Note in Figure 4 that four classes of high boiling compounds are synthesized, in addition to the synthesis of high and low boiling point alkyl silane monomers discussed above.

disilanes	$\equiv$ Si - Si$\equiv$	(1)
siloxanes	$\equiv$ Si -O- Si$\equiv$	(2)
disylmethylenes	$\equiv$ Si (CH_2) Si$\equiv$	(3)
cyclics	[-Si Me_2 O-]$_{n}$ $n = 3, 4, 5$	(4)

In all of these classes, the $\equiv$Si radical can have Cl, Me, or H on each valence.

It was found during the optimization of the SiC filament formation process that if the high boiling species with boiling points over 100° C were not rejected from the main silane supply, the filament quality was degraded, and

---

[1]K. R. Burson and C. P. Kenner, Gas Chromatograph Separation of Chlorosilanes, Methylchlorosilanes, and associated siloxanes, Anal Chem 41 (6) pp. 870-872, 1969.

difficulties were experienced with tungsten substrate burnout shortly after an attempted reactor startup. It has not been determined which one or several of these high boiling point species are responsible for the quality degradation and startup difficulties.

SEMICONDUCTOR PHENOMENA OF SiC MONOFILAMENT

Early in the program of the development of SiC monofilament several phenomena were observed related to the semiconduction properties of the SiC layer of the filament. On occasion it was found that the reactor running time was short, a minute or less, because of filament burnout at the bottom or exit electrode. When this difficulty became severe, the filament was watched closely near the exit electrode. Prior to burnout, the filament was seen to become very hot (estimated temperature about 1500° C) just above or below the mercury surface. As illustrated in Figure 5, the filament overheated below the mercury surface if the atmosphere over the mercury were hydrogen, and above the mercury if the atmosphere over the mercury were an argon-hydrogen mixture. Subsequently, the reactor was equipped with instrumentation to measure the voltage drop across the SiC sheath at the bottom electrode as shown in Figure 6. An impedance of the measurement apparatus as low as 20,000 ohms was adequate to obtain an accurate measurement of this voltage, the instrumentation consisting of a combination of an oscilloscope, Sanborn recorder, and a portable Simpson volt-ohmmeter. The voltage was observed to vary with reactor conditions from millivolts up to about 35 volts DC, at which level a breakdown conduction phenomenon occurred, yielding a noisy high frequency voltage fluctuation at the electrode, varying between zero and 35 volts at frequencies in the KHz range. The filament made during breakdown was always very weak, as a result of structural damage observed as surface pitting shown in Plate 1.

The reactor operating condition found to bring this voltage under control was the presence of air in the reactive gas mixture, in the concentration range of a few hundred PPM. Once it was determined that the voltage could be varied from the breakdown level down to millivolts by adding air, experiments were conducted to determine whether it was the O_2 or N_2 responsible for the effect, since it is well known that nitrogen doping of SiC will increase its conductivity. Addition of pure N_2 was found to have no effect, while addition of a mixture of O_2 and argon had the same effect as addition of a mixture of O_2 and N_2. It was therefore concluded that O_2 in the vapor deposition gas mixture can control the conductivity of the SiC sheath. The relationship between the voltage drop and O_2 addition to the gas mixture is shown in Figure 7. It was found that nitrogen doping could also control conductivity, but at the deposition temperatures used the nitrogen doping had to be done with NH_3 rather than N_2, and the effect of NH_3 was found to be about the same as that of O_2 as shown in Figure 7. The addition of NH_3 was not a practical method for long term operation because NH_4Cl buildup quickly clogged the system. The results obtained with O_2 addition have not yet been adequately explained; H. Gatos[2] has indicated that oxygen does not dope the SiC lattice as does nitrogen so as to increase its conductivity.

E. Kerns[3] observed during independent work on SiC filament production that phosphorous would serve to control the conductivity of the SiC sheath in the same way as does nitrogen. He added PCl_3 in low concentrations to the reactive mixture in order to dope the SiC with phosphorous.

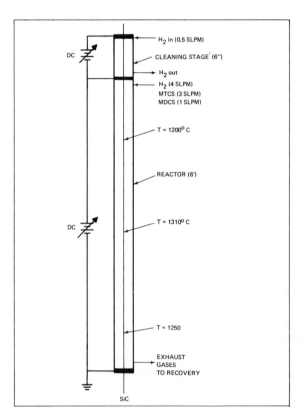

Figure 1 FLOW RATES AND REACTOR CONFIGURATION
FOR DEPOSITING SILICON CARBIDE

[2]H. Gatos, private communication.

[3]E. Kerns, private communication.

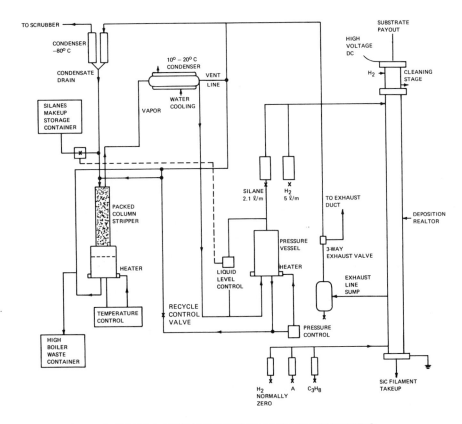

Figure 2 SiC FILAMENT PRODUCTION PROCESS SCHEMATIC

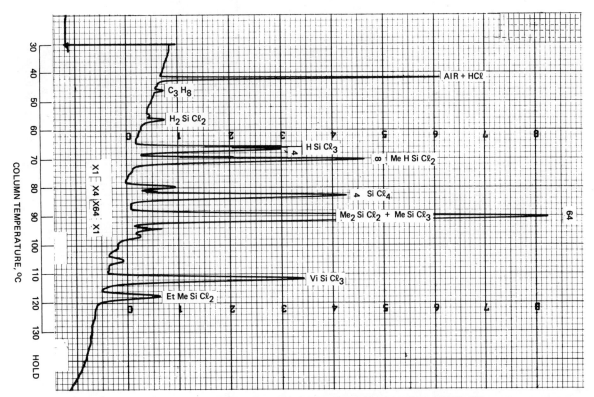

**Figure 3 GAS CHROMATOGRAM OF LIQUID FROM PRESSURE VESSEL OF
SILANES RECOVERY SYSTEM**

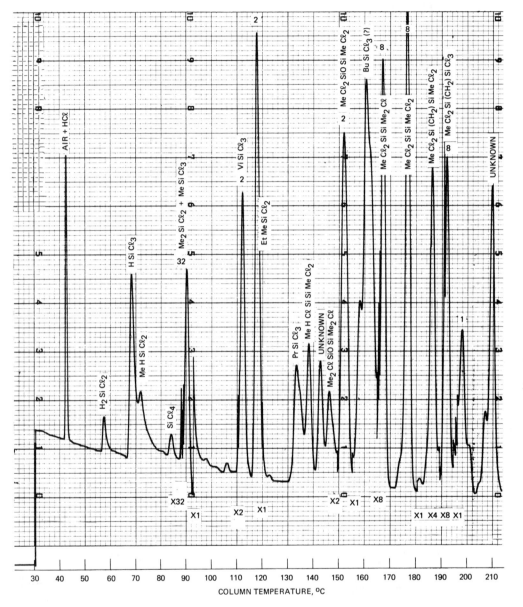

Figure 4 GAS CHROMATOGRAM OF LIQUID REJECTED FROM STILL POT
OF SILANES RECOVERY SYSTEM

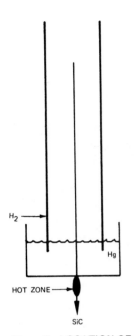

Figure 5a LOCATION OF
HOT ZONE
WHEN PURGE GAS IS H$_2$

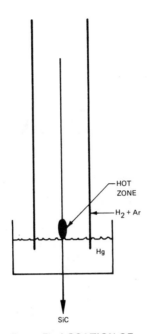

Figure 5b LOCATION OF
HOT ZONE
WHEN PURGE GAS IS H$_2$ + A

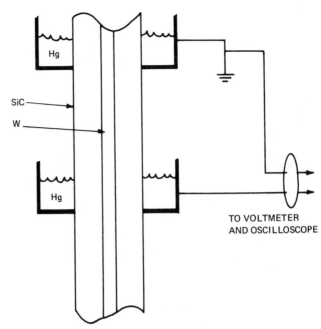

Figure 6 SCHEMATIC DIAGRAM OF VOLTAGE
MEASUREMENT CIRCUIT

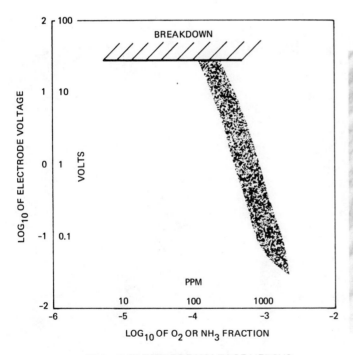

Figure 7 ELECTRODE VOLTAGE VERSUS
CONCENTRATION OF O$_2$ OR NH$_3$
ADDITION

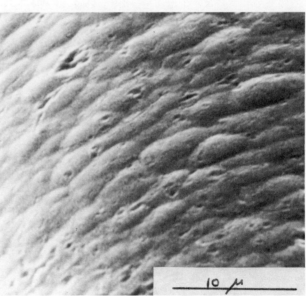

Plate 1 SCANNING ELECTRON MICROGRAPH
OF SiC SURFACE PITTED BY ARCING

PART II
POLYTYPISM

Speculations on the Origins of the Polytypism of SiC

Arrigo Addamiano

The existence of the phenomenon of "polytypism" in silicon carbide was reported 61 years ago by Baumhauer (1) following a morphological investigation of various carborundum crystals. Baumhauer introduced the words "polytypism" and "polytype" after he observed three different modifications of carborundum crystals grown "in equilibrium" with each other. The three crystal phases, SiC I, SiC II, and SiC III (or, in modern terminology, 15R SiC, 6H SiC and 4H SiC) did not show any tendency to transform into one another and were all equally stable. Here, then, was an apparent violation of Gibbs's phase rule and a system with a negative variance.

Simultaneously with the work of Baumhauer, the phenomenon of X-ray diffraction in crystals was discovered and techniques became immediately available for the study of the structure of crystalline solids. So Baumhauer in a second paper on the polytypism of silicon carbide (2) could report that SiC I, SiC II, and SiC III gave different X-ray diffraction patterns, therefore their "molecular or atomic" structure had to be different. Hence the suggestion that the polytypes are, in effect, different molecules, is over half a century old. In this period of time numerous observations of carborundum crystals and of high purity, laboratory made, SiC crystals have been reported, with the result that some 200 polytypes are now known. Indeed it appears that the theoretical limit to the number of polytypes that can exist is infinite. For the polytypes constitute series wherein longer and longer zig-zag sequences of atoms can be accommodated in correspondingly longer and longer unit cells. This feature is so striking that Ramsdell and Kohn in 1952 (3) suggested that the polytypes must originate by condensation of different radicals or molecules performed in the gas phase. In many respects, therefore,

the silicon carbide polytypes could be regarded as the inorganic analogs of
organic polymers, e.g. polyethylene, where different chain lengths can be obtained
by varying the degree of polymerization. This belief was fortified by the discovery of the existence of two different polytypes, 51 R_a and R_b, having identical
c axis length but different zig-zag sequence. This type of polymorphism, though
common for organic molecules, was new in inorganic chemistry.

Ramsdell and Kohn did not go so far as to propose a mechanism for the formation of the gaseous polymers envisaged by them. However, Adamsky and Merz (4)
reporting on the discovery of the wurtzite form of silicon carbide, which had
never been identified before in Lely-type furnaces, suggested that the temperature
in the cavity of the Lely furnaces is too high for the formation and existence
of the radicals or molecules from which the 2H SiC crystals originate. Once
again, therefore, the "polymer theory" of polytypism allowed a simple explanation
of an apparently intriguing question.*

In order to develop further the ideas of Ramsdell and Kohn, we shall make
two assumptions:

a) the polymers which generate the polytypes are formed by reaction of the
gaseous species present in the gas phase.

b) there is a 1:1 correspondence between the size of the condensing polymers
and the length of the zig-zag chain observed in the crystals.
In other words, the 2H SiC structure, with the ABAB... sequence, implies the
existence of Si_2C_2 polymers, the cubic ABCABC... structure derives from Si_3C_3
molecules, and so on.

Let us further elaborate these ideas and see whether they are in agreement
with the experimental facts. The gaseous species in equilibrium with hexagonal
silicon carbide at different temperatures are known from the work of Drowart and
De Maria (5). Si, Si_2C and SiC_2 are dominant over the interval of temperatures
of interest to us. The slopes of the logP vs 1/T lines are such that at low
temperature Si is in strong excess relatively to SiC_2 and Si_2C. At the highest
temperatures, which include the temperatures of operation of the Lely furnaces,
2500 - 2700°C SiC_2 and Si_2C become more important.

The process of dissociation of the silicon carbide charge, therefore, can be
formulated as follows**:

_____

* The wurtzite form of SiC is often present in the form of a white fluffy powder
on the inner lid of the Lely furnaces. This is one more indication that the
2H SiC modification is a low temperature form of SiC.

(1) $SiC \overset{\rightarrow}{\leftarrow} Si_{(g)} + C_{(s)}$

(2) $2\,SiC \overset{\rightarrow}{\leftarrow} Si_2C_{(g)} + C_{(s)}$

(3) $2\,SiC \overset{\rightarrow}{\leftarrow} SiC_{2(g)} + Si_{(g)}$

(4) $3\,SiC \overset{\rightarrow}{\leftarrow} SiC_{2(g)} + Si_2C_{(g)}$

where reaction (1) is favored, relatively to (2), (3) and (4), by low temperatures.

We must now consider the possible mechanism of the reactions, which produce the gaseous polymers that, by condensation, originate the polytypes. While all the polytypes can be described in terms of SiC groups repeated by symmetry elements, we shall keep in mind that we are actually dealing with chains of atoms of different lengths and geometries, i.e. crystal symmetry and molecular structure must be related. Thus, for the formation of the wurtzite structure we shall write:

(5) $Si + SiC_2 \overset{\rightarrow}{\leftarrow} Si_2C_2 \overset{\rightarrow}{\leftarrow} 2H\,SiC$

More generally, reactions of the type:

(6) $n\,Si + n\,SiC_2 \overset{\rightarrow}{\leftarrow} Si_{2n}C_{2n} \overset{\rightarrow}{\leftarrow} (2n)\,H\,SiC$

may be written for the series of which the 2H, 4H, 6H, 8H... types are members.

Another possible type of reaction involves interaction between the radicals SiC_2 and Si_2C. The simplest possible scheme involving these radicals corresponds to the formation of the cubic structure:

(7) $SiC_2 + Si_2C \overset{\rightarrow}{\leftarrow} Si_3C_3 \overset{\rightarrow}{\leftarrow} cubic\ SiC$

A double condensation

(8) $2\,SiC_2 + 2\,Si_2C \overset{\rightarrow}{\leftarrow} Si_6C_6 \overset{\rightarrow}{\leftarrow} 6H\,SiC$

can lead to the formation of the 6H SiC polytype, which, as we have seen, also can form from reaction (6).

Concurrent reactions of the type (6) and (8) can lead to the formation of other types. For instance, the 15R SiC structure, which requires Si_5C_5 polymers, can result from:

(9) $3(Si + 2\,SiC_2 + Si_2C) \overset{\rightarrow}{\leftarrow} (Si_5C_5)_3 \overset{\rightarrow}{\leftarrow} 15R\,SiC$

** These reactions, strictly speaking, apply only to the SiC used by Drowart and De Maria. Differences in vapor pressure and gas phase equilibria can be expected when different polytypes are considered. For instance Knippenberg (6) reported that the vapor pressure of cubic SiC is higher than that of the hexagonal polytypes at all temperatures.

$$(10) \qquad 3(Si + 3SiC_2 + Si_2C) \rightleftarrows (Si_7C_7)_3 \rightleftarrows 21R \ SiC$$

It can be expected that, in general, the higher the temperature, the higher the degree of polymerization. Thus, as the temperature increases, we expect the equilibria to produce, in the order, the 2H; 3C; 4H; 15R; 6H; 21R; 8H;... polymers. Knippenberg has given (6) the results of statistical observations on the relative abundance of different polytypes at various furnace operating temperatures. They fully agree with our thesis, except for the lack of data on the 21R polytype, which is not a rare polytype and can be expected to be "common" at about 2700°C.

Let us consider, now, the equilibria between different phases. In spite of the enormous number of polytypes known and of much experimental work (6-12) on the mechanism of phase transformation, the existence of true triple points have not yet been reported. So far as we know all the phase transformations are mono-tropic, the low temperature forms going irreversibly into the high temperature forms, either through the gas phase or by transport of matter. If we were dealing in all cases with monomers arranged in different geometrical patterns, one would expect to find some sharp, well defined, transformation point in an interval that extends from, say, 1400 to 2800°C. Reversible equilibria should also exist. If, on the other hand, we are dealing with polymers of different complexity, true triple points cannot exist and the transformations must be irreversible. Apparently, therefore, the experimental picture, as it stands today, supports the polymer theory of the polytypism.

As for the effects of an external pressure of nitrogen (13, 14) or of argon (10) on the transformation of cubic into hexagonal SiC, this, also, is not in disagreement with the polymer theory. An external pressure reduces the dissocia-tion and evaporation rate of the cubic SiC and favors the occurrence of reaction (7) (a two-body collision) over reaction (8) (a four-body collision). In a sense, since the neutral gases, N_2 and Ar, do not enter in the composition of the reacting system, their behavior is essentially that of a negative catalyst of reaction.

In conclusion, the existence of a very large number of "polytypes" and the phenomenon of "polytypism" itself can best be understood if the polytypes corres-pond to molecules of different complexity, i.e. to polymers of different molecular weight, different chain lengths and bond angles and different thermal stability. A possible (although highly speculative) mechanism of formation of the polymers has been suggested. According to this mechanism the different polymers form according to reactions (5-10) and the like for the higher polymers. The occurrence of polymers of higher and higher complexity at higher and higher temperatures is in agreement with what one would expect if polymer formation does occur. Similarly, the absence of well defined transformation points and the effect of an external pressure on the cubic to hexagonal transformation, support the "polymer theory of polytypism".

(1) H. Baumhauer, Z. Kristallorgr. 50, 33 (1912).

(2) H. Baumhauer, idem 55, 249 (1915-20).

(3) R. S. Ramsdell and J. A. Kohn, Acta Crystallorgr. 5, 215 (1952).

(4) R. F. Adamsky and K. M. Merz. Z. Kristallogr. 111, 350 (1959).

(5) J. Drowart and G. De Maria, Proc. Conf. on Silicon Carbide, Boston, Mass.,
        1959 (Edited by J. O'Connor and J. Smiltens, Pergamon Press, New York
        1960) p. 16-23.

(6) W. F. Knippenberg, Philips Research Reports 16, 161-274 (1963).

(7) P. Krishna and R. C. Marshall, J. Crystal Growth 11, 147 (1971).

(8) P. Krishna and A. R. Verma, Ind. J. Pure and Appl. Phys. 1, 242 (1963);
        Acta Crystallogr. 15, 383 (1962).

(9) P. Krishna, R. C. Marshall and C. E. Ryan, J. Crystal Growth 8, 129 (1971).

(10) C. E. Ryan, R. C. Marshall, J. J. Hawley, I. Berman and D. P. Considine,
        AFCRL-67-0436 (August 1967), Phys. Sci. Res. Papers No. 336.

(11) G. A. Bootsma, W. F. Knippenberg and G. Verspui, J. Crystal Growth 8, 341
        (1971).

(12) P. Krishna and R. C. Marshall, idem 9, 319 (1971).

(13) A. Addamiano and L. S. Staikoff, J. Phys. Chem. Solids 26, 669 (1965).

(14) A. R. Kieffer, P. Ettmayer, E. Gugel and A. Schmidt, Mat. Res. Bull. 4,
        153 (1969).

On the Growth Mechanism of Silicon Carbide Polytypes

U. S. Ram, M. Dubey and G. Singh

ABSTRACT

A lot of effort has been directed to the growth of a desired polytype of silicon carbide under controlled conditions, without much success due to the complex nature of the growth mechanism. In order to study the growth of SiC polytypes, we have conducted experimental studies on SiC crystals by X-ray diffraction, chemical etching and optical microscopy. The implication of these observations on the growth mechanism of silicon carbide polytypes is discussed.

INTRODUCTION

Polytypism is a well-known phenomenon in which the same chemical compound crystallizes into different crystallographic modifications. These modifications known as polytypes are all similar along a and b axes but differ from each other along the c axis. Silicon carbide is a prominent polytypic substance. A fairly satifactory explanation for the growth of SiC polytypes was given by Frank (1951) based on the presence of screw dislocations. Later experimental observations, however, raised certain doubts regarding the applicability of this theory. We have conducted experimental studies of silicon carbide crystals by X-ray diffraction, chemical etching and optical microscopy which has been reported in this paper.

EXPERIMENTAL OBSERVATIONS

Etching techniques have been utilized earlier (Amelinckx & Strumann 1960, Patel & Mathai 1972, etc.) to investigate defects in SiC. Since silicon carbide

184

polytypes are known to exist in syntactic coalescence in most of the crystals, we have tried to study their characteristics by successive etching to remove thin layers of the crystals. The chemical etching was carried out using a fused salt mixture of NaOH and KNO_2 (2:1 by weight) at about 600^oC. After each etching run, Laue photographs were taken along the c-axis to reveal the structural features. The observations of these studies are summarized below:

(1) In some crystals whose growth surfaces do not possess spiral features, it is observed that spiral markings appear after etching. Thus the absence of spiral features on the growth surface does not necessarily mean that such crystals do not contain screw-dislocations, which are supposed to be responsible for the growth of polytypic crystals. It has been pointed out earlier (Shaffer 1969) that deposition of oxide layers during later stages of growth or interactions of screw-dislocations may annihilate the spiral features on the growth surfaces. Figures 1(a) and 1(b) are optical micrographs of the growth surface and the surface obtained after etching respectively of a SiC crystal.

(2) Syntactically coalesced structures are usually situated one over the other in the form of thin slices of varying thicknesses having a common c-axis. This is revealed by Laue photographs taken after each run of etching. First, the X-ray diffraction spots corresponding to a particular structure near the growth surface become weaker and after successive etchings, we are left with the crystal containing a single polytypic structure. Figure 2(a) shows the Laue photograph of a SiC crystal taken with Cu radiation along c-axis. Figure 2(b) shows a similar photograph from the same crystal showing spots of a single polytypic structure (6H), obtained after etching out the other coalesced structures. In a few cases, the polytypic structures are found to be misoriented from each other i.e., their c-axis are inclined to each other.

(3) In some cases the interfaces between coalesced structures represent sharp transition from one structure to the other. This is indicated by sharp reflections of different polytypic structures lying on the same curves of constant ζ on c-axis oscillation photographs. However, in many cases these spots are accompanied by continuous streaks showing the presence of one dimensional disorder. In such crystals whose X-ray diffraction patterns contain sharp spots as well as streaks along h0·1 reflections, we have found that after etching the crystals to varying extents, the continuous streaks disappear leaving only sharp spots corresponding to some ordered structure. Figure 3(a) and (b) shows Laue patterns taken with Cu radiation along c-axis of the same crystal, (a) without etching

and (b) after etching. This clearly indicates that streaking is not necessarily
due to a disordered structure because every region of a disordered structure would
give rise to continuous streaks.

Therefore, if any disordered structure exists, it is likely to be in the form
of vary thin layers sandwitched by ordered polytypic structures. The existence
of such disordered layers at the interfaces of the syntactically coalesced polytypes has also been reported earlier (Nishida 1971).

(4) On minute inspection of X-ray diffraction patterns obtained from silicon
carbide crystals under magnification, one often finds that each spot consists of
a number of closely spaced sharp spots. Ordinarily this feature is not evident
when diffraction patterns are seen by the naked eye. These fine spots are actually
the spots of some polytype of very high periodicity whose structure is based on
some polytype other than 6H, 15R and 4H. When a structure of such a high periodicity is weakly based on a parent structure like 21R, 33R, etc., its spots would
be relatively stronger (Mitchell 1954) in intensity near the spot positions of
the parent structure but would also have appreciable intensities throughout the
$h0 \cdot 1$ reciprocal lattice rows. Since these spots are very closely spaced, at each
spot position of the parent structure, more than one comparatively strong spots
appear to form a single spot of the parent structure. In between these parent
spot positions, since the spots are weak in intensity, they merge to form feable
continuous streaks. When X-ray diffraction patterns showing no streaks along $h0 \cdot 1$
rows were viewed minutely under magnification, in some of them it was observed
that each spot consists of more than one fine spot. These are the cases where
a high period structure is strongly based on some parent structure and consequently
its closely spaced spots have appreciable intensities only in the vicinity of spot
positions of the parent structure unless seen minutely under magnification.
Figure 4(a) shown $10 \cdot 1$ reflexions of 147R and Figure 4(b) shows the same under
magnification. It may be noted that finer spots are only visible in Figure 4(b).

Thus, the continuous streaks along $h0 \cdot 1$ reflections are quite often a consequence of sharp spots arising due to very high periodicities which are so closely
spaced that they merge to appear as a streak (Ram et al. 1972).

(5) Polytypic structures of higher periodicities have been found whose structures are based on 21R, 33R, 147R etc. indicating thereby that a screw-dislocation
is likely to be created in any polytype (not necessarily 6H, 15R or 4H as is
usually found) to give rise to new polytypic structures (Ram et al. 1972, Dubey
et al. 1973). The crystal structures of two such polytypes viz. 147R and 189R

have been worked out. The structure of 147R which is based on 33R was found to be $[(3332)_4 32]_3$ in Zhdanov notation. The structure of 189R based on 21R was found to be $[(34)_8 43]_3$.

DISCUSSION

It is a well-known fact that silicon carbide can prove to be a very useful substance for solid state devices provided its desired single crystals could be grown under controlled conditions, which has not been possible as yet. However, good crystals having polytypic structures in syntactic coalescence can be grown. As we have seen, the coalesced structures of SiC have usually a common c-axis, it seems possible to obtain a desired single crystal out of a composite crystal by etching out undesired portions under precisely controlled conditions. This, however, requires the knowledge of location and thickness of the desired structure in the composite crystal.

The experimental observations reported here are also interesting from the point of view of understanding the growth mechanism of polytypic crystals. As mentioned earlier, one dimensional disorder in silicon carbide as inferred from the continuous streaks along $h0 \cdot l$ reflections is considered to be contrary to the predictions of screw-dislocation theory of polytypism. However, we have seen that continuous streaks may also arise due to very high periodicities. Since continuous streaks arise from some localised thin layers as revealed from our etching studies, the thin regions at the interfaces of the coalesced structures seem to be ordered structures of very high periodicities. This, along with the fact that continuous streaks do not arise due to disordered structure extending throughout the crystal, shows that the observed X-ray diffraction patterns viz. sharp spots accompanied by continuous streaks are expected from such crystals as well which have grown by a screw-dislocation mechanism. In fact we have shown that only in very rare cases, a slightly modified version of the screw-dislocation mechanism operates in the growth of silicon carbide polytypes (Ram et al. 1973).

(1) Amelinckx, S. and Strumann, G., J. Appl. Phys. 31, 1354 (1960).

(2) Dubey, M., Ram, U. S. and Singh, G., Acta Cryst. B29, 1548 (1973).

(3) Frank, F. C., Phil. Mag. 42, 1014 (1951).

(4) Mitchell, R. S., J. Chem. Phys. 22, 1977 (1954).

(5) Nishida, T., Miner. Journal 6, 216 (1971).

(6) Patel, A. R. and Mathai, John K., J. Phys. D. Appl. Phys. 5, 390 (1972).

(7) Ram, U. S., Dubey, M. and Singh. G., Z. Krist, (1972), To be published.

(8) Ram, U. S., Dubey, M. and Singh, G.,J. Appl. Cryst. (1973), To be published.

(9) Shaffer, P. T. B., Acta Cryst. B25, 477 (1969).

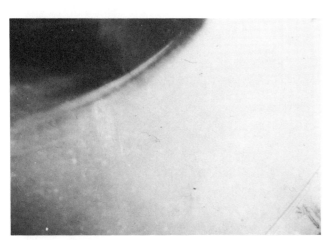

Figure 1(a)

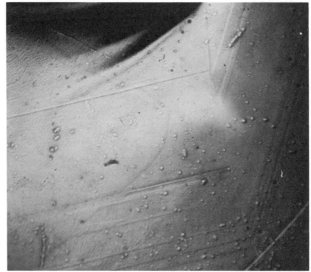

Figure 1(b)

Figure 2(a)

Figúre 2(b)

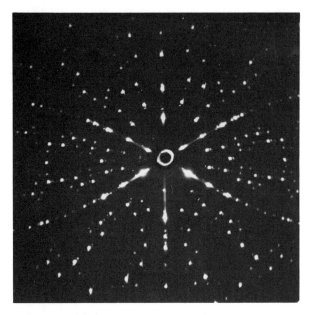

Figure 3(a)

Figure 3(b)

Figure 4(a)

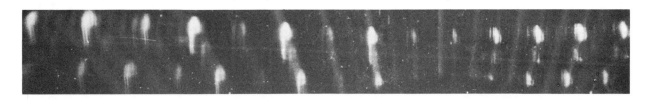

Figure 4(b)

A New Polytype of SiC, 21T

Z. Inoue, H. Komatsu, H. Tanaka and Y. Inomata

A new 21-layer trigonal polytype have been found in SiC crystals which were grown in the reaction of a graphite crucible with the molten silicon at 2200°C for 5 hours. The layer structure of this polytype was determined on the assumtion that the stacking element '1' in Zhdanov symbol shall not appear in SiC crystals grown above 2000°C. It was found that the stacking sequence of this polytype is (534333) and its space group belongs to P3ml. In connection with such internal trigonal structure, this polytype showed the trigonal spiral pattern on the (0001) surface and its step height coincided with c-period determined by the x-ray method.

INTRODUCTION

Since Baumhauer first reported the polytypism in SiC, many polytypes have been discovered up to now(1, 2, 3). In addition to those, a new polytype of 21T is revealed here.

For the analysis of this polytype, a simple assumption(4) was introduced ; SiC polytypes of high temperature phases will not have such stacking element as represented by the Zhdanov symbol '1'. As a result of this assumption, the layer structure of this polytype was easily analysed to be (534333).

In order to describe the various kinds of polytypes, Ramsdell has indicated a polytype by the number of layers in the unit cell with the letter H or R denoting hexagonal or rhombohedral symmetry respectively(5). Although his notation has been extensively used, it can not discriminate, from hexagonal structures, a trigonal structure having a primitive hexagonal lattice, both of them being designated by the letter H. Therefore, in order to avoid such an ambiguity, we adopt a symbol T to designate polytypes of trigonal symmetry(6, 7).

EXPERIMENTAL

A new polytype of 21T was grown together with many SiC crystals, mostly

191

6H and 15R, under the rich silicon vapour condition (8) at 2200°C for 5 hours. The ash content of the graphite crucible used for the experiments was less than 20 ppm and the purity of powder silicon used for the starting material was better than 99.999%.

The external form of this crystal is hexagonal prism and is 0.5 mm in diameter and 0.5 mm in length along the c-axis. The colour of this crystal was pale green in the transmitted light.

In order to examine the layer structure of this polytype, x-ray oscillation photogrpah about the c-axis were taken with Cu Ka radiation. The x-ray reflection spots are shown in Plate-1. As will be obvious from Plate-1, the structure of this polytype is trigonal one and consists of 21 Si-and C-layers along to the c-axis in the unit cell.

STACKING SEQUENCE DETERMINATION

In order to determine the layer structure of SiC polytypes, the trial-and-error method has been often used. For the cases of long-period polytypes, however, this method does not work so effectively because of formidable numbers of possible stacking sequence models, the larger the unit cell is, the greater are the difficulties in structure determination.

On the other hand, it has been noticed that no polytypes of SiC bearing the stacking element of '1' have been discovered except the 2H type (9, 10). Moreover, according to recent investigations, 2H type is supposed to be a low temperature modification grown below 1400°C or 1600°C (11, 12, 13) and transformed into 3C, 4H and 6H as the rise of temperature (14, 15, 16). Therefore, we many safely assume that the element '1' will not appear in the Zhdanov symbols for SiC polytypes belonging to high temperature phases grown over 2000°C. In fact, we have succeeded (4) in the analysis of the layer structure of 45R, $(223323)_3$, by using a simplified method based on this assumption. We applied the same method to the determination of the 21T structure.

With the aid of this assumption, the number of possible stacking models for 21-layer trigonal structure could be reduced to 109 from about 4000. Moreover, the x-ray diffraction pattern of this polytype shows the extinctions in 10·l, 20·l and 21·l rows when $l = 21 \cdot N$, where N is interger. 21T, therefore, has an equal number of Si (or C) atoms in the unit cell on each of the three symmetry axis, namely A:$[001]_{00}$; B:$[001]_{1/3,-1/3}$; and C:$[001]_{-1/3,1/3}$ as shown by Niggli's

notation.   Considering such criteria,   We could further reduce the number of the
possible stacking models from 109 to 44.   These 44 stacking models are given in
Table-1.   The true structure of 21T polytype should be one of these 44 models.

Table-1

The 44 possible stacking models for 21-layer trigonal structure, 21T

(18·3), (15·6), (13·323), (12·9), (12·432), (12·333), (11·343), (10·623),
(10·353), (9732), (9462), (9633), (9363), (9534), (923232), (8373), (8643),
(832323), (7626), (7653), (724242), (743223), (732432), (733323), (6663),
(6564), (626232), (653232), (623532), (644322), (642342), (643332),
(623433), (643233), (632343), (633333), (535323), (542424), (523443),
(543243), (534333), (42422322), (43223232), (33323232)

As the next step to the analysis of the structure, we calculated the periodic
intensity distribution function S which is characteristic of the layer sequence of
SiC polytype (17).   S is given as

$$S(hkl) = \frac{F(hkl)}{f_0(hkl)} \quad ,$$

where F(hkl) is the structure factor of the SiC polytype, and $f_0(hkl)$ is the
Fourier transform of the single Si-C layer.   Each of $S_{c·1\sim44}$, namely, S values
calculated for 10·l row with each of the 44 stacking models mentioned earlier,
was compared with S_0 values observed over the 10·l row of 21T.

We here define the reliability index R as follow:

$$R_n (\%) = \frac{\sum \left| |S_0| - |S_{c·n}| \right|}{\sum |S_0|} \times 100 \quad ,$$

where n is the reference number for one of the 44 stacking models.   These R_n
values are presented in Table-2.

Table-2

The R values for 44 possible stacking models

Models	R(%)	Models	R(%)	Models	R(%)
(18·3)	52.3	(15·6)	55.9	(13·323)	59.4
(12·9)	57.1	(12·432)	59.8	(12·333)	46.3
(11·343)	38.2	(10·623)	46.8	(10·353)	45.5
(9732)	55.1	(9462)	72.7	(9633)	43.6
(9363)	52.4	(9534)	67.5	(923232)	72.6
(8373)	50.0	(8643)	51.2	(832323)	70.2
(7626)	67.8	(7653)	45.4	(724242)	84.2
(743223)	60.8	(732432)	67.0	(733323)	42.9
(6663)	70.4	(6564)	93.8	(626232)	91.4
(653232)	72.9	(623532)	89.7	(644322)	67.8
(642342)	72.4	(643332)	65.8	(623433)	74.0
(643233)	61.2	(632343)	44.2	(633333)	42.1
(535323)	41.8	(542424)	98.1	(523443)	50.2
(543243)	48.6	(534333)	9.8	(42422322)	80.9
(43223232)	55.7	(33323232)	80.7		

As will be obvious from Table-2, the stacking model of (534333) gives a minimum R value. Therefore, the layer structure of 21T have been successfully deduced to be (534333). The observed S_O and calculated S_C values of 21T are tabulated in Table-3 for reference.

OBSERVATION

In order to determine the surface structure of 21T polytype, the differential interference contrast microscopic observation and multiple-beam interferometric measurement were made on the (0001) surface. A differential interference contrast photograph is shown in Plate-2. This polytype showed a trigonal spiral pattern on the (0001) surface as a result of the internal trigonal structure of (534333).

The step heigh of this trigonal spiral was measured by the multiple-beam interferometric method using a interference filter of 5460Å wave length. As shown in Plate-3, the shifted width is about 1/6 of the interference fringe interval along the direction of the arrow where nine mono-layers are bunched together (compare with Plate-2 at same position).

Therefore, the step height of the mono-layer of this polytype was measured to be about 52Å, whose value coincides with c-period determined by the x-ray method. This single-centered spiral with the height of the unit cill size evidences that the origin of the polytype is due to a screw dislocation (18).

CONCLUSION

A new 21-layer trigonal polytype have been grown in a rich silicon vapour condition. Its layer structure has been easily determined to be (534333) by using a simplified method. This new polytype is a trigonal structure and its surface shows a trigonal symmetry as a reflection of internal trigonal structure. In order to discriminate, from hexagonal structure, a trigonal structure having a primitive hexagonal lattice, this new polytype had better be designated as 21T instead of 21H.

Table-3

The observed S_O and calculated S_C values for $10 \cdot l$ row of 21T

$10 \cdot l$	S_C	S_O
−21	0.0	0.0
−20	1.1	0.8
−19	1.4	1.2
−18	1.7	1.3
−17	5.3	6.1
−16	1.4	1.0
−15	0.8	0.7
−14	12.0	11.8
−13	1.0	0.7
−12	2.0	4.1
−11	9.5	9.7
−10	3.9	3.9
− 9	4.6	4.9
− 8	6.5	6.9
− 7	5.2	5.4
− 6	5.0	4.9
− 5	2.8	3.2
− 4	3.9	4.6
− 3	3.0	3.4
− 2	1.2	1.8
− 1	1.0	1.3
0	0.0	0.0
1	1.1	1.2
2	1.4	1.4
3	1.7	2.1
4	5.3	5.3
5	1.4	1.5
6	0.8	0.5
7	12.0	12.0
8	1.0	0.6
9	2.0	2.2
10	9.5	10.1
11	3.9	3.7
12	4.6	4.1
13	6.5	7.1
14	5.2	4.6
15	5.0	4.6
16	2.8	2.2
17	3.9	4.2
18	3.0	3.6
19	1.2	1.5
20	1.0	1.4
21	0.0	0.0

REFERENCES

1. VERMA, A.R. and KRISHNA, P. Polymorphism and Polytypism in Crystals, John Wiley and Sons, Inc., New York (1966).

2. SHAFFER, P.T.B. (1969).   Acta Cryst., B25, 477.

3. TRIGUNAYAT, G.C. and CHADHA, G.K. (1971).   Phys. Stat. Sol., (a)4,9.

4. INOUE, Z., INOMATA, Y. and TANAKA, H. (1972).   Miner, Journ., 6,486.

5. RAMSDELL, L.S. (1947). Amer. Miner., 32,64.

6. YUASA, T., TOMITA, T. and TOKONAMI, M. (1967). Journ. Phys. Soc. Japan, 23,136.

7. INOUE, Z., SUENO, S., TAGAI, T. and INOMATA, Y. (1971). Journ. Cryst. Growth, 8,179.

8. INOMATA, Y, (1972). Journ. Cryst. Growth, 12,57.

9. MERZ, K.M. and ADAMSKY, R.F. (1959). Journ. Amer. Chem. Soc., 81,250.

10. ADAMSKY, R.F. and MERZ, K.M. (1959). Zeit. Krist., 111,350.

11. INOMATA, Y., INOUE, Z., MITOMO, M. and SUENO, S. (1969). Jour. Ceram. Assoc. Japan, 77,143.

12. INOMATA, Y. and INOUE, Z. (1970). Journ. Ceram. Assoc. Japan, 78,133.

13. SOKHOR, M.I. and GLUKHOV, V.P. (1965). Soviet Phys. Cryst., 10,341.

14. TAGAI, T., SUENO, S. and SADANAGA, R. (1971). Miner. Journ., 6,240.

15. KRISHNA, P., MARSHALL, R.C. and RYAN, C.F. (1971). Journ. Cryst. Growth, 8, 129.

16. KRISHNA, P. and MARSHALL, R.C. (1971). Journ. Cryst. Growth, 11, 147.

17. TAKEDA, H. (1967). Acta Cryst., 22, 845.

18. FRANK, F.C. (1951). Phil. Mag., 42, 1014.

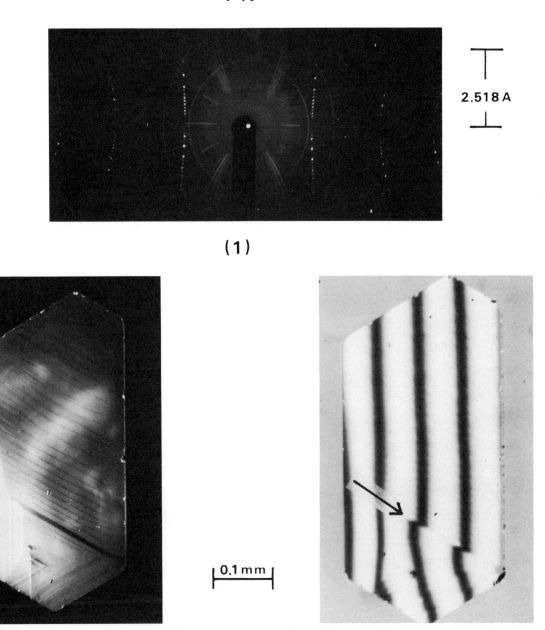

2.518 A

(1)

0.1 mm

(2)

(3)

Plates

(1) A oscillation photograph of 21T taken with Cu Kα radiation.

(2) A differential interference contrast photograph taken on the (0001) surface of 21T.

(3) A multiple-beam interference photograph taken on the (0001) surface of 21T. Nine mono-layers of 21T are bunched together along the direction of the arrow.

Mechanism of Solid State Transformations in Silicon Carbide

D. Pandey and P. Krishna

In recent years several workers (1-6) have reported independent investigations of phase transformations in silicon carbide from the 2H or ABAB... structure to the cubic (ABCABC...) and other small-period structures of SiC. Krishna and Marshall (2,3) performed an X-ray study of the structural changes produced in single crystals of 2H SiC by thermal annealing at different temperatures and reported (1) that the 2H structure transforms into the cubic structure when the transformation nucleates at temperatures between $1400^{o}C$ and $1600^{o}C$ and proceeds further towards the 6H structure on annealing at higher temperatures (ii) that the 2H structure transforms into a disordered intermediate state, which is not cubic, when the transformation nucleates at temperatures between $1600^{o}C$ and $2000^{o}C$, and then proceeds further towards the 6H structure on annealing at higher temperatures, and (iii) that the 2H structure transforms directly into the 6H structure when the transformation nucleates at temperatures above $2000^{o}C$. In all cases the transformation was found to commence with the insertion of random stacking faults and then proceed towards the new ordered arrangement. No change in the shape of the crystals was observed as a consequence of the transformations and it was suggested that they occur through layer-displacements in the solid state caused by the nucleation and propagation of stacking faults.

Tagai et al. (4) performed a similar investigation of the structural changes produced by annealing crystals of 2H SiC. Their crystals behaving somewhat differently from those of Krishna and Marshall. On their X-ray diffraction photographs the reflections corresponding to the cubic, 4H and 6H phases appeared at temperatures above $1600^{o}C$, $1900^{o}C$ and $2000^{o}C$ respectively, without the 2H reflections disappearing completely. This implies that some microscopic regions of the crystals transformed while others remained unchanged. Apparently nucleation of the second phase could only occur over certain regions of the parent crystal for some unknown reason. The transformation was reported to occur by the insertion of stacking faults and no change in the shape of the crystals was found.

On the other hand Bootsma et al. (5) have reported from a study of the structural changes produced by thermal annealing of polycrystalline samples of 2H SiC (containing about 5% cubic SiC) that these transform rapidly to the cubic structure at temperatures above 1500°C with a marked change in the morphology of the individual crystallites in the sample. Annealing at higher temperatures caused the structure to transform, comparatively gradually, to the 6H structure, with further changes in the size and shape of the grains in the sample. The powder method of X-ray diffraction was used to identify the structure but disorder effects due to stacking faults were not measured. Transformations from the 4H and 15R structure to the 6H structure were also observed in polycrystalline samples annealed at temperatures above 2000°C. It was concluded that the cubic structure is a metastable intermediate state in the 2H to 6H transformation and that all transformations occur by a process of surface diffusion.

More recently, Powell and Wills (6) have studied the transformation behaviour of thin single-crystal flakes of 2H SiC prepared from the as-grown crystals by grinding. They find that these flakes transform to the cubic structure even at temperatures as low as 400°C and the change is accompanied by visible deformations of the platelets. They have, therefore, suggested that the transformation proceeds by a deformation process of periodic slip caused by the rotation of a partical dislocation around a suitable screw dislocation parallel to c. Such a process is known to be responsible for the hexagonal close-packed to cubic close-packed transformation in certain metals (7) and in ZnS (8).

This paper examines in detail the layer displacement mechanism and the basal slip mechanism as applied to transformation in SiC and compares the results with experimental findings with a view to determine which of the two mechanisms is more likely for the transformations.

THE BASAL SLIP MECHANISM

The changes in the layer sequence required for a structural transformation in SiC can be brought about by slipping parts of the crystal past each other along the basal plane by partial slip vectors $\pm s_1 = \pm 1/3 (a + 2b)$, $\pm s_2 = \pm 1/3 (2a + b)$

and $\pm \underline{s}_3 = \pm 1/3 (\underline{a} - \underline{b})$. Of these slip-vectors only three are possible on a particular layer if the laws of close-packing are not to be violated since no two successive layers in the structure can be in the same orientation A, B or C. Since the three slip vectors $+ \underline{s}_1$, $+ \underline{s}_3$ and $-\underline{s}_2$ lead to the same structural configuration they will be denoted as $\underline{S}_1$ and the other 3 vectors $- \underline{s}_1$, $- \underline{s}_3$ and $+ \underline{s}_2$ as $\underline{S}_2$. $\underline{S}_1$ causes a cyclic shift of the layers (A to C, B to A, C to B).

A periodic basal-slip mechanism is known (8-10) to operate in ZnS which is isostructural with SiC and, like the latter, exhibits a large number of polytypic structures. The 2H to 3C transformation in ZnS is believed to occur through periodic slip caused by the rotation of a partial dislocation around an axial screw dislocation. This mechanism is also believed (10,11) to be responsible for the formation of polytypes in this material. The basal plane in a close-packed structure containing a screw dislocation along $\underline{c}$, is not perfectly flat but spirals round the screw dislocation with a pitch equal to its Burger's vector. A partial dislocation gliding on such a plane would therefore rotate round the screw dislocation and cause slip to occur periodically. If the screw dislocation in a 2H crystal has a Burger's vector equal to two times the separation between successive close-packed layers and a partial dislocation moving round it causes slip by an amount $\underline{S}_2$, the structure would transform into the cubic sequence as shown below. (The vertical lines represent slip-planes across which the layers on the right shift past those on the left by $\underline{S}_2$).

Initial Structure (2H):

```
A B|A B A B A B A B A B A B ...
    C A|C A C A C A C A C A ...
        B C|B C B C B C B C ...
            A B|A B A B A B ...
```

Resulting Structure:

```
A B C A B C ... = 3C
```

Consider now the 3C to 6H transformation in SiC in terms of the basal slip mechanism. The kind of basal slips required to transform an ABCABC... (3C) sequence into an ABCACB (6H) sequence are depicted below.

Initial Structure (3C):

```
A B C A|B C A B C A B C A B C A B C ...
        C|A B C A B C A B C A B C A ...
          B|C A B C A B C A B C A B ...
              A B C A|B C A B C A B C ...
                      C|A B C A B C A ...
                        B|C A B C A B ...
```

Resulting Structure: A B C A C B A B C A C B ... = 6H

The transformation would require slip by an amount $\underline{S}_1$ to occur on three successive layers with this pattern repeating indefinitely every 6 layers. This is only possible if there is in the crystal a screw dislocation of Burger's vector equal to 6 layer separations with 3 partial dislocations, on successive layers, rotating round the screw.

The basal slips required for transforming the 2H structure directly into the 6H structure are represented below:

Initial Structure (2H):   A B|A B A B A B A B A B A B ...
                                            C A C|A C A C A C A C A ...
                                              B A B|A B A B A B A B ...
                                                C A C|A C A C A ...
                                                  B A B|A B ...

Resulting Structure: A B C A C B A B C A C B.... = 6H

It is clear that the transformation would require basal slips of $\underline{S}_1$ and $\underline{S}_2$ to occur alternately, every three layers. This is possible only if there is a screw dislocation of Burger's vector equal to 6 layer separations with two partial dislocations rotating round it in slip planes separated by 3 layers.

THE LAYER-DISPLACEMENT MECHANISM

The structural transformations in SiC discussed above can also result from suitable layer displacements caused by the nucleation and expansion of stacking faults in individual close-packed double-layers (of Si and C). Such a process would be governed by thermal diffusion since the nucleation of a stacking fault would require the migration of atoms inside the crystal. The nucleation of the stacking faults and the consequent layer displacements would be a statistical phenomenon but there would be a tendency for the layer-displacements to occur in such a manner as to minimize the free energy and take the structure towards the stable state. If the stable state happens to be one with a different order such an order will tend to result. Such a mechanism was also suggested by Jagodzinski (12) for the formation of SiC polytypes.

Consider first the nucleation of a fault in a close-packed layer in the 2H structure. Figure 1a depicts an idealized A layer with all atoms in position. If now the temperature is raised, vacancies are created and atoms become free to migrate within the crystal and to the surface by a process of diffusion, the layer can be visualized with some vacancies as shown in Figure 1b. If a number of these vacancies come close together it will become possible for neighboring atoms to move into C-sites (indicated by a cross x) thereby nucleating a fault as shown in Figure 1c. Within the faulted region the atoms are in C-sites while the rest of the atoms are in A-sites causing a partial dislocation to bound the fault as shown in Figure 1d. This partial dislocation can now glide, causing the fault to expand until the entire layer of atoms is displaced into the C-sites and the partial dislocation moves out of the crystal. The entire layer is then displaced from the A to the C orientation. By the same mechanism displacements would occur for other layers througout the structure in such a manner as to result in the new structure. Since this is a statistical process it is not possible to represent it diagramatically. However, the layer displacements that would be required for each transformation, and would therefore be the most probable, are depicted below in rectangular blocks.

2H to 3C Transformation:

Initial Structure (2H):   A B A B A B A B A B A B A B

Resulting Structure (3C)   A B C A B C A B C A B C A B

3C to 6H Transformation:

Initial Structure (3C):   A B C A B C A B C A B C A B C

Resulting Structure (6H):   A B C A C B A B C A C B A B C

2H to 6H Transformation:

Initial Structure (2H):   A B A B A B A B A B A B A B A

Resulting Structure (6H):   A B C A C B A B C A C B A B C

According to this mechanism the transformation would proceed through one-dimensional disorder caused by the nucleation of stacking faults. Since the rise in entropy with disorder makes it increasingly difficult to arrive at the stable ordered configuration the final structure is expected to exhibit considerable one-dimensional disorder.

DISCUSSION AND CONCLUSIONS

It is evident from the above analysis that the transformations 2H to 6H and 3C to 6H by the periodic slip mechanism require more than one partial dislocation to operate around an axial screw dislocation at suitable layer intervals. Such a configuration of dislocations is very unlikely to occur. A single partial

dislocation operating round an axial screw dislocation in the 2H structure would
result in the cubic structure if the Burger's vector of the screw dislocation
equals two layer spacings. If the Burger's vector is larger, a rhombohedral
structure would be generated as shown below for a Burger's vector of 4 layer
spacings.

Initial Structure (2H):

```
A B A B|A B A B A B A B A B A B ...
        C A C A|C A C A C A C A ...
                B C B C|B C B C B C ...
                        A B A B|A B ...
```

Resulting Structure (12R):   A B C A C A B C B C A B

Zhdanov Symbol: $(31)_3$

A periodic slip mechanism involving a screw dislocation with a Burger's
vector equal to an odd number of layer spacings is not possible in a 2H structure
as it will violate the laws of close-packing and cause successive layers to be in
the same orientation. Since the periodic slip is caused by the same partial, it
has to occur in the same direction after every pitch of the screw. The layers
cannot slip cyclically (A to B, B to C, C to A) across one slip plane and anti-
cyclically (A to C, B to A, C to B) over the next. The result of periodic slip
in 2H after every 3 layers is illustrated below.

Initial Structure (2H):

```
A B A|B A B A B A B A B A B...
      C B C|B C B C B C B C B...
            C A C|A C A C A...
                  B A B|A B A B...
```

Resulting Structure: A B A C B C C A C A B A B - not possible.

It follows that hexagonal structures cannot result from the 2H structure by
the operation of a single partial around an axial screw. This fact was overlooked
by Daniels (9) and Mardix et al. (10,11,13) while proposing such a mechanism for
the creation of long-period polytypes in ZnS, because they employed Zhdanov symbols
in their deductions instead of the actual ABC sequence of layers. As pointed out
by Krishna and Verma (14) this can lead to incorrect results. From the foregoing
it is evident that the direct transformation from 2H to 6H structure, as observed
in SiC (3) cannot occur by the above mechanism.

The 2H to 3C transformation is the only transformation that may occur by
the periodic slip mechanism and this is known to happen in ZnS which isostructural
with SiC (8). However, such a mechanism caused deformation, producing easily

measurable kinks in the external shape of the crystal. Such kinks have been observed (9,10) in ZnS crystals after the transformation but in SiC neither Krishna et al. (2,3) nor Tagai et al. (4) could notice any visible changes in the external shape of the 2H single crystals after transformation. Since SiC is a very hard and brittle material it is not easy for a deformation process to occur as it is in ZnS. Only in the crystal platelets used by Powell and Wills (6) does it appear likely that the periodic slip mechanism has operated. In their experiments the transformation was observed at too low a temperatue (400°C) to permit layer-displacement by a process governed by diffusion. Their crystals showed visible signs of deformation. Perhaps the process of grinding the crystals into thin platelets introduces dislocations in the crystal which enable the periodic slip mechanism to operate.

In all other cases of solid state transformations observed in SiC, the layer-displacement mechanism governed by diffusion appears to operate. The single-crystal X-ray diffraction photographs recorded by Krishna and Marshall (2,3) and Tagai et al. (4) at different stages of the transformation clearly indicate that the transformation commences with the insertion of random stacking faults. Since this involves a process of nucleation, the behaviour of individual crystals is very different. Some crystals transform readily even at 1400°C while others do not transform even at 2000°C for a long time (2). As the concentration of faults increases a new periodicity becomes apparent and the resultant structure contains a large concentration of random faults. There is no change in the external shape of the crystal. All this is in full agreement with the layer-displacement mechanism discussed above. In the case of polycrystalline specimens the changes in the shape of the grains with transformation, as observed by Bootsma et al. (5) are not due to deformation but due to surface-flow and sintering of crystallites caused by grain boundary diffusion at elevated temperatures.

The fact that 2H crystals have been observed to transform to the cubic phase at temperatures as low as 400°C confirms the view (2) that the 2H structure in SiC is a metastable phase and the cubic modification is the stable low temperature modification of SiC. The possibility (4) that the 6H structure is the stable structure at all temperatures and the cubic structure results only as an intermediate metastable state in the transformation of the 2H to the 6H structure appears remote since there is no kinetic reason why the 2H structure should prefer to transform via the cubic phase instead of transforming directly into the 6H structure.

REFERENCES

(1) P. Krishna, R. C. Marshall and C. E. Ryan, J. Cryst. Growth, $\underline{11}$, 129, (1971).

(2) P. Krishna and R. C. Marshall, J. Cryst. Growth, $\underline{9}$, 319 (1971).

(3) P. Krishna and R. C. Marshall, J. Cryst. Growth, $\underline{11}$, 147 (1971).

(4) Tokuhei Tagai, Shigeho Sueno, and Ryoichi Sadanage, Min. Journal, $\underline{6}$, 240 (1971).

(5) G. A. Bootsma, W. F. Knippenberg and G. Verspui, J. Cryst. Growth, $\underline{8}$, 341 (1971).

(6) J. Anthony Powell and Herbert A. Will, J. Appl. Phys., $\underline{43}$, 1400 (1972).

(7) A. Seegar, Z. Metallk., $\underline{4}$, 247 (1953).

(8) F. S. D'Aragona, P. Delavignette and S. Amelinckx, Phys. Stat. Sol., $\underline{14}$, K115, (1966).

(9) B. K. Daniels, Phil. Mag., $\underline{14}$, 487 (1969).

(10) S. Mardix, Z. H. Kalman and I. T. Steinberger, Acta Cryst., $\underline{A24}$, 464 (1968).

(11) E. Alexander, Z. H. Kalman, S. Mardix and I. T. Steinberger, Phil. Mag., $\underline{21}$, 1237 (1970).

(12) H. Jagodzinski, Acta Cryst. $\underline{7}$, 300 (1954).

(13) S. Mardix, E. Alaxander, O. Brafman and I. T. Steinberger, Acta Cryst., $\underline{22}$, 808 (1967).

(14) P. Krishna and A. R. Verma, Z. Krist., $\underline{121}$, 36 (1965).

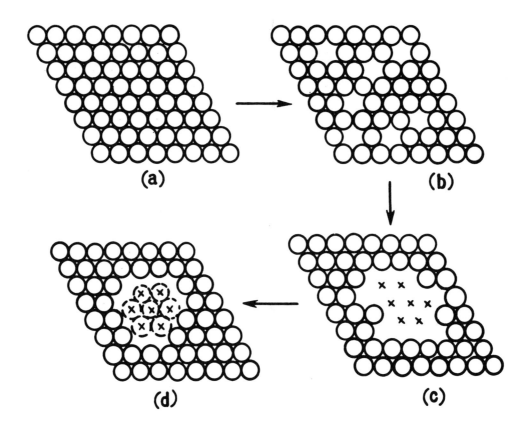

Figure 1. Nucleation of a stacking fault in a
           close-packed layer by thermal diffu-
           sion. Crosses (x) indicate sites for
           atoms in the faulted layer. A partial
           dislocation surrounds the faulted region.

Natural Superlattice in Silicon Carbide Crystals

G. B. Dubrovskii

ABSTRACT

The electron energy spectrum of α-SiC is shown to be the result of a perturbation of the spectrum of cubic SiC by the natural one-dimensional superlattice potential arising from additional interaction between atoms in hexagonal h-layers. The energy band structure along the c-axis of the crystal is calculated on the basis of a simple physical model and compared with experiment.

INTRODUCTION

Over the last ten years there has been an increasing interest in the investigation of artificial superlattices in semiconductors. It has been shown theoretically (1-3) that the negative differential conductivity is attainable in the narrow energy bands resulting from the superlattice perturbation of semiconductors. At present, however, there are no reliable experimental data on artificial superlattices.

In this paper we consider a natural superlattice in SiC crystals and show that the conduction bands in these crystals consist of alternating allowed and forbidden bands. The width of the lowest allowed band is found to vary from a few millielectronvolts to about one electronvolt depending on the superlattice period.

THEORY

All the known modifications of SiC have identical tetrahedral configurations of neighboring Si and C atoms and differ only in the packing of subsequent layers of atoms along the c-axis of the crystal. In 3C SiC all layers are in cubic

207

positions whereas in 2H SiC in hexagonal. All the multilayer modifications of SiC are characterized by their own sequence of c- and h-layers. In comparison with the cubic layer there is additional interaction between the third nearest neighbors in h-layers. This interaction is directed along the c-axis of the crystal and is the result of 13% decrease in the distance between the respective atoms. Thus, h-layers in multilayer SiC structures may be considered as regular stacking faults superimposed onto the basic cubic SiC structure producing a one-dimensional superlattice potential of the period equal in hexagonal crystals to one half and in rhombohedral to one third of the unit cell length along the c-axis.

A superlattice in SiC is quite evident from the point of view of crystallography. Qualitative analysis of a superlattice has been successfully used, for instance, in the investigation of the phonon dispersion curves of SiC crystals in the standard large zone (4). However, energy aspects of a superlattice in SiC have not been analysed so far and the role of a superlattice in defining the electron energy spectrum has been neglected.

Periodic perturbation introduces additional energy gaps into the electronic spectrum of the crystal and changes the electron energy and the effective mass at the band extrema. In the case of SiC all the characteristics of the valence band do not seem to depend on a superlattice, whereas the electron mobilities and donor activation energies show a strong polytype dependence (4). From this we can conclude that a superlattice in SiC affects the electron properties in the conduction band only and does not influence the valence band. Thus, the whole variation in the band gap, E_g, from 2.4 eV for 3C SiC to 3.3 eV for 2H SiC (5) may be ascribed to the change in the electron energy at the bottom of the conduction band.

The period of a superlattice and relative position of h-layers in any SiC modification are known exactly from crystallography. Since h-layers are not situated very close to each other, their potentials may be presumed to be the same in all modifications. This assumption enables us to represent the potential of individual h-layer in any appropriate form, bypassing its actual form. Then the electron energy spectrum of any α-SiC crystal may be obtained from the solution of the Schrodinger equation for an electron of the effective mass M_β, moving in the field of a superlattice formed by the regular array of h-layers. The only two unknown parameters of the equation, the electron effective mass, M_β, in the unperturbed β-SiC structure along [111] direction, and the magnitude of the individual h-layer potential, V, can be estimated from the combined solution of at least two equations for different periods, a, of the superlattice, by fitting it with experimental data on the dependence of E_g on a.

Using the simplest model of a potential of the form $U = V (1 + \cos 2\pi x/a)$, where x is the coordinate along the c-axis, we have obtained the best fit with experiment for SiC modifications 4H, 6H and 8H with the parameters V = 0.96 eV and M_β = 1.65 M_o. Calculated with these parameters electron energy spectra in the x-direction for 6H and 8H SiC are shown in Figure 1. In the same figure $E(k_x)$ dependence for unperturbed cubic SiC crystal is given with M_β = 1.65 M_o. Calculated values of the effective mass of an electron at the bottom of the conduction band are 2.0 M_o and 2.2 M_o for 6H and 8H SiC, respectively.

Fig. 1. Calculated energy bands along the c-axis of the crystal for a) 6H, and b) 8H SiC.

The conduction bnad of cubic SiC is shown by the dotted line.

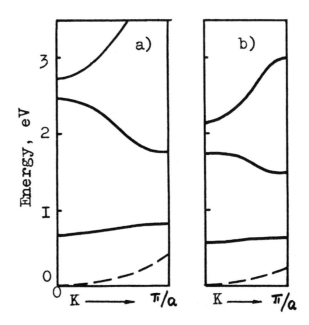

It can be noticed from the data of Figure 1 that the energy gaps for 6H and 8H SiC between the first and subsequent subbands at $K = (k - k_{min}) = 0$, where k_{min} is the wave vectors of electrons at the absolute minimum of the conduction band, are very close to the energies of the strong polarized absorption bands observed in n-type crystals. The nature of these bands remain unexplained so far.

Owing to the presence of a superlattice of a period, a, the wave vectors of electrons at the points $2n\pi/a$ are equivalent. Hence, direct transitions of electrons between subbands are possible. Selection rules for these transitions are determined by the relative symmetry of the wave functions in the initial and final states. In the tight-binding approximation, the wave functions of odd and even-numbered bands are $\cos K_x x$ and $\sin K_x x$, respectively. Thus, for the light polarization $E||c$ transitions from the first subband into the second and successive even-numbered subbands are allowed and transitions into the odd-numbered bands are forbidden. The wave functions of electron in the directions perpendicular to the c-axis of the crystal in all the bands are simply $\exp(i\,k_r\,r)$, i.e., all transitions in the light polarization $E\perp c$ are forbidden.

We have concluded (6) from the consideration of the energy position, the polarization and the relative intensity of the absorption bands in $E\|c$ and $E\perp c$ for several modifications of SiC that the bands in $E\|c$ are due to electron transitions into the second subband and weaker bands in $E\perp c$ are due to forbidden transitions into the third subband.

Thus, further refinement of the electron energy spectra of SiC becomes possible utilizing experimental data on the energy gaps between subbands at $K = 0$.

It is evident that the superlattice potential of the form $\cos(2\pi x/a)$ is valid only for a range of a-values in hexagonal crystals and fails to describe the superlattice potential in rhombohedral structures. Another simple but somewhat more versatile model is that of Kronig and Penney with rectangular barriers of finite height (7). Using this model we have calculated the height and the width of the barriers located at h-layers and the position of the bottom of the potential well between the barriers (which reflects the overlapping of the actual potentials of neighboring h-layers) for 6H and 8H SiC modifications by fitting the following experimental data on the electron energies at $K = 0$ in three lowest subbands (6): 0.6, 2.0 and 2.6 eV for 6H SiC and 0.4, 1.55 and 2.0 eV for 8H SiC (all the energies are reckoned from the bottom of the conduction band of cubic SiC). It was found that to satisfy these experimental data the following parameters of the barrier must be chosen the height $U_o = 2.4\,eV$, the width $\beta = 0.8 - 1.0\,Å$, and the bottom of the potential well $U_1 = 0.4 - 0.5$ eV.

Taking these parameters as the conventional form of the potential of an individual h-layer, we can calculate the electron energy spectrum of any rhombohedral SiC modification. Agreement of the calculated band picture with experiment would confirm the validity of the model.

Fig. 2. Model of the superlattice potential for 21R SiC.

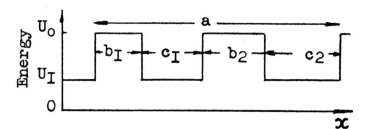

We have calculated the electron energy spectrum of 21R SiC using the model shown in Figure 2. No experimental data were used in these calculations. The expression for the spectrum was obtained from the solution of the Schrodinger equation for an electron of the effective mass M_β moving in the field of the potential in Figure 2. Calculated energies of the lowest 5 allowed subbands at K = 0 are given in Table 1.

Table 1. Energies of the lowest subbands at K = 0 in eV from the bottom of the conduction band of cubic SiC.

Structure of SiC	Number of the subband				
	1	2	3	4	5
6H	0.6	2.0	2.6	-	-
8H	0.4	1.55	2.0	-	-
21R	0.45	0.65	0.95	1.7	2.0

The calculated value of the electron energy at the bottom of the conduction band of 21R SiC agrees well with the experimental value of 0.463 eV (5). Absorption bands observed in n-type 21R SiC at 1.2 eV in the light polarization E||c and at 1.85 eV in E⊥c seem to correspond to allowed transitions into the 4th subband and to forbidden transitions into the 5th subband, respectively. Transitions into the 2nd and 3rd subbands have not yet been observed.

It is of interest to note that the calculated width of the first allowed band in 21R SiC is only 12 meV. Related values in 6H and 8H SiC are about 100 meV and 30 meV, respectively.

An important conclusion on the location of the conduction band minima in multilayer SiC modifications can be deduced from the consideration of a superlattice in these crystals. The perturbing superlattice potential is evidently weaker than that associated with the interaction between neighboring atoms generating the crystal field of the cubic SiC matrix. Hence, we can suppose that the conduction band minima in all multilayer SiC modifications must be localized at the same points in the reciprocal space as in cubic SiC matrix. Absolute minima in cubic SiC are known to be situated at the points X (8). Thus the minima in multilayer SiC structures are to be near the point M at the boundary of the large zone, and the number of equivalent valleys in the conduction bands of these crystals is to be the same as that in cubic SiC, i.e., three. This conclusion is confirmed by the identity of the principal phonons observed in the edge absorption of many SiC modifications except for 2H (5). Theoretical calculations also give the conduction band minima in 4H and 6H SiC near the point M (9).

The electron energy spectra obtained in our calculations can be used to give a qualitative explanation of the stability of complex modifications of SiC, based

on the comparison of the average energies of free electrons at the crystal-growth
temperature. It was found that because of the presence of a forbidden band inside
the conduction band the average energy of electrons in a crystal with a superlattice can be less than that in unperturbed cubic SiC at the appropriate temperature.
This confirms Knippenberg's hypothesis on the significant role of the energy of
free electrons in the stabilization of complex SiC modifications.

EXPERIMENT

We have measured the optical absorption in n-type SiC crystals of seven
modifications: 4H, 6H, 8H, 15R, 27R, 33R, and 21R containing no interlayers of
heterogeneous structures. All the samples showed clearly resolved polarized
absorption bands at wavelengths characteristic for each crystal structure. Most
of the structures have a single strong band in each polarization, $E||c$ and $E_\perp c$.
The spectral dependence of the reduced absorption coefficients, α/α_{max}, for these
bands in seven SiC modifications investigated is shown in Figure 3 for the light
polarization $E||c$ and in Figure 4 for $E_\perp c$.

The concentration dependence of the absorption at the peaks of all bands is
linear.

In all SiC crystals measured the absorption band in $E||c$ is much stronger
than that in $E_\perp c$ and is situated at lower energies. The oscillator strength values
calculated for 6H and 15R SiC absorption bands are 0.72 and 0.61 for $E||c$ and
0.031 and 0.059 for $E_\perp c$, respectively.

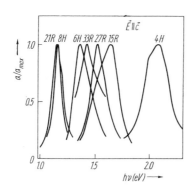

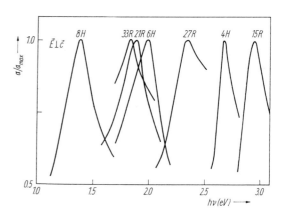

Fig. 3. Spectral dependence of the re-
duced absorption coefficient, α/α_{max}, for
seven SiC modifications in the light
polarization $E||c$; T = 300°K.

Fig. 4. Spectral dependence of the re-
duced absorption coefficient, α/α_{max}, for
seven SiC modifications in the light polarization $E_\perp c$; T = 300°K.

The temperature dependence of absorption in the bands measured from 77°K to about 800°K is very similar in all the crystals. The magnitude of absorption at the peak of the band changes slightly with temperature. The oscillator strength values practically do not depend on temperature.

From the similarity of the main characteristics of absorption under investigation it may be concluded that the mechanism of transitions found for 6H SiC (10) is valid for all the crystal modifications measured. Electrons are excited at low temperature from donors and at high temperature from the bottom of the conduction band into some higher bands. The energy position of the final states for transitions in both polarizations depends upon the crystal structure. A simple unique relationship may be obtained from the consideration of separate groups of modifications having common structural characteristics.

One can distinguish hexagonal crystals, structural formulae of which are (11), (22), (33), and (44), and several groups of rhombohedral crystals: $(2_{2n + 1}3)_3$ $(3_{2n + 1}2)$, $(3_{2n + 1}4)$, etc. It is seen from the data presented that within any group of structures determined in this way, the transition energies monotonously decrease with increasing length of the unit cell along the c-axis of the crystal or, in our definition, period of superlattice, a.

We have made an attempt to observe transitions into the 2nd and 3rd subbands in 21R SiC, predicted by the above theory, and have found in E$||$c the absorption band at 0.26 eV in agreement with the energy gap of 0.2 eV between the 1st and 2nd subbands.

No direct measurements of the width of the allowed bands were reported for SiC. We believe, however, that the anomaly in the edge absorption of 21R SiC (11) is due to two-dimensional density of states in the forbidden interval between 1st and 2nd subbands.

Conductivity measurements in high electric fields are complicated in SiC because of low resistivity of the crystals. Using short pulses we have observed nonlinearity of the I-V characteristic along the c-axis in the crystal 297R SiC due, apparently, to tunneling of electrons between subsequent subbands (12).

It may be concluded that there is both theoretical and experimental evidence that electronic spectrum of SiC is associated with a superlattice. Quantitative estimations indicate narrowing of allowed bands to some few millielectron-volts. This determines the possibility of obtaining negative differential conductivity due to Bragg reflections and resonant tunneling of electrons.

REFERENCES

(1) L. V. Keldysh, Fiz. Tverd. Tela, $\underline{4}$, 2265 (1962).

(2) L. Esaki and R. Tsu, IBM J. Res. Dev., $\underline{14}$, 61 (1970).

(3) R. F. Kasarinov and R. A. Suris, Fiz. Techn. Polupr., $\underline{5}$, 797 (1971).

(4) L. Patrick, Mat. Res. Bull. (SiC issue), $\underline{4}$, S129 (1969).

(5) W. J. Choyke, Mat. Res. Bull. (SiC issue), $\underline{4}$, S141 (1969).

(6) G. B. Dubrovskii, Fiz. Tverd. Tela, $\underline{13}$, 2505 (1971); G. B. Dubrovskii,
       A. A. Lepneva and E. I. Radovanova, Phys. Stat. Sol., (b) $\underline{57}$, 423 (1973).

(7) R. A. Smith, in "Wave Mechanics of Crystalline Solids", New York (1961),
       p. 134.

(8) F. Herman, J. P. van Dyke and R. L. Kortum, Mat. Res. Bull. (SiC issue),
       $\underline{4}$, S167 (1969).

(9) H. G. Junginger and W. van Haeringen, Phys. Stat. Sol., $\underline{37}$, 709 (1970).

(10) G. B. Dubrovskii and E. I. Radovanova, Fiz. Tverd. Tela, $\underline{II}$, 680 (1969).

(11) D. R. Hamilton, L. Patrick and W. J. Choyke, Phys. Rev., $\underline{138}$, A1472 (1965).

(12) G. B. Dubrovskii and A. A. Wolfson, JETP Letters, $\underline{17}$, 22 (1973).

A Study of Etch Pits on Pure Polytypes of SiC

J. W. Faust, Jr., Y. Tung and H. M. Liaw

Several techniques are available for the determination of polytypes in SiC. One very simple technique is to observe the luminescence under ultraviolet light at liquid nitrogen temperature and then again as the sample warms up (thermoluminescence). This technique gives only the predominant polytype in a mixed polytype crystal. It is, however, very useful in screening a large number of crystals in a short time for use as such or for more detailed structural investigation.

More exact information can be obtained using electron or x-ray diffraction techniques. Using these techniques one can not only identify the polytypes but also can determine whether a crystal is a single polytype or contains several polytypes. Reflection electron and x-ray diffraction techniques will only give information about the polytype or polytypes on the surface being studied. Transmission techniques are necessary to determine if there are buried layers of other polytypes and if so to identify them.

Several attempts have been made to utilize optical techniques to determine polytypes. Wolff (1) using a low power laser as a light source for an optical goniometer suggested that one could differentiate between polytypes by examining the light figures created by growth steps on unetched surfaces of SiC. Shaffer (2) studied refractive index and birefringence of several polytypes of SiC to determine if such techniques could be used to differentiate mixed polytypes in a crystal. He found an approximate relationship but not one of sufficient accuracy to be useful. With a polarizing microscope he was able to show layers and inclusions of other polytypes but could not identify them.

Interferometric techniques have been used by several investigators, to
measure the step height of growth spirals at screw dislocations on a number of
polytypes of SiC (this has been summarized by Verma and Krishna (3)). It has been
found that the step height of screw dislocations with unit Buergers vector is
equal to the lattice constant in the c-direction, within experimental error, for
hexagonal polytypes while it is one-third of the lattice constant in the c-direction for rhombohedral crystals. A few examples are given in Table I. A word of
caution is necessary; in many cases, especially in the crystals that were grown

Table I

Polytype	Interferometric data in Å	x-ray data in Å
6H	15 $\pm$ 2	15.08
15R	12 $\pm$ 2	37.5
33R	27 $\pm$ 3	82.5
126R	102.8 - 109 $\pm$ 5	371.4
66H	168 $\pm$ 3	165.88

earlier, screw dislocations with multiple Buergers vectors were quite common.
Thus a large step height could be mistaken for a higher polytype when in reality
it is a multiple of a lattice parameter for a lower polytype.

Numerous etchants have been reported for SiC. The most useful and widely
used etchants are chemical, both molten salts and gases, and electrolytic. The
etchants and a discussion of the etching processes have been reviewed by Faust (4)
and by Jennings (5). In these papers pit shapes and some etching rates were also
given and discussed. Much of the early work was done on crystals that were of
unknown polytype; although the platelets were generally 6H as the host matrix
with other polytypes as layered inclusions. It was shown that β-SiC could be distinguished by its triangular etch pits from α-SiC with hexagonal etch pits.
Faust (4) also showed a photomicrograph of pits that were "distorted" hexagons
(i.e., three alternating long edges and short edges). At that time no explaination
could be given for these pits. In this study, pure polytypes and some mixed polytypes of SiC have been etched and the etch pits studied by optical microscopy,
replica electron microscopy (REM) and scanning electron microscope (SEM) to see if
one could differentiate between the polytypes by simple microscopic examination of
etch pits. Also some work was done on etch rates.

EXPERIMENTAL PROCEDURES

Platelets of pure polytypes of 4H, 6H, and 15R and also known mixed polytypes were studied. The samples were etched in the molten salt etchant $NaOH:Na_2O_2$ (3:1) at $500^{\circ}C$. The etching time was varied from 2 to 30 minutes so that the shapes of the pits could be studied as they evolved. A three minute etch was used as a standard for studying the shapes and details of the etch pits for comparison between the different polytypes. Etching action was stopped by decanting the molten salt and washing the sample in boiling water and then in dilute HCl. Sometimes it was necessary to rinse the samples in aqua regia. The samples were then examined first with the optical microscope, then with the SEM and occasionally with REM. The electron microscope used was a Siemens Elmiskop 1A, and the SEM was a JOEL SMU-3.

Etching was also studied as a function of temperature to determine the etch rates. For etch rate studies a special furnace was constructed that allowed a more precise control over the time of etching without changing the temperature of the molten etchant. The apparatus consisted of a vertical tube furnace over which a rod was positioned. A tungsten wire fastened to the rod was used to hold the samples. A one mil diameter hole was drilled in the corner of each sample to attach the sample to the wire. The rod could be rotated by a motor drive mechanism to raise or lower the sample and could be moved horizontally to position the sample. The temperature was measured directly by means of a protected thermocouple.

RESULTS & DISCUSSION

The platelets of the pure polytypes 4H, 6H, and 15R were first etched and examined under the optical microscope. The carbon face of the platelets of the various polytypes all gave a similar appearance which depended on the etching time and temperature. As was reported previously (4) no pits were found on this face. The face changed from smooth to a needle like structure, and then to a "wormy" appearance with increasing temperature. On the silicon face, several shapes and sized of pits were found on any given platelet. The size of the pits depended mainly on the time and temperature of etching. The higher the temperature and/or the longer the time of the etching, the larger the pits. The actual pit shape did not change as the etching time was increased (i.e., sequential etching); it did however, increase in size. Pits were formed both at the sites of dislocations and at other sites on the surface; however the pits that were not at the sites of

dislocations could be easily distinguished from the dislocation pits by the fact
that they were truncated (4, 6). The larger pits were randomly distributed over
the surface while the smaller ones were aligned in lineage lines in the <11$\bar{2}$0>
directions.

The random etch pits on the 15R crystals were "distorted" hexagons; that
is they had three long edges alternating with three short edges as shown in
Figure 1. The etch pits on the 4H and 6H crystals were perfect hexagons as seen
in Figures 2 and 3; however, "distorted" hexagons were also found. The "distorted"
pits were generally larger than the perfect hexagons. Examination of all pits at
higher magnification (including REM and SEM) showed that the majority of the
larger pits contained more than one apex indicating several dislocations very close
together whose pits interacted to form one large pit with the "distorted" hexagon
shape. The etch pits containing only one apex, then, were studied on all samples.
All such pits in the 4H and 6H platelets were perfectly hexagonal. (See Figures
2 and 3). The sides of the pits in the 6H polytypes were found to be stepped.
Examination of the steps, using SEM showed, however, that the steps were not perfectly straight and uniform as had been found earlier (4). The steps were often
wavy as seen in Figure 4. This wavyness quite possibly comes from jogs in the
steps. In some cases, one step appeared to merge with another. The sides of the
hexagonal pits on the 4H polytypes were generally devoid of step; perhaps because
the steps were too small to be resolved. (In Figure 2 one can see a faint indication of steps). Examination of the sides of the pits in 15R platelets showed
that only the long sides were stepped; the short sides were smooth as shown in
Figures 1 and 6. One can occasionally see one or two large steps on the short
faces as seen on one of the pits in Figure 1. From these studies, then, one can
say that the "distorted" hexagonal pits unexplained by Faust (4) were from 15R
platelets.

Platelets that contained one polytype in a 6H matrix, as determined by transmission x-ray Laue techniques, were etched and their pits examined. In several
cases the second polytype shared some of the surface with the 6H matrix; it was
possible to see this by examining the etch pits. Figure 5 shows an example where
15R shared part of the surface with the 6H matrix. The shapes of the pits can be
clearly seen. Figure 3 is a SEM micrograph of one of the 6H pits on this sample.
Figure 6 shows a 15R pit on the same sample. These pits had a bright outline
around them other pits on the same sample did not as seen in Figure 5. Thus far
we have no plausible explaination for this bright outline phenomenon. In the case

of a 3C partially covering a 6H crystal, it was possible to determine this by the trangular etch pits characteristic of the 3C (4) and the hexagonal pits on the 6H part of the crystal.

Mitchell (7) reported that the Buergers vector of a screw dislocation is n(4d) for a 4H, n(5d) for a 15R, and n(6d) for a 6H polytype. Where n is an integar, and d is the c-component of the two nearest Si-Si interatomic distance in the various polytype unit cells. It is not inconceivable to expect that the heights of the steps in the pits would be a multiple of the basic Buergers vector for each polytype (at least in pits associated with screw dislocations) or are a multiple of the c-lattice parameter. If such is the case one could conceivably distinguish between the polytypes by measuring the step height on the sides of the etch pits. This is a very difficult measurement, and thus far it has not been possible to measure the height of the steps.

The kinetic studies of etching showed that the etch rates did not increase exponentially with temperature for a single charge of molten salt. This is not too surprising because of the decomposition of the Na_2O_2 with temperature and time. Etch rates for mixtures of $NaOH:Na_2O_2$ (3:1) made up "fresh" for each time and temperature were 0.35μ/min at 360°C and 26μ/min at 685°C. These values are higher than those reported by Jennings (5). The discrepancy could be due in part to the age of the Na_2O_2 used in making the etchants.

CONCLUSIONS

These studies show that at least some polytypes can be determined by studying etch pits under an ordinary metallurgical microscope. The 3C pits are triangular with smooth sides; the 4H are perfectly hexagonal with no, or few, steps visible on the sides of the pits; the 6H are perfectly hexagonal with steps on all sides of the pits; and the 15R are "distorted" hexagons with steps only on the three long sides. This technique will show only pits on polytypes at the surface and will give no data as to any buried polytypes.

This work was partially supported by the Air Force Cambridge Research Laboratories under Contract #F19628-72-C0136.

REFERENCES

(1) Cooks, F. H., Da, B. N., and Wolff, G. A., J. Materials Soc. 2, 470 (1967).

(2) P. T. B. Shaffer, The Microscope 18, 179 (1970).

(3) Verma, A. R., and Krishna, P., "Polymorphism and Polytypism in Crystals"
     John Wiley and Sons, Inc., New York (1966).

(4) Faust, Jr., J. W., in "Silicon Carbide" (Smiltens and O'Connor, eds.) p. 403
     Pergamon Press, 1960.

(5) Jennings, V. J., Materials Res. Bull. $\underline{4}$, S206 (1969).

(6) Amelinckx, S. and Strumane, G., J. Appl. Phys. $\underline{31}$, 1359 (1960).

(7) Mitchell, R. S., Zeit. f. Krist. $\underline{109}$, 1 (1957).

Figure 1. "Distorted" hexagons in the
15R polytype. SEM 1×10^4 X

Figure 2. Pit in 4H polytype.
SEM 5×10^3 X

        SEM micrographs are shown here for best detail; however, an optical microscope
is adequate for identifying polytypes from etch pits.

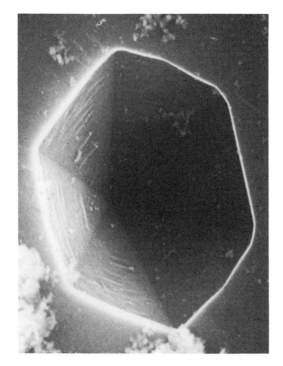

Figure 3. Pit in 6H polytype from 6H
part of Figure 5. SEM 5×10^3 X

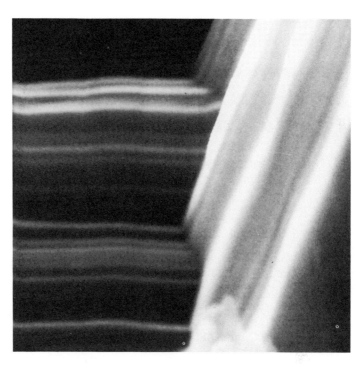

Figure 4. Steps on two sides of a pit
in 6H polytype. SEM 3×10^4 X

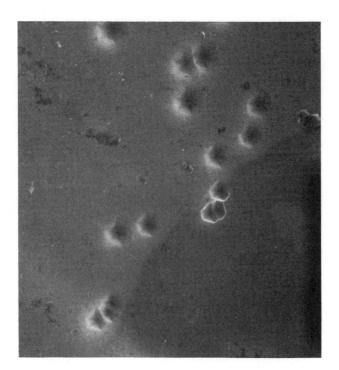

Figure 5. Pits on a surface that is
part 6H and part 15R. SEM 1×10^3 X

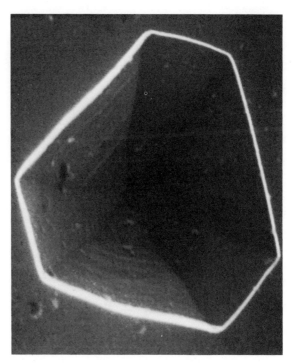

Figure 6. One pit of 15R polytype
section of Figure 5.

Transmission Electron Microscopy on Growth Characteristics of SiC Crystals during Reaction Sintering

H. Sato, S. Shinozaki, M. Yessik and J. E. Noakes

Newly formed SiC crystals during reaction sintering are exclusively of 3C structure and found to grow epitaxially on the surface of original α-SiC crystals. Original α-SiC crystals are full of stacking faults while newly formed 3C crystals are heavily twinned. 3C crystals independently grown in the space surrounded by original SiC particles expose the (111) and the (100) surface and bond each other only when they are of the same orientation.

INTRODUCTION

Reaction-sintered SiC materials are generally produced from a compacted mixture of silicon carbide particles and carbon by introducing molten silicon to convert the carbon into SiC. The newly formed SiC crystals then bond the original SiC crystals and themselves together (1). Therefore, the growth characteristics of the newly formed SiC crystals eventually determine the intrinsic strength of the silicided materials. In particular, it is important to determine the nature of the bonding of the newly formed SiC particles to the original SiC particles and to each other. The aim of this paper is the application of transmission electron microscopy to the observation of the boundaries of crystallites to investigate the bonding characteristics in detail and also the defects associated with both the original and the newly formed SiC crystals.

SiC is known to have an extensive polytypical behavior (2). This means that SiC can take many different crystal structures each of which is a stacking variant of a close packed structure. In order to identify each structure, the Ramsdell notation (3) is used throughout this paper.

EXPERIMENTAL RESULTS

An investigation was made on a reaction-sintered material fabricated in this laboratory. The material is characterized by a fairly large original

SiC particle size (maximum 40 $\sim$ 50 μm in diameter). The use of such a large original SiC grain size, although not favorable with regard to mechanical properties, has the advantage of allowing the original SiC particles to be distinguished visually from the newly formed SiC particles.

A microstructure of this material taken by optical microscopy is shown in plate 1. Here, large, irregularly shaped particles are the original SiC particles, whereas white portions indicate the existence of unreacted Si metal and the remainder is the newly formed SiC crystals. It is easy to see that some newly formed SiC particles directly cover the original SiC particles, indicating that they grow on the surface of the original SiC particles. This is more clearly seen in plate 2 where Si is removed by a selective etching, leaving only original SiC covered with newly formed β-SiC on the surface. On the other hand, in the space surrounded by the original SiC particles, new SiC crystals are found to grow independently in a rather equi-axed shape. The etching pattern of the original SiC particles indicates that these are mostly single crystals. X-ray examinations of individual particles confirmed this observation.

Since the specimens must be thinned down for observation in the electron microscope, the morphological distinction between the original SiC crystals and the newly formed SiC crystals is eventually lost. However, the possible difference in crystallographic structure between these two species can be utilized as an indicator of old and new.

A neutron-diffraction study of the bulk of the original SiC powders and an X-ray diffraction study of individual particles from the same batch show that the structure of the original SiC particles used for the preparation of the specimens is either 4H or 6H and the possible existence of other structures is found to be negligible. In fact, there are two possible 6H structures ($6H_1$: ABCACB$\cdots$ and $6H_2$: ABCBCB$\cdots$), and what we find in SiC is $6H_1$, while, $6H_2$, which is rather common in structures of long period stacking order in alloys (4,5), has never been reported in SiC. On the other hand, a neutron-diffraction study of reaction-sintered specimens shows that the newly formed SiC crystals are exclusively 3C and that the existence ratio of 4H and 6H does not change during the reaction. It has been customary that SiC with 3C structure is called β-SiC while SiC of all other structures, whether it is hexagonal or rhombohedral, is called α-SiC. This means, according to whether the structure of the grain we observe in the electron microscope is α or β, we can identify it as an original SiC crystal or a newly formed one.

The ion beam thinning technique is utilized to thin the specimens to the desired thickness after they have been mechanically ground to a thickness of ∿ 75-100 μm. Our specimens with a large initial SiC grain size were found to be far easier to thin down by this technique than REFEL or hot-pressed material which has far smaller grain size. A Philips 200 electron microscope with an accelerating voltage of 100 kV was used for the observations. With this accelerating voltage, however, only a very limited area around the hole created by the thinning could be observed.

The polytypism of SiC indicates that the energy required to change the stacking order and, hence, the energy of formation of stacking faults, is very small and a high density of stacking faults is expected. Further, the stacking fault is not generally expected to end inside the crystal, since this means the existence of dislocations which requires a certain amount of energy. Such faulting should occur on the basal plane perpendicular to the c-axis and hence creates one-dimensional disorder in the c-direction. This one-dimensional disorder tends to create streaks in the diffraction pattern in the c-direction. In the case of the 3C structure, a twin boundary also coincides with one of the cubic {111} planes. Since the energy required to create the twin boundary is very small in such a case, the existence of a large number of twins is expected in β-SiC.

In plate 3, an example of a heavily faulted 6H crystal is shown. The corresponding diffraction pattern also shows streaks along the c-axis as a result of these faults. Generally, however, the density of stacking faults in α-SiC is found to be less than expected. The density of dislocations is also low as expected and the SiC crystals are otherwise relatively perfect and Kikuchi lines are often observed (not shown).

Crystals with the cubic structure are those grown during the reaction-sintering process. These crystals are found to be very heavily twinned. The twins are clearly seen by diffraction contrast as stripes perpendicular to one of the <111> axes. In plate 4, such micrographs taken with both bright and dark field techniques are shown. These crystals are expected to grow parallel to a {111} plane. Therefore, the existence of twins indicates that the stacking order changes from time to time during the growth, but, nevertheless the cubic stacking order continues. This indicates that the cubic structure is the most stable state under this condition of growth, which is very important in discussing the origin of the polytypism. Stacking faults exist also in cubic crystals.

When the stacking faults do not end inside the crystal, it is very difficult to distinguish the image from those of twin boundaries. In plate 5, the image of the twin boundaries is observed in the same crystal shown in plate 4.

Optical micrographs (plates 1 and 2) show that β-SiC grows on the surface of a α-SiC. Plate 5 shows an example of such boundaries between 4H crystal and a 3C crystal. The 4H crystal (A) shows the image of stacking faults whereas the 3C crystal (B) shows the image of twin boundaries. The diffraction pattern obtained at the boundary, which shows the superposition of the pattern from the two crystals, is also included in plate 5. Both the electron micrograph and the diffraction pattern show that the (111) axis of the 3C crystal and the c-axis of the 4H crystal makes an angle of about 70°. The same relation is found whenever a boundary between α- and β-SiC is observed. This means that one of the {111} planes of 3C crystal matches the basal plane of an α-crystal and indicates that β-SiC grows epitaxially on the surface of α-SiC. This also means that the bonding between the newly formed β-SiC crystal and the α-SiC crystal is intrinsic and not due to the presence of some other material like Si or SiO_2 as a glue at the boundary. In plate 6, another example of such boundaries between a 4H crystal and a 3C crystal is shown.

The above conclusion does not necessarily mean that the intrinsic bonding between β-SiC and α-SiC always occurs over the entire surface of α-SiC crystals. Rather, it is likely that the intrinsic bonding occurs at limited areas where the condition is right as schematically shown in figure 1. Such a situation is actually observed in plate 7. However, by optical microscopy, it is possible to confirm that the composite crystals consisting of newly formed β-SiC on the original α-SiC crystals behave as if they were single crystalline in fracture (not shown). Therefore, it is a reasonable guess that such intrinsic bonding occurs generally on wide areas of the surface of original α-SiC particles. It has been confirmed that this is especially true if the crystal growth occurs at high temperatures.

The existence of an epitaxial relation between the β-SiC and the α-SiC, however, raises a reasonable question as to whether newly formed β-SiC crystals of arbitrary orientations can bond together properly. In order for these crystals to bond together properly, a regular array of dislocations of proper Burgers vector should be introduced at the grain boundary. The following observations indirectly indicates that this is unfavorable. The density of dislocations in α-grains is found to be extremely low so that the energy of formation of dislocations is high. Also, initially prepared α-SiC grains are found to be all single crystals and polycrystalline grains are not found. The same thing then can be expected in the case of β-SiC. In fact, we could not find a boundary where two β-SiC grains touch each other with an arbitrary angle in the thinned specimens. On the other hand, we found a case where possibly two β-SiC crystals with the same orientation touch each other as is shown in plate 8. In plate 9 a scanning electron micrograph of β-SiC crystals grown independently in the space surrounded by large α-SiC crystals is shown. These crystals are rather equiaxed and expose only the (111) and the (100) surface. Morphology of the crystals indicate that these crystals grow from the liquid phase by precipitation rather than from a vapor phase during the reaction sintering process. Some of these crystals seem to bond to each other with a very limited area where their orientations are the same. In order to improve the strength of reaction sintered SiC materials of this kind, it seems necessary to devise a way to increase the area of contact of β-SiC crystals.

REFERENCES

(1) P. Popper, Special Ceramics, ed. P. Popper, British Ceramics Association. London (1960) p. 209.

(2) A. R. Verma and P. Krishna, Polymorphism and Polytypism in Crystals, John Wiley and Sons, Inc., New York (1966).

(3) L. S. Ramsdell, Am. Mineralogist 32, 64 (1947).

(4) H. Sato, R. S. Toth and G. Honjo, J. Phys. Chem. Solids 28, 137 (1967).

(5) M. Hirabayashi, K. Hiraga, S. Yamaguchi and N. Ino, J. Phys. Soc. Japan 27, 80 (1969).

Plate 1. Microstructure of a reaction
sintered material (200X).

Plate 2. Removal of Si by a selective
etchings (375X).

Plate 3. High density of stacking
faults in a 6H grain.

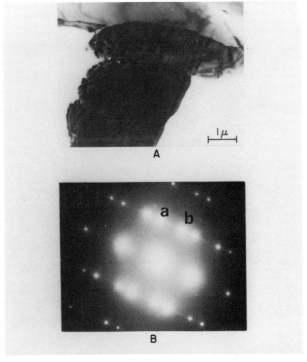

Plate 4a. Existence of twin bands in a
3C grain.
  A) Bright-field image.
  B) Diffraction pattern.

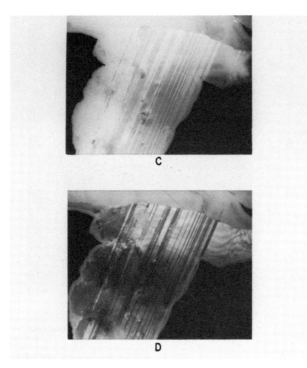

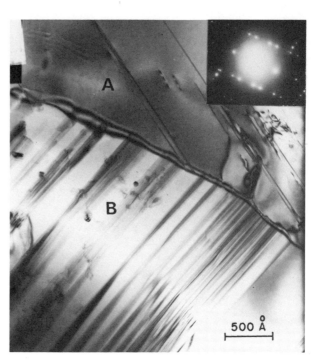

Plate 4b. C) Dark-field image (with the diffracted beam a in B).
D) Dark-field image (with the diffracted beam b in B).

Plate 5. A grain boundary between a newly formed 3C grain (B) and an original 4H grain (A).

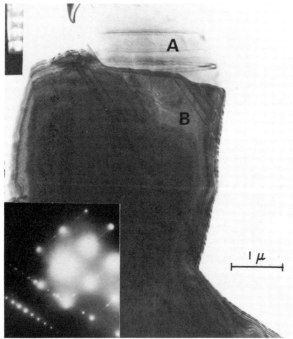

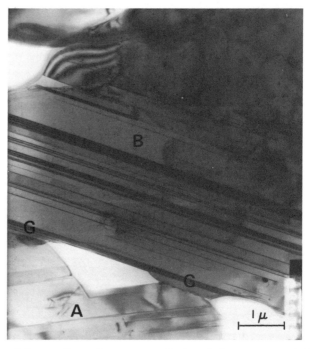

Plate 6. Overlapping of a 4H grain (A) and a 3C grain.

Plate 7. Contact of 6H (A) and 3C (B) at limited area (G).

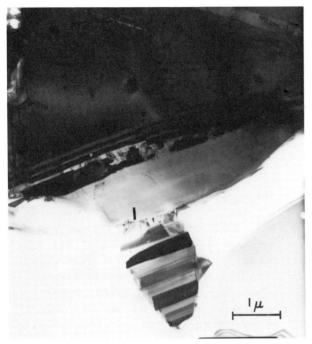

Plate 8.   A boundary (I) of two 3C
grains at the same orientation.

Plate 9.   Independently grown β-SiC
crystals.

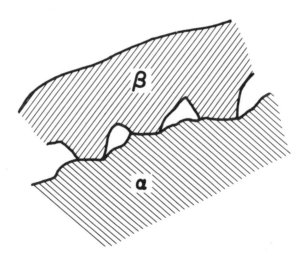

Figure 1.   Contact of α-SiC and β-SiC
crystals at limited area (schematic)

Direct Observation and Identification of Long Period Structures of SiC by Transmission Electron Microscopy

H. Sato, S. Shinozaki and M. Yessik

Lattice periods of long period structures in SiC are directly observed by transmission electron microscopy and their structures are analysed. Among these, 222R, 303R, etc. have been identified. The long distance information concerning the stacking order which leads to the stabilization of long period structures seems to be conveyed externally by means of an effect of environments.

INTRODUCTION

It has been known for some time that SiC shows extensive polytypical behavior and takes many different forms of stacking variants of close packed structure (1). However, at the moment, no single theory can account satisfactorily for all the observed facts on polytypism in SiC. One of the most puzzling features is the existence of extremely long periods. Here, we present an attempt to utilize transmission electron microscopy to identify the structures and the distribution of such long period structures as an approach toward understanding the problem. Although it is commonly thought that electron diffraction is not preferable for structure determination because of extensive dynamical effects, electron microscopy has some definite advantage over the standard X-ray diffraction. First, structures can be observed both in the real space as a direct resolution of the lattice period and in the reciprocal space as the diffraction pattern under the same condition. Also very small grains can be observed as single crystalline specimens so that the problem of syntactic coalescense on a very fine scale, for example, can be easily avoided. In fact, an alternating growth of 6H and 15R structures in bands of several hundred angstroms has been found (2). Such application of transmission electron microscopy to the study of the structures with long period stacking order has been commonly used (3,4). The technique, however, has not been utilized yet in the study of polytypism in SiC. Here, we present several examples in identifying structures which have not been reported and then give a tentative explanation for the origin of the long periods in SiC.

STRUCTURAL DESCRIPTION AND POLYTYPISM IN SiC

Before we discuss the problem of polytypism, it might be worthwhile to describe the structural characteristics of SiC in a way convenient to our later discussions. A SiC molecule can be described as a tetrahedron in which a carbon atom is surrounded by four silicon atoms, or in the same way, a silicon atom is surrounded by four carbon atoms. These tetrahedra are arranged in such a way that all atoms lie in parallel planes on the nodes of regular hexagonal networks. The Si-C bond is mostly a covalent bond which is a short range interaction. The characteristic of the stacking can then be considered as a stacking of spheres of radius a/2, where a is the edge length of the unit tetrahedron, centered either at Si atoms or at C atoms, with nearest neighbor central force interaction. In other words, the SiC structure is represented conveniently by a close packing of spheres with nearest neighbor interactions, and any structure can be specified simply by the stacking order of the close packed planes on three possible stacking positions, A, B and C.

As is well known, in the case of close packing of spheres with nearest neighbor interactions, the total energy of the structure does not depend on its stacking order. This roughly explains the origin of the polytypism in SiC where many different stacking variants exist, since any close packed stacking variant is energetically equivalent and hence equally possible in the ideal case of nearest neighbor interactions. Ideally, the most probable structure would then be a non-periodic, one-dimensionally disordered structure. Although such a one-dimensionally disordered structure is actually found in CVD SiC material (5), in most cases, SiC crystallizes into a small number of definite structures of short period. This is due to the deviation from the ideal condition and in this deviation our main interest lies. For the specification of the structure, the Ramsdell notation (6) and the Zhdanov notation (7) are used throughout this paper.

EXPERIMENTAL RESULTS

Three types of SiC materials are used for observation. These are a hot pressed SiC made by Norton Co., a reaction-sintered material, REFEL, made by U.K.A.E.A. and a reaction-sintered material made in our laboratory using α-SiC particles supplied from Norton Co. as the original SiC particles (8) (referred to hereafter as FORD). The latter two differ in the processing method and in the grain size of the original SiC particles. Long period structures are found in the hot pressed material and very frequently in REFEL but not at all in FORD even after a very extensive examination. A Philips 200 electron microscope with 100 kV acceleration with large angle goniometer stage has been used. A Hitachi 650 kV high voltage electron microscope at the Case Western Reserve University was also used to supplement the data. An ion beam thinning technique has been utilized for the preparation of specimens. Although the existence of a large number of long period structures was observed by the high voltage electron microscope, a reliable intensity distribution in the diffraction pattern could not be obtained. Therefore, the identification of long period structures was achieved only on those specimens investigated with the Philips 200 microscope.

In plate 1, an observation of a long period structure in a REFEL specimen as an assembly of stripes of about $255\overset{\circ}{A}$ in width is shown. This was taken by a dark field technique in order to observe these stripes more clearly (9). These stripes correspond to a direct resolution of the lattice image (long period) and correspond to the period of about 100 close-packed layers. The diffraction pattern of this grain is basically that of 6H. However, in between two of the spots of the underlying 6H structure in the $(10 \cdot \ell)$ series, seventeen additional spots are clearly observed as shown in plate 1. The intensity distribution of the seventeen spots and that for the entire unit reciprocal layer spacing (3) taken by a densitometer is shown in plate 2 and 3. Although an extensive refinement in structural analysis has not been made, the intensity distribution can be well reproduced by the calculated structure factor of a 101 layer structure $[(33)_{16}32]_3$ or 303R. The comparison of the calculated intensity distribution (thin vertical lines) and that measured is also shown in plate 3. In fact, this belongs to a well known family of structures in SiC, $[(33)_n32]_3$.

In plate 4, a lattice image of another long period structure in a REFEL specimen is shown. The width of the stripes is about $185\overset{\circ}{A}$ which corresponds to a structure with about a 75 layer period. The intensity distribution of the diffraction pattern and the comparison with that of the analysed structure is shown in plate 5. The intensity distribution in the reciprocal space is also basically that of the 6H structure but the fine detail is a little more complicated than the 303R structure (11). The stacking order of the most probable structure thus analysed is $[(33)_6 34(33)_4 34]_3$ or 222R with a 74 layer period. This structure also is a variant of a well known family of structures in SiC, $[(33)_n 34]_3$. These two structures, 303R and 222R, have not been previously observed. In plate 6, the distribution of grains with long period structure in one REFEL specimen as observed by the Hitachi 650 is shown. Although the exact structures of each grain were not analysed, the periods can be obtained from the widths of the stripes and these are all different depending on the grains. However, this indicates the very frequent existence of long period structures in SiC and the advantage of using transmission electron microscopy in searching for the new long period structures.

DISCUSSIONS

Although the dynamical effect is strong in electron diffraction, the intensity distribution seems to be accurate enough for a reliable analysis of the structure as long as good diffraction patterns can be taken. This is, however, not easy, since the penetration of the electron beam is small and hence the intensity is low. In fact, in the analysis of the structures with long period stacking order in alloys where homogeneous thinning of specimens is easy and enough intensities in the diffraction patterns can easily be obtained, a unique determination of the structure could be made without any trouble (3,4). Also, the direct observation of the lattice images proves beyond doubt that these are real crystallographic structures and any possible confusion from the syntactic coalescence, etc., can be avoided.

As is clear from the lattice image, the actual lattice period is one third of the number of layers in one period specified by the Ramsdell notation in the case of R symmetry. This is because the size of the "hexagonal" unit cell is taken as the period in the Ramsdell notation. Although quite satisfactory as a notation, we find it more convenient to use the number of spots in the unit reciprocal layer spacing as the number of layers of close-packed planes in one period in physical arguments like the width of stripe patterns, etc. (3,4). However, here, the Ramsdell notation is used in order to avoid the confusion.

As pointed out by Frank (10) and others (11), both 303R and 222R structures can be derived by introducing simple screw dislocations from the 6H structure. Therefore, it has seemed quite attractive to explain the occurrence of these structures as a result of the growth of 6H structure via screw dislocation, assuming the stability of the 6H structure. However, the screw dislocation theory has difficulties in explaining, for example, the existence of extensive stacking faults often found in long period structures and the fact that the lattice period and the growth step height on the surface of the crystal do not usually match (1). The existence of one-dimensionally disordered structure, on the other hand, supports the concept that any close-packed structure is almost equally possible. Here, however, it is necessary to understand how the information concerning the stacking period can be conveyed for such a long distance.

In the case of metallic alloys, the occurrence of a series of structures of long period stacking order can be qualitatively understood as an effort of the conduction electrons to bring the Brillouin zone boundaries into contact with Fermi surface (3,12). In a semiconductor like SiC, however, it is hard to accept a similar interpretation. Further, in SiC, many structures can appear at a fixed electron-atom ratio. However, information concerning the stacking period can be conveyed <u>externally</u> in a crystal like SiC where the change in the stacking order does not cause any appreciable change in energy. If a crystal is placed or grows under a constraint from the surroundings, internal strain will be created. This strain can be accommodated by a series of appropriate stacking shifts, or preferably in the form of a structure with a regular stacking shift. In figure 1, the comparison of the stacking order of 6H and that of $[(33)_n 32]_3$ structure at the stacking shift is shown in a zig-zag stacking diagram. It is clear from the figure that a horizontal shift indicated by an arrow is involved at the stacking shift and, hence, a homogeneous shear strain can be absorbed by a transformation from 6H into a long period structure as shown in figure 2. In other words, the existence of such long period structure is interpreted as a result of a "strain-induced transformation". Whether or not such transformation can occur, or what type of structure should be produced under such a condition, should then depend on whether or not such a transformation process is energetically favorable. The problem can then be reduced to one like that discussed by Mitchell (11) concerning the energy required to introduce an appropriate screw dislocation in the basic structure necessary to describe the transformation.

The discussions given above as to the origin of the long period structures are yet highly speculative and, at the moment, there is no direct experimental support. However, we might mention the fact that many long period structures are found in hot pressed material and REFEL but not in FORD. The former two materials use far finer starting SiC grains and a larger strain would be involved in these grains due to the pulverization process. Further, a higher external pressure is involved in processing. Whereas REFEL was formed by an extrusion process, FORD was fabricated by a different technique in which considerably lower pressures were used. The difference in the frequency of observing long period structures in these specimens might then be taken as an indirect support of our statement.

REFERENCES

1. A. R. Verma and P. Krishna, Polymorphism and Polytypism in Crystals, John Wiley and Sons, Inc., New York (1966).

2. S. Shinozaki and H. Sato, unpublished.

3. H. Sato, R. S. Toth and G. Honjo, J. Phys. Chem. Solids 28, 137 (1967).

4. H. Sato and R. S. Toth, J. Phys. Chem. Solids 29, 2015 (1968).

5. S. Shinozaki and H. Sato, unpublished.

6. L. S. Ramsdell, Am. Mineralogist 32, 64 (1947).

7. G. S. Zhdanov, Dokl. Akad. Nauk. SSSR 48, 39 (1945).

8. H. Sato, S. Shinozaki, M. Yessik and J. E. Noakes, previous paper.

9. R. S. Toth and H. Sato, Appl. Phys. Letters 9, 101 (1966).

10. F. C. Frank, Phil. Mag. 42, 1014 (1951).

11. R. S. Mitchell, Zs. f. Krist. 109, 1 (1957).

12. H. Sato and R. S. Toth, Bull. Soc. Franc. Miner. Crist. 91, 557 (1968).

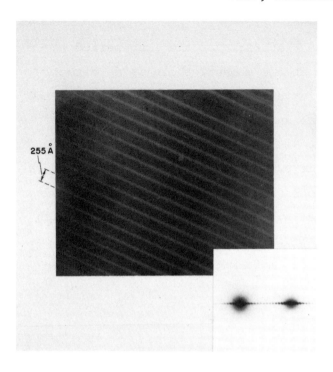

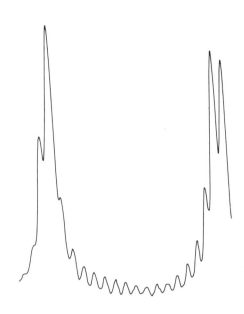

Plate 1. A dark-field micrograph indicating the direct resolution of the lattice period of 303R.

Plate 2. The intensity distribution of the seventeen additional diffraction spots shown in plate 1.

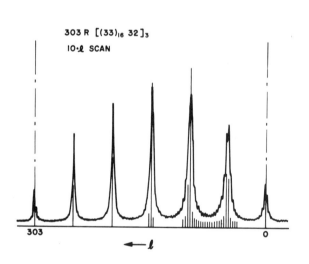

Plate 3. The intensity distribution in the reciprocal space of 303R.

Plate 4. A bright-field micrograph indicating the direct resolution of the lattice period of 222R.

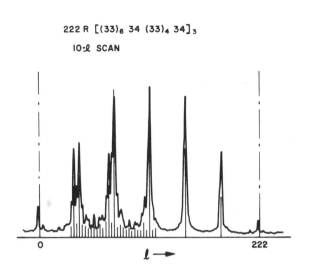

Plate 5.   The intensity distribution in the reciprocal space of 222R.

Plate 6.   Electron micrograph of a REFEL specimen.   Grain with long period structures are indicated by arrows.

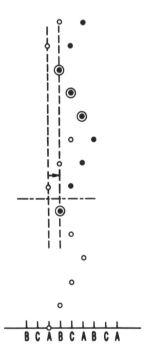

Figure 1.   Comparison of the stacking order of 6H (open circles) and that of $[(33)_n \ 32]_3$ (filled circles).

Figure 2.   Accomodation of a homogeneous shear strain by periodic step shifts.

Observation of Slip Distributions in α-SiC

H. Posen and J. A. Bruce

INTRODUCTION

The growth of large single crystals of silicon carbide has been the objective of many laboratories throughout the world, and in fact has become somewhat of an international contest. However, semiconductor device applications are less dependent on size than on defect distributions, and understanding when such defects occurr in the growth process may be important in the obtention of defect free silicon carbide.

In this paper we report on slip distributions in three α-SiC crystals of the same polytype but of different national origin and of different dopant.

The primary investigative tool was x-ray topography with ancillary optical evaluations.

EXPERIMENTAL

A. Crystals

Crystal I is an American crystal, grown by the General Electric Company, is green in color and is doped with nitrogen. Crystals II and III are Russian in origin, and grown presumably, as is the American crystal, by the Lely technique. Crystal II is also greenish and is doped with nitrogen. Crystal III is opaque grey in colour with a metallic lustre and nominally doped with beryllium.

Both crystals I and II exhibit the six-sided plate-like morphology,

238

characteristic of hexagonal crystals. Crystal III has a highly distorted plate-like habit.

Differential interference contrast microscopy of crystals I and II show apparent growth spirals and step-like thinning of the crystallite edges.

Microscopy of crystal III shows a rough, pebbly surface.

B. X-Ray Topography

Although the defect structure of SiC has been studied by a variety of methods, the large size and national origin of these crystals precluded, at least initially, the use of a destructive method to evaluate thin defect structures.

Polarized light studies, and Normarski differential interference contrast can distinguish the presence of different polytypes or twins in SiC, but cannot delineate the defect structure. X-ray rocking curves of selected reflections can indirectly give dislocation densities, and in some instances the proportion of screw to edge type dislocations, but cannot give a detailed picture of the dislocation network. Transmission electron micrographs could give direct information about type and configuration of defects for selected small areas, if thinnable, and representative sections were readily available.

X-ray diffraction topography offers the desiratum of non-destructive defect evaluation, coupled with a direct visual display of the defect configuration. Some applications of this technique to the study of SiC have been made by others.[1, 2, 3]

The x-ray camera used in this study is a Schwuttke design[4], incorporating a long collimator, crystal goniometer and both specimen rocking and scanning facilities. The x-ray technique is that developed by Lang[5] and Schwuttke[6]. The topographs are made with $MoK\alpha_1$ radiation on Ilford Nuclear Emulsion Plates, Type G5, with exposure times of 1 to 2 hrs, depending on the viewing Bragg reflection and specimen thickness.

The enhancement of diffracting power associated with imperfect crystal regions results in diffraction contrast, which when recorded on a

photographic plate delineates defect regions of the crystal. Optimum diffraction contrast is obtained for crystal thickness t and linear absorption coefficient μ, such that $\mu t \sim 1$. In these topographs, the unitary absorption path
criterion was not always observed in order that the topographs reflect the
as-received defect configuration. Thus the crystal thicknesses varied from
0.23 mm to 0.69 mm, whereas the optimum thickness was 0.64 mm. Such
deviations from the optimum thickness are reflected in a slight diminution in
diffraction contrast and shorter exposure times.

　　　Topographs of the same crystal, when viewed by different reflections,
allow the approximate determination of the Burgers vectors of those defects
exhibiting contrast variation by using the criterion that $\underline{b} \cdot \underline{g} = 0$ when the
diffraction contrast is a minimum; $\underline{b}$ is the Burgers vector of the dislocation,
and $\underline{g}$ is the diffraction vector of the viewing Bragg reflection.

　　　A comparison of x-ray transmission Laue patterns for each of the crystals with simulated patterns obtained from the Atlas of SiC Patterns[7] gives
a one-to-one correlation with the (0001) orientation of the 6H polytype for
crystals I, II and III.

　　　The transmission pattern for Crystal III, has an additional Laue spot
on each of the six-fold reciprocal lattice lines at slightly varying distances
from the pattern's center of symmetry. These spots can be ascribed to β-SiC,
but may be associated with a lattice rotation.

RESULTS AND DISCUSSION

　　　Topographs of crystal I are shown in Plate 1 for the $(11\bar{2}0)$ type planes,
and in Plate 2 for the $(30\bar{3}0)$ type planes. The topographs are divided into five
zones by dislocation bundles originating from a point on the crystal edge that
had been attached to the furnace wall. Plate 3(a) is a line drawing of the
prominent contrast features of crystal I. Zones 1 and 2 although dominated by
feature D, a multiple dislocation loops, has significantly fewer dislocation
trails.

　　　Features A and B of crystal I can be identified as dislocations of $\langle 11\bar{2}0 \rangle$

Burgers vectors, whereas feature C of the same crystal can be identified as a dislocation pencil of Burgers vector $\langle \bar{1}2\bar{1}0 \rangle$.

Features D and F are dislocation loops which do not appear to be directly associated with a crystal growth mechanism, although the loop designated as F does appear to originate at the nucleation edge. These loops are not correlateable with features on optical micrographs.

The spaghetti-like trails of contrast associated with zones 3, 4, and 5 may be ascribed to screw-like components of dislocations, which do not lie in the base plane.

Topographs of crystal II are shown in Plate 4 for $(11\bar{2}0)$ type planes. and in Plate 5 for the $(30\bar{3}0)$ type planes. Features A and B (plate 3b) can be identified as dislocations with Burgers vectors $\langle 2\bar{1}\bar{1}0 \rangle$ and $\langle \bar{1}2\bar{1}0 \rangle$. Note that feature A, consists of two bundles, A_1 originating at the nucleating edge and A_2 at the boundary of feature D respectively, whereas feature C are dislocations of Burgers vector $\langle 11\bar{2}0 \rangle$. Feature D appears to be a hexagonal progenitor of the final platelet. The boundaries of feature D do not go in and out of contrast, regardless of the viewing reflection. This suggests that the permanent contrast associated with the boundaries of D is due to differential absorption, arising from a physical discontinuity in the crystal. Nomarski microscopy of this crystal does not delineate feature D. There is no evidence of the spaghetti-like contrast trails or zones of varying perfection as evidenced in crystal I.

Topographs of crystal III are shown in Plate 6 for the $(11\bar{2}0)$ type planes. The quality of this crystal was so poor that it was necessary to oscillate the specimen $\pm 1/8^\circ$ while obtaining the scanning topograph. Again varying contrast bands can be ascribed to the presence of dislocations of the $\langle 2\bar{1}\bar{1}0 \rangle$ type Burgers vectors. Burgers vectors of the type $\langle \bar{1}2\bar{1}0 \rangle$ and $\langle 11\bar{2}0 \rangle$ were not observed, and again the spaghetti-like contrast of crystal I was not seen.

Thus crystals I, II and III exhibit the basal slip, characteristic of hexagonal materials - slip in the $\langle 11\bar{2}0 \rangle$ direction on the (0001) plane.

The topographs for crystal II suggest an interrupted growth history with the boundary for feature D of this crystal, representing some thermo-dynamic discontinuity. As the crystal grows beyond this boundary, stresses arise sufficient to create plastic deformation, and dislocations represented by features C and part of feature A are generated. At a later time, after the platelet is fully grown, another stress relieval occurrs and the rest of the dis-locations represented by feature A are generated from the nucleating wall, throughout the crystal.

SUMMARY

The slip mechanism for 6H-SiC crystals has been confirmed by x-ray topography.

The thermal history of a particular crystal has been inferred from ex-amination of its topographs.

REFERENCES

1. Amelincyx, S., Strumane, G., & Webb, W. W., "J. Appl. Phys." 31, 1359(1960)

2. Ohata, K., Tomita, T., & Watanbe, "Japan J. Appl. Phys."4, 652 (1965)

3. Bartlett, R. W., & Martin, G. W., "J. Appl. Phys."39, 2324 (1968)

4. Schwuttke, G. H., "J. Electrochem. Soc.", 109, 27, (1962)

5. Lang, A. R. "J. Appl. Phys.", 597, (1958)

6. Schwuttke, G. H., "J. Appl. Phys.", 36, 2712, (1965)

7. Posen, H., "An Atlas of Laue Patterns for SiC Polytypes" - this conference

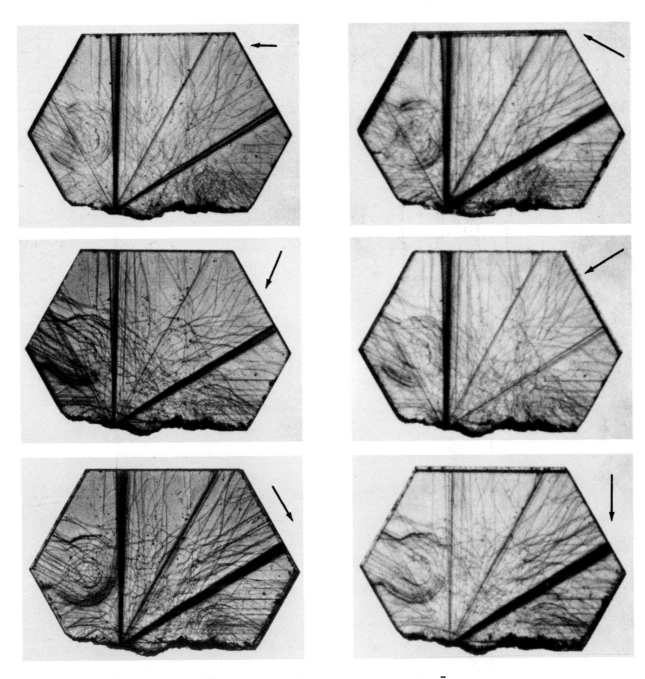

Plate 1.   (11$\bar{2}$0) - Type X-ray Trans-
           mission Topographs of SiC,
           Crystal I.
           Magnification is 10x.

Plate 2.   (30$\bar{3}$0) - Type X-ray Trans-
           mission Topographs of SiC,
           Crystal I.
           Magnification is 10x.

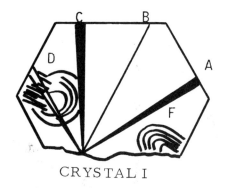 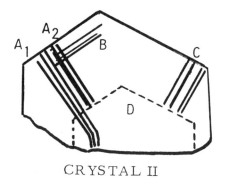

Plate 3. Line drawing illustrating features of x-ray topographs of
          Crystals I(3a) and II(3b).

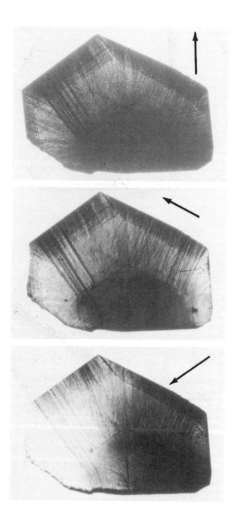 

Plate 4. (11$\bar{2}$0) - Type X-ray Trans-   Plate 5. (30$\bar{3}$0) - Type X-ray Trans-
          mission Topographs of SiC,   mission Topographs of SiC,
          Crystal II.   Crystal II.
          Magnification is 5x.   Magnification is 5x.

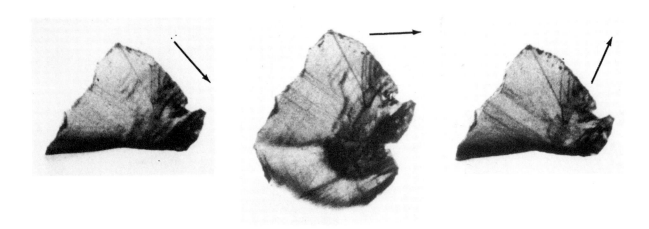

Plate 6. $(11\bar{2}0)$ - Type X-ray Transmission Topographs of SiC, Crystal III.
          Magnification is 10x.

An Atlas of Laue Patterns for SiC Polytypes

H. Posen

INTRODUCTION

Laue patterns have been used to orient single crystals since the beginnings of x-ray science. These patterns are obtained by recording on film, x-rays diffracted from the various atomic planes of the single crystal. The patterns consist of spots of diffracted intensity on the film, and the arrangement of such spots on the film reflect the orientation of the crystal vis-a-vis the incident x-ray beam, and the intrinsic symmetry of the crystal.

Normally the analyst identifies prominent [hkl] zones of the diffracting crystal from the Laue pattern, and then using either a Greninger (back-reflection) or a Leonhardt chart (transmission) determines the stereographic latitude and longitude of a particular diffraction spot for plotting the pole of the diffracting plane on a stereogram. Then, in conjunction with several standard stereographic projections, the orientation of the crystal is identified.

Alternatively, a one-to-one comparison is made with a library of Laue patterns obtained for that particular crystal from a series of oriented samples. Dunn and Martin[1] have published such a library of actual Laue photographs for iron, and Wood [2] in her manual of x-ray orientation has published a few selected patterns.

Unfortunately, the orientations of non-cubic crystals, or for that matter, cubic crystals of non-simple orientation are not easily assigned because "standard" stereographic projections for the crystal of interest don't exist, and

246

libraries of appropriate Laue patterns are not available.

For crystal systems exhibiting a variety of polytypes, such as SiC, in which different polytypes arise from small differences in atom stacking along the c-axis, analysts have abjured Laue techniques as non-diagnostic and have turned to x-ray Weissenberg methods for distinguishing between the various polytypes of SiC.

In this paper we report on a computer program called STEREO [3] which can simulate the x-ray Laue experiment and thus produce an atlas of Laue patterns for the various polytypes of SiC. We also demonstrate how such an atlas of patterns can be used to rapidly distinguish between the various polytypes of SiC.

THE SIMULATION PROGRAM

The program STEREO is a digital computer program written originally for an IBM 7094 machine, but now running routinely on a CDC 6600 machine.

The program produces both Laue back-reflection and transmission patterns, and also stereographic projections for any specified crystal orientation and set of crystal parameters. Patterns are presented in both annotated and unannotated form to enhance pattern recognition.

The following facts are pertinent:

1) All plots are accurate to 0.01 inches.

2) The stereograms are based on a 10 cm Wulff net (simple scaling of listed values will provide any other desired diameter).

3) Specimen-to-film distance, x-ray wavelength, x-ray excitation voltage are free parameters, to be supplied by the experimentalist.

4) The Laue spot is scaled in proportion to the relative intensity of the particular (hkl) reflection, the number of reflection orders superimposed and polarization factor. The spot sizes should only be considered as guides, rather than exact reproductions of true intensity values, since extinction effects and other crystal dependent perfection parameters have been ignored.

5) Bragg spots or spots corresponding to diffraction from those particular

planes, whose orientation is such that they select and diffract the characteristic x-ray wavelength, are particularily intense and are designated by an asterisk symbol on the simulated patterns.

6) Only those (hkl) values consistent with the space group vanishings for the particular crystal are plotted.

7) Laue spots or poles, whose coordinates are computed, but whose plotted position would overlap previous points are listed, but not plotted.

8) A simulated fiducial mark (upper right hand corner) on the Laue patterns corresponds to a similar mark on an experimental plate or film.

9) For hexagonal crystals, the three symbol - (hkl) notation is used. Conversion to the four symbol - (hkil) notation is accomplished by using the algorithm i =-(h + k).

10) Programming restrictions exclude examination and plotting of those (hkl) triads which contain an index greater than nine.

11) The printout lists for each plot the pole, its latitude and longitude in degrees, its Cartesian coordinates in inches and its relative intensity.

RESULTS AND EXAMPLES

Stereograms and Laue patterns were generated for the following SiC polytypes; β, 2H, 4H, 6H, and 15R. The lattice parameters were taken from Shaffer [4].

All simulations were based on sample exposure to CuK x-rays excited at 23 KV, with a specimen-to-film distance of 3 cm.

For β-SiC, (100), (110), (111), (210), (211) and (221) patterns have been generated.

For the hexagonal polytypes, (001), (100), (110), (111), (210), and (223) patterns have been generated.

Typical annotated Laue patterns, obtained by simulation are illustrated by plate 1 and plate 2 for the back-reflection and transmission patterns expected from a (223) oriented 4H-SiC crystal. Note that the variation in spot size, and the density of Laue spots is much smaller in the transmission patterns than in the back reflection patterns. Note also that the (104) and the (014) spots in the transmission pattern are Bragg spots. There has been no photo reduction of these patterns.

An example of how the atlas of generated patterns can be used to quickly identify the different SiC polytypes is illustrated in plate 3 and plate 4. Samples of SiC, independently identified as 4H and 6H polytypes by x-ray Weissenberg methods were examined by transmission Lue methods using Cu x-rays and a 3 cm specimen-to-film distance. The generated patterns and the corresponding Laue pattern are displayed in the same figure for comparison.

Although both polytypes are hexagonal with the same a_o spacing, the difference in the c_o lattice parameter, between the two polytypes, predicates a clearly identifiable difference in Laue patterns, and this difference is accurately portrayed by the simulated patterns.

SUMMARY

An atlas of Laue patterns and stereograms have been generated by a digital computer for polytypes of SiC.

Application of such an atlas to the determination of unknown orientations and to the distinguishing between various polytypes has been demonstrated.

My sincere thanks go to J. Bruce for the careful x-ray work, represented in plates 4 and 5 and for making critical and helpful comments on the manuscript.

REFERENCES

1. Dunn, G. C. & Martin, W. W. , "Metal Trans."(1949), 185, pg. 471

2. Wood, E. A. , "Crystal Orientation Manual", Columbia, N. Y. (1963)

3. Posen, H., "STEREO - A Computer Program", - to be published

4. Shaffer, P. T. B. , "Acta Cryst.", B25, part 3 (March 1969)

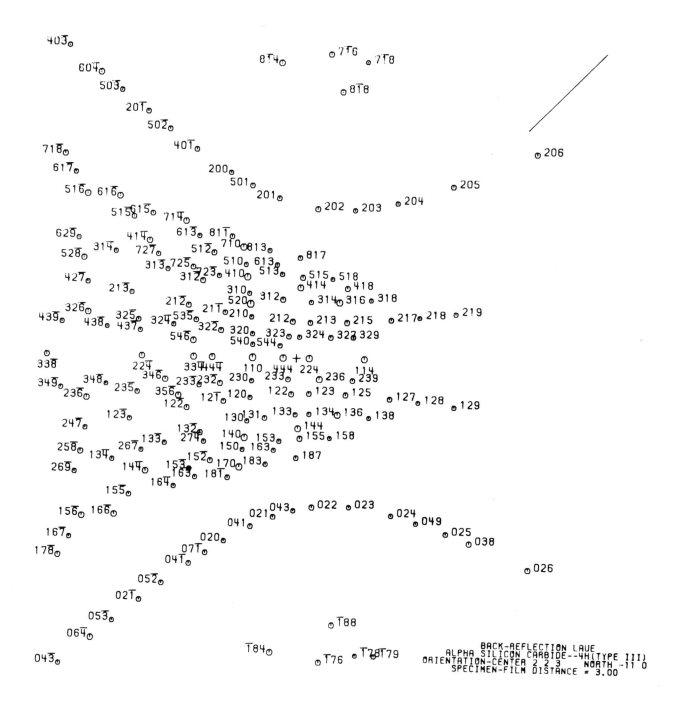

Plate 1.   Simulated Laue Back Reflection Pattern (223) oriented 4H-SiC crystal.

Plate 2.   Simulated Laue Transmission Pattern (223) oriented 4H-SiC crystal.

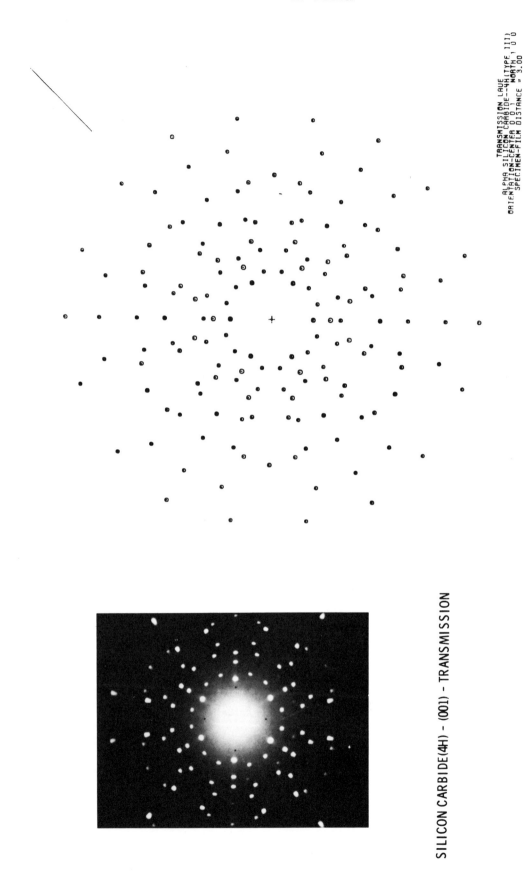

ALPHA SILICON CARBIDE
TRANSMISSION LAUE
ORIENTATON-CENTER 0 0 1   NORTH 1 0 0
SILICON CARBIDE-4H(TYPE 111)
SPECIMEN-FILM DISTANCE = 3.00

SILICON CARBIDE(4H) - (001) - TRANSMISSION

Plate 3.   Comparison of experimental Laue Pattern with simulated Laue Pattern for an (001) oriented 4H-SiC crystal.

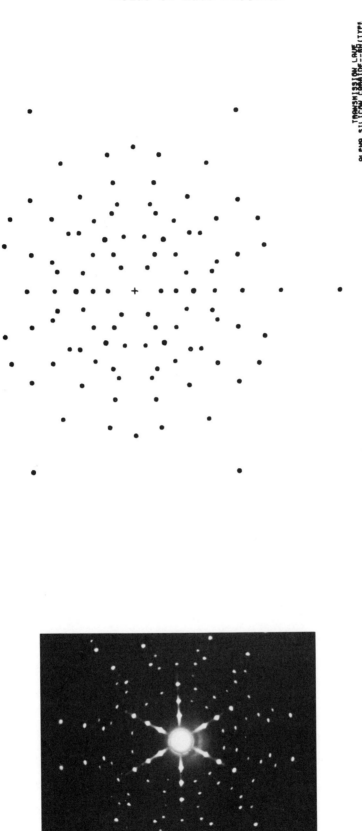

SILICON CARBIDE(6H) - (001) - TRANSMISSION

Plate 4.   Comparison of experimental Laue Pattern with simulated Laue Pattern for an (001) oriented 6H-SiC crystal.

Standard Transmission Laue Patterns for Polytype Determination

Y. Tung and J. W. Faust, Jr.

There are a number of techniques for differentiating between the various polytypes of SiC, at least as far as the matrix is concerned. These techniques include measuring of the angles between external faces using an optical goniometer, luminescence (both at liquid nitrogen and the thermoluminescence upon its warming up), etch pit studies, and various microscopic techniques. These determinations give generally, only the main polytype and no information about any buried polytypes. More detailed information can be obtained by use of electron and x-ray diffraction. Reflection techniques in either electron or x-ray diffraction give information mainly about the polytype at the surface. Transmission techniques are necessary to get full information about the sample. X-ray diffraction techniques that are used most often for polytype determination of SiC are the oscillation, rotation, Weissenberg and Debye-Scherrer. These techniques are slow and only a few crystals a day can be studied.

The Laue transmission patterns for the common polytypes differ considerably from one another making this a very attractive technique not only for the identification of a single polytype but also for identifying the indiviual polytypes in a mixed crystal. With conventional x-ray film, one crystal can be processed in approximately one hour. This time is, however, still too long for identification of many samples to be used for measuring properties or for device processing.

Hamilton (1) modified a Polaroid Land camera back to include a DuPont CB2 intensifying screen as reported by Schmidt and Spencer (2). This shortened the exposure time allowing him to examine many more crystals an hour by comparing them with standard patterns he had taken. Unfortunately there are no standard patterns in the literature to give this technique more widespread use. In this volume Posen (3) has reported on a computer program that gives printouts of patterns of pure polytypes making available for the first time a way of easily obtaining standard patterns for many polytypes. At the start of our program at USC, we had obtained pure polytypes of 4H, 6H, and 15R and a mixed crystal of 4H-6H polytypes from the Astronuclear Division of the Westinghouse Electric Corporation. We purchased a Polaroid system for x-ray crystallography and made standard patterns. Since then we have added to our collection and used this technique to identify many crystals of SiC from various sources. It is the purpose of this note to report on the use of this technique.

EXPERIMENTAL PROCEDURES

The Polaroid XR-7 System was mounted on one port of a standard Philips 35kv x-ray generator with a Cu target. The specimen to film distance was 3 cm. Polaroid film type 57 (ASA 3000) was used. Since the main faces of the platelets are {0001}, they can be quickly and easily attached directly to the front of the x-ray collimator by plasticene or some similar material. Exposure time depended on the thickness of the samples. Times of a few minutes were used for ordinary thickness while longer times were needed for very thin samples. Usually no surface preparation was needed; although samples were put in HF and then rinsed in distilled water to remove any possible oxide layers thicker than the normal one.

DISCUSSION AND RESULTS

The resolution of the Polaroid film is not quite as good as conventional x-ray film, making it more difficult to index the spots with a Dunn-Martin chart. This, however, is not necessary with a set of standard patterns, and the Polaroid film is quite adequate for this purpose. One merely takes his exposure and compares the pattern with the standards. By comparing all spots -- those along both the 6 principle poles and also the spots in between these poles, it is relatively easy to determine if the sample is a single polytype, and if so which one. If the sample is not a single polytype, the other polytype or polytypes can be identified, and if necessary the relative proportions of the polytypes can be estimated. Figure 1 shows a standard pattern of a 4H; Figure 2, a 6H and Figure 3, a 15R. One can easily see the marked difference in these patterns. A person can quickly learn to identify these patterns by their typical grouping of spots. Platelets of usual thickness, about a few hundred microns, could be processed in a few minutes; however, platelets that were thinner than 100 microns needed a longer exposure (the order of 30 to 60 minutes) to bring out the spots between the 6 principle poles.

Crystals that contain a second polytype result in the superposition of the two polytype patterns. This is illustrated in Figure 4 which gives the pattern for a crystal containing 6H and 15R. One can see, by comparing with the standard patterns, the superposition of the two standard patterns. Mixtures of the following polytypes: 3C, 4H, 6H and 15R can easily be distinguished. It is not possible to detect a 2H polytype in a matrix of 4H or 6H because of the high number of coincident spots; however, the 2H is rarely found in these platelets. Higher polytypes can be seen as "extraneous" spots but cannot be easily identified unless one has standard patterns for the higher polytypes. It is difficult to get a standard pattern for a 3C crystal because these crystals are usually twinned.

REFERENCES

(1) Hamilton, D. R., private communication.

(2) Schmidt, P. H., and Spencer, E. G., Rev. Sci. Instr. <u>35</u>, 957 (1964).

(3) Posen, H., This volume, page 246.

This work was partially supported by the Air Force Cambridge Research Laboratories under contract #F19628-72-C0136

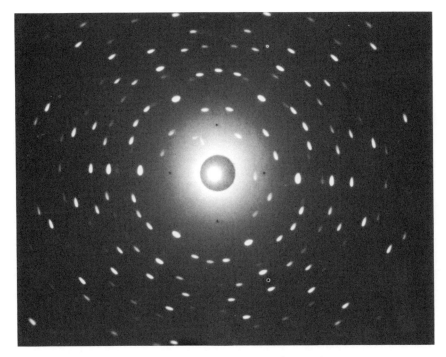

Figure 1.   Transmission Laue Patterns for pure 4H polytype.

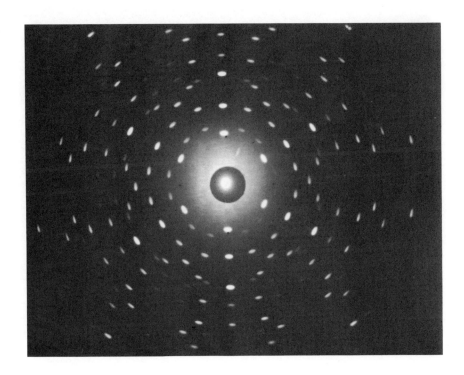

Figure 2.   Transmission Laue Pattern for pure 6H polytype.

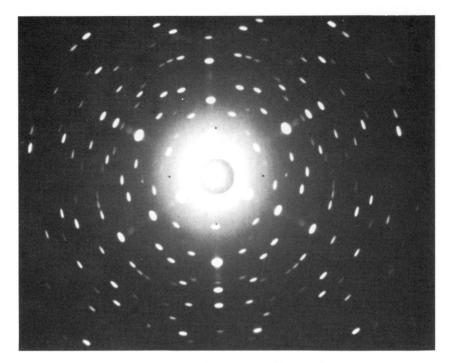

Figure 3.   Transmission Laue Pattern for pure 15R polytype.

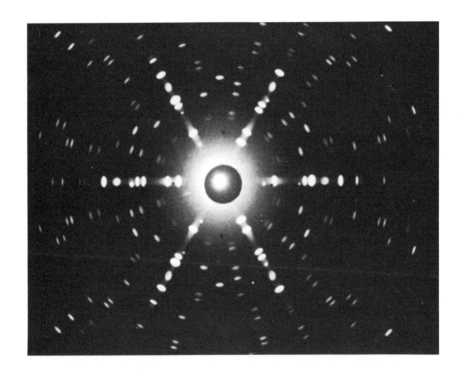

Figure 4.   Transmission Laue Pattern for a platelet containing both 6H and 15R.

PART III
PHYSICAL PHENOMENA

Review of Optical Work in SiC since 1968[*]

W. J. Choyke and Lyle Patrick

This review discusses a selected group of papers on SiC published since 1968. The topics are optical absorption, Raman scattering, dielectric constants, mobilities, and the luminescence of donor-acceptor pairs, radiation defects, isoelectric Ti and other impurities.

INTRODUCTION

This review is not intended to be comprehensive, but is limited to papers that have been of special interest to us. Thus, many good papers will go unmentioned. The eight section headings are the following:

1. Optical Absorption

2. Raman Scattering

3. Defect Luminescence

4. General Luminescence

5. Donor-Acceptor Spectra

6. The Spectrum of Isoelectronic Ti

7. Dielectric Constants

8. Polytype Dependence of Mobility

1. OPTICAL ABSORPTION

Only a few of many papers on the absorption of light in SiC are mentioned here. They are divided into three categories: (a) intrinsic electronic transitions, (b) intrinsic lattice absorption, and (c) impurity absorption. In category (c) we have restricted our review largely to absorption by N donors.

---

[*]Work supported in part by the U.S. Air Force Office of Scientific Research (AFSC) under Contract F44620-70-C-0111.

Intrinsic Electronic Absorption

Some very good work on the absorption edge of polytype 3C was done by two groups at the A.A. Zhdanov State University at Leningrad [1,2]. Their crystals were of good enough quality to permit the observation of much fine structure, and a modulation technique was used by the second group. They were able to report accurate values for the energy gap, the exciton binding energy, and the binding energy of an exciton to the donor N. A mechanism was suggested to explain dips in the absorption edge called "transparency lines".

Electro-absorption measurements of the indirect edge in 6H SiC were also done in Leningrad [3]. In this case the complexity of polytype 6H made it considerably more difficult to interpret all the structure observed.

A number of absorption measurements were made on thin samples in an attempt to find higher absorption edges. In 3C indirect edges were found at 3.55 and 4.2 eV [4a], and it was concluded that the first direct transition was near 6 eV. The 4.2 eV transition has been identified by Hemstreet and Fong [4b] as $\Gamma_{15v} - L_{1c}$. In 6H SiC, measurements to 4.9 eV showed three additional indirect transitions and a fourth that could not be positively identified as indirect [5]. Makarov extended the absorption in 6H to 5.8 eV by using a surface-barrier diode as detector [6].

Lattice Absorption

Raman scattering measurements showed that a single set of dispersion curves for the axial direction could be used for all SiC polytypes [7]. The crystal structure of a particular polytype determines which points on the dispersion curves are accessible to experimental observation. Many of the accessible phonons are weakly infrared active, and in the last few years many of them have been observed, as shown by a review of the Russian literature. In one paper [8], however, the phonon energies were incorrectly compared with phonon energies reported in luminescence measurements. It should be remembered that the latter reflect the positions of the conduction band minima, which do not fall on the axis of the Brillouin zone, hence phonons found in the luminescence have no place on the axial dispersion curves.

The most complete work was that by Dubrovskii and Radavanova [9], who made measurements on polytypes 4H, 6H, 15R and 8H. Phonon energies found in the first three agreed with the Raman results. No Raman measurements had been made on 8H,

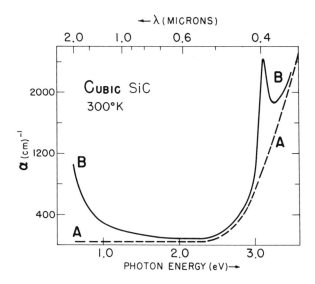

Fig. 1. Absorption spectra of 3C SiC crystals A (dashed) and B (solid) at
         300°K. Crystal A is relatively pure; crystal B is strongly n-type
         (perhaps 10^{19}/cm^3 donors). From [12].

but the infrared results for this polytype also fitted the set of axial
dispersion curves given in [7].

Impurity Absorption

The color of a nitrogen-doped SiC crystal depends on its polytype (e.g.,
green 6H and yellow 15R) because of dichroic absorption bands that were attributed to direct electronic transitions from the lowest conduction band to higher
bands [10]. The positions of the conduction band minima and the energies of the
higher bands both have a strong dependence on polytype, for the minima lie at the
zone boundaries [11]. Consequently, the Biedermann bands vary considerably from
one polytype to another, and their structure reflects the complexity of the
polytype band structure.

Recently the corresponding absorption band has been found in N-doped cubic
SiC [12] as shown in Fig. 1. Because 3C polytype has a relatively simple band
structure the N donors induce a single narrow isotropic absorption band at
3.1 eV, just within the weak indirect absorption edge. The transition can be
identified as $X_{1c} - X_{3c}$, and the energy is in good agreement with the calculated
value [4b,13]. Such a comparison between experimental and theoretical values is
not yet possible for the other polytypes.

The electronic transitions in the Biedermann bands may originate on free
electrons in the conduction band or on electrons bound to N donors. If these

two components could be separated, one could deduce the ionization energies of N donors. A new optical determination of these energies is highly desirable, for the optical values previously given are invalid, now that the spectrum once attributed to ionized N has been assigned instead to Ti (Sec. 6). A number of authors have recently attempted to separate the bound and free carrier peaks. Dubrovskii and Radovanova [14] studied the temperature dependence of a band at about 1.9 eV and saw evidence of the change from absorption by bound electrons at low temperatures to absorption by free electrons at high temperatures. Three other groups [15-17] have preferred to work with a different band in the range 1.33 to 1.39 eV. This band shows a structure that depends on temperature and on UV irradiation [17]. The possible temperature dependence of the interband separation may be a problem in interpreting the data, but all the papers give some evidence of one or more N binding energies of the order of 0.1 eV.

Finally, absorption measurements on B- and Al-doped 6H samples [18] showed rather broad absorption bands that were fitted to give $E_A(Al) = 0.27$ eV and $E_A(B) \sim 1.0$ eV, the latter being considerably higher than the values of E_A derived from electrical measurements.

2. RAMAN SCATTERING

Colwell and Klein [19] have investigated Raman scattering by electronic excitations in n-type SiC. The room-temperature Raman scattering of polytype 6H, reported earlier [20], was interpreted entirely in terms of lattice phonon scattering. However, at 6.4°K, as shown in Fig. 2, there are four new peaks in the 6H Raman data in N-doped crystals, and all have E_2 symmetry. Three of the peaks, at 13.0, 60.3, and 62.6 meV are evidently due to the $1s(A_1)$ to $1s(E)$ valley-orbit excitations of the three inequivalent N donors. The fourth was attributed to scattering by a localized mode.

The set of $1s(A_1)$ to $1s(E)$ energy values, one small and two large, is analogous to the set of exciton binding energies at N-donor sites, which are 16, 31, and 32 meV [21]. It suggests that there may be a similar set of N ionization energies, i.e., one small and two large. If so, the value of ~ 0.1 eV usually found in electrical measurements [22] would be largely due to the N site with the small ionization energy. Although a simple classification of the three sites describes one as having hexagonal symmetry and two as having cubic symmetry, it is not clear that this has any relationship to the ionization

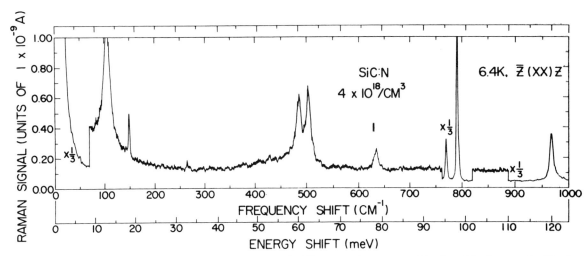

Fig. 2.   Raman spectrum of 6H SiC:N at 6.4°K. Lines at 13.0, 60.3, and 62.6 meV
are due to valley-orbit transitions of the N-donor electron at the
three inequivalent sites. From Colwell and Klein [19].

energies. A different mechanism, the Kohn-Luttinger interference effect, seems
to offer a better explanation of the differences [23].

The Raman results throw some light on the location of the 6H conduction-band
minima, for the E_2 symmetry of the excitations is predicted only for minima on
the M-L symmetry lines of the Brillouin zone. This is consistent with a
suggestion of Herman et al [13] at the last SiC conference, and also with the
calculations of Junginger and van Haeringen [24]. This placement on a line
leaves one undetermined parameter k_z. It has been suggested [23] that k_z might
be determined by ENDOR measurements, as a similar parameter was in Si, for the
Kohn-Luttinger interference effect, in SiC as in Si, results in a complex donor-
electron density pattern. Because good enough ENDOR results have not yet been
reported, an interpretation of the exciton binding energies was used to suggest
a position of the minima in [23]. The suggested k_z is one for which 6H SiC has
six conduction-band minima.

Colwell and Klein [19] also observed an interference between the 13 meV
electronic transition and an E_2 lattice mode at 18.6 meV. From a theory of the
interference effect they calculated an electron-phonon coupling constant. In a
sample with 6×10^{19} N/cm^3 they found plasmon coupling with the LO phonon. This
phenomenon was the subject of a theoretical and experimental study in another
paper [25].

3. DEFECT LUMINESCENCE

In the last five years the luminescence of defects in SiC has become a
subject of much interest because of the increasing use of ion implantation for
introducing donors and acceptors [26-28]. During implantation each heavy ion
displaces several thousand lattice atoms, and the lattice structure must be
restored by annealing if the electrical or luminescent effects of the defects
are not to overshadow those of the implanted ions. Hart et al [29] found, by
He or H backscattering, that the lattice damage is largely repaired by a 1200°C
anneal. However, Marsh and Dunlap [27] found that electron mobility in implanted
layers continues to increase with annealing up to 1700°C. The effect of residual
defects turns out to be even greater in the photoluminescence. After implanting
a variety of ions and then annealing at increasing temperatures, we found that
two defects, called D_1 and D_2, continue to dominate the luminescence up to our
highest annealing temperature of 1700°C.

We have been especially interested in detecting implanted ions by their
luminescence. After implantation of the light ions H and D, the low-temperature
luminescence is strong, and it is quite efficient because of the absence of
Auger recombination. It is the subject of a separate paper at this conference
[30]. However, after implanting heavy ions we have so far been unable to detect
luminescence due to the implanted ions themselves, perhaps because of the
dominance of the defect spectra. It may be possible to identify implanted
impurities indirectly through the luminescence of impurity-defect (ID) pairs
that anneal out below 1500°C, using a method to be described later. The D_1 and
D_2 spectra are independent of sample and implanted ion, whereas the ID pairs
are sample dependent. Thus, when the active impurities in the pairs have been
identified, these spectra may be used for impurity analysis. The following
three sections will give brief descriptions of the three kinds of defect
luminescence, D_1, D_2, and ID pairs.

The D_1 Spectrum

The severe lattice damage caused by the implantation of <u>heavy</u> ions broadens
the luminescence. Recovery of good line structure requires an anneal of perhaps
1000°C. The D_1 spectrum is already observed after the 1000°C anneal but
increases in intensity with annealing temperature to about 1300°C and persists
even after a 1700°C anneal [31,32]. The growth of a spectrum at high temperature

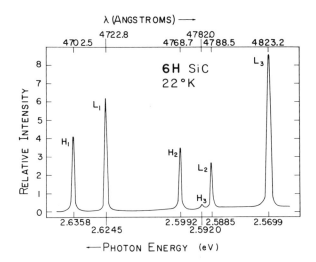

Fig. 3.   No-phonon lines of the 6H D_1 spectrum at an intermediate temperature
(22°K), showing both L and H, the low- and high-temperature forms.
The energy and wavelength are shown for each of the six lines.
From [32].

is evidence that it is not due to a <u>simple</u> lattice defect. Its independence of
implanted ion and sample show that no impurity is involved. A plausible model
for the D_1 center is some form of divacancy.

The D_1 spectrum has an unusual temperature dependence, with two abrupt
changes below 13°K in polytype 3C [31] and one change in 6H [32]. Figure 3
shows the no-phonon lines in 6H at 22°K, when both low-temperature (L) and high-
temperature (H) forms appear together. Because of the three inequivalent sites
in 6H, there are three lines for each form of the spectrum, as indicated by
subscripts 1 to 3. The shift in line positions with temperature is too large
for exciton j-j coupling (22.1 meV for L_3 to H_3), and it is thought that the
center undergoes a low-temperature distortion, sometimes called a pseudo-Jahn-
Teller effect. In both 3C and 6H polytypes the D_1 spectrum has a strong line
due to an 83 meV localized mode and much structure due to resonant modes, all
of which appears to be independent of temperature.

The high-temperature form of the D_1 spectrum was observed in 3C SiC in
electron-irradiated samples by Geiczy et al, who called it the "A" spectrum [33].
This spectrum was also reported in proton-bombarded samples and in crystals that
were quenched from 3000°C [34]. Makarov reported the D_1 spectrum in polytypes
4H, 6H, 21R, and 3C after neutron irradiation, and he commented on the phonon

structure [35]. Again, only the high-temperature form was found, at 80°K, because of experimental limitations.

The D_2 Spectrum

The D_2 center reported in polytypes 3C [36] and 15R [37] is another intrinsic defect for, like D_1, it is independent of implanted ion and sample. It is also a complex defect, for its intensity increased with annealing temperature to about 1550°C, and it persisted to our highest annealing temperature of 1700°C. It differs from D_1 in having a number of high-energy localized modes, ranging up to 164.7 meV, much higher than the lattice limit of 120.5 meV.

High-energy localized modes are generated by light atoms, and require the lattice force constants to be stiffened by interstitial atoms rather than to be weakened by vacancies. Thus, a plausible model for the D_2 center is the carbon di-interstitial. The D_2 spectrum requires a higher annealing temperature than does D_1, and our model suggests that this is because its appearance depends on the diffusion of interstitial C atoms. The D_2 spectrum does not have a high-temperature form and is thermally quenched below 77°K. It was therefore not observed by those who investigated radiation damage only above 77°K [33-35].

Impurity-Defect Pairs

Impurity-defect (ID) pairs are observed in implanted samples but are best examined in electron irradiated samples, where the minimal lattice damage allows the omission of the annealing operation. One then observes a number of lines that undergo progressive changes with annealing below 1000°C, where one has no interference from the D_1 and D_2 spectra. The original ID spectra anneal out at various temperatures while others grow with increasing annealing temperature or, in some cases, first appears only above a certain temperature. Very little remains of the ID spectra after annealing at 1200°C.

The ID spectra are sample dependent and are sometimes quite different in samples from different furnace runs. Both the original sample impurities and the electron-produced defects are needed to form the centers, hence the name impurity-defect pairs. The ID spectra can also be observed after proton bombardment, which does not damage the lattice enough to require an anneal.

An important problem is to identify the impurity in each ID spectrum. A possible solution is to introduce heavy ions by ion implantation, followed by

high-temperature annealing to repair the lattice, and then to reintroduce the
ID spectra by electron or proton bombardment. Thus, it may be possible to
identify an impurity by its ID luminescence even though it does not produce a
spectrum of its own.

Balona and Loubser [38] found seven electron-paramagnetic-resonance (EPR)
spectra in electron irradiated 3C and 6H SiC and studied their changes with
annealing. They proposed tentative models for the seven centers, of which only
one involved an impurity. The large number and the complexity of both the EPR
and optical spectra make it difficult to assign an EPR and an optical spectrum
to the same center. If the impurities play as big a role in these centers as
we believe, both EPR and optical experiments should be done on the same samples.

4.   GENERAL LUMINESCENCE

Work has continued on the use of SiC in electroluminescent diodes [39,40],
but it will not be discussed here, for it is the topic of separate papers at
this conference.

Cathodoluminescence was used to investigate the luminescence of surface
layers, for low-energy electrons have a small penetration distance. Vodakov
et al [40] cut diffused samples at a small angle to a p-n junction plane, thus
exposing several different regions for bombardment one at a time. In this way
they could examine the roles played by the several impurities, Al, O and B in
the broad-band luminescence.

A number of recent papers in the Russian literature report the luminescence
of samples activated by new impurities. Sc was introduced by a traveling-solvent
method [41] and imparted room-temperature luminescence to the grown crystal.
A p-n junction could also be formed by growing the n-SiC(Sc) on a p-SiC seed.
Further investigations have been carried out on the luminescence of Be [42], and
the role of O in p-n junction electroluminescence is the subject of current
research [43].

An interesting 3C SiC spectrum was reported for crystals under high
excitation by an Ar ion laser [44]. Many new lines were observed, in addition
to the usual N donor spectrum that appeared in the same crystals under low
excitation. One new series of lines had the same phonon structure as the N
spectrum but was displaced 1.5 meV to higher energy. The intensity of this
series increased with decreasing temperature, and it had a superlinear dependence

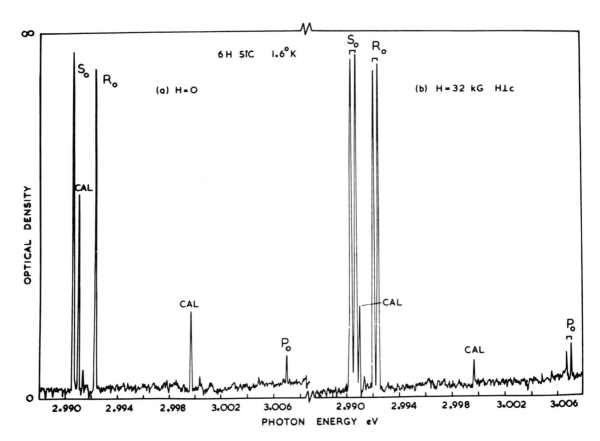

Fig. 4.   Portions of the low-temperature photoluminescence of 6H SiC close to
          the exciton energy gap (3.024 eV), (a) at zero magnetic field, (b) at
          H = 32 kG. For $\vec{H} \perp$ c each of the three no-phonon, bound-exciton lines
          P_o, R_o, and S_o splits into two. Lines marked CAL are from an Fe
          calibration lamp. From Dean and Hartman [45].

on the excitation level. It was attributed to the recombination of one of the
excitons in an exciton molecule (biexciton) bound to the N donor. Thus, the
final state of this transition is a single exciton bound to the N. It is also
possible for the recombining exciton to transfer enough energy to free and
dissociate the second exciton, thereby reducing the photon energy by a known
amount. Lines corresponding to such a process were also identified.

Dean and Hartman [45] did a magneto-optic study of the PRS spectrum in
6H SiC, a spectrum attributed to exciton recombination at neutral N donors on the
three inequivalent C sites [21]. Figure 4a shows these lines at 1.6°K in zero
magnetic field, and Fig. 4b shows them split in a field of 32 kG, with $\vec{H} \perp$ c.
Using the Thomas and Hopfield model developed for CdS [46], Dean and Hartman
showed that the g value of the electron can be derived from the H $\perp$ c data, and

that $g_e \sim 2$ for all three lines in Fig. 4b. A different splitting is found with H $\parallel$ c, from which a value of $g_h \sim 3.2$ can be deduced for the g value of the hole. The magneto-optic data is consistent with the postulate that the PRS lines arise from the decay of excitons bound to neutral donors, and Dean and Hartman conclude that it is reasonable to identify N as the optically active donor.

The two strongest lines, R and S, had previously been detected in absorption [47]. Wecker et al [48] more recently measured the absorption of R and S in a magnetic field, with H $\perp$ c, and found $g_e \sim 2.1$, in satisfactory agreement with the results of Dean and Hartman.

Hartman and Dean [49] had earlier made magneto-optic measurements on the luminescence spectrum attributed to exciton recombination at a neutral N donor in polytype 3C [50]. The experiment was motivated in part by the absence of an identifiable nitrogen electron-paramagnetic-resonance (EPR) signal in crystals with the N luminescence, including their own samples. However, the EPR spectrum of N in 3C SiC, showing the expected hyperfine splitting, has now been reported [51]. Hartman and Dean concluded that the luminescence spectrum was due to exciton recombination at a point defect consistent with a neutral donor. In view of the more recent EPR and 6H magneto-optic results, there seems little doubt that the donor is N. The measured g values were $g_e = 1.96$, $g_h(1/2) = 1.12$, $g_h(3/2) = 1.10$. Uniaxial stress measurements were made on the no-phonon line to obtain a deformation-potential constant, and the luminescence decay time was found to be 390 nsec at 5.5°K.

5. DONOR-ACCEPTOR SPECTRA

Donor-acceptor (DA) pairs were observed in 3C SiC as early as 1963 [50], but it appeared impossible to give a satisfactory analysis at that time. By 1970 experimental techniques had improved so that the fine structure was well resolved [52], as shown in Fig. 5, and the analytical techniques had improved so that the multipole splittings and the intensity cutoffs and enhancements could be interpreted [53].

Figure 5 shows part of a Type II spectrum, attributed to nitrogen on C sites and aluminum on Si sites. The lines were resolved, far beyond the portion shown, up to shell m = 80, and an extrapolation to infinite separation yielded an accurate value of the minimum photon energy, $h\nu_\infty = 2.0934$ eV. For the 3C

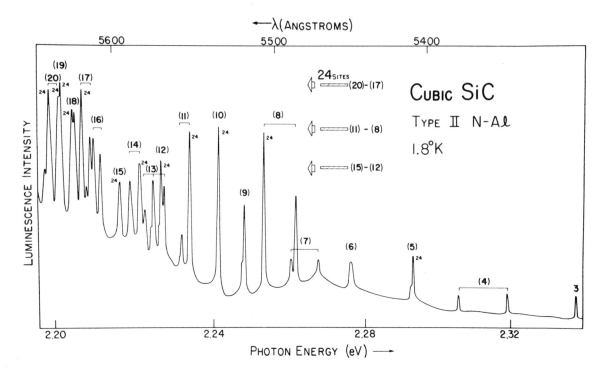

Fig. 5.   Portion of the Type II donor-acceptor spectrum of 3C SiC with N-Al
           pairs. Only some of the close pair lines are shown, with the bracketed
           shell number m ≤ 20. Many of the shells have well-resolved substructure
           components. Lines to which 24 equivalent pairs contribute are marked.
           Intensity levels for groups of these lines are shown, indicating where
           the intensity anomalies appear. From [52].

energy gap we can use the value E_G = 2.403 eV at 4.2°K [2]. From the difference
in these energies we obtain

$$E_G - h\nu_\infty = E_D(N) + E_A(Al) = 310 \text{ meV} ,$$

where E_D and E_A are the donor and acceptor ionization energies.

     The division of the 310 meV between E_D and E_A is open to two interpretations.
In [52] we analyzed the two intensity anomalies shown in Fig. 5 to obtain a
value of $E_D(N)$ = 118 meV. However, in 1968 Zanmarchi [54] had shown that above
77°K the DA spectrum is weak and the luminescence is largely due to what he
called the A spectrum. This he attributed to free-to-bound transitions, i.e.,
free electrons recombining with holes bound to the deep Al acceptors. We also
saw the A spectrum, with its no-phonon peak A_0 at 2.153 eV, but we remarked
that Zanmarchi's interpretation was inconsistent with ours. However, recent
work [55] has verified the free-to-bound mechanism in GaP, which, like 3C SiC,
has the zincblende structure and conduction band minima at X. The various

luminescence mechanisms in GaP and 3C SiC are very similar, and the temperature dependence of the free-to-bound transitions in GaP is like that observed for the A_0 peak. Furthermore, a similar peak has now been reported in 3C SiC containing nitrogen-boron DA pairs [56], apparently due to recombination of free electrons with holes bound to the deep B acceptors at 77°K. The N-B pair spectrum is discussed below. Because of these developments, we now believe the free-to-bound interpretation of peak A_0 is correct, even though it leaves the intensity anomalies of Fig. 5 unexplained.

The energy of the A_0 peak is expected to increase with temperature up to 100°K or more, for

$$E(A_0) = E_G - E_A(A1) + ckT \tag{1}$$

and the kinetic energy of the electrons (ckT) increases with temperature faster than E_G decreases at low temperatures. The constant c in Eq. 1 is determined by the energy dependence of the capture cross-section. For an energy independent cross-section c = 1, and for the GaP example reported in [55] the Fig. 12 plot showed c = 1.37. With E_G = 2.402 eV at 77°K for 3C SiC, the deduced value of $E_A(A1)$ is 256 meV if c = 1, and 258 meV if c = 1.37. We shall take the average value of 257 meV for $E_A(A1)$, and by subtraction from 310 meV we obtain $E_D(N)$ = 53 meV.

An increase in the energy of the A_0 peak was found by Long et al [57] in the temperature range 77 to 105°K. However, their correction for the change of E_G with temperature was excessive, for this correction is itself temperature independent and is quite small near 77°K. More accurate values of the dependence of E_G on temperature are needed before the constant c of Eq. 1 can be evaluated.

In an earlier paper [58] Long et al also showed a line spectrum in 3C SiC that they attributed to N-Al pairs. Their measurements had to be made above 4.2°K to avoid interference with a strong N spectrum, hence thermally excited lines were present and positive identification of the shell numbers was difficult. We have not been able to compare it with our spectrum in detail.

The DA spectrum also has a broad peak of unresolved lines due to distant pairs, which we called the "band" spectrum in [52]. The shell number at the maximum of this spectrum depends on excitation intensity, falling in the range 360 > m > 210 in our work. This band is often observed in samples in which the impurity density is too great for the appearance of well-resolved lines at small m. It was observed by Zanmarchi [54], and has since been studied by Long et al

[57], who showed how the band width and peak wavelength depend on the delay time of the measurement following excitation.

In the DA spectrum due to N-B pairs in 3C SiC [56] no resolved line structure was observed, but the band of unresolved lines was attributed to DA pairs because it shifted to higher energy with increased excitation intensity. A lower limit of 639 meV was deduced for the ionization energy of the boron acceptor $E_A(B)$. If Kuwabara et al had used $E_D(N) = 53$ meV instead of our earlier value of 118 meV, they would have obtained a lower limit of more than 700 meV for $E_A(B)$. This value seems high, but, as mentioned in Sec. 1, an absorption experiment gave $E_A(B) \sim 1$ eV in 6H SiC [18].

In 6H SiC, DA spectra due to N-Al pairs have been observed, and are frequently of good quality, with many lines. However, the reduced symmetry of 6H leads to additional shell splittings, and the presence of three inequivalent sites for each donor and acceptor leads to a multiplicity of nine in the number of DA spectra. A complete analysis of such a spectrum is a difficult task.

6. THE SPECTRUM OF ISOELECTRONIC Ti

One of the most prominent spectra in 6H SiC, the ABC spectrum, was attributed to exciton recombination at ionized N donors [59]. Much work has been done on this spectrum in the last five years, and we now think it should be attributed to Ti as an isoelectronic substituent for Si [60]. A similar spectrum was also found in polytypes 15R, 4H, and 33R, and we now attribute all of these to Ti. The earliest doubts about the role of N followed the discovery of the spectrum in 4H SiC [61], for the deduced value of the N donor ionization energies E_i were unreasonably high, yet there was little doubt that the spectrum was due to the same center as the ABC spectrum in 6H [62]. For example, a characteristic 90 meV localized mode is a prominent feature of both.

The most revealing experimental work was the magneto-optical study of the ABC spectrum by Dean and Hartman [45]. Their conclusion was that the luminescence center was complex. That view was disputed by Patrick [63], who interpreted their data to show that the luminescence could be attributed to a substitutional atom, although the center was not a donor, hence not N. Another observation by Dean and Hartman has proved valuable in identifying the center. While making the magneto-optical study, using very high resolution, they observed two satellite lines on either side of the A, B, and C lines. They wrote -- "An

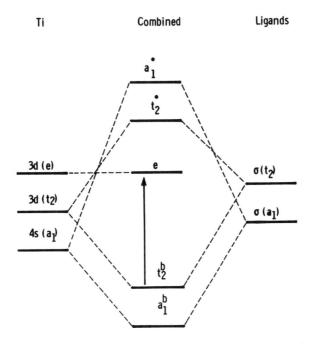

Fig. 6.   Schematic drawing, showing the combination of Ti and ligand states to
          form bonding (a_1^b and t_2^b), non-bonding (e), and antibonding states
          (a_1^* and t_2^*). Degenerate levels are shown separated for clarity. In
          the ground state the bonding levels are all occupied. It is suggested
          that exciton capture is equivalent to the transition indicated by the
          arrow, and luminescence to the reverse transition. From [60].

obvious possibility is that the satellites arise from isotope shifts of the
no-phonon line. However, we have not been able to find a satisfactory interpretation in terms of isotope shifts for plausible impurities". We now attribute
these lines to the presence of the five Ti isotopes. Ti is a common impurity in
SiC.

     Ti can be considered as an isoelectronic substituent for Si. Ligand field
theory shows that the d^2s^2 configuration of Ti is capable of combining with the
sp^3 bonding structure of the four C neighbors. The resulting bonding, anti-
bonding, and non-bonding levels are shown schematically in Fig. 6. It is doubtful
that Ti differs enough from Si in size and electronegativity to be a hole trap in
the usual sense of an isoelectronic center [64]. However, as Fig. 6 illustrates,
the d levels split, in T_d symmetry, into t_2 and e levels. It is the t_2 levels
that join in forming the bonding structure, leaving the non-bonding e levels
empty. If the e levels fall within the energy gap, they make the center an

electron trap. After an electron is captured, a hole may then be bound by the
Coulomb force, as in the conventional isoelectronic center. The most compelling
reason for choosing Ti is, of course, that Ti has five isotopes whose abundances
are consistent with the intensities of the main line and the four satellite lines
observed by Dean and Hartman.

In our model an exciton capture is equivalent to promoting a bonding
electron into the empty e level, as indicated by the arrow in Fig. 6, and the
luminescence is due to the reverse transition. The electron is tightly bound in
the d shell, in a state that is not derived from the conduction band. Thus, the
level is nearly independent of polytype, unlike the energy gap differences that
depend on the positions of the conduction band minima. That is why the luminescence lines fall in the same energy range in 4H and 6H, even though the 4H energy
gap is 0.24 eV greater than that of 6H. It also explains the absence of the
spectrum in the small-gap polytypes 3C and 21R, for if the energy gap is too
small the e levels are resonant with the conduction band levels, and the Ti atom
cannot then function as an electron trap.

The assignment of the ABC spectrum to Ti removes a conflict between optical
and electrical values of the N donor ionization energies E_i, for the former no
longer exist. Hagen and Kapteyns [22] made careful measurements on good n-type
6H samples and found $E_i < 0.1$ eV, considerably less than any of the three E_i
attributed to N donors in [59]. Although three different values of E_i must
coexist in polytype 6H, it is understandable that they are not resolved in
electrical measurements. Later Hagen and van Kemanade [65a] grew 6H SiC crystals
doped with the isotope N^{15} and found that the ABC spectrum was unchanged, thereby
showing that it is not a N spectrum. Still more recently, trap depths near
0.1 eV were found in thermoluminescence, and were attributed to N [65b].

In addition to resolving the conflict with electrical measurements, the
assignment of the spectrum to a center other than an ionized donor resolves a
conflict with Hopfield's theory of exciton binding [66]. Hopfield showed that an
exciton is bound to an ionized donor only when the electron-hole mass ratio is
small, and then only with a very small binding energy. Since no other spectrum
has been found that can be attributed to the ionized N donor, it seems likely
that the bound state does not exist.

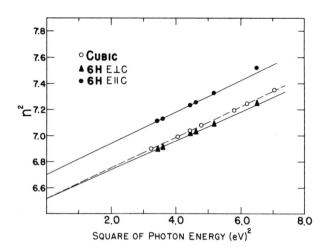

Fig. 7. Extrapolation of the square of the refractive index (n^2) to zero

photon energy to obtain ε_∞.

Refractive index values are from [68] for 3C, and from [69] for both

ordinary ($\vec{E} \perp c$) and extraordinary ($\vec{E} \parallel c$) rays of 6H. From [67].

7. DIELECTRIC CONSTANTS

More accurate values of 3C and 6H static dielectric constants were obtained
[67] by using extrapolated refractive index measurements and the Lyddane–Sachs–
Teller (LST) relation. Figure 7 shows the square of the refractive index plotted
against the square of the photon energy and extrapolated to zero photon energy
to obtain values of ε_∞, the "optical" dielectric constant. This extrapolation
is based on a simple one-oscillator model that works well in practice. These
values of ε_∞ and certain phonon energies measured by Raman scattering [7,20] were
then used in the LST relation to derive the static dielectric constants ε_s. The
phonon energies and dielectric constants are listed in Table I.

TABLE I. Extrapolated ε_∞ from Fig. 7, Raman
phonon energies from [7] and [20], and calculated
static dielectric constants ε_s for 3C and 6H SiC.

Polytype	ε_∞	$\hbar\omega_L(cm^{-1})$	$\hbar\omega_T(cm^{-1})$	ε_s
3C	6.52	972	796	9.72
6H($\perp$)	6.52	970	797	9.66
6H($\parallel$)	6.70	964	788	10.03

The 3C refractive index values in Fig. 7 were from recent measurements by Shaffer and Naum [68]. The 6H values were from old measurements by Thibault [69], but the same extrapolated ε_∞ is obtained by using the more recent measurements by Shaffer [70], who also reported refractive index values for polytypes 15R and 4H.

Golightly [71] measured the birefringence of twelve SiC polytypes and plotted birefringence against hexagonality, as was done earlier for ZnS polytypes [72]. He also observed the dichroism due to the Biedermann bands in n-type crystals and listed the colors for all twelve polytypes.

Powell [73] measured the refractive indices and birefringence of polytype 2H. He was able to make the tiny 2H crystals into wedges and he succeeded in obtaining accurate values of both ordinary and extraordinary refractive indices over a range of wavelengths. He also discussed the polytype dependence of the birefringence and showed that 2H did not fall on the line plotted by Golightly.

Singh et al [74] reported values of the second harmonic coefficients, and discussed the possibility of using hexagonal SiC in optical parametric oscillators. Their measurements were on "hexagonal crystals of unknown polytype".

8. POLYTYPE DEPENDENCE OF MOBILITY

Much of the optical work has been concerned with discovering and correlating polytype differences. Thus, we are very interested in the topic of this section, although the results are derived from electrical rather than optical measurements.

At the last SiC conference it was suggested that the electrical properties of SiC should be polytype dependent for n-type samples, because the conduction band minima are at the zone edge [11]. On the other hand, these properties should be largely independent of polytype for p-type samples, for the valence band maximum is at the zone center. Barrett and Campbell [75] had found considerable difference in the electron mobilities of three polytypes, with increasing mobility in the order 6H, 15R, and 4H. Now Lomakina et al [76] have confirmed this polytype order for mobility and have found the reverse order to hold for N donor activation energies. Thus, the order of increasing mobility is a consequence of the order of decreasing electron mass in these three polytypes. The same paper also reported that there were no differences between p-type samples of these polytypes in the Al acceptor activation energy or in the hole mobility.

Later, the effect of annealing the crystals at 2300°C on electron mobility
was investigated by a group that included most of the same authors [77]. Some
of the crystals showed significantly increased electron mobility after annealing,
reaching values of 1800, 2400, and 3100 cm^2/V sec at 100°K for 6H, 15R and 4H
respectively. These are the largest SiC mobility values published to date. At
300°K the 6H, 15R and 4H mobilities were 330, 500, and 700 cm^2/V sec, which may
be regarded as near optimum room temperature values, for in these samples the
lattice scattering is dominant at 300°K. The increase in mobility with high-
temperature annealing was attributed to the formation of neutral complexes from
pairs of donors and acceptors, for the number of compensating acceptors was found
to decrease. Thus, the introduction of a neutral center by annealing reduced
the amount of charged impurity scattering.

REFERENCES

1. D.S. Nedzvetskii, B.V. Novikov, N.K. Prokof'eva, and M.B. Reifman, Fiz.
    Tekh. Poluprov. 2, 1089 (1968). [Sov. Phys. Semiconductors 2,
    914 (1969)].

2. V.A. Kiselev, B.V. Novikov, M.M. Pimonenko, and E.B. Shadrin, Fiz. Tverd.
    Tela 13, 1118 (1971). [Sov. Phys. Solid State 13, 926 (1971)].

3. G.B. Dubrovskii and V.I. Sankin, Fiz. Tverd. Tela 14, 1200 (1972). [Sov.
    Phys. Solid State 14, 1024 (1972)].

4a W.J. Choyke and Lyle Patrick, Phys. Rev. 187, 1041 (1969).

4b L.A. Hemstreet, Jr., and C.Y. Fong, Solid State Comm. 9, 643 (1971); Phys.
    Rev. B 6, 1464 (1972).

5. W.J. Choyke and Lyle Patrick, Phys. Rev. 172, 769 (1968).

6. V.V. Makarov, Fiz. Tekh. Poluprov. 6, 1805 (1972). [Sov. Phys. Semi-
    conductors 6, 1556 (1973)].

7. D.W. Feldman, J.H. Parker, Jr., W.J. Choyke, and Lyle Patrick, Phys.
    Rev. 173, 787 (1968).

8. M.A. Il'in, E.M. Karshtedt, and E.P. Rashevskaya, Fiz. Tekh. Poluprov. 6,
    2230 (1972). [Sov. Phys. Semiconductors 6, 1877 (1973)].

9. G.B. Dubrovskii and E.I. Radovanova, Fiz. Tverd. Tela 14, 2456 (1972).
    [Sov. Phys. Solid State 14, 2127 (1973)].

10. E. Biedermann, Solid State Comm. 3, 343 (1965).

11. Lyle Patrick, Materials Res. Bull. 4, S129 (1969).

12. Lyle Patrick and W.J. Choyke, Phys. Rev. 186, 775 (1969).

13. F. Herman, J.P. Van Dyke, and R.L. Kortum, Mat. Res. Bull. 4, S167 (1969).

14. G.B. Dubrovskii and E.I. Radovanova, Fiz. Tverd. Tela 11, 680 (1969). [Sov. Phys. Solid State 11, 545 (1969)].

15. O.V. Vakulenko and O.A. Govorova, Fiz. Tverd. Tela 12, 1857 (1970). [Sov. Phys. Solid State 12, 1478 (1970).

16. I.S. Gorban', V.P. Zavada, and A.S. Skirda, Fiz. Tverd. Tela 14, 3095 (1973). [Sov. Phys. Solid State 14, 2652 (1973)].

17. M.P. Lisita, O.V. Vakulenko, Yu.S. Krasnov, and V.N. Solodov, Fiz. Tekh. Poluprov. 5, 2047 (1971). [Sov. Phys. Semiconductors 5, 1785 (1972)].

18. O.V. Vakulenko and O.A. Govorova, Fiz. Tverd. Tela 13, 633 (1971). [Sov. Phys. Solid State 13, 520 (1971)].

19. P.J. Colwell and M.V. Klein, Phys. Rev. B 6, 498 (1972).

20. D.W. Feldman, J.H. Parker, Jr., W.J. Choyke, and Lyle Patrick, Phys. Rev. 170, 698 (1968).

21. W.J. Choyke and Lyle Patrick, Phys. Rev. 127, 1868 (1962).

22. S.H. Hagen and C.J. Kapteyns, Philips Res. Repts. 25, 1 (1970).

23. Lyle Patrick, Phys. Rev. B 5, 2198 (1972).

24. H.G. Junginger and W. van Haeringen, Phys. Status Solidi 37, 709 (1970).

25. M.V. Klein, B.N. Ganguly, and P.J. Colwell, Phys. Rev. B 6, 2380 (1972).

26. H.L. Dunlap and O.J. Marsh, Appl. Phys. Letters 15, 311 (1969).

27. O.J. Marsh and H.L. Dunlap, Rad. Effects 6, 301 (1970).

28. A. Addamiano, W.H. Lucke, and J. Comas, J. Luminescence 6, 143 (1973).

29. R.R. Hart, H.L. Dunlap, and O.J. Marsh, in Radiation Effects in Semiconductors, (Gordon and Breach, N.Y., 1971) p. 405.

30. Lyle Patrick and W.J. Choyke, this conference.

31. W.J. Choyke and Lyle Patrick, Phys. Rev. B 4, 1843 (1971).

32. Lyle Patrick and W.J. Choyke, Phys. Rev. B 5, 3253 (1972).

33. I.I. Geiczy, A.A. Nesterov, and L.S. Smirnov, in Radiation Effects in Semiconductors, (Gordon and Breach, N.Y., 1971) p. 327.

34. A.A. Nesterov, V.D. Gurko, and L.S. Smirnov, Fiz. Tekh. Pol. 6, 1292 (1972). [Sov. Phys. Semiconductors 6, 1130 (1973)].

35. V.V. Makarov, Fiz. Tverd. Tela 13, 2357 (1971). [Sov. Phys. Solid State 13, 1974 (1972)]; Fiz. Tverd. Tela 9, 596 (1967). [Sov. Phys. Solid State 9, 457 (1967)].

36. Lyle Patrick and W.J. Choyke, J. Phys. Chem. Solids 34, 565 (1973).

37. W.J. Choyke and Lyle Patrick, in Proc. International Conference on Defects in Semiconductors, Reading, 1972, p. 218.

38. L.A. DeS.Balona and J.H.N. Loubser, J. Phys. C 3, 2344 (1970).

39. R.M. Potter, J.M. Blank, and A. Addamiano, J. Appl. Phys. 40, 2253 (1969).

40. Yu.A. Vodakov, G.F. Kholuyanov, and E.N. Mokhov, Fiz. Tech. Poluprov. 5, 1615 (1971). [Sov. Phys. Semiconductors 5, 1409 (1972)].

41. Kh. Vakhner and Yu.M. Tairov, Fiz. Tverd. Tela 11, 2440 (1969). [Sov. Phys. Solid State 11, 1972 (1970)].

42. A.A. Kal'nin, B.I. Seleznev, and Yu.M. Tairov, Fiz. Tverd. Tela 12, 3215 (1970). [Sov. Phys. Solid State 12, 2599 (1971).

43. E.E. Violin, Yu.M. Tairov, and O.A. Fayans, Fiz. Tech. Poluprov. 6, 2301 (1972). [Sov. Phys. Semiconductors 6, 1941 (1973)].

44. B.V. Novikov and M.M. Pimonenko, Fiz. Tverd. Tela 13, 2777 (1971). [Sov. Phys. Solid State 13, 2323 (1972)].

45. P.J. Dean and R.L. Hartman, Phys. Rev. B 5, 4911 (1972).

46. D.G. Thomas and J.J. Hopfield, Phys. Rev. 128, 2135 (1962).

47. I.S. Gorban' and A.P. Krokhmal', Fiz. Tverd. Tela 12, 905 (1970). [Sov. Phys. Solid State 12, 699 (1970)].

48. C. Wecker, M. Certier, S. Nikitine, and L. Dietrich, Phys. Stat. Solidi (b) 50, K81 (1972).

49. R.L. Hartman and P.J. Dean, Phys. Rev. B 2, 951 (1970).

50. W.J. Choyke, D.R. Hamilton, and Lyle Patrick, Phys. Rev. 133, A1163 (1964).

51. Yu.M. Altaiskii, I.M. Zaritskii, V.Ya. Zevin, and A.A. Konchits, Fiz. Tverd. Tela 12, 3036 (1970). [Sov. Phys. Solid State 12, 2453 (1971)].

52. W.J. Choyke and Lyle Patrick, Phys. Rev. B 2, 4959 (1970).

53. P.J. Dean, in Progress in Solid State Chemistry, (Pergamon, New York, 1973), Vol. 8, p. 1.

54. G. Zanmarchi, J. Phys. Chem. Solids 29, 1727 (1968).

55. R.Z. Bachrach and O.G. Lorimor, Phys. Rev. B 7, 700 (1973).

56. H. Kuwabara, S. Shiokawa, and S. Yamada, Phys. Stat. Solidi (a) 16,
     K67 (1973).

57. N.N. Long, D.S. Nedzvetskii, N.K. Prokofeva, and M.B. Reifman, Opt.
     Spektrosk. 30, 306 (1971). [Optics and Spectroscopy 30, 165 (1971)].

58. N.N. Long, D.S. Nedzvetskii, N.K. Prokofeva, and M.B. Reifman, Opt.
     Spektrosk. 29, 727 (1970). [Optics and Spectroscopy 29, 388 (1970)].

59. D.R. Hamilton, W.J. Choyke, and Lyle Patrick, Phys. Rev. 131, 127 (1963).

60. Lyle Patrick and W.J. Choyke, unpublished.

61. W.J. Choyke, Lyle Patrick, and D.R. Hamilton, in Proceedings of the
     Seventh International Conference on the Physics of Semiconductors,
     (Dunod, Paris, 1964), p. 751.

62. I.S. Gorban' and V.A. Gubanov, Fiz. Tverd. Tela 13, 2076 (1971). [Sov.
     Phys. Solid State 13, 1741 (1972)].

63. Lyle Patrick, Phys. Rev. B 7, 1719 (1973).

64. A. Baldereschi and J.J. Hopfield, Phys. Rev. Letters 28, 171 (1972).

65a S.H. Hagen and A.W.C. van Kemanade, J. Luminescence 3, 131 (1973).

65b A. Halperin, E. Zacks, and E. Silberg, J. Luminescence 6, 304 (1973).

66. J.J. Hopfield, in Proceedings of the Seventh International Conference on
     the Physics of Semiconductors, (Dunod, Paris, 1964), p. 725.

67. Lyle Patrick and W.J. Choyke, Phys. Rev. B 2, 2255 (1970).

68. P.T.B. Shaffer and R.G. Naum, J. Opt. Soc. Am. 59, 1498 (1969).

69. N.W. Thibault, Am. Mineral. 29, 327 (1944).

70. P.T.B. Shaffer, Appl. Optics 10, 1034 (1971).

71. J.P. Golightly, Can. Mineral. 10, 105 (1969).

72. O. Brafman and I.T. Steinberger, Phys. Rev. 143, 501 (1966).

73. J.A. Powell, J. Opt. Soc. Am. 62, 341 (1972).

74. S. Singh, J.R. Potopowicz, L.G. Van Uitert, and S.H. Wemple, Appl. Phys.
     Letters 19, 53 (1971).

75. D.L. Barrett and R.B. Campbell, J. Appl. Phys. 38, 53 (1967).

Review of Optical Work

283

76. G.A. Lomakina, Yu.A. Vodakov, E.N. Mokhov, V.G. Oding, and G.F. Kholuyanov, Fiz. Tverd. Tela 12, 2918 (1970). [Sov. Phys. Solid State 12, 2356 (1971)].

77. G.A. Lomakina, G.F. Kholuyanov, R.G. Verenchikova, E.N. Mokhov, and Yu.A. Vodakov, Fiz. Tekh. Poluprov. 6, 1133 (1972). [Sov. Phys. Semiconductors 6, 988 (1972)].

Recent Band Structure Calculations of Cubic and Hexagonal Polytypes of Silicon Carbide

L. A. Hemstreet and C. Y. Fong

The purpose of this paper is to discuss recent investigations of the energy band structure and optical properties of cubic and hexagonal polytypes of silicon carbide. Since the large number of atoms per unit cell of the more complex structures exceedingly complicates the problem, we shall focus our attention on the two simplest modifications, 3CSiC and 2HSiC, which exhibit zincblende and wurtzite symmetry, respectively, with the hope that any insight gained from their study will carry over to the more complicated cases.

The calculations to be reported here were done using the formalism of the empirical pseudopotential method (EPM). Since an excellent review of this method already appears in the literature,[1] we shall merely outline the procedure. The weak pseudopotential $V_p(\vec{r})$ is expanded in terms of a finite set of reciprocal lattice vectors $|\vec{G}| < G_o$, where G_o is some cut-off reciprocal lattice vector (RLV) length chosen to insure adequate convergence of the eigenvalues:

$$V_p(\vec{r}) = \sum_{j\lambda} v_p(\vec{r} - \vec{R}_j - \vec{\tau}_\lambda) = \sum_{|\vec{G}| < G_o} V(\vec{G}) e^{i\vec{G}\cdot\vec{r}} \qquad (1)$$

For the case of binary compounds the Fourier coefficients can be factored as follows:

$$V(\vec{G}) = V^S(|\vec{G}|)S^S(\vec{G}) + iV^A(|\vec{G}|)S^A(\vec{G}) \qquad (2)$$

where

$$S^S(\vec{G}) = \frac{1}{n} \sum_\lambda e^{-i\vec{G}\cdot\vec{\tau}_\lambda} \qquad (3)$$

and

$$S^A(\vec{G}) = \frac{-i}{n} \sum_\lambda P_\lambda e^{-i\vec{G}\cdot\vec{\tau}_\lambda} \qquad (3a)$$

are called the symmetric and antisymmetric structure factors, respectively, and
depend only on the positions of the atoms in the unit cell, given by the
$\vec{\tau}_\lambda$, and where

$$2v^S(|\vec{G}|) = \frac{n}{\Omega_o} \int e^{-i\vec{G}\cdot\vec{r}} [v_1(\vec{r}) + v_2(\vec{r})]_d{}^3r = v_1(|\vec{G}|) + v_2(|\vec{G}|) \qquad (4a)$$

and

$$2v^A(|\vec{G}| = \frac{n}{\Omega_o} \int e^{-i\vec{G}\cdot\vec{r}} [v_1(\vec{r}) - v_2(\vec{r})]_d{}^3r = v_1(|\vec{G}|) - v_2(|\vec{G}|) \qquad (4b)$$

are the symmetric and antisymmetric form factors, which are independent of the
positions of the atoms. In these equations, $P_\lambda = +1$ if $\vec{\tau}_\lambda$ locates an atom of
Type 1, $P_\lambda = -1$ for atoms of Type 2, n represents the number of atoms per unit
cell, and Ω_o is the unit cell volume.

Once the form factors $v^S(G)$ and $v^A(G)$, $G < G_o$, are specified, the energy
bands of the solid are determined by the secular equation

$$\det |\tilde{H}_{gg'} - E\,\delta_{gg'}| = 0 \qquad (5)$$

where

$$\tilde{H}_{gg'} = H_{gg'} + \sum_{G_1 \leq |\vec{h}| < G_2} \frac{H_{gh}H_{hg}}{E - H_{hh}} \qquad (6)$$

and

$$H_{gg'} = \frac{p^2}{2m} + v^S(|\vec{g} - \vec{g}'|)S^S(\vec{g} - \vec{g}') + iv^A(|\vec{g} - \vec{g}'|)S^A(\vec{g} - \vec{g}') \qquad (7)$$

and where we have expanded the pseudowave-function $\phi_{nk}(\vec{r})$ over a basis of plane
wave states $|\vec{k} + \vec{g}\rangle$, with $|\vec{g}| < G_1$, and have applied a form of pertubation theory
first introduced by Lowdin.[2]

In the EPM the form factors are determined by requiring that the resulting
energy band structure of the solid be consistent with the available experimental
data (primarily optical data in the case of semiconductors). This is accomplished as follows: one first determines an initial set of form factors by some
means and uses these to calculate the energy levels at selected points of high
symmetry in the Brillouin zone (BZ). Various gaps at these points are then
compared to those values extrapolated from the experimental data. If the agreement is not considered totally satisfactory, the initial set of form factors is
adjusted and a new set of gaps are calculated. This process is repeated until a
consistent set of theoretical and experimental energy gaps emerges. Clearly,
the success of this method depends on the reliability and availability of the
experimental data. In those cases where good experimental data exists, the

EPM should provide very accurate band structures. In addition, the EPM can be used very successfully to calculate the electronic structure of many solids for which there is insufficient experimental information available, but which contain component atoms for which pseudopotential form factors have previously been determined. With proper scaling to account for differences in lattice constant and screening between the various structures, the pseudopotentials can be thought of as properties of the atoms involved, and the form factors can be transferred from one structure to the next.

A case in point are the polytypes of silicon carbide. In this study the cut-off RLV length of 3CSiC was chosen to be $G_O = \sqrt{12}$ (in units of $2\pi/a_{ZB}$, where $a_{ZB} = 4.35\mathring{A}$ is the lattice constant of 3CSiC). This choice requires three symmetric and three antisymmetric form factors, which were determined by scaling the silicon form factors of Brust[3] and the carbon form factors of the present authors[4] and combining them according to eq. 4. These six parameters were then allowed to vary slightly in order to yield gaps in better agreement with the experimental data.

The case of 2HSiC is more difficult. Since there are more atoms per unit cell it was necessary to choose $G_O = \sqrt{8}$ (in units of $2\pi/a_w$, $a_w = a_{ZB}/\sqrt{2} = 3.08\mathring{A}$). In units of $2\pi/a_{ZB}$, this is equivalent to $G_O = \sqrt{16}$. This choice requires that ten symmetric and nine antisymmetric form factors be specified. At the same time there is only one reliable piece of experimental datum available, a value of 3.33 eV for the indirect gap determined via luminescence and lattice absorption measurements.[5] Clearly, one cannot hope to use one piece of datum to adjust nineteen parameters. However, one can use the similarity in the two structures to get around this problem.

Figure 1 shows the arrangement of the atoms in the zincblende and wurtzite structures. Here the z axis is directed along the c axis of wurtzite and the [111] direction in zincblende. The wurtzite structure can be derived from zincblende by rotating the tetrahedra in the zincblende layers by 60° about the [111] direction relative to one another. Each atom retains its tetrahedral coordination and the nearest neighbor distance remains the same. In addition nine of the twelve next nearest neighbors remain the same. Because of this close similarity of the two structures one can, as a first approximation, assume that the form factors $V^S(|\vec{G}|)$ and $V^A(|\vec{G}|)$, as a function of $|\vec{G}|$, will be the same for the two polytypes. Since the wurtzite lattice has different

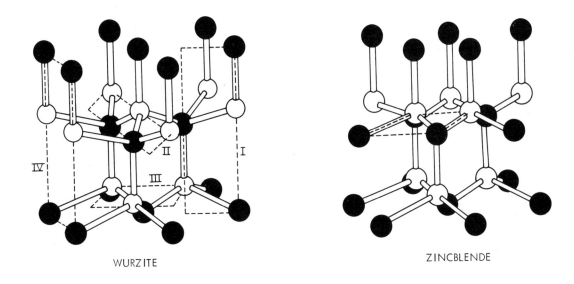

WURZITE   ZINCBLENDE

Fig. 1 Arrangement of the atoms in the (a) wurtzite and (b) zincblende
crystal structures. The z direction is taken to be along the c axis in
wurtzite and along the (111) direction in zincblende.

RLV's than does the cubic structure it is necessary to interpolate (extrapolate)
between (beyond) the zincblende values. This was accomplished by drawing a
smooth curve through the zincblende values of $V^S(G)$ and $V^A(G)$ as a function of
G and reading the intermediate values directly from this curve. Using this
procedure it was possible to use the experimental data of 3CSiC to calculate
the properties of both 2HSiC and 3CSiC.

The energy band structures of 3CSiC and 2HSiC derived in this manner are
shown in Figures 2 and 3, respectively. Also shown in Figure 2 for comparison,
is the band structure calculated by Herman et al.[6] using a first principles
OPW technique. As one can see, the two cubic band structures are qualitatively
similar but differ in quantitative detail. Important energy gaps are summarized
in Tables 1 and 2 for 3CSiC and 2HSiC, respectively. We also list results from
previous calculations[6-8] as well as those values extrapolated from the experimental absorption and reflectivity data.[9-13]

It should be noted that the 3CSiC results reported here are not identical
to those published previously by the authors.[14-15] That study was based on
the use of a nonlocal pseudopotential. We have just recently discovered a
numerical instability in the routine used to evaluate the matrix elements of
the nonlocal potential which casts some doubt on the validity of that evaluation.

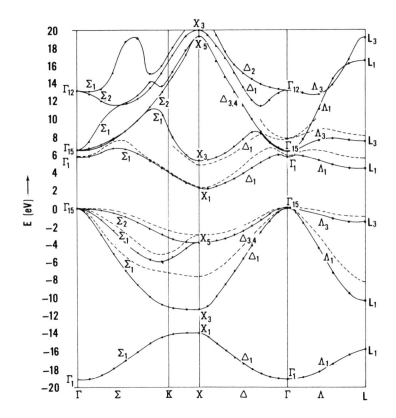

Fig. 2 Energy band structure of 3CSiC. Also shown (dashed curve) are results reported in reference 6.

	$\Gamma_{15v} - \Gamma_{1c}$	$\Gamma_{15v} - \Gamma_{15c}$	$X_5 - X_{1c}$	$X_{1c} - X_{3c}$	$L_{3v} - L_{1c}$	$\Gamma_{15v} - L_{1c}$	$\Gamma_{15v} - X_{1c}$	$L_{3v} - X_{1c}$
PRESENT CALCULATION	5.92 eV	6.49 eV	6.36 eV	3.08 eV	6.02 eV.	4.38 eV.	2.35 eV.	3.90 eV
HERMAN ET AL.	5.9	7.8	5.3	2.6	6.4	5.5	2.3	3.1
JUNGINGER ET AL.	5.14	10.83	5.27	3.24	6.75	5.93	2.40	3.26
BASSANI ET AL.	6.8	8.6	5.8	3.2	9.9	6.8	2.7	6.0
EXPT.	6.0			3.10		4.20	2.39	3.55

Table 1   Important energy gaps in 3CSiC.

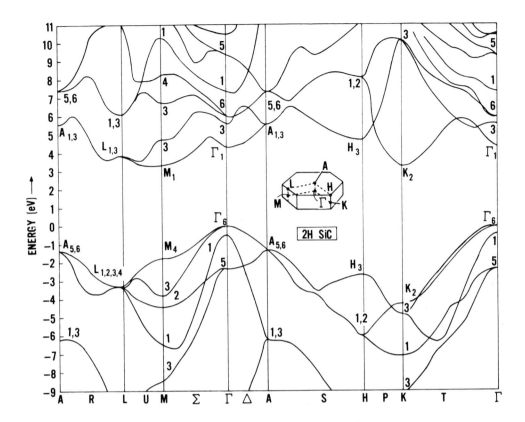

Fig. 3 Energy band structure of 2HSiC, following the symmetry notation of Rashba (reference 20).

ENERGY GAP	PRESENT CALC.	JUNGINGER ET AL (REF. 8)	HERMAN ET AL (REF. 8)
$\Gamma_{6v} - K_{2c}$	3.30 eV	4.02 eV	3.3 eV
$\Gamma_{1v} - K_{2c}$	3.77	3.35	3.7
$\Gamma_{6v} - M_{1c}$	3.40	4.42	4.0
$\Gamma_{1v} - M_{1c}$	3.87	3.75	4.4
$\Gamma_{6v} - L_{1c}$	3.81	4.49	3.3
$\Gamma_{1v} - L_{1c}$	4.28	3.82	3.7
$\Gamma_{6v} - \Gamma_{1c}$	4.39	5.09	6.0
$\Gamma_{1v} - \Gamma_{1c}$	4.86	4.46	6.4

Table 2. Important energy gaps in 2HSiC.

We present here results obtained using only the local form factors reported in references 14-15. However, the differences are quite small for most points in the Brillouin zone. It is only for those relatively few points with large $|\vec{k}|$ that discrepancies as large as .1 - .2 eV are found in the calculated energy states. Since these affected points include regions near X and K, some shifts are introduced into the positions of the optical structure from those values reported earlier, but the effect is small and essentially uniform.

We have also calculated the density of valence band states for 3CSiC and we compare the result for the three highest valence bands to the soft x-ray emission spectrum of Wiech[16] in Figure 4. The width of the three valence bands is found to be 11.25 eV, while the total valence bandwidth is 19.0 eV. These numbers compare reasonably well to the experimental values[16-17] of 9-10 eV and

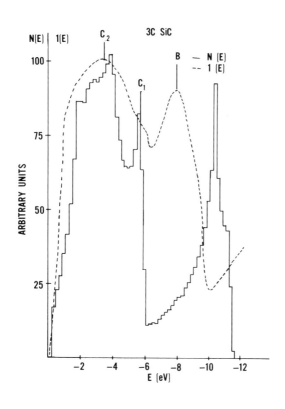

Fig. 4 Comparison of the calculated density of states of 3CSiC (full line) to the soft x-ray emission spectrum of Wiech (dashed curve).

18.2 - 19.2 eV, respectively.

It is interesting to compare the band structure of 2HSiC to that of 3CSiC. Birman[18] has pointed out that due to the similarity of the Brillouin zones of the two structures, there should exist a correspondence between energy states along various symmetry directions. If the two zones are aligned as shown in Figure 5, with the cubic Λ axis parallel to the hexagonal Δ axis, each vector

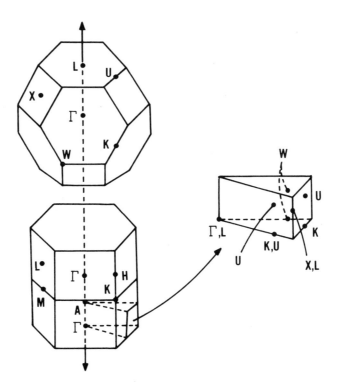

Fig. 5 Comparison of the zincblende and wurtzite Brillouin zones. Positions in
the wurtzite zone of special symmetry points of the cubic zone are indicated.

$\vec{k}$ along the Γ-Λ-L direction in zincblende corresponds to an identical $\vec{k}$ vector
along the Γ-A-Γ direction in wurtzite. I.e., the cubic Λ axis is "mapped" onto
the hexagonal Δ axis from Γ to A and back to Γ, with the cubic point L being
"folded back" onto the hexagonal point Γ. This mapping is indeed reflected in
the two band structures. The 2HSiC direct gap of 4.39 eV, identified with the
$\Gamma_{6v}-\Gamma_{3c}$ transition, compares very nicely to the value of 4.38 eV for the
corresponding cubic $\Gamma_{15v}-L_{1c}$ transition. The crystal field splitting near the
valence band edge ($\Gamma_{6v}-\Gamma_{1v}$) is about .45 eV. The differences between the
wurtzite energy levels at Γ and those of the corresponding zincblende states
along Γ and L are less than this value for all states up to and including the
second Γ_1 state in the conduction band.

It is also possible to make some comparisons between $\vec{k}$ vectors and states
along directions perpendicular to the "polar" direction discussed above. However,
there is no longer a unique one-to-one correspondence between zincblende and
wurtzite. For example, the cubic point K is mapped onto two regions at the
wurtzite zone - a region around M, extending along the Σ direction to Γ, and a

region between the wurtzite points M and K and extending upward along Z. Such mappings manifest themselves in the optical constants of the two solids and are generally indicated by the mappings of cubic zone points onto the hexagonal zone shown in Figure 5, taken from the work of Bergstresser et al.[19]

Instead of comparing calculated energy gaps at isolated $\vec{k}$ points in the BZ to values extracted from experiment, a more stringent test of any band structure would be to use the calculated energy bands and wavefunctions to determine the optical constants of the solid, as this calculation tests the band structure over a much wider range of energies. The imaginary part of the dielectric constant $\varepsilon_2(\omega)$ can be determined from the equation[1]

$$\varepsilon_2(\omega) = \frac{e^2}{\pi m^2 \omega^2} \sum_{vc} \int_{BZ} |\vec{\xi} \cdot \vec{P}_{vc}|^2 \frac{dS}{|\nabla_{\vec{k}} E_{vc}(\vec{k})|} \tag{8}$$

where

$$\vec{P}_{vc} = -i\hbar \langle \vec{k}c | \vec{\nabla} | \vec{k}v \rangle \tag{9}$$

is the matrix element of the momentum operator, $\vec{\xi}$ is the polarization vector of the photons, $E_{vc} = E_c - E_v = \hbar\omega$ is the interband energy, and S is a surface of constant interband energy. Here the sum is over all valence and conduction band states, the integral is over the Brillouin zone, and $|\vec{k}c\rangle$, $|\vec{k}v\rangle$ represent the wavefunctions of the conduction and valence bands states, respectively. Once $\varepsilon_2(\omega)$ is determined, $\varepsilon_1(\omega)$, the real part of $\varepsilon(\omega)$, is given by the Kramers-Kronig relation

$$\varepsilon_1(\omega) = 1 + \frac{2}{\pi} \int_0^\infty \frac{\omega' \varepsilon_2(\omega')}{\omega'^2 - \omega^2} \, dw' \tag{10}$$

Given $\varepsilon_1(\omega)$ and $\varepsilon_2(\omega)$ other optical constants can be determined. In particular, the reflectance is given by

$$R(\omega) = \frac{(\varepsilon_1^2 + \varepsilon_2^2)^{1/2} - [2\varepsilon_1 + 2(\varepsilon_1^2 + \varepsilon_2^2)^{1/2}]^{1/2} + 1}{(\varepsilon_1^2 + \varepsilon_2^2)^{1/2} + [2\varepsilon_1 + 2(\varepsilon_1^2 + \varepsilon_2^2)^{1/2}]^{1/2} + 1} \tag{11}$$

We have performed calculations of $\varepsilon_2(\omega)$ for both 3CSiC and 2HSiC using our EPM band structures and wavefunctions (without core corrections). The results are shown in Figures 6 and 7. Since 2HSiC has a preferred direction, Figure 7 contains two functions; $\varepsilon_2^\perp(\omega)$ (full line) and $\varepsilon_2^\parallel(\omega)$ (dashed line), where the superscript denotes the direction of the electric field vector $\vec{E}$ measured with

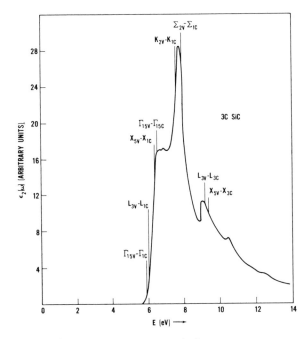

Fig. 6 Calculated $\varepsilon_2(\omega)$ spectrum of 3CSiC, with important interband transitions indicated.

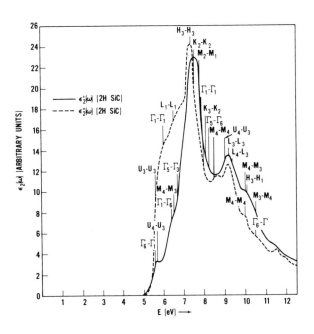

Fig. 7 Calculated $\varepsilon_2(\omega)$ spectrum of 2HSiC, with important interband transitions indicated.

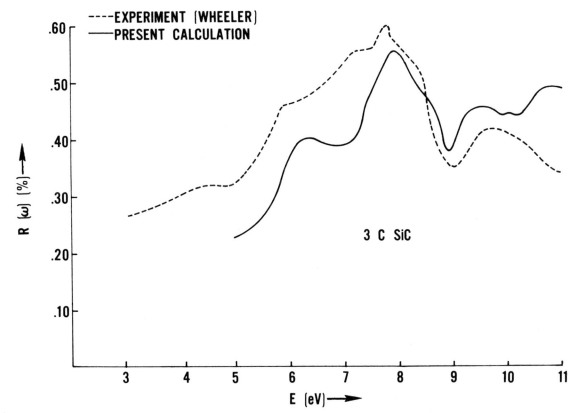

Fig. 8 Reflectance spectra of 3CSiC. The full line shows the calculated result, while the dashed curve indicates the experimental results reported in ref. 13.

respect to the c axis of the crystal. As one might expect because of the structural similarities of the two solids, the cubic and hexagonal spectra are quite similar in shape. Both are dominated by a single large peak between 7.5 - 8.0 eV, both show peaks or shoulders near 9 eV and 10 eV, and both exhibit shoulders near 6.5 eV.

We have also computed $R(\omega)$ for 3CSiC and compare it to the experimental measurements of Wheeler[13] in Figure 8. The experimental curve shows structure at 4.6 eV, 6.0eV, 7.1 eV, 7.8 eV, 8.3 eV, and 9.7 eV. It should be noted, however, that Choyke and Patrick[11] were unable to reproduce the weak peak at 4.6 eV in independent measurements, while Belle et al.[12] report a doublet structure in the range 9-10 eV. In light of these experimental uncertainties, one must consider the agreement between theory and experiment shown in Figure 8 to be quite satisfactory.

In order to determine the interband transitions responsible for the experimental structure in the case of 3CSiC, it is more convenient to analyze $\varepsilon_2(\omega)$, which depends directly on such transitions. In so doing one should remember that corresponding structure in the $R(\omega)$ and $\varepsilon_2(\omega)$ spectra will be shifted slightly (.1 - .5 eV) with respect to each other because of the Kramers-Kronig transformation.

ENERGY OF OPTICAL STRUCTURE			ASSOCIATED CRITICAL POINTS	
EXPT.	THEORY	INTERBAND TRANSITION	SYMMETRY	ENERGY
6.0 eV.	6.4 eV.	$\Gamma_{15_v} \cdot \Gamma_{1c}$	M_0	5.90 eV.
		$L_{3v} \cdot L_{1c}$	M_0	6.02
7.1 eV.	7.3 eV.	VOLUME EFFECT		
7.8	8.1	$\Sigma_{2v} \cdot \Sigma_{1c}$	M_2	7.88
8.3	8.4	VOLUME EFFECT		
9.7	9.5	$L_{3v} \cdot L_{3c}$	M_0	9.18
		$X_{5v} \cdot X_{3c}$	M_0	9.40
	9.9	VOLUME EFFECT		

Table 3 Summary of optical structure and associated critical points for 3CSiC. The energies listed in columns 1 and 2 refer to the experimental and calculated reflectance spectra, respectively.

POSITION OF STRUCTURE (eV.)	REGION	TRANSITION	ENERGY
PARALLEL TRANSITIONS:			
5.7 eV	M_2 cp AT	$U_3 \cdot U_3$	5.68 eV.
6.5	M_1 cp AT	$L_1 \cdot L_1$	
		$L_3 \cdot L_3$	6.46
7.4	TRANSITIONS AROUND AND INCLUDING		
	M_1 cp AT	$H_3 \cdot H_3$	7.41
	M_2 cp AT	$K_2 \cdot K_2$	7.53
8.6	M_1 cp AT	$M_4 \cdot M_4$	8.50
9.2	M_0 cp AT	$L_3 \cdot L_3$	9.30
PERPENDICULAR TRANSITIONS:			
5.7 eV	M_2 cp AT	$U_4 \cdot U_3$	5.62 eV.
6.5	M_1 cp AT	$M_4 \cdot M_3$	6.51
6.7	M_0 cp AT	$\Gamma_5 \cdot \Gamma_3$	6.65
7.0	M_1 cp ALONG	$\Delta_6 \cdot \Delta_5$	6.93
7.6	TRANSITIONS AROUND AND INCLUDING		
	M_2 cp AT	$M_2 \cdot M_1$	7.78
8.1	M_2 cp AT	$K_3 \cdot K_2$	8.11
8.3	M_0 cp AT	$\Gamma_5 \cdot \Gamma_6$	8.25
9.2	M_0 cp AT	$U_4 \cdot U_3$	9.24
10.0	M_2 cp AT	$M_4 \cdot M_3$	10.00
10.3	M_0 cp AT	$H_3 \cdot H_{1,2}$	10.26

Table 4 Summary of the optical structure and associated critical points of 2HSiC. The energies listed in column 1 refer to the calculated $\varepsilon_2(\omega)$ spectrum.

The important interband transitions for 3CSiC are indicated on Figure 6 and summarized in Table 3. Structure below the main peak is due mostly to 4-5 transitions between the highest valence band and lowest conduction band, while 4-6 transitions become more important for energies above 8 eV.

We have also performed a critical point analysis for 2HSiC and we list these results in Table 4. It is interesting to compare these transitions with those responsible for the structure in 3CSiC. In the energy range 6-7 eV the cubic spectrum exhibits structure due to critical points at Γ, L, and X. There is also a singularity (M_1) at energy 6.10 eV along Λ, about 2/5 of the way from Γ to L. In this range the main contributions to $\epsilon_2(\omega)$ come from 4-5 transitions around Γ and extending out along Δ toward X and along Λ toward L. As discussed previously, the cubic points Γ and L are mapped onto the hexagonal point Γ. In addition, the cubic point L is also mapped onto the U axis about 2/3 of the way between M and L. The cubic points X, K, and U (K differs from U by a RLV) are mapped into several different locations in the hexagonal zone, as indicated in Figure 4. Hence, one would expect to find critical points in the wurtzite spectra along U, coming from the cubic L and X points, at Γ, coming from cubic Γ and L, and along Δ, corresponding to the cubic c.p. along Λ, in this energy range. This is precisely what is found.

In the case of the main peak, the major contributions to the cubic $\epsilon_2(\omega)$ come from transitions around K, extending down along Σ and then over toward the line WX. In the hexagonal case, one finds contributions from the regions around H, K, and M, as well as around the U axis and extending out toward Δ (hex). This is also consistent with the mappings indicated in Figure 4. Cubic structure in the 9-10 eV region is due to the 4-6 transitions near Γ, X, and L, which again show up as critical points for transitions along U and at M in the hexagonal spectrum.

To summarize, by taking advantage of the structural similarity between the wurtzite and zincblende crystal lattices, we have been able to investigate the properties of both 3CSiC and 2HSiC using data derived primarily from experiments on the cubic polytype. As a result of this investigation we have been able to identify the major interband transitions associated with the optical structure of these materials.

It is hoped that this analysis will also offer some insight into the nature of the optical properties of other hexagonal polytypes such as 4HSiC and 6HSiC. The reflectance spectrum of 6HSiC, for example, is found to be very similar to

that of 3CSiC, being dominated by a single large peak at 7.8 eV and showing a
broad shoulder at 8.4 eV, with additional structure near 5.5, 6.7, 7.1, and
9.6 eV. On the basis of our discussion here, one might expect to find critical
points near H, K, and M with energies in the range 7.5 - 8.0 eV, corresponding
to the main peak, with additional critical points expected near Γ, L, M and
along U with energies around 6.5 - 7.0 eV and near L, M, and H with energies
in the range 9-10 eV. It will be interesting to see if future analysis will
bear out this conjecture.

References

1. M. L. Cohen and V. Heine, Solid State Physics 24 (Academic Press, 1970).

2. P. Lowdin, J. Chem. Phys. 19, 1396 (1951).

3. D. Brust, Phys. Rev. 134, A1337 (1974).

4. L. A. Hemstreet and C. Y. Fong, Phys. Rev. B 2, 2054 (1970).

5. L. Patrick, D. R. Hamilton, and W. J. Choyke, Phys. Rev. 143, 526 (1966).

6. F. Herman, J. P. Van Dyke, and R. L. Kortum, Mat. Res. Bull. 4, 5167 (1969).

7. F. Bassani and M. Yoshimine, Phys. Rev. 130, 20 (1963).

8. H. G. Junginger and W. Van Haeringen, Phys. Stat. Sol. 37, 709 (1970).

9. W. J. Choyke, D. R. Hamilton, and Lyle Patrick, Phys. Rev. 133, A1163 (1964).

10. Lyle Patrick and W. J. Choyke, Phys. Rev. 186, 775 (1969).

11. W. J. Choyke and Lyle Patrick, Phys. Rev. 187, 1041 (1969).

12. M. L. Belle, N. K. Prokof'eva, and M. B. Reifman, Sov. Phys. Semiconductors,
    1, 315 (1967).

13. B. E. Wheeler, Solid State Comm. 4, 173 (1966).

14. L. A. Hemstreet and C. Y. Fong, Solid State Comm. 9, 643 (1971).

15. L. A. Hemstreet and C. Y. Fong, Phys. Rev. B 6, 1464 (1972).

16. G. Wiech, in Soft X-ray Band Spectra (Academic Press, 1968) p. 59.

17. I. I. Zhukova, V. A. Fomichev, A. S. Vinogradov, and T. M. Zimkina, Sov.
    Phys. - Solid State 10, 1097 (1968).

18. J. L. Birman, Phys. Rev. 115, 1493 (1959).

19. T. K. Bergstresser, M. L. Cohen, Phys. Rev. <u>164</u>, 1069 (1967).

20. E. I. Rashba, Sov. Phys. Solid State <u>1</u>, 2569 (1970).

Photoluminescence of H- and D-Implanted SiC*

Lyle Patrick and W. J. Choyke

A strong new photoluminescence spectrum has been produced in SiC polytypes 4H, 6H, and 15R by implanting H or D ions, followed by annealing. The observation of CH and CD bond-stretching modes leads to a model of the luminescence center, namely, an H or D atom bonded to a C atom at a Si vacancy. Two kinds of spectra are observed. They are called primary and secondary, and may be due to exciton recombination at two different charge states of the center. Missing 15R spectra are attributed to the failure of some centers to bind excitons, and the quenching of a 4H spectrum is described. Several strong spectral lines are due to localized vibrational modes that are thought to be introduced by the Si vacancy.

I. INTRODUCTION

An efficient photoluminescence in SiC is produced by implantation of H ions (protons), followed by annealing at 700°C or so [1],[2]. The most remarkable feature of the spectrum is a line showing the excitation of a phonon with an energy of 369 meV, more than three times the 120 meV lattice limit. This phonon can be identified as a CH bond-stretching mode. Most spectral line energies remain about the same when the sample is implanted with D instead of H, but there is a large change in the bond-stretching mode, for the energy of the CD mode is only 273 meV. The CH and CD vibrational frequencies are very close to those of CH and CD bonds in organic molecules [1].

First, we show how the observation of CH and CD bond-stretching modes and the need for annealing lead to a model of the luminescence center that comprises an H or D atom bound to a C atom at a Si vacancy. After showing a typical spectrum we compare H spectra in polytypes 4H, 6H, and 15R. The constancy of the CH mode energy is contrasted with the variability of the exciton binding energy

---
* Work supported in part by the U.S. Air Force Office of Scientific Research (AFSC) under Contract F44620-70-C-0111.

E_B, with a suggestion that E_B may be negative at some of the many inequivalent sites, thus explaining certain "missing" spectra. In polytype 4H one spectrum is rapidly quenched under the uv exciting light at 1.3°K. Finally, we give a short comparison of three kinds of phonon modes that contribute to the luminescence spectra, namely, bond-stretching, localized, and momentum-conserving.

Recent high-resolution experiments [3] show that each no-phonon line can be resolved into two independent components. This discovery leads to an interpretation and a nomenclature different from that given in [1] and [2]. The differences are explained in Section II.

II. MODEL OF LUMINESCENCE CENTER

The observation of the CH bond-stretching mode shows that within the luminescence center an H atom is bonded to a C atom at a site free from perturbations by near neighbors. The need for annealing at 700°C indicates that interstitial H (as implanted) cannot form such a bond. This is understandable, for each neighboring C atom is fully bonded to Si atoms. The growth in intensity of the spectrum with annealing shows that the H atom diffuses to the active site, the obvious choice being one of the Si vacancies created during the implantation. At such a site there are four dangling C bonds, and the H atom becomes strongly attached to <u>one</u> of them, for the CH bond length is only 1.12 Å, much less than the C-Si distance of 1.89 Å. Thus, the H atom is attached to a single C atom and lies at a considerably greater distance from the other three.

In uniaxial crystals one of the four possible CH bonds is parallel to the crystal c axis and gives rise to what we call an axial center. The three other possible CH bonds are equivalent and give rise to nonaxial centers. It is now thought that the axial and nonaxial centers generate the two independent components of each no-phonon line that are observed when high enough resolution is employed. The 1.3°K spectra shown in this paper were formerly called nonaxial [1,2], but are now called the primary spectra, for the discovery of the two components shows that both axial and nonaxial centers contribute. At 4°K and higher another kind of center produces what were formerly called axial, but are now called secondary spectra. The secondary spectra will not be discussed here. They may be due to exciton recombination at a different charge state of the same center.

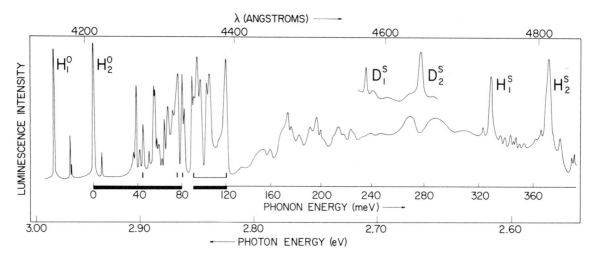

Fig. 1. Photoluminescence spectrum of H-implanted 15R SiC at 1.3°K. CH bonds
at two of the five inequivalent lattice sites in polytype 15R generate
overlapping spectra whose no-phonon lines are marked H_1^0 and H_2^0.
Energies of H_2 phonons may be read from the special phonon energy scale,
H_2^S being 368.7 meV. The insert shows the bond-stretching mode lines
in D-implanted 15R SiC.

III. EXPERIMENTAL RESULTS

Figure 1 shows the luminescence spectrum of a 15R SiC sample implanted with
3×10^{14} H/cm^2 at 150 keV, and then annealed at 900°C. The spectrum was photo-
graphically recorded on a Kodak 103F plate, and Fig. 1 is a copy of a densitometer trace. The 15R polytype has five inequivalent Si or C sites, but only two
distinct spectra are observed, with no-phonon lines marked H_1^0 and H_2^0. Thus,
excitons are bound at two of the five sites with binding energies of 4 and 43 meV,
the 15R exciton energy gap being 2.986 eV [4]. The axial and nonaxial components
of H_1^0 and H_2^0 are not resolved here.

The H_1 and H_2 spectra are similar except for the energy shift and the greater
intensity of H_2. For convenience, a phonon energy scale is included in Fig. 1,
with its origin at H_2^0. Lattice phonon energies, with a limit at 120.5 meV, are
shown by the dark strip with a gap between acoustical and optical branches from
80 to 91 meV. The strongest one-phonon lines in Fig. 1 belong to the H_2 spectrum,
and are indicated by the markers above the scale, i.e., lines at 46 and 77 meV,
a gap mode at 81 meV, and all the lines between 91 and 120.5 meV, the last being
the LO mode at the zone center [5]. Most of the phonons can be identified as
momentum-conserving (MC) by a comparison with those observed in the N spectra [4].

Beyond the lattice limit are some two-phonon bands and, finally, the lines corresponding to the CH bond-stretching vibrations, marked H_1^S and H_2^S. The latter falls at 368.7 meV on the scale.

When 15R SiC is implanted with deuterons instead of protons, a similar low-temperature luminescence is observed. There are again two spectra, called D_1 and D_2, isotope shifted from H_1 and H_2 by less than 1 meV for D_1 and about 2 meV for D_2. The one-phonon spectrum is also similar except for the absence of the 81 meV gap mode. However, the lines D_1^S and D_2^S are very different from H_1^S and H_2^S, as shown by the insert in Fig. 1. These lines represent the stretching vibrations of the CD bonds, and the phonon energies are 272 and 273 meV, approximately 74% of the corresponding CH mode energies [6].

IV. MISSING SPECTRA

In Table I we have collected the CH mode energies and exciton binding energies E_B for spectra in three polytypes. The variation in CH mode energy is less than 1%, whereas E_B runs from 4 to 125 meV. The small CH mode variation is understandable, for the H atom is strongly bound to a single C atom at a distance of 1.12 Å and lies considerably farther from any other atom. It therefore is not sensitive to the spatial patterns of atoms that distinguish polytypes or the

TABLE I. CH bond-stretching mode energies and exciton binding energies E_B from seven spectra in three H-implanted polytypes. The subscript on H is a serial number to distinguish the inequivalent centers in each polytype.

Spectrum	CH(meV)	E_B(meV)
4H - H_1	367	110
4H - H_2	370	125
6H - H_1	370	12
6H - H_2	367	19
6H - H_3	369	70
15R - H_1	367	4
15R - H_2	369	43

inequivalent sites within a polytype. On the other hand, an exciton is loosely bound and therefore <u>does</u> sense the pattern of neighboring atoms, and its binding energy is strongly influenced by this pattern. We wish to apply this observation to the problem of missing spectra.

It is likely that at 900°C the migrating H atoms cannot distinguish between the five inequivalent sites in polytype 15R and therefore occupy all of them equally. However, Fig. 1 shows that only two of these five sites give rise to luminescence spectra under uv excitation. We conclude that excitons can be bound at only two sites and that the missing spectra are to be attributed to zero or negative values of E_B at the other three sites.

Some of the expected secondary spectra are also missing [7]. In all cases we attribute the missing spectra to the failure of the CH or CD centers to bind excitons. In cubic SiC (3C) there is only a single site. In view of the fact that implantation of H or D in cubic SiC does not produce any luminescence, we conclude that the single 3C site fails to bind an exciton.

V. QUENCHING IN POLYTYPE 4H

A very unusual quenching is found in the H_2 primary spectrum of 4H SiC [7]. This spectrum is efficient during the first minute of exposure to the uv exciting light at 1.3°K but soon fades out, only to be restored by a room-temperature anneal. There is some evidence that a 50°K anneal might suffice.

In 4H SiC there are two inequivalent Si sites. Two primary spectra are observed, H_1 and H_2, but no secondary spectra. We assume that neither secondary center binds an exciton. We have tentatively attributed both primary and secondary spectra to the same center, H bonded to C at a Si vacancy. It is possible that the quenching of H_2 results from a change in the center's charge state, thus changing it from a primary center to a secondary center that does not bind an exciton. Trapped charge could be thermally released by a room-temperature anneal. However, any explanation of the quenching is confronted by two experimental facts. H_1 does not quench, although it is attributed to a center that differs from the H_2 center only with respect to the configuration of next-nearest neighbors. Also the D_2 spectrum does not quench, although the D_2 center differs from the H_2 center only in the replacement of H by D.

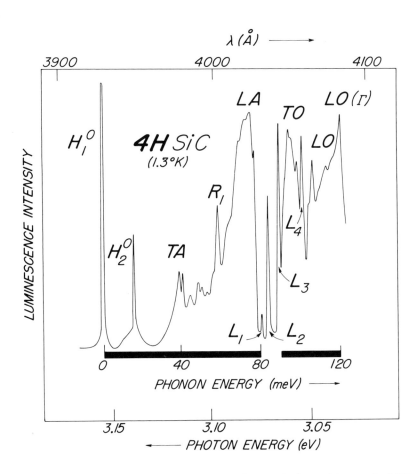

Fig. 2. Photoluminescence spectrum of H-implanted 4H SiC at 1.3°K. Only the
no-phonon and one-phonon regions are shown here. The phonon lines
identified by letters all belong to the H_1 spectrum. The relatively
weak H_2^0 and its associated phonon lines are partially quenched by a
ten-minute exposure to the uv exciting light.

VI. ISOTOPE DEPENDENCE OF PHONONS

Figure 2 shows the one-phonon structure in the spectrum of an H-implanted
4H sample, except for the CH modes that fall far off-scale on the right. H_2^0 is
weak because the sample has been exposed to the uv exciting light for about
10 min, hence the phonon spectrum is largely associated with H_1^0, from which line
the energies are measured on the special phonon-energy scale. Momentum-
conserving (MC) modes for this indirect-gap material are denoted TA, LA, TO, and
LO. These modes have the same energies in D-implanted samples, for the energies
of MC modes depend only on the band structure. A resonant mode is labeled R_1.
The resolved localized modes are more numerous in 4H SiC than in 6H or 15R and

they are labeled L_1 to L_4. The first three fall in the gap between acoustic and optic branches and L_4 falls between TO and LO modes. The H atom by itself can only introduce bond-stretching and bond-bending modes, and the latter are not observed. Thus, these localized modes must be associated with the Si vacancy. They are only slightly changed in energy by the substitution of D for H [2,7], indicating that the motion of the H or D atom is only a small part of the total vibrational mode.

Thus, we find three scales of size, or degrees of localization of the phonon, and we observe three degrees of dependence of the energy on the substitution of D for H. The correlation of size and energy change may be summarized as follows: (a) strongly localized on CH or CD bond: large change, (b) weakly localized on Si vacancy: small change, (c) not localized (MC modes): no change.

REFERENCES

[1] W.J. Choyke and Lyle Patrick, Phys. Rev. Letters 29, 355 (1972).

[2] Lyle Patrick and W.J. Choyke, Phys. Rev. B8, 1660 (1973).

[3] W.J. Choyke, P.J. Dean, and Lyle Patrick, unpublished.

[4] Lyle Patrick, D.R. Hamilton, and W.J. Choyke, Phys. Rev. 132, 2023 (1963).

[5] D.W. Feldman, J.H. Parker, Jr., W.J. Choyke, and Lyle Patrick, Phys. Rev. 173, 787 (1968).

[6] W.J. Choyke and Lyle Patrick, in Proceedings of the Eleventh International Conference on the Physics of Semiconductors, (PWN - Polish Scientific Publishers, Warsaw, 1972), p. 177.

[7] W.J. Choyke and Lyle Patrick, to be published.

Photoluminescence of β-SiC Doped with Boron and Nitrogen

Shoji Yamada and Hiroshi Kuwabara

Photoluminescence spectra are measured on β-SiC crystals doped with boron and nitrogen. There are two different kinds of emission series; A and B series. The mechanism responsible for the B series emission is the electronic transition between nitrogen donor and boron acceptor. The A series emission may be attributed to the transitions between conduction band tail states and boron acceptor.

INTRODUCTION

The role of acceptors which have important effects upon the luminescence of SiC has been studied by many investigators (1-3), and especially G. Zanmarchi (4) and W. Choyke et al. (5) have investigated the luminescence mechanisms of β-SiC doped with aluminum in detail. Two series are reported to be present in the photoluminescence spectra, one of which is assigned to be caused from the transition between nitrogen donors and aluminum acceptors. Another series of the luminescence is located at about 40 meV higher photon energies than the series of the donor-acceptor pair type, but the process responsible for the luminescence has not been understood.

The luminescence in SiC crystals doped with gallium, indium or aluminum is strong only at low temperatures. On the other hand, boron-doped crystals show rather bright luminescence even at room temperature (6). Only few investigations have been made on the luminescent properties of SiC doped with boron (6,7), and, moreover, the recombination mechanisms of which have not been clarified.

This paper presents the experimintal data on the characteristic luminescence of β-SiC doped with boron. Two series of emission quite similar to those of β-SiC doped with aluminum are found. The properties of the luminescence are studied as functions of boron concentration, the amount of intentionally added nitrogen, temperature, and excitation intensity.

EXPERIMENTAL PROCEDURES

The β-SiC crystals used were grown at 1550°C from the silicon solution in a high-density, high-purity graphite crucible (8). As source materials of silicon, high purity single crystals, boron-doped crystals and crystals grown in a mixture gas of 720 Torr helium and 40 Torr nitrogen were used, and the SiC crystals containing various amounts of boron and nitrogen were prepared. The boron contents in these SiC crystals were estimated from the dopant concentration in the silicon melt using the published data (9). The impurity contents of the samples are summarized in Table 1.

Photoluminescence intensities were measured as a function of emitted photon energy under the excitation by the 488 nm emission from an argon laser. The samples were mounted in a double Dewar vessel. The emitted light from the sample was passed through an optical system composed of a grating spectrometer and filters, and was detected by a photomultiplier or by a photographic plate.

TABLE 1

Impurities in the samples

Sample number	Boron concentration in SiC crystal (cm^{-3})		Weight percent of nitrogen-doped Si in silicon melt
1	2	$\times 10^{16}$	0
2	3.5	$\times 10^{16}$	0
3	9	$\times 10^{16}$	0
4	1.5	$\times 10^{17}$	0
5-1	2.5	$\times 10^{17}$	0
5-2	2.5	$\times 10^{17}$	0.3
5-3	2.5	$\times 10^{17}$	3.3
5-4	2.5	$\times 10^{17}$	13.6
6	4.5	$\times 10^{17}$	0
7-1	1	$\times 10^{18}$	0
7-2	1	$\times 10^{18}$	0.2

EXPERIMENTAL RESULTS

Typical luminescence spectra are shown in Fig. 1. The broken line represents the 77 K spectrum of a sample doped with both boron and nitrogen. This spectrum consists of a peak series, whose individual peaks are designated as B_0, B_1, B_2, ... as shown in the figure. The solid line is the 77 K spectrum of a sample doped with boron but without intentional nitrogen doping. The B series emission is very weak in this curve. So, the emission of the B series is believed to be concerned to the presence of nitrogen. The humps in the solid line may be attributed to the presence of residual nitrogen in the sample. There is another peak series of A_0, A_1, A_2, ... in the solid line. Similar spectra are observed in all the samples containing boron in the concentration range from 10^{16} to 10^{18} cm^{-3}, but are hardly observed in the concentration below 10^{16} and above 10^{18} cm^{-3}. Therefore, both the A and B series of emission are concerned to the presence of boron.

The B_1 and B_2 peaks are apart from the B_0 peak by about 69 and 116 meV, respectively. These energies well correspond to those of LA and LO phonons found in Al-doped crystals (4,5). Then it is reasonable to assign B_0 as a zero-phonon peak, and B_1 and B_2 as its phonon replicas. The B_3 and B_4 peaks can also be interpreted as the phonon replicas concerning to the emission of LA + LO and LO + LO phonons. Each peak in the A series is located at about 40 meV higher energy than the corresponding peak of the B series. Therefore, the method of assignment of phonon replicas used in the B series can be also employed to the assignment in the A series.

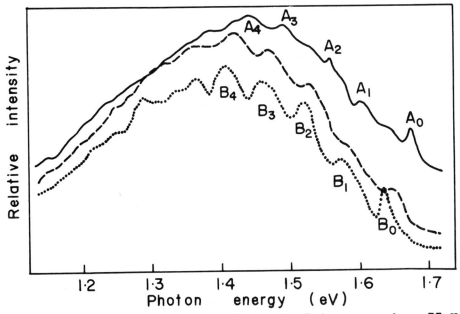

Fig. 1 Luminescence spectra of sample 7-1 measured at 77 K (solid line) and at 4.2 K (dotted line), and sample 5-4 measured at 77 K (broken line).

The dotted line in Fig. 1 shows the spectrum at 4.2 K of the same sample as that by the solid line. It is seen that the intensity ratio between the A_0 peak and the B_0 peak decreases with decreasing temperature.

The luminescence spectra corresponding to the A_0 and B_0 peaks of two samples with different boron concentrations are shown in Fig. 2. The maximum of the

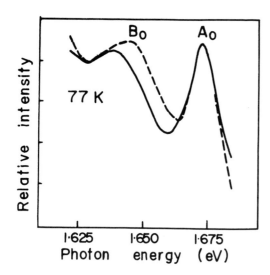

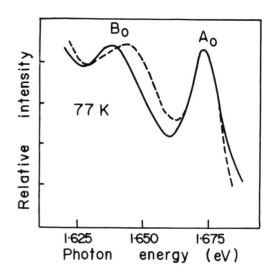

Fig. 2 Luminescence spectra
measured at 77 K for sample 6 (solid
line) and sample 7-1 (broken line).

Fig. 3 Luminescence spectra of sample
7-1 measured at 77 K, at the relative
excitation rates of 1 and 0.02 for the
broken and the solid lines, respectively.

B_0 peak of the sample with higher boron concentration locates at higher energy
than that of the sample with lower concentration. On the other hand, energy
change of the A_0 peak can not be observed as a function of boron concentration.
With increasing excitation intensity, the B_0 peak shifts toward higher energies,
and at the same time becomes broader as shown in Fig. 3. But the shape and the
position of the A_0 peak are practically unchanged with excitation intensity.

At low temperatures, fine structures are observed in the higher energy
portion of the B_0 peak. Figure 4 shows a densitometer trace of the luminescence
spectrum measured at 1.6 K. The spectrum is rather complicated owing to the
presence of isotopes of boron (10). There are two kinds of boron atoms with mass
numbers 10 and 11 whose percent abundances are 19.2 and 80.8, respectively.
There are many pairs of emission lines which show the intensity ratio of approximately 1 versus 4, and the energy separation in the pair is about 1 meV. We
have selected only the emission lines corresponding to boron of mass number 11
for the analysis.

The width of the A_0 peak increases with temperature as shown in Fig. 5.
At high temperatures, the A_1 peak broadens and overlaps with the A_0 peak. At low
temperatures the intensity ratio of the A_0 peak to the B_0 peak is decreased, and
the A_0 peak is difficult to be recognized. Therefore, the half-width of the A_0
peak is determined in rather narrow temperature range from 65 K to 160 K. The

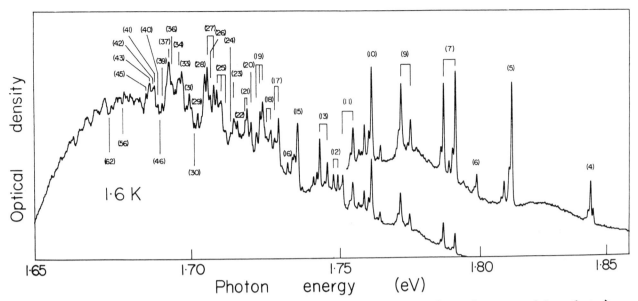

Fig. 4 Luminescence spectra (densitometer trace of a photographic plate) of sample 7-2 measured at 1.6 K. The integer in the bracket is the shell number of equivalent lattice sites.

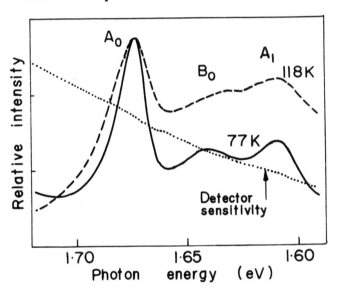

Fig. 5 Broadening of the A_0 peak with temperature for sample 3.

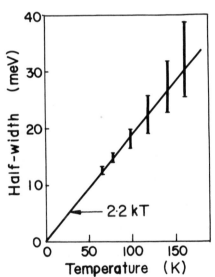

Fig. 6 Temperature dependence of half-width of the A_0 peak of sample 3.

results are shown in Fig. 6. The width of the peak is proportional to the absolute temperature, and is about 2.2 kT. Here, k is the Boltzmann's constant and T the absolute temperature.

DISCUSSION

The behavior of the B_0 peak is associated with donor-acceptor pairs.

According to Hopfield et al. (11), pair spectrum is described as follows:

$$h\nu = E_g - (E_D + E_A) - E_C + E_{vdW}, \qquad (1)$$

where $h\nu$ is the emitted photon energy, E_g the energy gap, E_D and E_A the ionization energy of donors and acceptors, respectively, and E_{vdW} the interaction energy between neutral donor and acceptor. The Coulomb interaction energy between ionized donor and acceptor is assumed that $E_C = - e^2 / \varepsilon r$, where e is the electronic charge, ε the static dielectric constant, and r the donor-acceptor separation in the pair.

With increasing boron concentration, the mean separation in the pairs decreases and the mean energy of emitted photons increases owing to the Coulomb term. Therefore, the shift of the B_0 peak with the change in the boron concentration is reasonably interpreted.

For pairs with large separations, the transition probability is small and the emission intensity has a tendency to be saturated with increasing excitation. For pairs with smaller separations, which correspond to higher emitting photon energies, the emission intensity is difficult to be saturated even at high excitation levels. This results in the shift and broadening of the B_0 peak as shown in Fig. 3.

Fine structures shown in Fig. 4 are originated from the discrete values of separation r. From the considerations on the correlation between the luminescence intensity and the number of the equivalent lattice sites, and also on the absence of lines in the spectrum at the special shell numbers, it is concluded that the measured spectrum is classified as Type 1: i.e. both the donor and the acceptor atoms substitite the same kind of host lattice atoms. The experimental results on ESR showed that both the boron and the nitrogen atoms substitute the carbon sites in SiC lattice (12, 13). The conclusion obtained from the ESR data agrees with the result of our analysis.

The integer in brackets in Fig. 4 is the number of a shell composed of equivalent lattice sites thus assigned. This assignment is somewhat tentative in the low energy region. The main reason for the difficulties in the assignment is that the energy separation between adjacent number shells becomes smaller with respective to the isotopic shift in the low energy region.

The photon energy corresponding to the infinitely large separation is expressed as $h\nu_\infty = E_g - (E_D + E_A)$, and is estimated to be 1.614 eV. The exciton energy gap of β-SiC is 2.390 eV at low temperatures, and exciton binding energy

is estimated to be 13.5 meV(14). Then $E_D + E_A$ may have a value of about 0.790 eV.
This value is fairly larger than the sum of E_D(0.118 eV) determined from the
luminescence spectra on β-SiC doped with aluminum (5) and E_A (0.39 eV) estimated
from the Hall data (15).

Figure 7 shows the emitted photon energy as a function of separation r.
Under the assumption that E_{vdW} = 0 and ε = 9.7, the experimental points are
fitted well to Eq. (1) for large values of r. Therefore the interaction between
neutral donor and acceptor is supposed to be negligible for large separations.

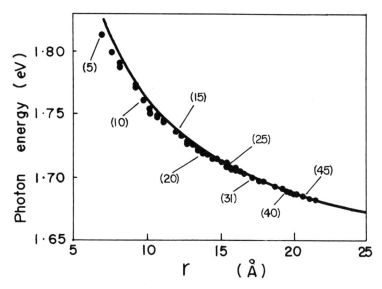

Fig. 7 The energies of the descrete pair lines as a function of the pair
separation r. The curve represents the calcurated values from Eq. (1)
under the assumption that E_{vdW} = 0.

At present, definite conclusions can not be obtained on the A series
emission. The energy difference between the A and B series is about 40 meV, and
this value is nearly equal to the corresponding energy separation in the luminescence spectra of β-SiC doped with aluminum. This energy difference is smaller
than the ionization energy of nitrogen donor. Therefore, it is probable that the
A series emission is caused by the electronic transitions between some kinds of
defect states located just below the bottom of the conduction band and the boron
acceptors. This unknown defect states must not be charged localized centers when
unoccupied, because the maximum and the half-width of the A_0 peak are unchanged
by the variation of both excitation intensity and boron concentration. The fact
that the half-width of the A_0 peak is of the order of thermal energy means that
the defect states are not strongly localized. The energy distribution of the

state density may be an approximately linear function of energy. There seems to be some kinds of tail states of the conduction band in the β-SiC crystals and these tail states are considered to be responsible for the A series emission.

There may be other possibilities of interpreting the A series emission; such as recombination of excitons bound to complex centers containing boron.

ACKNOWLEDGMENT

The authors wish to express their sincere thanks to Dr. K. Mizuma of Nippon-Denshi-Kinzoku Co. Ltd. for supplying silicon. They are also indebted to Mr. K. Urabe and Mr. S. Shiokawa for their assistances. This work is partly supported by Grant-in-Aid for Scientific Research from the Ministry of Education.

REFERENCES

1. G. F. Kholuyanov, Soviet Phys.-Solid State, 7, 2620 (1966).

2. A. A. Kal'nin, V. V. Pasynkov, Yu. M. Tairov, and D. A. Yas'kov, Soviet Phys.-Solid State, 8, 2381 (1967).

3. A. Addamiano, J. Electrochem. Soc. 113, 134 (1966).

4. G. Zanmarchi, J. Phys. Chem. Solids, 29, 1727 (1968).

5. W. J. Choyke and L. Patrick, Phys. Rev. B 2, 4959 (1970).

6. R. M. Potter and D. A. Cusano, J. Electrochem. Soc. 114, 848 (1967).

7. H. Kuwabara, S. Shiokawa, and S. Yamada, phys. stat. sol. (a) 16, K67 (1973).

8. S. Yamada, T. Kawai, and M. Kumagawa, J. Cryst. Growth, 19, 74 (1973).

9. W. E. Nelson, F. A. Halden, and A. Rosengreen, J. Appl. Phys. 37, 333 (1966).

10. P. J. Dean, C. H. Henry, and C. J. Frosch, Phys. Rev. 163, 812 (1968).

11. J. J. Hopfield, D. G. Thomas, and M. Gershenzon, Phys. Rev. Letters, 10, 162 (1963).

12. H. H. Woodbury and G. W. Ludwig, Phys. Rev. 124, 1083 (1961).

13. G. E. G. Hardeman, J. Phys. Chem. Solids, 23, 1223 (1963).

14. D. S. Nedzvetskii, B. V. Novikov, N. K. Prekf'eva, and M. B. Reifman, Soviet Phys.-Semicond. 2, 914 (1969).

15. G. A. Lomakina, Soviet Phys.-Solid State, 7, 475 (1965).

Some Aspects of Cathodoluminescence of β-SiC Grown by the Gasphase Method

I. I. Geiczy and A. A. Nesterov

GREEN BAND OF β-SiC LUMINESCENCE

A. Non-Doped or "Pure" Crystals

Investigation of β-SiC "pure" crystals at the electron beam excitation showed that the radiation is made of the main (green) band (5200 - 5400 Å) and two more long-wave ones: 5500 - 5750 Å and 5800 - 6000 Å correspondingly (Figure 1). In their turn each of these bands consists of three more narrow lines (T = 120°K).

It was established that with temperature increase the band is broadened, the structure is smoothed, the peak I_3 intensity is reduced (Figures 1,2) and this causes shifting of the green band maximum to the peak I_2. The latter is being most precisely expressed in the range of 120 - 290°K was defined to have the shift temperature coefficient equal to -1.1×10^{-4} eV/°K.

The band intensity decreased 1.5 - 2 times in the same temperature range. At 290°K the structure was presented as weak bendings near maxima (Figures 1,3), at higher temperatures the structure disappeared.

The investigation of the intensity of radiation main band on the excitation density showed that it can be approximated with degree dependence $I \sim j^a$ where "a" equals to 1.2 - 1.5 at $10^{-5} < j < 1$ a $\cdot$ cm^{-2} and of the order 0.6 for $j > 1$ a $\cdot$ cm^{-2}.

The band characteristics (let us call it the excitation density effect) for $j > 1$ a $\cdot$ cm^{-2} were the structure disappearance, broadening, shifting to the field of more long-waves (20 Å for $j = 1$ a $\cdot$ cm^{-2}, T = 100°K), Figure 2a. The mentioned effects are not connected with the specimen heating as shifting to lower quantum energies is connected with the peak I_3 intensity increase which as it was shown for $j = 0.1$ ma $\cdot$ cm^{-2} decreases as temperature increase, Figure 1b.

313

The spectra analogous to those we obtained at $j = 1a \cdot cm^{-2}$ and in the paper (1) were obtained for light-diodes on β-SiC (300 and 120°K) i.e. the structure in the main band was not discovered.

The radiation of the comparatively high intensity that permitted us to resolve the structure in the bands as a rule could be observed from the natural mirror side B ($\overline{1}\overline{1}\overline{1}$). The opposite natural side A (111) radiated much weaker, the radiation band shift and disappearance of the structure up to $2a \cdot cm^{-2}$ was not observed.

B. Crystals Doped with Nitrogen and Boron

To obtain additional information on β-SiC luminescence in the green band crystals specially doped with nitrogen and boron were investigated. It was found that with increasing impurity concentration (apart from the doping element) the radiation intensity decreased, the structure disappeared, the band was widened, and the maximum was shifted to the field of lower energies, Figure 2b. The half-width of the band changed from 90 to 200 Å (100°K) and from 240 to 350 Å (300°K) in the impurity concentration range 10^{17} to 10^{20} cm^{-3}. The band maximum shift was reached 80 Å in cooled specimens. More long wave bands decreased with doping but the character of the intensity dependence on the excitation density did not change with temperature and doping degree and coincided with the analogous dependence for "pure" crystals.

As the zone in β-SiC is an indirect one the line maxima in the green band (I_1, I_2, I_3) can be connected with phonon-assisted transitions: TA (46 meV), LA (79 meV) and TO (94 meV). As the following bands are connected with the main one we explain them as its phonon-assisted components.

Estimations showed that the most reasonable mechanism of radiation (T $\simeq 300^{o}$K) is the annihilation of free excitons (E_i = 10 meV).

The intensity decrease with temperature increase in the range of 100 - 290°K with activation energy of 10 meV is well coordinated with dissociation energy of exciton (2).

Changes of the green band for "pure" crystals at $j \geqslant 1a \cdot cm^{-2}$ are in quality similar to those that have been observed for exciton radiation in silicon under electron excitation (3). The authors explained them as transition from the exciton mechanism to band-to-band.

The criterion of exciton decay requires equality of Debye screening length 1 to Bohr radius of exciton r_{ex} (4), where

$$1 = \left(\frac{EkT}{4\pi e^2 n}\right)^{1/2}, \quad r_{ex} = \frac{Eh^2}{e^2 m^*}$$

For β-SiC $m_e = 0.4\ m_o$ (5) $m_h = 0.7\ m_o$ (6)

$$= 9.7\ (7)$$

calculation shows that with the used concentrations of non-equilibrium charge carriers ($\sim 5 \times 10^{17}$ to 10^{18} cm^{-3}) exciton decay does not occur and radiation mechanism remains the same. Observed changes in the green band are evidently connected with splitting levels in the exciton band (8).

The availability of additional recombination centres that reduce concentration of non-equilibrium charge carriers and shift the effect to high levels of excitation is responsible for the absence of the effect of excitation density for the specimens with the weak green band.

The results of investigating specimens doped with N and B over a wide range of concentrations proved the explanation of the β-SiC green band behaviour depending on the excitation density to be correct.

The results obtained with light-diodes show that at large injection, exciton interaction with free charge carriers takes place as well. In paper (1) this effect could increase also due to the large concentration of impurities.

LONG-WAVE RADIATION

Long-wave (defect) radiation of initial specimens was observed independently in "pure" crystals and in specially doped (grown in the atmosphere with carbon or silicon excess) as well.

In all observed bands the depended $I = f(j)$ was of sublinear character and radiation intensity decreased as temperature increased. The band shape was not dependent on the excitation density.

The most intense luminescence was produced by "pure" crystals from natural carbon side B ($\overline{111}$) i.e. bands G and F. Band G (Figure 3) consists of the head line G_o (6274 Å, T = 100°K) with the halfwidth 0.3kT and following more wide lines with peculiar distribution of intensity that are identified as its phonon-assisted transitions. Temperature coefficient of the line G_o shift $\frac{\Delta h\nu}{\Delta T}$, makes -3×10^{-5} eV/°K (100 to 300°K). Band F (Figure 4) consist of the head line F_o (7840 Å T = 100°K) with halfwidth 0.4kT and $\frac{\Delta h\nu}{\Delta T} = -3 \times 10^{-5}$ ev/°K in the temperature range, 100 to 300°K. The following long-wave lines more weak in intensity accompany line F_o.

The third variety of defect radiation in initial crystals grown in an atmosphere with excess carbon was band E, Figure 5a. It was weak in intensity but with a distinctly expressed line E_o (7140 Å, T = 100°K) with halfwidth less than kT. More long-wave structure corresponded to its phonon-assisted transitions. At T = 290°K the structure in band E was not available. The spectrum shown in Figure 5b, was registered at the sensibility limit of the installation and was characteristic of the crystals grown in an atmosphere with excess silicon.

The bands considered could be available in combination F and G, E and F. Crystals specially doped with N and B radiated only in band F.

The given cathodoluminescence of β-SiC with narrow (halfwidth less than kT) major lines show the recombination of nearly immobile charge carriers, i.e. not connected with free zones. The major lines represent zero-phonon transitions. As paper (9) shows the weaker the defect connection with lattic oscillations and the greater the value of mean oscillating quantum, the greater is the probability of zero-phonon transition. The more long-wave part of bands is completely determined by the interaction character with the lattice oscillations.

Bands E, G, and F are an example how the radiation spectrum changes depending on the defect nature: from a single line (F_o) (weak interaction with lattice oscillations) up to wide E-band, upon which zero-phonon line E_o, and phonon-assisted transitions are visible.

The greatest degree of electron localization which decreases successively for G- and E- bands corresponds to F- band.

Quasi-line spectrum behind the line F_o, weak in intensity, is due to either local oscillations of the defect, the narrow maxima in the distribution function, or losses by the frequencies of oscillations of the crystal lattice (10). The more long-wave part of band F is probably connected with other radiation centres.

Afterglow time of radiation in bands F and G which equals several microseconds (comparing to $\sim 10^{-8}$ sec. for the green band) shows the availability of electron transitions in isolated systems.

Band F was the most peculiar defect band in most initial crystals. All crystals investigated had twin lamellae.

That the band F is connected with the packing defect can be demonstrated by the fact that it was observed, as a rule, from the natural side (111) in parallel to which at the depth of 0.5 to 20μ the twin lamella was localized. The removal of the layer from the surface of the specimen took this band of luminescence away.

The same regulation in the behaviour was characteristic of band G as well.

It is known that after β-SiC crystals heating in air at T = 1100 to 1200°C weak band A (11) appears. Its intensity is 1.5 to 2 orders less than after annealing of earlier irradiated crystals (dose equal to 10^{16} e · cm^{-2}, E = 3.5 meV). We happened to obtain intense band A in initial crystals with the help of thermal quenching at T = 2500 - 3000°C. The effect did not depend on whether it took place in air (electric arc), in vacuum or in inert atmosphere, and was of volume character.

This result allows us to somewhat specify the nature of the defect connected with band A. Thermal quenching results in "freezing" some part of the vacancies that are in equilibrium in the material at higher temperatures. In such experiments no complications arise connected with the availability of interstitials (12). It is believed that the defects due to this quenching are vacancy complexing with impurities(13). Which of the vacancies is dominating this complex: V_{Si} or V_C? Taking into account a big coefficient of silicon self-diffusion in SiC it will apparently be V_{Si}. We are inclined to believe that centre A is the complex-centre containing a vacancy (silicon vacancies) and an impurity atom of oxygen or nitrogen. Nitrogen is always present in SiC in large quantities. Information of this kind on oxygen is not available so far. However in paper (14) there are data on possible non-controlled "contamination" of SiC with oxygen in the process of growing.

(1) A. I. Luk'yanova, Sh. A. Mirsagatov, V. V. Morozkin and V. M. Rubinov, Sg. III Vsesoyuzn. Konf. po poluprovodn. karbidu kremnia, str. 213, M., 1970.

(2) M. L. Belle, N. K. Prokofieva and M. B. Reifman, FTP, 1, 383 (1967).

(3) V. S. Vavilov and E. L. Nolle, FTP, 2, 742 (1968).

(4) B. M. Ashkinadze and I. D. Iaroshetskii, FTP, 1, 1706 (1967).

(5) Proc. Int. Conf. on Silicon Carbide, University Park, Pennsylvania, October, 1968, (Pergamon Press, New York, 1969). pg. 639.

(6) M. S. Saidov, Kh. A. Shamuratov and A. S. Saidov, Sb. III Vsesoyuzn konf. po poluprov. karbidu kremnia, str. 191, M., 1970.

(7) L. Patrick, W. J. Choyke, Phys. Rev., 2, B 2255 (1970).

(8) Eksitoni v poluprovodnikakh, Izd. "Nauka", M., 1971.

(9) E. D. Trifonov, DAN SSR, 147, 826 (1962).

(10) G. S. Zavg, FTP, 5, 1954 (1963).

(11) I. E. Geiczy, A. A. Nesterov and L. S. Smirnov, Rad. Effects, 9, 243 (1971).

(12) A. Damask, G. Dins, Tochechnie defekti v metallakh, izd. "Mir", M., 1966.

(13) M. L. Swanson, Phys. Stat. Sol. 33, 721 (1969).

(14) Yu. M. Tairov and V. F. Tsvetkov, FTP, 5, 317 (1971).

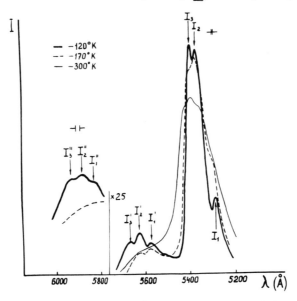

Figure 1. Green band of β-SiC Cathodoluminescence and its phonon-assisted transitions for three ranges of temperature, T°K: 1-120, 2-170, 3-290; j = 0.1 mA.cm².

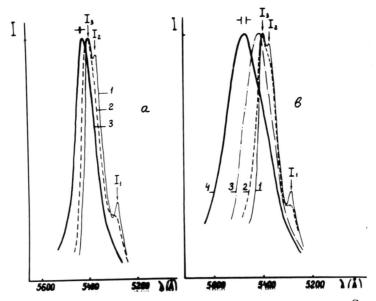

Figure 2. Green band of β-SiC Cathodoluminescence (at T = 100°K) depending on: a) excitation density ("pure" crystals) mA/cm²: 1-1, 2-10, 3-1000; b) degree of nitrogen doping, cm⁻³; 1x10¹⁶ to 10¹⁷, 2,3,4x 10₂¹⁸ to 10¹⁹ (N₂<N₃<N₄; j=0.1mA/cm²)

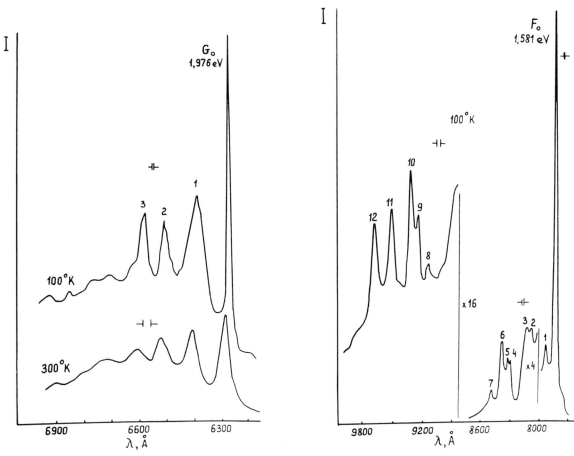

Figure 3. G-band of β-SiC Cathodoluminescence.

Figure 4. F-band of β-SiC Cathodoluminescence.

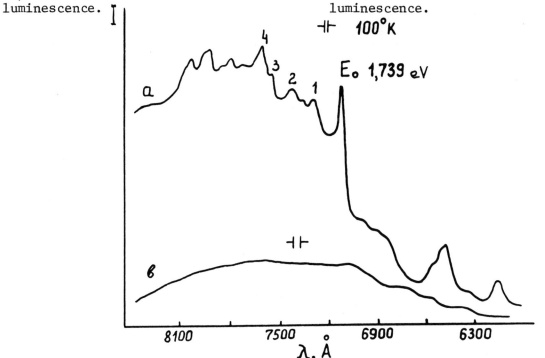

Figure 5. Long-wave spectra of β-SiC crystals luminescence grown in atmosphere:
     a) with carbon excess
     b) with silicon excess

Kinetic Studies on the Oxidation of Silicon Carbide

E. Fitzer and R. Ebi

This paper describes investigations on the high temperature oxidation of SiC. As is known, the excellent oxidation resistance of SiC at temperatures between 1,000 and 1,600°C is caused by the formation of a protective SiO_2 layer in an oxidizing atmosphere. In this paper pure oxygen, air as well as CO_2 and CO in various concentrations were used as oxidizing agents. Also the additional influence of water vapour was tested. All experiments were performed at a total pressure of 1 atm. This experimental study was performed in order to get basic information on the oxidation behaviour of SiC-heating elements and refractories in industrial atmospheres.

Figure 1 compares the free energies of the reaction SiC + O_2 related to 1 mol SiC at various O_2 partial pressures. One can recognize that at a very low oxygen pressure, i.e. in CO-atmospheres, volatile SiO is formed without SiO_2 formation and a catastrophic oxidation results. For example, at 1 atm. O_2 partial pressure, SiO is not formed below 2100°K. However, at an O_2 pressure of only 10^{-5} atm. the SiO formation is favoured as low as 1,350°K. Of even more practical importance is the solid state reaction between the previously formed SiO_2 surface layer and the SiC-substrate. This reaction can occur because of the low transport rates of O_2 through the SiO_2 layer and the very low O_2 activity at the interface between the SiO_2 and the SiC.

There is a lot of literature on the oxidation behaviour of Si, SiC and other refractory Si-compounds. In 1963 Motzfeld (1) compiled these data as shown in Figure 2 in an Arrhenius plot. The oxidation of elemental Si is indicated as diamonds. Full circles show the oxidation behaviour of SiC. In this diagram also the premeation data of molecular O_2 through SiO_2-glass-walls are included as open circles. Most of the values lie near a single line indicating not only the same activation energy but also a comparable rate constant. Motzfeld has concluded that the oxidation rate of SiC as well as of Si is controlled by the transport rate of O_2 through the reaction product, that is the SiO_2 layer.

Figure 3 shows our results of the SiC oxidation in pure oxygen and in air. Data on Si_3N_4 are included for comparison. Contrary to the former literature, three temperature ranges with different activation energies were found. The activation energy of 36 kcal/mol in the low temperature range up to $1200^{\circ}C$ agrees well with literature data. In the temperature range between 1200 and $1400^{\circ}C$ an activation energy of only 21.6 kcal/mol is measured. Above $1400^{\circ}C$ the activation energy increased again. If one regards carefully the old literature as shown in Figure 2, its scattering values fit the stepline quite well found in this paper. X-ray studies of the SiO_2 layers between 1200 and $1400^{\circ}C$ show that the SiO_2 layers consist of 10-70% cristobalite. This cristobalite content decreases with increasing temperature. By these experiments the lower activation energy in the temperature range between 1200 and $1400^{\circ}C$ can be correlated with cristobalite formation. The facilitated oxygen transport can be explained by an increased number of defects in the SiO_2-glass between the crystals, or by a preferred oxygen diffusion at the cristobalite/glass-interface. The diffusion of oxygen through the cristobalite is more difficult than through the silicon glass layer.

Concerning the transport of the volatile species involved in the oxidation reaction, further conclusions can be drawn from the temperature dependence of the oxidation kinetics. As confirmed by gas chromatographic results, CO is formed as a product during the SiC-oxidation in oxygen and air. For this gaseous by-product a similar transport mechanism as for the oxidizing gas must be assumed, that is permeation through the silica-glass as indicated in the upper left corner of Figure 4. Dietzel (2) has estimated that the width of SiO_2-rings is large enough to allow the permeation of O_2 molecules. The table in Figure 4 compiles the kinetic gas diameters of oxygen and CO for various temperatures, compared with the width of SiO_2-rings. N_2 is included because of the similar oxidation behaviour of Si_3N_4. The comparison of all these data shows that the transport of the products CO and also N_2 in the case of Si_3N_4 is rate controlling and not the transport of the oxygen molecule. This assumption, based only on a very rough estimation, seems to be proved by experiments.

Figure 5 shows the ratio of the oxidation rates in pure oxygen and in air for SiC and Si_3N_4. These data are presented in the diagram in the ordinate as ratio k_A over k_{O_2}. If the O_2 transport controls the oxidation rate, k-ratios below 1 must indicate a distinct influence of the O_2 partial pressure. If the oxidation rate is controlled by the transport of the oxidation products, the O_2-pressure influence will decrease and ratios near 1 are to be expected. This is the case for SiC above 1400°C. For Si_3N_4 this influence of the oxidation by-product-transport is even larger.

The experiments with CO_2 as oxidizing agent were carried out in N_2 atmospheres containing 2.5 and 17% CO_2 and in 100% CO_2 in the temperature range between 1200 and 1600°C. Figure 6 shows the weight changes of the SiC samples in dependence of time - in the lower diagram with pure CO_2, in the two upper diagrams with different CO_2 contents in nitrogen. One can recognize that the parabolic oxidation behaviour is overlapped by a weight decrease, which does not occur before some hours oxidation time. This weight loss can be explained by the secondary reaction between the performed SiO_2 layer and the SiC substrate. This reaction causes the formation of volatile SiO and CO by which the sample is destroyed. As mentioned in the beginning this secondary solid state reaction depends upon the O_2 activity in the interface. Therefore, it will occur at low CO_2 pressure earlier, that means at lower temperatures compared with pure CO_2.

Figure 7 shows the experimental results of the heat treatment of SiC in CO. Even under these conditions of very low O_2 activity a weight increase is found by thermal analysis, indicating that thin SiO_2 - layers are formed in the beginning. The overlapping secondary reaction between preformed SiO_2 and SiC is found in all cases again after some hours. At temperatures above 1450°C only the volatile SiO is formed. The linear time dependence in this range indicates that only the chemical reaction at the gas solid interface is the rate controlling step. An amelioration of the oxidation resistance can be obtained by an additional water content in the gas. The temperature limit for the stability of SiC in CO is then shifted up to 1450°C. More systematic studies about the influence of additional water vapour during the oxidation of SiC were performed in the case of oxygen and CO_2. The right diagram in Figure 8 shows the oxidation behaviour of SiC in a CO_2 atmoshpere with a water vapour content. The behaviour in the same CO_2 atmosphere without water is shown in the left diagram for comparison. One can recognize that the weight increase, which means the SiC oxidation is increased, if water pressure is present.

Also the overlapping secondary solid state reaction between the SiO_2 layer and the SiC substrate is retarded because of the higher O_2 activity in the interface. Presumably, the structure of the SiO_2-glass is influenced and therefore the diffusion of the volatile species is promoted. These experimental results agree with the industrial experience that SiC is quite resistant in CO_2 atmospheres even at higher temperatures. Contrary to this beneficial effect of water in the case of CO and CO_2, water vapour diminishes the oxidation resistance of SiC in air. The influence of water on the oxidation of SiC is of technical interest as moisture contents in air accelerate the oxidation of SiC. Figure 9 shows oxidation in dry air as dashed lines and full lines indicate the thermogravimetric results in air with additional water vapour. One can clearly recognize that the reaction

is increased, especially at the lower temperatures. During the first 20 hours, a linear time dependence is observed. As described with experiments in CO atmosphere this increased oxidation rate is caused by the easier transport of the volatile species. However, in the case of carbon oxides the necessary O_2 activity at the interface is guaranteed by this easier transport. In the case of air attack this easier transport accelerates only the oxidation rate of SiC. These explanations are compiled in Figure 10. The experimental results of the laboratory tests are in agreement with industrial experience.

SUMMARY

The oxidation resistance of SiC is caused by the formation of a protective SiO_2 layer. Further oxidation is controlled by the permeation of oxygen or the diffusion of the gaseous reaction product CO. In different temperature ranges different temperature dependences of the oxidation rate were found. This was explained by a change of the diffusion mechanism. The stability of the SiC substrate depends mainly on the O_2 activity at the surface of the substrate, that means, at the interface between the SiO_2 layer and the SiC substrate. At a low O_2-potential at the SiO_2/SiC interface, the reaction of SiC with SiO_2 leads to the formation of volatile SiO and CO. At a low partial pressure in the primary stage SiO is formed at once which causes catastrophic oxidation. By water vapour contents in the reaction gases an amelioration of the high temperature resistance can be obtained in carbon oxide gases. However, in O_2 and air the attack is increased by water.

REFERENCES

(1) K. Motzfeld: Acta Chem. Skand. _18_ (1964) p. 1596-1606.

(2) A. Dietzel and F. Oberlies: Glastechn. Ber. 30 (1957), 37.

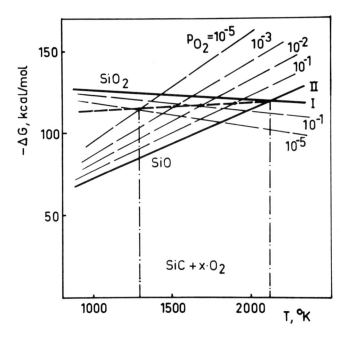

Figure 1.   Free energy of reaction for the oxidation of SiC by oxygen, related to 1 mol SiC.

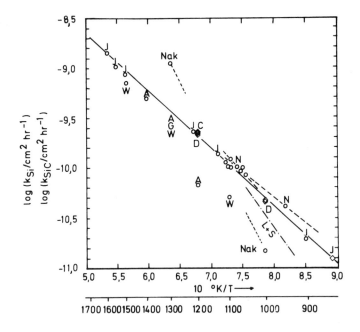

Figure 2.   Literature data on the oxidation kinetic of Si and SiC (1).

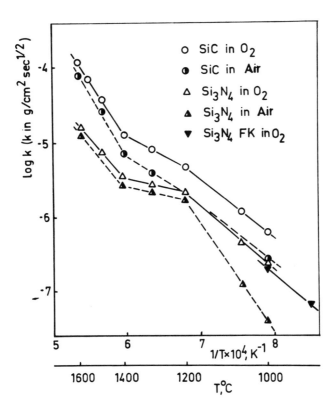

Figure 3. Experimental results on the oxidation kinetic of SiC and Si_3N_4 in pure oxygen and air.

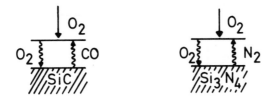

Gas	20°C	1000°C	1500°C
O_2	2,98	2,58	2,53
N_2	3,15	2,7	2,65
CO	3,19	2,73	2,68
CO_2	3,34	3,0	2,95
Width of SiO_2 ring	2,5	2,5	2,5
Width of Cristobalite	2,2	2,25	2,3

Figure 4. Molecule diameters of gases and ring widths within the solid SiO_2.

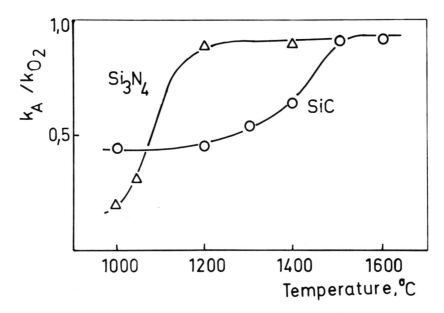

Figure 5.   Ratio of oxidation rates in pure oxygen and air.

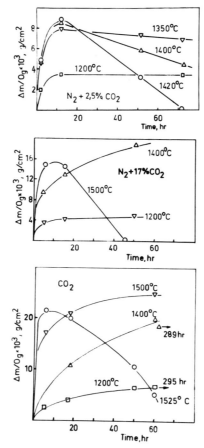

Figure 6.   Oxidation isotherms of SiC in CO_2.

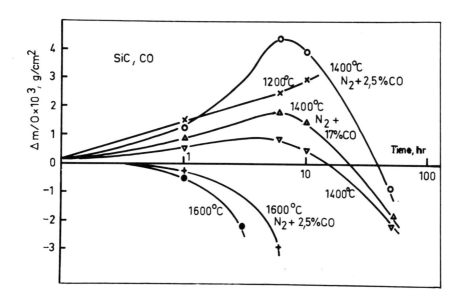

Figure 7. Oxidation isotherms of SiC in CO.

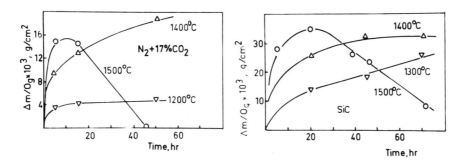

Figure 8. Oxidation isotherms of SiC in CO_2 with additional water vapour pressure (right-hand diagram) compared with that in dry CO_2 gas (left-hand).

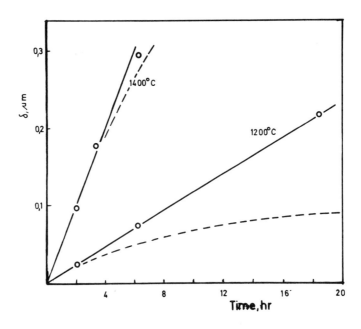

Figure 9.   Influence of water vapour in air on the oxidation behaviour of SiC (full lines); isotherms in dry air (dashed lines).

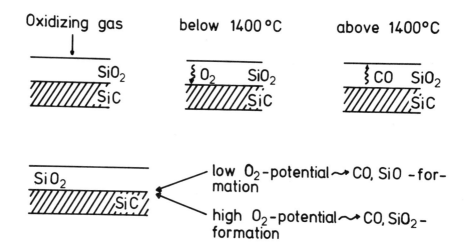

Figure 10.   Competive primary and secondary oxidation reactions of SiC.

Oxidation of 6H Alpha–Silicon Carbide

R. C. A. Harris and R. L. Call

The oxidation on the two crystal faces of SiC platelets was studied. The oxidation of one face was found to follow a simple parabolic law which was oxide diffusion controlled with an activation energy of 47 KCal/mole. The oxidation of the opposite crystal face followed a linear growth rate which was surface reaction controlled with an activation energy of 85 KCal/mole.

Previous investigators have determined the oxidation rate of SiC by making weight change measurements, and have found the oxidation rate to follow a simple parabolic law at the lower oxidation temperatures and times. At higher temperatures and times, these investigators found an increase in the oxidation rate which they have attributed to a change in state of the oxide. This investigation, however, shows that this change is due to the linear oxidation rate becoming the dominant rate.

INTRODUCTION

Silicon carbide crystallizes in polar form and so exhibits a unique oxidation rate on each of its faces. Oxidation experiments to date have been performed by monitoring weight changes of the oxidizing semiconductor, or of its reaction by-products, (1,2,3). These experimental methods result in only one oxidation rate which is for the carbon face since the oxidation of this face dominates at lower temperatures and short oxidation periods used. In this work, the oxidation rate of both crystal faces is determined.

The oxidation of silicon carbide follows the equation

$$X^2 + AX = Bt \tag{1}$$

which gives the oxide thickness X as a function of oxidation time t.

In the limit of large oxidation times, which results in thick oxides, Eq. (1) yields the familiar parabolic law of oxidation growth:

$$X^2 = Bt \tag{2}$$

where B is proportional to the diffusion coefficient of oxygen through SiO_2 and is termed the diffusion rate constant.

329

For small t, which yields thin oxides, Eq. (1) can be approximated by the linear equation

$$X \approx \frac{B}{A} t \qquad (3)$$

where B/A is proportional to the surface chemical reaction rate.

Experimental Procedure

Oxidations were carried out in an open-ended system. The Furnace was a barrel type with three zones heated with SiC rods. The center zone's temperature was automatically controlled by a thermocouple inserted into the side of the furnace. A quartz tube was inserted into the furnace. The tube was necked down on one end and connected to flow meters which were used to control the flow of appropriate gases through the tube. The other end of the tube was open. A quartz slab was used as the vehicle for inserting crystals into the hot zone. The slab was moved in the furnace with a quartz push rod.

Silicon carbide crystals were given an initial cleaning in hot trichloroethylene and isopropyl alcohol, blown dry with nitrogen, and then etched in a dilute HF solution to remove any oxide initially present. The crystals taken from the acid solution were rinsed in six megohm water, blown dry with nitrogen, loaded onto the quartz slab, and inserted into the furnace tube. The crystals preheated in the hot zone for five minutes with 500 cc/min of N_2 flowing through the tube. After the preheat cycle, a flow of 500 cc/min of O_2 was established. Oxidation times ranged from approximately thirty minutes to two weeks. The crystals were moved from the hot zone to the cool end zone in one rapid motion, and allowed to cool for five minutes before they were removed.

During all oxidations, the temperature of the crystals was monitored by a platinum vs. platinum 10% rhodium thermocouple inserted from the open end of the tube and placed directly over the crystals.

Oxide thickness measurements were made by etching a groove completely through the SiO_2 layer using standard photoresist and etching techniques. The entire crystal face was then aluminized and the step height measured using a Normarski Polarized Interferometer. Approximately ten depth measurements were made on each crystal and the readings averaged to give the measured oxide thickness. With careful measurements, readings could be held to within $\pm$ 200 Å of the average. This gives accuracies ranging from $\pm$ 20% for thicknesses of 1000 Å to $\pm$ 1% for thicknesses of 20,000 Å.

An indication of the difference in oxidation rates on the two crystal faces was verified by oxidizing both faces of a piece of SiC for 70 hours at $1060^{\circ}C$. The thickness on one side was 900 Å, while on the opposite face, the oxide thickness measured 6500 Å. This is a 7 to 1 difference in oxide thickness. For future discussion, the side that had the thin oxide at low oxidation temperatures will be referred to as the "Thin Oxide Side," while the other side will be called the "Thick Oxide Side."

A catalog was made of ten silicon carbide crystals which could be consulted to determine which crystal face was the Thick Oxide Side and which was the Thin Oxide Side. These crystals were used repeatedly to obtain the oxidation data plotted if Figures 1 and 2.

Experimental Results

From Figure 1, which gives the oxidation data for the Thick Oxide Side, it is seen that above an oxide thickness of approximately 2500 Å the curves have a slope of one half. This means that in this region, the oxidation on the Thick Oxide Side follows the simple parabolic law. Below 2500 Å the curves increase in slope and tend to approach a linear oxidation rate. Taking the log of Eq. (2) and letting t = 1, we obtain

$$\text{Ln} \ (x) = \text{Ln} \ (B^{\frac{1}{2}}) \tag{4}$$

If the curve which lies above 2500 Å is extended to the left, B can be read directly as the intercept at t = 1. When this is done, the data in Table I are obtained. In Figure 3 are plotted the data for B given in Table I vs. temperature. A best straight line fit is made for the data points.

The curve of Figure 3 fits the equation

$$B = K \ \text{exp} \ (-\epsilon_a/kT) \tag{5}$$

where ϵ_a is termed the oxide diffusion activation energy, and is defined by Adamsky (2) as the energy necessary to initiate diffusion through the oxide and is computed from Figure 3 as approximately 2ev or 47 KCal/mole.

Figure 2 gives the experimental data taken for the oxidation on the Thin Oxide Side of the crystal. These curves have unity slope for all oxidation temperatures and times, and correspond to the region governed by chemical surface reaction. Taking the log of Eq. (3), we obtain

$$\text{Ln} \ (x) = \text{Ln} \ (B/A) + \text{Ln} \ (t) \tag{6}$$

By letting t = 1, we may obtain the values of B/A directly from Figure 2. The data obtained when this is done are tabulated in Table II and plotted in Figure 4 which yield on activation energy of approximately 3.7ev or 85 KCal/mole.

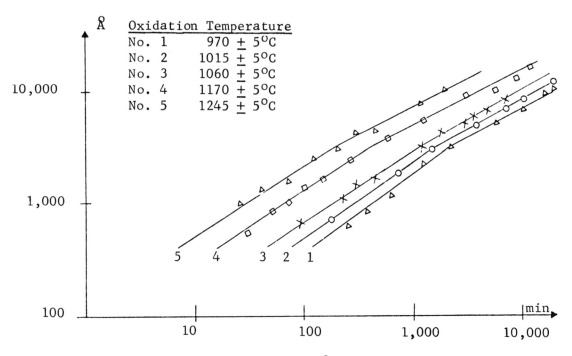

Figure 1. Oxide Thickness (Å) vs. Oxidation Time
(min.) on Thick Oxide Side

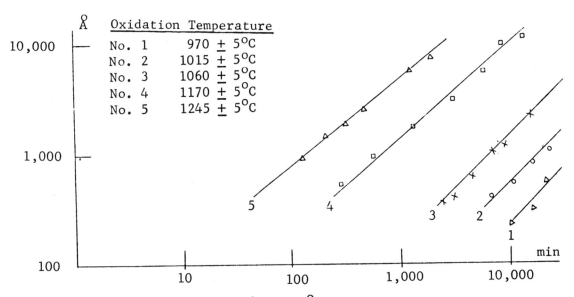

Figure 2. Oxide Thickness (Å) vs. Oxidation Time
(min.) on Thin Oxide Sides

TABLE   I

DIFFUSION RATE DATA

Curve No.	Temperature ^oC	$B^{\frac{1}{2}}$ $\overset{o}{A}/min^{\frac{1}{2}}$	B $\overset{o2}{A}/min$
1	970	49.5	2450
2	1015	56.0	3140
3	1060	68.0	4625
4	1170	150.0	22500
5	1245	190.0	36100

Discussion and Conclusions

One face of the crystal oxidizes at a linear rate for a period of time, indicating that a surface chemical reaction is the governing force. The oxidation rate then changes and follows a parabolic law meaning that the diffusion of oxidant through the oxide is the rate determining factor. The diffusion activation energy on this crystal face is found to be 47 KCal/mole. This figure compares well with values found for oxidation studies carried out by weight change measurements on powdered samples, (4,5).

The Thin Oxide crystal face oxidizes at a linear rate for all times and temperatures used, showing that it is the chemical surface reaction rate that dominates this oxide growth.

Examination of the oxidation curves in Figures 2 and 3 shows that for shorter oxidation times or lower temperatures the oxide on the Thick Oxide Side is indeed the thicker of the two. Thus, if one were measuring a weight change, this growth would dominate and the results would show a parabolic growth.

For longer oxidation times or higher temperatures, the oxide thickness on the Thin Oxide Side will eventually become thicker than on the Thick Oxide Side due to the linear growth rate. Under these conditions, oxide growth rate curves would show an increase after some time period.

Other investigators (5) have attributed the observed change in slope of such curves to a change of state of the oxide--the oxide changing from amorphous in the parabolic region to cristobalite in the linear region. It has been shown in this work that the rate changes due to the linear oxidation rate of the Thin Oxide Side becoming the dominant rate.

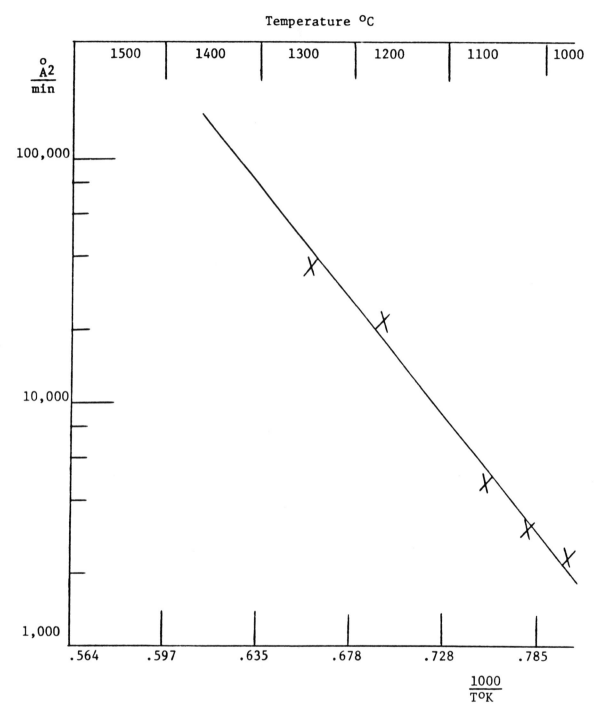

Figure 3.   Diffusion Rate Constant B vs. Oxidation Temperature for Thick Oxide Side

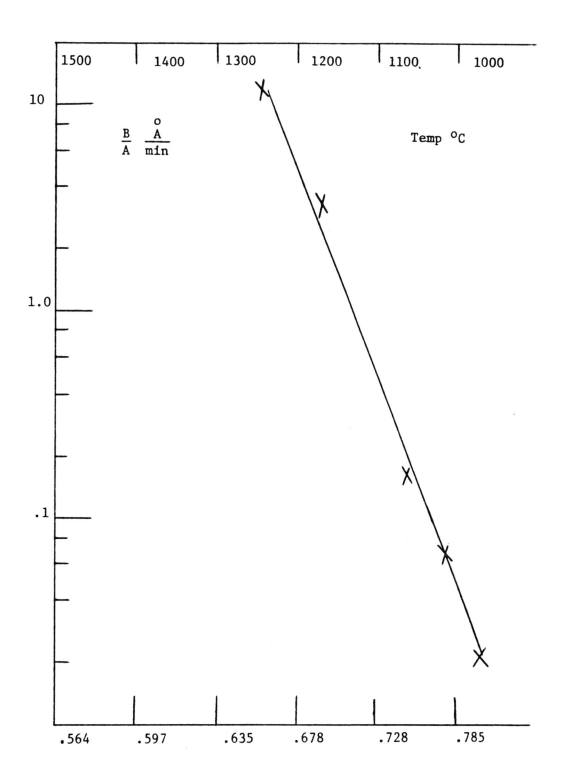

Figure 4.   Reaction Rate Constant B/A vs. Oxidation
Temperature for Thin Oxide Side

TABLE II

REACTION RATE DATA

Curve No.	Temperature $^\circ$C	B/A $\mathring{A}$/min.
1	970	.022
2	1015	.07
3	1060	.17
4	1170	3.50
5	1245	11.50

ACKNOWLEDGMENTS

The authors would like to thank Messrs. Ryan and Potter for their generosity in supplying silicon carbide crystals for this work.

REFERENCES

1.   P.J. Jorgensen, M.E. Wadsworth, and I.B. Cutler, "The Kinetics of the Oxidation of SiC," Pergamon Press, 1959.

2.   R.F. Adamsky, "Oxidation of Silicon Carbide in the Temperature Range 1200° C to 1500°C," J. Phys. Chem. 63, 1959.

3.   E.A. Gulbransen, K.F. Andrew, and F.A. Brassart, "Oxidation of Silicon Carbide at 1150° to 1400°C and at 9 X 10^{-3} to 5 x 10^{-1} Torr Oxygen Pressure," J. of Electrochemical Soc., Vol. 113, No. 12, December 1966.

4.   P.J. Jorgensen, M.E. Wadsworth, and I.B. Cutler, "Oxidation of Silicon Carbide," J. Am. Ceramic Soc., Vol. 42, No. 12, December 1959.

5.   G. Ervin, Jr., "Oxidation Behavior of Silicon Carbide," J. Am. Ceramic Soc., Vol. 41, No. 9, September 1958.

Charge Carrier Concentration and Mobility in n-Type 6H Polytypes of SiC

R. B. Hilborn, Jr., and H. Kang

Although electrical properties such as resistivity and mobility have been reported for SiC for many years, the strong dependence of these properties on crystal structure (other than between α and β) has only been distinguished since the late 1960's[1,2]. As such, considerable work still remains to be done in the area of characterizing the polytypes of SiC by their electrical properties. This paper represents an endeavor in this direction, in that we wish to present a set of curves which can be used to correlate the impurity concentration in uncompensated n-type 6H SiC with Hall mobility and resistivity at 300°K. Conversely these curves can be used to determine whether or not a sample is of the 6H polytype.

The electrical characteristics of materials are most generally described by magneto-electric transport measurements from which one is able to ascertain values of resistivity, mobility and carrier concentration. The results of such measurements taken at a temperature of 300°K on a number of different specimens of n-type 6H SiC are shown in figure 1. Here we have made a composite log-log plot of the resistivity versus charge carrier concentration and Hall mobility versus charge carrier concentration. The carrier concentration was determined from the relation: $n = 1/(eR_H)$ where R_H is the measured Hall coefficient. The curves shown resulted from data taken from measurements of our own and published data[1,2] on uncompensated or slightly compensated samples. Nitrogen impurities were the donors for all samples. The maximum deviation from the curves for data taken from any one sample was 15%.

It should be noted that the abscissa of figure 1 represents the charge carrier concentration, n, and not the density of donor impurities, N_D. The conduction being predominantly extrinsic at 300°K, the charge carrier concentration can be

taken as being equal to the concentration of ionized donors, N_D^+.
This quantity can then be related
to the donor concentration. The
donor levels for nitrogen in SiC
are sufficiently deep that not
all of them will be ionized at
$300°K$. If we assume the ionization mechanism for nitrogen donors
in 6H SiC to be the same as that
for singly charged donors such as
phosphorous, in silicon, then the
concentration of ionized donors
can be calculated in the standard
manner[3].

The relation between N_D^+ to
N_D is dependent on the ionization
energy of the donor, E_D. As the
ionization energy for nitrogen in

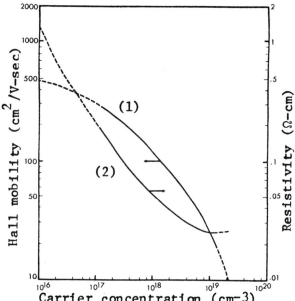

Figure 1:   Hall mobility[1] and resis-
            tivity[2] vs carrier
            concentration for n-type
            6H SiC at $300°K$.

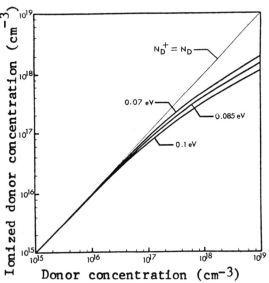

Figure 2:   Ionized donor concentra-
            tion vs donor concen-
            tration for nitrogen
            in 6H SiC at $300°K$.

SiC varies from sample to sample (between
0.07 eV to 0.10 eV), a set of curves, one
for each E_D, is required to adequately
relate N_D^+ to N_D for this impurity. Such
a set of curves is shown in a log-log
plot in figure 2, where the parameter
indicated is the ionization energy used
for the computation of the curves associated with it.

In order to obtain the greatest accuracy in relating the nitrogen concentration to mobility or resistivity, it is
necessary to determine the ionization
energy associated with the nitrogen impurity in each sample considered. This
typically requires the measurement of the
charge carrier concentration as a function

of temperature. If this information is not readily available, however, one can see from figure 2 that using a value of $E_D = 0.085$ eV will enable the determination of N_D to within less than one-third of an order of magnitude for a value of $N_D^+ = 10^{18}$ cm^{-3} and to considerably greater accuracy for smaller values of N_D^+.

The curves presented here should be of some help to researchers in SiC by providing a ready source for correlating basic room temperature electrical properties of n-type 6H SiC to the concentration of nitrogen donors in the crystal. The advent of similar data for other polytypes of both n- and p-type material will be of considerable importance in the future to engineers as they try to incorporate the use of SiC for the practical design of solid state electronic devices.

Acknowledgments--We wish to thank the Westinghouse and General Electric Corporations for providing us with samples and Mr. Y. Tung of our laboratory for his identification of the polytype of each sample. This work was partially supported by the Air Force Cambridge Research Laboratories under contract #F19628-72-C0136.

(1)   D. L. Barrett and R. B. Campbell, J. Appl. Phys., 38, 53 (1967).

(2)   S. H. Hagen and C. J. Kapteyns, Philips Res. Repts., 25, 1 (1970).

(3)   R. A. Smith, Semiconductors, p. 90, Cambridge University Press, 1968.

PART IV
NON-ELECTRONIC APPLICATIONS

Non-Electronic Applications of SiC

Peter T. B. Shaffer

The electronic applications, single crystal growth, dislocation studies, crystal structure determinations, electroluminescence represent the glamour of silicon carbide. The tonages, profits and mass production fall into the areas of abrasive, refractories, and chemical uses. These are the every day meat and potatoes of the SiC industry.

To attempt to do more than to skim over these uses very briefly would be impossible in the time allotted. Not too surprisingly the amount of SiC used increases as the required purity decreases and also as the ease of fabrication increases. According to the 1970 Minerals Yearbook the annual production of SiC in the U.S. and Canada was 150,000,000 KG (3.3 x 10^8 lbs) at a total value of $24,000,000. This averages out to 16¢ per KG (7¢ per lb).

The largest single use for SiC is as an additive in ferrous metallurgy, to deoxidize the steel and to make it more readily machinable. While SiC is quite inert under many conditions, it is extremely reactive toward many metals, such as the transition metals of Groups IV through VIII (Ti, V, Cr, Mn, Fe, Co, Ni groups). It dissolves as carbides and silicides, both of which react with dissolved oxygen. Purity required for this use is about one nine, 90%.

Next in importance are the refractory and abrasive applications. These require purities of the order of two nines or slightly less (95-99%). The next major use, and on a tonnage basis it is almost negligible, is in the manufacture of electrical heating elements and resistors. The purity required is slightly less than three nines, 99.5-99.9%.

As yet there are no major commercial applications of SiC requiring three nines or higher purity. Single crystal thermistors and nuclear applications will probably fall in this range.

As an indication of some current interests in SiC for non-electronic uses, I went through three volumes of <u>Chemical Abstracts</u> and noted the frequency of occurrence of several non-electronic applications. I fully expected the relationship to hold that the most common usage would be the least commonly published but what I found was not so. Roughly in order of frequency the uses mentioned were --

Whiskers - Composites	20
Refractories	10
Ferrous Metallurgy	7
Nuclear	7
Abrasives	6
Pyrolytic	5
Wear Resistant Surfaces, Friction Coating	3
Heaters, Resistors	4
Magnetic Tape	1
Filters	1

Publications follow in order of frequency both the most highly established and most far out uses with the commonly used and immediately pending applications receiving less attention.

Consider then, the various uses of SiC arranged by the mode of fabrication used.

Grain, Powders
Bonded (by a second phase)
Recrystallized
Self Bonded - Reaction Bonded
Hot-Pressed
Pyrolytic
Single Crystals (sublimed)

The degree of difficulty of producing shapes and structures falls roughly in this order with the Acheson process the simplist best understood and most easily controlled. Nobody here will dispute the fact that single crystal growth is the most difficult method of fabricating masses of SiC.

GRAINS, POWDERS

As manufactured commercially in the Acheson Furnace a mixture of sand and

coke is arranged around a packed graphite resistor placed between two end terminals.
Brick work in movable steel frames contain the mix in the reaction zone. This
method permits the production of huge quantities of a useful material. The
unreacted and partially reacted charge becomes the insulator to retain the heat
within the reaction zone. During recrystallization the heat of crystallization
is liberated within the unreacted charge and is not lost. The unreacted, and
partially reacted charge are quite effective insulators.

After the passage of current, equivalent to 100,000 to 200,000 kwh, the
power is turned off and the sides are removed letting the unreacted material fall
away and increasing the rate at which the reacted charge cools. What is left is
a long cylinder of SiC with a semi hollow core in which is found the graphite left
behind as part of the SiC was decomposed. Crystallite size decreases from inside
to outside.

After cooling the cylinder is broken up and the product is hand sorted to
specifications depending on the final intended use. Such sorting can be on the
bases of apparent bulk density, lack of contaminating phases, or color for example.
The material is then crushed, screened and cleaned by roasting, magnetic separation and acid leaching, again dependent on the final intended use.

As produced, the crude grain as it is called is ready for use in ferrous
metallurgy as a steel additive, as a BOF fuel, in ceramics or abrasives when a
binder will be supplied or in chemical processes such as in chlorination to produce silicon tetrachloride.

A second, similar process for producing loose powders is in the tube furnace.
This process is chemically essentially the same except that heat is provided from
the outside of a crucible in which the reactant powders are retained. Not only
is the method inefficient due to heat losses outward, but the amount of material
present in the reaction zone at any one time is small. On top of this the SiC
powder first formed inside the crucible is an efficient insulator which prevents
the rapid conduction of heat to the unreacted charge. Certain specialty items
such as fine powders and whiskers have been made by tube furnace processes. None
has become commercially significant, however.

BONDED

The second most common way of preparing SiC into a form for use is by the
addition of a bonding phase. Here powders, prepared by crushing and screening

the product from the Acheson Furnace are agglomerated by means of a bond phase, usually a material which will melt, cure or sinter at lower temperatures. For high temperature uses the low temperature bond phase limits the high temperature applicability of the resulting products. Such products typically are refractories or abrasives.

Certain powders bonded by glassy phases are used for nonlinear resistors (varistors) and for lightning arrestors.

RECRYSTALLIZED

Silicon carbide does not melt congruently but dispropertionates into a silicon rich vapor and a carbon-rich residue. Lacking a true molten phase, and high temperature creep properties it will not sinter in the usual ceramic sense. When a 1 inch cube of powder containing a fugitive binder is heated it becomes stronger. It does not shrink or densify but forms a skeleton of inter granular bridges. The final dimensions after recrystallization will be 1 inch.

Typically, items made in this way are heating elements and refractories. Such recrystallized bodies may be impregnated at a later time to improve certain properties.

SELF BONDED

If a cold pressed or extruded compact of SiC and carbon in the proper proportions is exposed to a vapor containing silicon, the carbon is converted to SiC which bonds the original SiC crystallites. Since silicon carbide formed from carbon occupies greater volume than the original carbon, a degree of densification occurs. Judicious choice of ratios of SiC to carbon could theoretically lead to a fully dense body. In practice, this is not possible since the formation of a dense SiC layer on the outside would prevent the diffusion of additional silicon in to convert the interior carbon. Diffusion through SiC is prohibitively slow.

In practice a density of 98-99% is attempted, with the remaining 1-2% of the pores filled with silicon. Depending on the final use this free silicon can be removed by heating under carbon or in vacuum.

Dense, self bonded SiC parts, as they are called, made by this process include ceramic armor, (Figures 1 and 2) wear resistant parts such as suction box covers used in paper making machines, (Figure 3) dense SiC heat exchange pipes, muffles, and even a pre-shaped electric heater! In most cases the green, unfired body is

shaped to the final dimensions before it is "shot" with silicon. Where tolerances
or surface finishes are critical final diamond grinding is used.

This process lends itself to large relatively complicated shapes. Its main
restriction is in thickness. It becomes difficult to obtain complete penetration
and conversion if any point in the structure is more than an inch or two from the
surface.

A modification of this process involves the direct conversion of a pure carbon
precursor to SiC by exposure to silicon or silicon monoxide vapor. Depending on
the density of the carbon precursor the final SiC body could be a foam, a relatively dense structure or a SiC coated carbon composite.

Recent developments in the field of automobile antipollution measures are
moving toward two possible applications of self bonded SiC. The first of these
is the automobile exhaust post burner. The engine heat released with the hot
exhaust gasses into a SiC exhaust manifold would cause secondary combustion with
injected air to remove unburned hydrocarbons and carbon monoxide. Similarly,
studies are underway to determine whether SiC can be used as a catalyst support
for the catalytic post-combustion of automobile exhaust gasses.

Gas turbine engines using self bonded SiC parts are being studied. As is
so often the case with refractory materials, engineers are unable to design around
the brittle, catastrophic fracture inherent with SiC. Many parts of self bonded
SiC are in use for pump parts to move liquid metals.

HOT PRESSED

Not only will silicon carbide not sinter, it will not yield and densify even
under the usual hot-pressing conditions. In order to obtain the near theoretical
density in hot pressed SiC it is necessary to add a second material, one which
will react with SiC and/or form a separate liquid phase. Typically, either MgO,
iron or aluminum is added.

As in the case of bonded SiC, the addition of a densifying agent results in
a premature decrease on the high temperature properties, notably strength and
elastic modulus of SiC.(1, 2)

As yet no major market for hot pressed SiC has developed. This is no doubt
due to the relative versatility of the self bonding process, and to the high cost
and limited geometry of hot-pressed pieces.

During hot pressing care must be taken to minimize grain growth to achieve maximum strengths in the final products. Notably, small additions of boron cause many orders of magnitude growth in crystallites from submicron to many millimeters in a period of minutes (3). Silicon in a free state has much the same effect but to a lesser degree. The presence of free carbon has the opposite effect, in that it largely prevents the growth of grains. Free carbon however has a deleterious effect on other properties, especially strength and elastic modulus.

PYROLYTIC

When certain silicon and carbon containing vapors come in contact with a hot substrate under proper conditions of stoichiometry, silicon carbide is formed. Over a prolonged period the SiC will continue to deposit theoretically to any final thickness desired. Depending on numerous conditions of time, temperature, vapor concentration, velocity, nucleation, etc., such pyrolytic or GVD coatings can be made over a range of crystallite sizes (100's of Angstroms to millimeters) and with varying degrees of crystal orientation.

Typically, $SiCl_4$ and toluene or a substituted chlorosilane are the reactants. Care must be taken to avoid the deposition of second phases, either Si or C. This can to a large extent be done by using a silicon rich system with excess hydrogen in the carrier gas.

Depending on the choice of substrate on which the SiC is deposited, the final structure can be a layered composite, or the conditions in the apparatus can be changed and the substrate removed to yield a free standing SiC structure.

The most notable example of pyrolytic SiC is the coaxial filament produced for reinforcing purposes. One could argue the fact that this product has not yet become truly a commercial one, but there seems little doubt that it is on the way to becoming one.

Pyrolytic coatings seem destined to find a place as coatings on nuclear fuels to prevent oxidation and to retain gaseous fission products. SiC has a low expansion on exposure to nuclear radiation which strongly favors it for these uses.

SINGLE CRYSTALS (SUBLIMED)

The final method of fabrication I'd like to mention and the one with which you are probably most familiar is sublimation. This method involves extremes in temperature ($\sim 2500^{\circ}$C) and time (many hours) and yields products more readily measured in grams than in pounds. It can and has been used to produce dense

(99+% of theoretical), pure, single phase bodies (4) (see Figure 4) whose mechanical properties (elastic modulus, strength) remain practically unchanged to temperatures in excess of $1500^{\circ}C$ (2). It is doubtful if such a process will ever become commercially feasible.

From the foregoing, it becomes evident that to no small degree the uses of silicon carbide are restricted by the difficulty in fabricating it into useful geometries, and by the temperature limitations of the phases used to bond it into useful shapes. Consider the properties of SiC, other than its difficulty in fabrication and determine how these find applicability.

Mechanically SiC is a hard, rigid, brittle solid which does not yield to applied stresses even at temperatures approaching its decomposition temperature. Its hardness and conchoidal fracture mode have made it one of the most useful synthetic abrasives. There are several harder materials but none that combine both hardness and mode of fracture. The latter is most important in producing a "self-sharpening" abrasive, as opposed to one which tends to round and become dull on use.

Its high strength and high elastic modulus make it a natural for use as a pyrolytic coaxial reinforcing filament, especially at high temperatures. The single crystal whisker forms represent the ultimate in strength and elastic modulus. Their use in practical applications in composites is far in the future not only due to cost but to inability to fabricate useful composites consistently.

SiC does not deform at high temperatures. As a result it is not capable of hot pressing without additives. At the same time it provides a heating element which will operate for extended periods unsupported without physical deformation.

Silicon carbide has a thermal conductivity comparable to that of metals (5). This alone suggests its excellence as a muffle or a heat exchanger pipe. Its thermal expansion coefficient is much lower than that of other typical ceramics and refractory metals. These two properties combine to give it outstanding thermal shock resistance.

Chemically SiC is resistant to aqueous acids, alkalis and to oxidation. Such properties lead to its use in fractories, pumps for corrosive fluids and metals, etc. On the other hand, it is vigorously reactive toward certain metals and metal oxides, which lend to its use as an additive in metallurgy. It is used to deoxidize ferrous alloys and as a fuel in the B.O.F. (basis oxygen furnace) process for steel manufacture.

MAJOR APPLICATIONS OF SiC

Present

    Additive to deoxidize steel

    Abrasive

    Refractory, ceramic

    Heating element

    Raw material for $SiCl_4$ manufacture

    Armor

    Electrical resistor, varistor, lightning arrestor

    Wear resistant parts

Future

    B.O.F. fuel

    Nuclear fuel cladding

    Gems

    Pigment, opacifier

    Reinforcement; whisker or filaments

    Protective coatings

As is typically the case with SiC, it is a paradox. It is either very reactive or it is inert; it is either a good thermal conductor or it is a thermal insulator. None of its properties are average. The only thing standing in the way or order of magnitude increase in applications is the difficulty in fabricating it to a useful geometry. Therein, like the problem of growth of perfect, pure, predictable crystals, lies the challenge and the promise of the future of SiC.

REFERENCES

(1) P. L. Farnsworth and R. L. Coble, J. Am. Ceram. Soc., 49, 264-8 (1966).

(2) P. T. B. Shaffer and C. K. Jun, Matl. Res. Bull, 7, 63-70 (1972).

(3) P. T. B. Shaffer and C. K. Jun, - unpublished results.

(4) P. T. B. Shaffer, Matl. Res. Bull., 4, 5-13-524 (1969).

(5) G. A. Slack, J. Appl. Phys., 35, 3460-6 (1964).

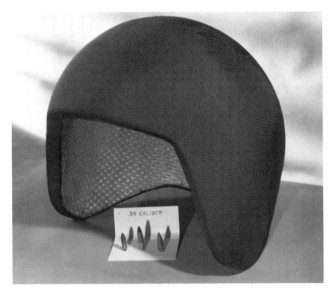

Figure 1. Ceramic Armor Helmet

Figure 2. Ceramic armor after stopping armor piercing bullet.

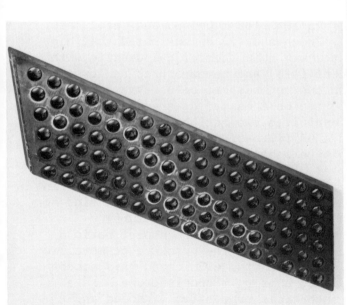

Figure 3. Dense silicon carbide suction box cover for paper making machine (length 18")

Figure 4. Dense, single phase body of silicon carbide prepared by sublimation.

The Effect of Carbon Additions on the Grain Growth and Strength of Hot Pressed SiC

R. L. Crane and G. W. Hollenberg

INTRODUCTION

Silicon carbide and silicon carbide based materials are being actively considered for use in structural components in the hot sections of air breathing propulsion systems. In this environment materials must possess high strength and impact resistance. It was with this in mind that the present investigation was initiated.

Recent emphasis on the synthesis of polycrystalline silicon carbide has contributed greatly to our understanding of its fabrication and mechanical properties. Alliegro et al. (1) showed that only a few elements could be used as densification aids in hot-pressing; and at this time aluminum (or alumina) and boron seem to be the most commonly used. More recently Prochazka (2) demonstrated that high purity, submicron SiC could be hot pressed to very near theoretical density with boron additions as small as 0.5 weight percent (wt%) and temperatures as low as 1950°C. The strength of SiC produced by various methods has been reviewed recently by Rice (3). He noted that, when plotted on a master diagram of strength versus square root of grain size, the data exhibited two regions of behavior. In the large grain size region the data showed a Griffith-Orowan type behavior, and in the small grain size region a Hall-Petch type behavior. The strength versus square root of grain size data of Forest et al. (4) similarly exhibited two regions of behavior. Their data was for material that consisted of α SiC and a fine particle size β SiC in a free silicon matrix. Gulden (5), however, could find no correlation between grain size and strength of β SiC at room temperature. Prochazka (2) observed that the growth of strength limiting,high aspect ratio α grains in a fine grain β matrix could be correlated with the presence of metallic silicon. The addition of small amounts of carbon (1 wt%) eliminated this discontinuous grain growth and yielded a fine grained, uniform microstructure. As a result, carbon rich specimens had superior mechanical properties. Rhodes and Cannon (6) hot-pressed billets of silicon carbide containing carbon fiber additions from 5 wt% to 50 wt% and observed a substantial increase in impact strength of samples containing 10 wt% fibers. They did not, however, note any decrease in grain size with the carbon additions.

EXPERIMENTAL PROCEDURE

Specimens used in this study were prepared from appropriate mixtures of silicon carbide and boron powders and carbon fibers. The silicon carbide powder was purchased from the Carborundum Co., Niagara Falls, New York with an average particle size of 7 μm. The weight percentage composition of the material according to the supplier was as follows: 99.00 SiC, 0.30 SiO_2, 0.27 Si, 0.04 Fe, 0.05 Al, 0.03 C. The boron powder was purchased from U.S. Borax Chemical Corp., Anaheim, California with a maximum particle size of 5.0 μm. Thornel 50 carbon fiber was purchased from Union Carbide Corp., Cleveland, Ohio. In its virgin state, the fiber has an average strength and modulus of 300 ksi and 50×10^6 psi respectively. Fiber tows were chopped into 1/4 inch (in.) long pieces before being added to a batch. After milling with SiC powder, it was noted that the fibers had been reduced to an aspect ratio of about 10 to 1.

Appropriate mixtures of the above materials were wet milled in a polyethylene bottle, containing two alumina balls and acetone, for 8 hours. After drying, batches of 150 grams each were hot pressed to produce nearly theoretically dense billets which were 2.5 in. in diameter and 0.75 in. high. A graphite die was lined with pyrolytic graphite sheet, before adding the powder, to facilitate pressing and reduce contamination. This assembly was placed in a Vacuum-Industries 15 kw hot press for consolidation in vacuum. The die and powder were first heated to 1000°C and held for 30 minutes to allow outgasing of the powder. Then maximum power was applied to bring the assembly to a temperature between 1950° to 2300°C where a pressure of 5000psi was applied. The specimen was pressure sintered for 15 minutes and then cooled to room temperature by reducing the power to zero over a 20 minute time span. The addition of carbon fibers significantly enhanced densification rate as Rhodes and Cannon (6) have also observed.

The silicon carbide billets were initially characterized by measuring their densities by the standard Archimedes method. Billets whose density was not at least 98.5% of theoretical were discarded; except for 0 wt% and 1 wt% boron specimens used in the strength characterization phase of the study. Assuming a carbon fiber density of 2.2 gm/cc, theoretical densities were calculated using the simple law of mixtures. X-ray diffraction analysis tentatively identified both the starting powder and hot-pressed billets as α SiC, type 6H. Although graphite was also noted, no boron containing compounds were indicated.

GRAIN GROWTH

It was noted during the initial phase of this study that a grain size discrepancy existed between billet that contained carbon fibers and those that did not. In billets with 0 and 1 wt% boron additions the grain size varied inversely with carbon fiber content; for example, from 20-23 μm for 0% or 1% fiber additions to 7-11 μm for 15-30 wt% additions. Although many attempts were made to reduce the grain size of 0 and 1 wt% carbon fiber billets without sacrificing density, all were unsuccessful. To document the affect of carbon fiber additions on the grain growth of SiC, the following study was conducted. All specimens contained 3 wt% boron and were hot pressed at 2250°C. Individual billets were cut into specimens measuring 0.25 x 0.25 x 0.5 in. These were annealed with a graphite resistance furnace in a purified argon atmosphere. Temperatures were measured and held to a precision of +10°C with a calibrated optical pyrometer. After annealing, individual samples were ground on progressively finer diamond impregnated laps; with the final polishing step being performed on a 1 μm lap. After being thermally etched at 1600°C for 15 minutes in vacuo, the average grain size was determined with the linear intercept method (7). This technique may incur significant bias errors when applied to the high aspect ratio grains encountered in

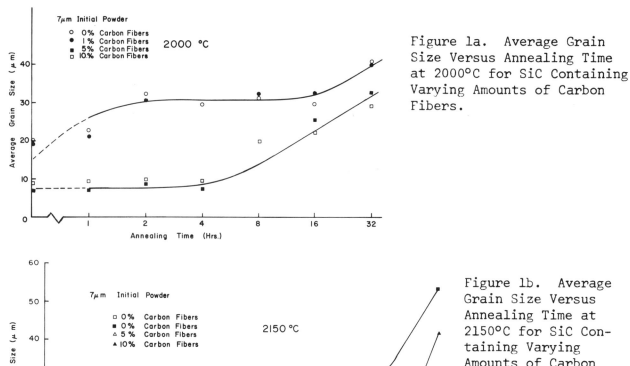

Figure 1a. Average Grain Size Versus Annealing Time at 2000°C for SiC Containing Varying Amounts of Carbon Fibers.

Figure 1b. Average Grain Size Versus Annealing Time at 2150°C for SiC Containing Varying Amounts of Carbon Fibers.

this study. However, the method was thought to be sufficiently accurate since comparative and not absolute measurements were sought. At least 250 grains were counted for each determination. Unavoidable grain pullouts were not included in the counting procedure.

Figures 1a and 1b demonstrate the dramatic effect that carbon fiber additions can have on the grain growth kinetics of SiC. It should be noted that at both temperatures an average grain size of less than 10 μm could be maintained for as long as 4 hours for 5 and 10 wt% fiber additions. On the other hand, specimens with only 1 or 0 wt% carbon fibers grew in a matter of one hour to a stabilized grain size of approximately 30 μm. It is in this first period of growth (or no growth) which determines the strength of the hot pressed material as will be shown later. However, at longer times (16-32 hours) average grain sizes became unstable and the average grain size began to increase again. In Plate 1 typical photomicrographs show the elongated grains that form in 0 wt% carbon fiber specimens during initial hot pressing, and the smaller isotropic grains of the specimens containing 5 wt% carbon fibers. It appeared that during annealing the carbon second phase coalesced into larger more equiaxed particles. Very large discontinuous grains typified the microstructures of the 16 and 32 hour specimens in which grain growth was reinitiated.

The concept of grain growth inhibition and limiting due to second phase particles is normally attributed to the "Zener Criterion" (8) although Haroun and Budworth (9) have attempted to improve upon it. Hillert (10) has recently presented an eloquent mathematically based interpretation of the same phenomenon. The Zener criterion for growth inhibited by spherical second-phase inclusions (carbon fibers in this case) is:

$$\text{Grain Size} = A\frac{r}{f} \tag{1}$$

where r is the average inclusion radius; f, the volume fraction of inclusion; and A, a constant of 0.15 for the Haroun and Budworth's modified criterion. For the case of 10% carbon fibers, limiting grain size of 10 μm would be predicted by the modified Zener criterion. Since the fibers possess an "effective diameter" of 7 μm, 10 μm is, of course, in good agreement with results. Specimens containing no fibers would be expected to grow continuously. However, the 3 wt% boron densification aid may have acted to limit growth to 30 μm. Notice that approximately the same limiting grain sizes were obtained at each temperature which is in fundamental agreement with the theory. Initial growth data was not sufficient to delineate a rate dependence.

The occurrence of the reinitiated region for growth can not be interpreted in terms of Zener's criterion. However, Hillert provided a simple interpretation for the data. First, normal or continuous grain growth occurs until a limiting grain size is reached due to the pinning constraint of second phase particles. By circumstances, a few particles are slightly larger than the normal distribution (for instance r=2r avg) at the point of limited growth, and these grains slowly begin to increase their rate of abnormal growth until they begin to control the overall grain size of the material. The experimentally observed coalescence of carbon into large particles was also predicged by Hillert.

MECHANICAL PROPERTIES

The following study was conducted to determine the affect of carbon fiber additions on the mechanical properties of silicon carbide. Specimens measuring 0.1 x 0.2 x 2.0 in. were cut from billets that ranged in composition from 0 to 30 wt% fibers and contained 0, 1 or 5 wt% boron. Specimens containing no fibers were polished on a 1 μm diamond lap to reduce the surface machining damage. No other bars were polished since the fibers represent much larger defects than any that resulted from machining. All samples were tested in four point bending on an Instron Testing machine at a crosshead speed of 0.005 in./minute. The load was applied by hardened steel pins 0.8 in. apart, while the bars were supported by pins 1.6 in. apart. At least four bars were tested for each composition.

The expected effect of carbon fiber additions on SiC was to increase its strength, by limiting grain growth, and to lower its modulus. Figure 2 illustrates that this did occur. Note that in the case of the 30 wt% fiber specimens, some "plastic" strain did occur before failure. Figure 3 shows the strength of silicon carbide as a function of fiber content at three different boron levels. The vertical bars at each composition represent the range of values observed during testing. The strength of the 0 and 1 wt% boron specimens increased to a maximum between 5 and 10 wt% carbon. This is the range of composition which limited grain growth during hot pressing. Examination of the 0, 5, 15 and 30 wt% fiber specimens revealed average grain diameters of 23, 9, 7, and 7 μm respectively. For compositions greater than 5 wt% to 10 wt% fibers, strength decreased rapidly.

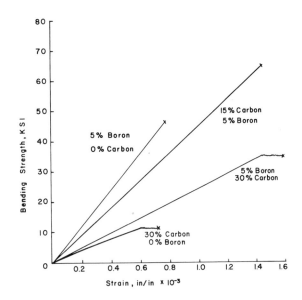

Figure 2. Typical Stress-Strain Curves for SiC Specimens Showing the Effects of Carbon Fiber and Boron Additions.

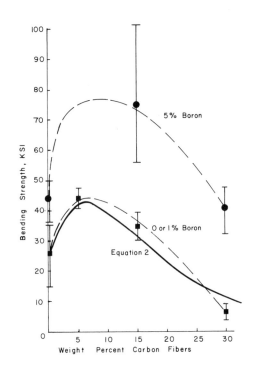

Figure 3. Four-Point Bend Strength of SiC Versus Carbon Fiber Content.

The affect of carbon fiber additions on the strength of SiC may be interpreted in light of well established theories relating strength to grain size and porosity. For simplicity it was assumed that the Knudsen equation (11) was applicable to this analysis. When combined with the Ryshkewitch (12)-Duckworth (13) porosity correction factor, the following expression was obtained:

$$\sigma = B(\text{Grain Size})^{-1/2} e^{-pC} \qquad (2)$$

where σ is the mean fracture strength, p is the volume fraction porosity, and B and C are constants. The porosity factor was included, based on the premise that the fibers act as pores in SiC. This is reasonable when the thermal expansion coefficients of the carbon fiber (α (axial) $\simeq 0$ and α (radial) $\simeq 6.5 \times 10^{-6}$°C^{-1} and polycrystalline α silicon carbide ($\alpha \approx 3.6 \times 10^{-6}$°C^{-1}) are compared. Thus, the large difference in expansions may cause separation of the fiber and matrix upon cooling. The size of any voids formed by this process would depend on the fiber-matrix bond strength and the thermal strains induced by the cooling process. Using the above grain size data for the four point bend specimens and a C=5, after Rice (3), equation 2 qualitatively predicts the strength-fiber content relationship as seen by the heavy, solid line in Figure 3. Above 10 wt% carbon the grain size is very nearly equal to that of the starting material and the grain size portion of the expression remains essentially constant. Fiber toughening effects were not taken into account since this would require considerably more data, such as, fiber matrix bond strength, etc., which are not available at this time.

There is a substantial difference between the strengths of specimens containing 0 and 1 wt% boron and those containing 5 wt% boron. A comparison of the scanning electron fractographs of two 30 wt% carbon specimens containing 0 and 5 wt% boron respectively indicates the reason for this behavior. Plate 2a shows the fracture surface of a 0 wt% boron specimen. Poor intergranular bonding is evident. In contrast, the fractograph of a 5 wt% boron specimen, Plate 2b, reveals better

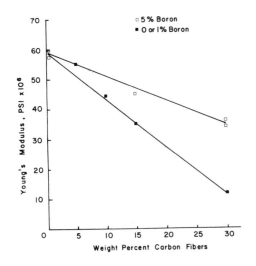

Figure 4. Young's Modulus of SiC Versus Carbon Fiber Content.

intergranular bonding. The fractograph of a 5 wt% boron, 15 wt% C Specimen that broke at 102 Ksi stress was included for comparison; Plate 2c.

The Young's modulus of the four point bend specimen is shown in Figure 4. The modulus of SiC decreases linearly with carbon content. For the same reasons outlined above, the 0 and 1 wt% boron specimens had lower moduli than the 5 wt% boron specimens.

CONCLUSIONS

Briefly, it has been demonstrated that carbon fiber additions can limit the grain size of hot pressed silicon carbide. Grain growth of α SiC containing various amounts of carbon was documented at 2000° and 2150°C. These data were interpreted using a modified Zener criterion. The strength and modulus of specimens containing from 0 to 30 wt% carbon were determined. Assuming that the fibers acted as pores, a model was presented to explain the increasing and then decreasing strength as a function of increasing carbon content.

REFERENCES

1. Alliegro, R.A., Coffin, L.B., and Tinklepaugh, J.R., J. Amer. Ceram. Soc. vol. 39, p. 386, 1956.
2. Prochazka, S., Final Report on Navy Contract N00019-72-C-0129, Dec. 1972.
3. Rice, R.W., Proc. Brit. Ceram. Soc., No. 20, p. 205, 1972.
4. Forrest, C.W., Kennedy, P., and Shennan, J.V., Proc. of the 5th Symposium on Special Ceramics, Ed. by P. Popper, Brit. Ceram. Res. Assoc., 1968.
5. Gulden, T.D., J. Amer. Ceram. Soc., vol. 52, p. 585, 1969.
6. Rhodes, W.H., and Cannon, R.M., NASA Rpt. No. NASA-CR 120966.
7. Mendleson, M.I., J. Amer. Ceram. Soc., vol. 52, p. 443, 1968.
8. Zener, C., private communication to C.S. Smith cited in Trans. AIME, vol. 175 p. 15, 1949.
9. Haroun, N.A. and Budworth, D.W., J. Mat'l. Sci., vol. 3, p. 326, 1968.
10. Hillert, M., Acta Met., vol. 13, p. 227, 1965.
11. Knudsen, F.P., J. Amer. Ceram. Soc., vol. 42, p. 376, 1959.
12. Ryshkewitch, E., J. Amer. Ceram. Soc., vol. 36, p. 65, 1953.
13. Duckworth, W., Ibid, p. 68.

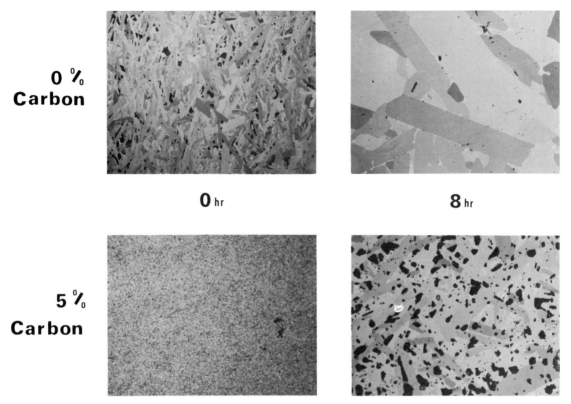

0 %
Carbon

0 hr   **8** hr

5 %
Carbon

Plate 1.   Photomicrographs of SiC Specimens Containing 0 and 5 wt% Carbon Fibers.
The Two Conditions Shown are as Fabricated and After an 8 hour Anneal at 2000°C. 200X

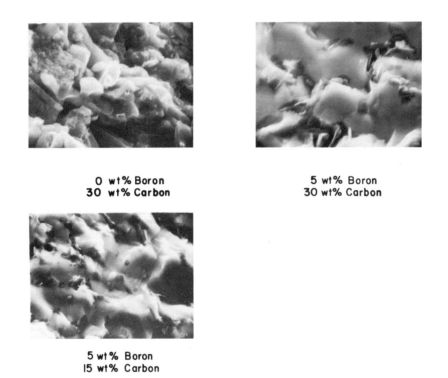

0 wt% Boron
30 wt% Carbon

5 wt% Boron
30 wt% Carbon

5 wt% Boron
15 wt% Carbon

Plate 2.   Scanning Electron Micrographs of Four Point Bend Strength Specimens. 1700X

Engineering Applications of REFEL Silicon Carbide

P. Kennedy and J. V. Shennan

Self-bonded silicon carbide was developed as a cladding for nuclear fuel in UK gas cooled reactors because of its potential for resisting irradiation damage, thermal stress and oxidation in CO_2 at high temperatures. Fabrication development, based on reaction sintering of SiC/graphite bodies produced a material with exceptional physical and mechanical properties - now designated REFEL silicon carbide. This material is fully dense, impermeable to gases and resistant to oxidation up to $1400^{\circ}C$. The combination of high strength, 500 MW/m^2, conductivity 50 W/cm at $1000^{\circ}C$, and low expansion, $4.3 \times 10^{-6}/^{\circ}C$ makes it suitable for operating under conditions of high thermal stress. With its additional property of high hardness REFEL silicon carbide is finding wide application in fields where abrasion resistance is important.

INTRODUCTION

For many years it has been known that SiC has outstanding mechanical and physical properties but it has also been appreciated that these properties are only realised in a continuous SiC matrix with little or no porosity. Such a material may be fabricated by reaction bonding, a process in which a SiC/graphite body is exposed to silicon and the graphite converted to bonding SiC. Thus the bond is formed in situ, and there is no dimensional change on firing. The 'green' material may be formed by a variety of routes, e.g. extrusion, isostatic compaction and warm moulding, and the green body may be machined before firing. Consequently the process is most versatile and a wide variety of shapes have been made. Plate 1. Because of its versatility the reaction bonding process was adopted by the UKAEA and a fine-grain homogeneous material, now designated REFEL SiC, was developed as a nuclear cladding material for use in high temperature gas cooled reactors.[1,2] The material is fully dense and impermeable to gases, resistant to oxidation and dimensionally stable under irradiation.

Thus with a low neutron absorption cross section it is ideally suited as a fission product retaining barrier. However, the performance of SiC cladding is limited by thermal stress, which for very highly rated fuel becomes excessive due to the five fold reduction in thermal conductivity caused by fast neutron irradiation.

It is in the non-nuclear engineering field that the major exploitation of REFEL SiC is currently being pursued within the overall programme of the UKAEA Ceramics Centre. Work comprising, a detailed evaluation of the mechanical properties, development of improved forming techniques for individual engineering components and assessment of component performance in collaboration with industrial users, has been in progress both at RFL Springfields and AERE Harwell since 1970.

This paper summarises the most recent results of property evaluation and outlines the most important engineering applications identified, together with relevant test data.

THE PROPERTIES OF REFEL SILICON CARBIDE

Many of the outstanding properties of REFEL SiC stem from the fact that it is fully dense - a characteristic conferred by the siliconising process. Plate 2. Some of the properties are shown in Table 1 and the corresponding properties of other engineering materials are shown for comparison.

COMPARATIVE VALUES FOR REFEL
SELF-BONDED SILICON CARBIDE AND OTHER MATERIALS

Material	Density g/cc	Hardness VPN	Rupture modulus '~' MN/m ($\times 10^3$ psi)	Young's modulus 'E' GN/m ($\times 10^6$ psi)	Poissons ratio	Thermal expansion coefficient '~' $\times 10^{-6}$/°C 0-1000°C	Thermal conductivity 'k' W/m°C (Cal/cm sec °C) 500°C	1200°C	Thermal shock parameter k°/E°' at 500°C (Cal/cm sec)
REFEL SiC	3.10	2500	525 (76)	413 (60)	0.24	4.3	83.6 (0.2)	38.9 (0.093)	59
Hot pressed Silicon Nitride	3.20	2500-3500	689 (100)	310 (45)	0.27	3.2	17.5 (0.042)	14 (0.033)	29
Reaction Bonded Silicon Nitride	2.60	900-1000	241 (35)	220 (32)	0.27	3.2	15 (0.036)	14.2 (0.034)	13
Hot pressed Beryllia	3.03	-	207 (30)	400 (58)	0.34	8.5	62.7 (0.15)	16.7 (0.04)	9
Hot pressed Alumina	3.90	2500	480 (70)	365 (53)	0.27	9.0	8.4 (0.02)	5.0 (0.012)	3
Tungsten Carbide(6% Co)	15.0	1500	1412 (205)	606 (88)	0.26	4.9	86.0 (~ 0.2)	-	-
Nimonic 105	8.0	350-400	-	220 (32)	-	18.8	-	-	-

Strength

For many applications the most important property of a ceramic is strength and extensive measurements have been made on REFEL SiC. The results of these measurements are presented in detail elsewhere[3] but in summary it may be stated that the observed mean strength is a function of surface finish, stressed volume and degree of biaxiality in the applied stress system, whilst the probability of failure at a given stress is a function of the flaw distribution. A Weibull modulus of 10 has been observed and the variation of mean bend strength with temperature, shown in Fig. 1, is invariant with temperature up to $1400^{\circ}C$ when the free silicon melts. It should be noted however that the residual strength above $1400^{\circ}C$ is still appreciable and that the free silicon could be removed for applications where it would prove embarrassing. The variation of Youngs Modulus with temperature is also shown in Fig. 1.[4]

Time dependence of strength

Rumsey and Roberts[5] and Marshall[6] have established a logarithric relationship between time to fracture and applied stress for self-bonded SiC. Using these workers results it has been deduced that failure would not occur in less than 10 years at a sustained loading equal to 75% of the instantaneous fracture stress.

Thermal stress and thermal shock resistance

The most important parameters affecting thermal stress and thermal shock are linear expansion, α and thermal conductivity 'k', $E\alpha/k\bar{\sigma}$ being a measure of the thermal stress or shock resistance, where $\bar{\sigma}$ is the mean strength. The variations of linear expansion and thermal conductivity with temperature are shown in Fig. 2 and it may be seen that α is small and varies little with temperature between $20-1000^{\circ}C$, whilst k is high, even at $1000^{\circ}C$. Consequently, the calculated value of $\frac{'E\alpha'}{k\bar{\sigma}}$ shown in Table 1 is higher than the corresponding values for the other materials cited.

Oxidation resistance

Bennet and Chaffey[7] have examined the oxidation of REFEL SiC at $950^{\circ}C$ and have shown that the rate decreases with time according to a parabolic rate law with a rate constant of 2.3×10^{-11} cm^2/hr. After the initial reaction, for all

practical purposes no further oxidation occurs and as the rate is governed by
diffusion through the silica layer, the SiC and silicon behave similarly.

At low oxidant pressure, volatile silicon monoxide is formed, the SiC
surface is no longer protected by a silica layer and a greatly enhanced corrosion rate is observed. However this only occurs below 10^{-6} atmos at $1000^{\circ}C$[8]
and most oxidation is protective.

Hardness and wear resistance

Microhardness measurements have been made in two laboratories at loads of
50 and 100 g. Good agreement has been obtained between measurements, the average
value being about 3000 kg/mm^2. Early measurements of wear resistance related to
the use of REFEL SiC as a fuel tube material, buttons of SiC were rubbed on SiC
plates at $650^{\circ}C$ in CO_2 and no wear was observed. Other tests have been made
under different sets of conditions, and similar results have been observed, but
wear rate is such a subjective property, depending on geometry, surface finish,
environment, etc, that it is practically impossible to generalise. Suffice it
to say that applications requiring high wear resistance are the ones which to
date have proved most successful.

APPLICATIONS

The key properties of SiC which place it in the field of engineering ceramics
are its high resistance to thermal stress and shock, its exceptional corrosion
resistance in high temperature oxidising atmospheres and its wear resistance.
Applications can be grouped into two general areas namely, components operating
at high temperature in oxidising atmospheres with varying degrees of stress and
components operating in abrasive environments. In addition there is the well
established use in heaters and a singular application which exploits the brittle
nature of the material, namely light armour. REFEL SiC is competitive with
boron carbide in this field, being slightly heavier but inherently cheaper.

High temperature applications

The major applications evaluated in this field comprise, nuclear cladding,
rocket nozzles, radiant heater tubes, gas turbine components and chemical plant
components.

Nuclear cladding

Detailed irradiation experiments have demonstrated that REFEL SiC cans are completely retentive to gaseous and solid fission products at temperatures of up to 1100°C in carbon dioxide and fuel burn-ups to 50,000 MWD/teU. A detailed assessment of the effect of fast neutron doses up to 2×10^{21} nvt on material properties confirm an overall probability of can failure of 1×10^{-6} can be achieved with peak fuel ratings up to 400 W/cm. At present, there is no intention in the UK of exploiting this cladding because the chosen coolant for HTRs is helium which permits the use of a graphite heat transfer interface.

Rocket nozzles

To guarantee performance of rocket motors low wear rate of the nozzle throat area is essential, particularly for long burns. REFEL SiC is currently being tested at the Rocket Propulsion Establishment, England, with most satisfactory results. Complete disclosure of information is not possible but 0.80 in. diam. experimental nozzles have shown no increase in area for firings up to 60 sec, at 340 psi throat pressure and 2760°K throat temperature.

Radiant heater tubes

Radiant tubes, heated by gas, are used in heat treatment furnaces where it is essential that the furnace atmosphere is uncontaminated by the products of fuel combustion. REFEL SiC is an ideal replacement for metals for extending the temperature range from 1100° to 1350°C, because of its impermeability, high conductivity and emissivity, oxidation and thermal shock resistance. The material has proved satisfactory under test conditions in typical gas atmospheres and is awaiting full scale evaluation for which tubes 3/5 ft in length will be required.

Gas turbine components

An assessment of suitable engineering materials has been made by Gostelow and Restall[4] principally on the basis of measurement of strength, creep and thermal fatigue. Above 1400°K the authors conclude that silicon nitride and silicon carbide are the only contenders for potential service in a gas-turbine and of these REFEL SiC is more likely to retain its strength above this temperature than any grade of nitride. A number of engine components have been fabricated and are currently on test at the National Gas Turbine Establishment.

Chemical plant components

As an example of the use of the material in a chemical plant, a major oil company has been using large diameter REFEL SiC tubes in "gasifiers" operating at about $900^{\circ}C$. In the course of normal operation a carbon deposit develops on the tube walls and must be periodically removed by oxidation. Temperatures as high as $1400^{\circ}C$ can be attained during this process. The REFEL SiC tubes have now been operating satisfactorily for about 1000 hours including several oxidation cycles. In a similar plant the same company has run larger components for some 500 hours to date in an atmosphere containing 8% SO_2 at a temperature of $1050^{\circ}C$.

Wear resistance applications

The important applications studied comprise mechanical seals, bearings, and spinnerets.

Mechanical seals

A major seal manufacturer has recently taken a license for the use of REFEL SiC and is engaged in field trials for a large number of high endurance applications. Two examples of these applications are crude sewage pumps and production grinding machines. Most sewage pumps are fitted with soft packing and grease lubricated seals on an automatic intermittent cycle which demands dry running at each start-up followed by the severe abrasiveness of unscreened sludge. The soft packing combined with abrasiveness wears the sleeve severely, and this must be replaced every 3-4 months. In addition, attempts to control the leakage lead to replacement of the soft packing every 5-6 weeks. REFEL SiC faces have run 8-9 months without attention, in approximately 40 cases to date. Rotary joints on the coolant system of production grinding machines originally consisted of carbon against ceramic wear faces, giving approximately 2-3 months life. REFEL SiC faces have been substituted and are still running after 9 months at 2000 rpm.

Bearings

REFEL SiC has been tested in hostile environments under conditions where severe abrasive forces are generated at start-up and shut-down. It has shown considerable promise but detailed performance figures are proprietary to users.

Spinnerets

In synthetic fibre spinning, it is essential that the hole geometry should not change over long periods. Also the spinneret material should be strong enough to withstand polymer pressure and electrically conducting to eliminate static problems and SiC fulfils both these requirements. A method of manufacturing spinnerets has been established and about 40 full size multihole components which have been subjected to accelerated tests on laboratory rigs have shown little or no wear with acceptable control of fibre diameter.

CONCLUSIONS

An improved form of self-bonded silicon carbide, REFEL SiC, has been developed at the UKAEA Laboratories, Springfields. The outstanding properties of this material place it in the field of engineering ceramics where it is currently being evaluated.

REFERENCES

1. J V Shennan. Dispersed ceramic fuels for the advanced gas-cooled reactor. Chemical Eng. Progress Symposium. Series 80, 63 (1967) Nucl. Eng. Part XVIII.

2. C W Forrest, P Kennedy and J V Shennan. The fabrication and properties of self-bonded silicon carbide bodies. Special Ceramics 5, Ed. P Popper, 1972.

3. P Kennedy, J V Shennan, P Braiden, J McLaren, R Davidge. An assessment of the performance of REFEL Silicon Carbide under conditions of thermal stress. Proc. Brit. Ceram. Soc., Vol 22, 1973.

4. C R Gostelow and J E Restall. Ceramics with potential for gas turbine applications. Proc. Brit. Ceram. Soc., Vol 22, 1973.

5. J Rumsey and A L Roberts. Delayed fracture and creep in silicon carbide. Proc. Brit. Ceram. Soc., Vol 7, 1967.

6. P Marshall. The relationship between delayed fracture, creep and texture in silicon carbide. Special Ceramics 4, Ed. P Popper

7. M J Bennett and G H Chaffey. The effect of fission fragment irradiation upon the oxidation of silicon carbide by oxygen at 950 C. J Nucl. Mat. 39, 1971, p 253-257.

8. J Antill. Active to passive transition in the oxidation of silicon carbide. Corrosion Science 11, 1971, p 337.

Plate 1.   Etched microstructure
of REFEL silicon carbide

Plate 2.   Some components fabricated
in REFEL silicon carbide

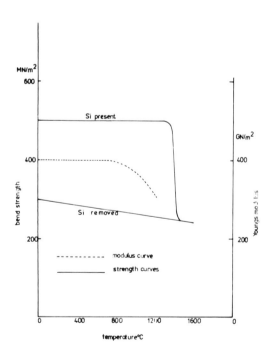

Figure 1.   Variation of the strength
and Youngs Modulus of REFEL silicon
carbide with temperature

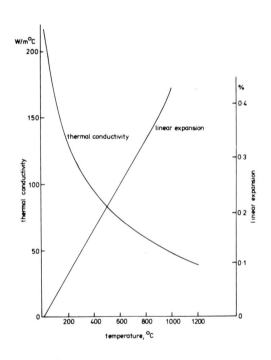

Figure 2.   Variation of linear
expansion and thermal conductivity
of REFEL silicon carbide with
temperature

High Strength Silicon Carbide for Use in Severe Environments

G. Q. Weaver and B. A. Olson

High strength SiC has been fabricated by hot pressing micron sized SiC powders with minor impurity additions. The resulting product has a density of 99-100 percent theoretical, an average grain size of 2-3 microns, and is 95-99 percent SiC with Al, Fe, and W as the major impurities.

The high strength and excellent oxidation and creep resistance of this material make it extremely important for high temperature mechanical applications in severe environments.

One of the areas of primary interests at this time is in both stationary and vehicular gas turbines. Silicon carbide is looked upon as providing the means of raising turbine inlet temperatures above 2500°F, without forced cooling of the turbine vanes. This would provide a marked increase in overall engine efficiency.

The use of ceramics in turbine engines is not a new concept, but efforts in the past using oxide ceramics and cermets have, in general, been unsuccessful. Most of these early failures were due to less than adequate materials properties combined with an attempt to make direct substitutions of ceramic components for metal parts already in use. In view of the experience gained from this early work the current approach is to obtain statistically significant mechanical and physical property data, and then by using computerized stress analyses, design each component to take advantage of the inherent strengths of the ceramic materials while at the same time minimizing their inherent weaknesses.

It is essential in performing such design analyses that one has sufficient and meaningful mechanical and physical property data.

The purpose of this paper is to present the physical and mechanical properties of one of the hot pressed silicon carbides that is now available commercially. This material is Norton NC203 HP SiC.

This material is fabricated from micron sized high purity SiC powder, which has been carefully selected and treated to minimize those impurities which would cause exaggerated grain growth. Minor amounts of selected impurities are then added to serve as both hot pressing aids and grain growth inhibitors. A typical chemical analysis of the material after hot pressing appears in Table I. The major impurities are oxygen, aluminum, tungsten and iron. The SiC is 85-100 percent 6H and amounts to 94-96 percent of the final composition. The microstructure of the NC203 material, as determined by SEM on a fracture surface, is shown in Figure 1. The average grain size of the material is approximately 1.5 microns with a maximum grain size less than 10 microns.

Al	1.48 %
Fe	0.24
B	0.005
W	2.5
O_2	1.6
Ti	0.01
V	0.02
Co	0.09
Ca	0.05
Mg	<0.05

TABLE I
Typical chemical analysis of
NC203 hot pressed silicon carbide

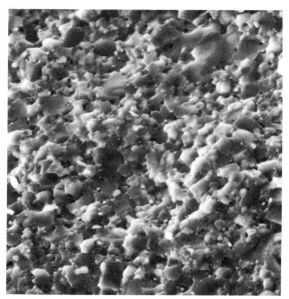

FIGURE 1
SEM of NC203 hot pressed
silicon carbide, fracture surface
at 2000X

This fine grain size combined with a well bonded structure imparts to the material one of its most outstanding mechanical properties; high strength at both room and elevated temperature. Figure 2 shows the strength from room temperature to 1500°C[1]. Samples were 1/8" x 1/8" in cross section and were measured on an Instron

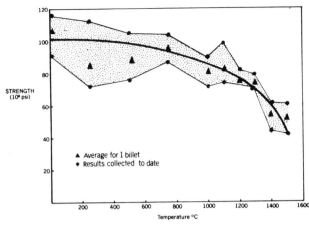

FIGURE 2
Cross bending strength measured
in 3-point loading as a function
of temperature

test machine using 3-point loading, a 1" span, and 0.02"/min. cross
head speed. The solid curve illustrates the data for one set of
samples while the shaded area encompasses all data gathered to date
on NC203.

The strength decreases slowly from 105,000 psi at room temperature to slightly over 70,000 psi at 1300°C. Above 1300°C the fall
off is more rapid, decreasing to 50,000 psi at 1500°C. While very
little work has been done above 1500°C there are indications that
the strength levels out again above this temperature.

Additional strength testing has been done by loading of samples
in tension[2]. Specimens were 0.145" gauge diameter by 1.5" gauge
length by 3" long and were tested in a helium atmosphere at a
strain rate of 0.001 inch per inch per minute. These results are
shown in Figure 3.

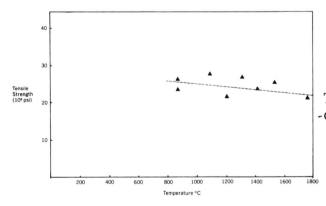

FIGURE 3
Tensile strength as a function
of temperature

Oxidation resistance of hot pressed SiC is excellent for a non-oxide ceramic. Specimens were prepared as bars 1/4" x 1/4" x 2". These were fired at 1440°C, in flowing air at approximately 60 percent relative humidity. The weight change was constantly recorded and is plotted in Figure 4 as percent weight gain and as weight gain per unit area.

Various physical properties have also been measured as a function of temperature. Thermal conductivity measurements were made using the comparative rod technique, a 1 inch diameter by 1 inch high specimen and an Armco Iron standard. These data are presented in Figure 5 together with several values calculated from thermal diffusivity measurements as a comparison[3].

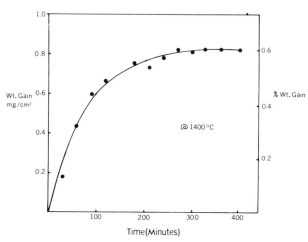

FIGURE 4
Oxidation (TGA) curves of NC203 SiC held at 1440°C in atmospheric air at approximately 60 percent relative humidity

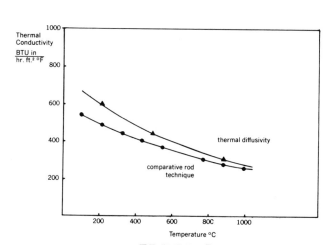

FIGURE 5
Thermal conductivity as a function of temperature up to 1000°C

Elastic properties were measured as a function of temperature using sonic techniques. Youngs modulus, shear modulus, and Poissons ratio are presented in Figures 6, 7, and 8[4].

Thermal expansion measurements were made on a 1/8" x 1/8" x 2" long bar heated at a rate of 1.5°C/min. The coefficient of expansion was found to increase significantly with increasing temperature ranging from 3.18 in/in/°C at 200°C to 5.81 in/in/°C at 1100°C. These data are shown in Figure 9.

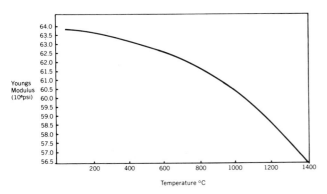

FIGURE 6
Youngs modulus, determined
sonically, as a function of
temperature

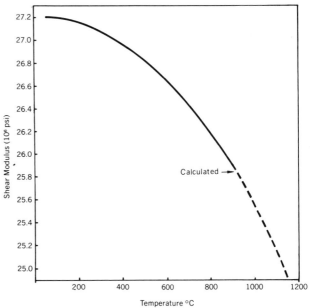

FIGURE 7
Sheer modulus, determined
sonically, as a function of
temperature

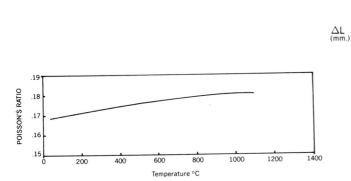

FIGURE 8
Poisson's ratio, determined
sonically, as a function of
temperature

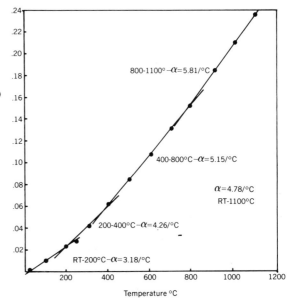

FIGURE 9
Thermal expansion curve showing
change in length as a function
of temperature up to 1100°C and
for various regions

Work of fracture measurements were made on 1/4" x 1/4" notched
samples, using an Instron test machine. More experimental details
are found in the paper by Bradt et.al.[5]. Little change with temperature was found below 1000°C. Above this temperature the work
of fracture appears to increase slightly and then to decrease below
1300°C. These data appear in Figure 10.

In addition to high strength at elevated temperatures, NC203
SiC also shows extremely good resistance to creep. Samples have
been tested in bending, compression and tension and in all cases
the amount of creep, even at loads near the breaking stress, has
been near the lower limits of detection. The results of this
testing has been summarized in Figure 11.

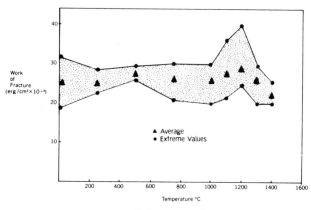

Loading	Temp °C	Load psi	Time hrs	Deformation	Comments
4 pt bend	1300	5,000	150	none	
4 pt bend	1300	7,500	150	none	
4 pt bend	1300	10,000	150	0.05%	possible knife edge deterioration
4 pt bend	1400	7,500	55	?	knife edges creeped-no measurable permanent deformation in samples
tensile	1260	10,000	1000	none detectable	lower limit of detection 0.1%
tensile	1260	20,000	1000	"	"
tensile	1371	10,000	1000	"	"
tensile	1371	20,000	8	"	sample broke
tensile	1315	16,000	500.6	0.02%	helium atmosphere

FIGURE 10
Work of fracture as a function
of temperature to 1500°C.

FIGURE 11
Summary of creep data collected
to date on NC203 SiC

The thermal shock behavior of hot pressed SiC appears at this
time to be its most limiting property. It is at best extremely
difficult to make predictions about thermal shock failure based on
mechanical properties or to generalize about thermal shock failures
from isolated tests. Our work does indicate that SiC shows excellent resistance to large steady state thermal gradients but is less
resistant to failure in rapid thermal quenching.

Preliminary analysis of the effects of rapid thermal quenching
has been made using a technique similar to that of Fleischner[6].
The specific damping capacity (Q^{-1}) of samples 1/8" x 1/8" x 4"
was measured (using a Nametre Acoustic Spectrometer) at room temperature before thermal shocking. Each bar was then heated slowly

to temperature, held for ten minutes and then quenched into water
at 20°C. The 25 gallon quench bath is located directly under the
furnace so that the test specimen has a free fall of approximately
9" before it hits the water. After shocking the specimen was dried
and the specific damping capacity remeasured.

Very little change in damping capacity was observed for
quenches having a ΔT <500°C. At ΔT ≃560°C a rapid increase in Q^{-1}
was observed, indicating significant micro damage in the sample.
Above ΔT = 600°C samples fractured into several pieces as they were
quenched. A summary of this data is given in Figure 12.

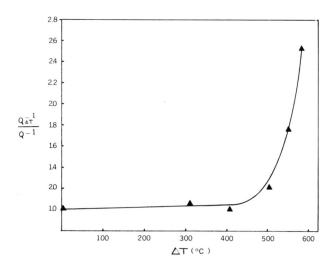

FIGURE 12
Effect of severity of thermal
quench on elastic modulus and
internal friction. Determined
by sonic techniques

Bibliography

1.   Personal communication R. C. Bradt, Dept. of Materials Science
       Ceramic Science Section, Pennsylvania State University,
       University Park, Pennsylvania.

2.   Personal communication D. D. Lawthers, Westinghouse Electric
       Corporation, Gas Turbine Systems Division, Philadelphia, PA

3.   Personal communication F. F. Lange, Westinghouse Electric
       Corporation, Research Laboratory, Pittsburgh, PA.

4.   Personal communication K. H. Styhr, Ford Motor Company, Detroit
       Michigan.

5.   R. C. Bradt, et.al. "Fracture of Commercial Silicon Carbides"
       to be presented at the International Conference on Silicon
       Carbide, 1973.

6.   P. L. Fleischner, "Devitrified Cordierite Compositions"; PhD
       Thesis, Rutgers, The State University, New Brunswick, NJ.

Infiltration of Silicon Carbide

A. R. Kieffer, W. H. Wruss, A. F. Vendl and R. M. Hubatschek

Summary

We tried to infiltrate porous silicon carbide specimens with silicon, metals and alloys containing silicon. We examined appearing
reactions by means of metallography as well as microanalyser.

We succeeded in producing a number of almost non-porous compound
materials by means of infiltration by laying the metal to be infiltrated on top of the porous SiC-samples. Subsequently the
effect of the metal contents and grain size of silicon carbide
upon mechanical strength was measured.

Infiltration of porous SiC-specimen

Because of its outstanding qualities silicon carbide is the most
frequently used hard material. It has technical applications owing
to its excellent hardness, chemical and oxidation resistance and
its semiconductor properties. It is used for organic or inorganic
bonded grinding wheels, heat resistant moldings in blast furnaces,
heating elements (i.e. Globar) and additionally as a component in
non abrasive concrete. Though the weight of silicon carbide used in
semiconductor devices is small, its value is of great importance.

Because of the easy decomposition of silicon carbide by some
metals experiments replacing other carbides in cutting materials
by SiC gave unsatisfactory results.
By means of hotpressing it is possible to produce almost dense
specimens from SiC and binder-alloys.

The production of porous bodies of SiC is quite easy. A subsequent
infiltration of the porous SiC-specimen with silicon by metal leads
to a technological interesting material for high temperature application with improved mechanical qualities compared with the porous
SiC specimen. Continuing previous experiments we tried to infiltrate SiC-bodies with metals and alloys. As silicon carbide is
generally not wetted by molten metals, the infiltration is fur-
thered by adding only a few parts of refractory metals, metal carbides or carbon to the SiC-samples. The bad wettability of silicon
carbide partly stems from the dense layer of SiO_2 surrounding the
single SiC grains. A treatment with hydrofluoric acid improved the
infiltration of various metals.

Production of SiC-bodies

Two different methods were tried in the production of SiC bodies
(1,3)

> a) cold pressing with organic binder
> b) hot pressing

In (a) SiC of various grain sizes (see table 1) was mixed with a
small quantity of an organic binder and pressed into mold of cylindrical shape in a steel die (Ø 20 mm, 10 mm high). Subsequently
these pills were heated in an argon atmosphere at $1800^{o}C$ for one
hour in a resistance heated furnace and strengthened. The resulting specimens have a porosity of 40-60 %. The measurements showed
that we were concerned with open porosity.

In (b) silicon carbide of different grain sizes was hotpressed in
graphite crucibles at temperatures of $1800-2000^{o}C$ and pressures
of 100-300 kp/cm^2 (see table 2). The advantage of this method is
that the porosity of the specimen can be varied to a somewhat
greater extent.

Table 1: Porosity of the SiC bodies (cold pressed)

SiC	grainsize μm	spec. surface m^2/g	pressure kp/cm^2	porosity %	binder %
K 46	800	–	500	38	2 starch
K 46	800	–	1000	36	2 starch
70 mesh	200	–	500	32	2 starch
70 mesh	200	–	1000	31	2 starch
120 mesh	125	0,05	500	33	2 molasses
120 mesh	125	0,05	1000	30	2 molasses
500 mesh	30	0,5	500	48	2 paraffin
500 mesh	30	0,5	1000	42	2 paraffin
1000 mesh	15	1,9	1000	56	2 paraffin
1000 mesh	15	1,9	2000	53	2 paraffin
1000 mesh	15	1,9	3000	51	2 paraffin

Table 2: Porosity of hot pressed SiC bodies ($200 \ kp/cm^2$)

SiC	grainsize μm	spec. surface m^2/g	temperature ^{o}C	porosity %
120 mesh	125	0,05	1800	42
120 mesh	125	0,05	1900	40
120 mesh	125	0,05	2000	36
500 mesh	30	0,5	1700	53
500 mesh	30	0,5	1800	49
500 mesh	30	0,5	1900	45
1000 mesh	15	1,9	1600	52
1000 mesh	15	1,9	1700	48
1000 mesh	15	1,9	1800	44

Methods of infiltration

There are several possibilities of infiltrating porous bodies with melts, principally two different methods can be distinguished:

1) Infiltration without pressure (method 1-3 in figure 1)

2) Infiltration with application of pressure (method 4 in figure 1)

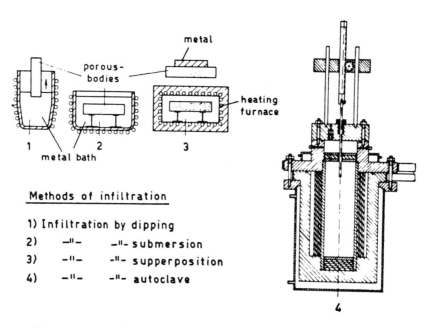

Figure 1: Methods of infiltration

In our experiments both methods were used but mainly the arrangement shown in figure 2 was chosen. The SiC specimens enclosed in
graphite crucibles coated with BN were put in a resistance heated
vacuum furnace which can also work under inert gas. The compacted
metal or alloy was laid upon the SiC body. Then the furnace was
evacuated, filled with argon and the temperature was raised. In
almost all experiments the temperature had to be raised at least
100-200°C above the melting point of the infiltrating material,
before infiltration occured. Partly because there were violent
reactions between the SiC-bodies and the materials used for infiltration.

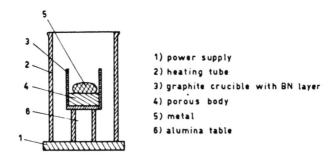

Figure 2: Preferred arrangement of Infiltration

Infiltration of SiC-specimens with silicon

The porosity of the SiC-specimens prepared for infiltration was measured and the quantity of silicon metal calculated that is needed to fill up all the pores. The quantity of impregnation material used was about 20 % higher, so even in the case of complete infiltration excess metal was left on the specimen. After the experiments the excess material on fully impregnated specimens was ground off and the density determined. Then the samples were cut into pieces, ground and polished and these specimens metallographically examined. The impregnation experiments of fine grained porous SiC-bodies with silicon (table 3) did not show positive results on samples without a hydrofluoric acid pretreatment. This treatment and the choice of a more coarse grained SiC-powder brought a full infiltration. Better results could be obtained by adding carbon. Here the infiltrating silicon metal reacts with the carbon which is included in the specimen as additional secondary SiC. The quantity of carbon can be varied from 1 to 12 % and the reaction proceeds without difficulty. A greater addition of carbon increases the quantity of secondary SiC while the remaining quantity of silicon is consequently diminished. A complete transformation of silicon to secondary SiC is impossible by the use of this method.

Table 3: Infiltration of SiC-specimen with silicon metal
Conditions: argon-atmosphere, graphite heater tube

SiC	addition of C %	temperature °C	remarks
K 46[+]	–	1500	partial infiltration
K 46[+]	1	1450	complete infiltration
K 46	3	1450	complete infiltration
K 46	5	1450	complete infiltration
K 46	10	1450	complete infiltration
1000 mesh	–	1700	wetting, no infiltration
1000 mesh[+]	–	1700	partial infiltration
1000 mesh[+]	1	1450	compl. infiltration, cracks
1000 mesh[+]	5	1450	compl. infiltration, cracks

[+]HF treated

However, SiC specimens with small contents of silicon can be obtained by Chemical Vapor Deposition. The infiltration technique we tried can be applied to specimens with a fine grained structure (30 μ) by sticking to exact conditions of infiltration. The results of these experiments are shown in table 3 (4,5).

Infiltration of porous SiC with metals and alloys

The series of experiments should explain the possibilities of infiltrating SiC with metals. The following metals were used as infiltration materials: Cu, Ag, Au, Fe, Co, Ni, Al, Sn (2,6,7).

In all these tests the temperature was some hundred degrees higher than the melting points of the applied metals. In most cases there were reactions between the SiC and infiltrating metal in which SiC was partly decomposed. We watched these processes in several cases and subsequently analysed the specimen with the microprobeanalyser. During the infiltration of a fine grained SiC specimen with a NiTi alloy (97 % Ni, 3 % Ti) the following things happened: 200°C above the melting point the surface of the SiC body was wetted, only a little infiltration occured. Increasing the temperature caused the molten metal to flow over the pill and infiltrated it from the side and bottom (figure 3).

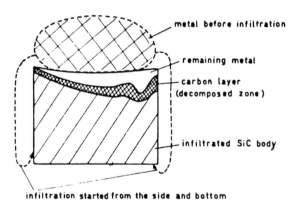

infiltration started from the side and bottom

Figure 3: Mechanism of infiltration

By means of the microprobeanalyser we found that the surface of the SiC body was initially decomposed by the NiTi alloy in the following way:

$$x \ Me \ + \ y \ SiC \longrightarrow Me_x Si_y + C$$

After this reaction the nickel alloy containing silicon infiltrated.

Metals such as copper, silver and gold especially with small amounts of 4a and 5a metals can be easily infiltrated with the SiC-grains being only slightly affected (plate 1,2). Aluminium and tin (8) infiltrated without additions, but the pretreatment with HF influenced the infiltration by lowering the infiltration temperature. Other metals show similar effects. The metals of the iron-group Fe, Co, Ni can be infiltrated after initial reactions with SiC, whereby silicon is alloyed with the infiltration metal. Sometimes the decomposition of the SiC progresses so fast that all the SiC is decomposed. These results are summarized in table 4.

Infiltration of SiC-specimens with small amounts of substances furthering wettability

We tried to improve the wettability of the SiC bodies adding 5-20 % of refractory metals or carbides. We found that these additions lowered the infiltration temperature. Therefore less secondary reactions were observed and almost dense and non-porous cermets could be obtained (table 5) (plate 3) (9,10).

Table 5: Infiltration of SiC with small contents of refractory metals and carbides

SiC mesh	additions wt %	infiltrating metal	temperature °C	remarks
120	5,10,20 Mo	Fe,Co,Ni,Cu	1600-1750	infiltration,decomposition begins
500	10,20 Mo_2C	Fe,Co,Ni	1750	infiltration,decomposition begins
1000	10,20 TiC	Fe,Co,Ni,Cu,Al	1500-1700	complete infiltration, only a little decomposition
1000	10,20 WC	Fe,Co,Ni	1700	decomposition on the surface, complete infiltration

Table 4: Infiltration of SiC-bodies with metals and alloys

SiC mesh	metal(alloy) wt.%	temperature of infiltration °C	remarks
120	Al	1400	complete infiltration
400	Al	1600	no porosity
1000	Al	1700	
120	Sn	1500	almost complete
1000	Sn	1700	infiltration
120	Fe,Co,Ni	1800	infiltration after decomposition
400	Ti,V	1800	decomposition
120	Cu,Ag	1650	infiltration, decomposition begins
400	Cr,Mn	1800	decomposition, infiltration
120	AlCu 10/90	1300	complete infiltration
120	AlCu 85/15	1250	
120	AlSi 50/50	1200	
120	AlFe 60/40	1450	decomposition, infiltration
400	CuSn 70/30	1700	small regions infiltrated
400	AgSn 70/30	1700	
120	CuMn 50/50	1600	complete infiltration
120	CuV 99/1	1600	
400	CuV 99/1	1600	infiltration, but remaining porosity
400	Cu(Nb,Ta) 99/1	1600	partial infiltration
120	FeCr 50/50	1600	reactions, infiltration
400	NiTi 99/1	1700	partial infiltration
120	AlSi 50/50	1350	complete infiltration
120	AlSi 90/10	1200	complete infiltration
400	NiSi 85/15	1500	complete infiltration
400	CuSi 90/10	1500	complete infiltration
400	FeSi 80/20	1700	decomposition, partial infiltration

Infiltration of SiC bodies with silicon containing alloys

Many of the above mentioned experiments showed that there are reactions before the infiltration starts. SiC was decomposed and silicon was alloyed with the infiltrating melts. After having gained these experiences we tried infiltration with silicon containing alloys. We found that these alloys attack SiC too, but the SiC grains were rounded (plate 4). In many cases these infiltrations have been successfull, producing a number of dense cermets (table 4) (11).

Examination of the infiltrated specimens

Metallographic examinations were made to judge the structure, the porosity and decrease of the grain size of SiC caused by the reactions. We measured the dependence of mechanical strength of the cermets on the content of metal as well as the influence of the grain size (figure 4). We found that in the examined range of porosity the strength improves with decreasing porosity and hence metal content. Fine grained cermets had a higher strength compared with coarse grained specimens (figure 5) (12).

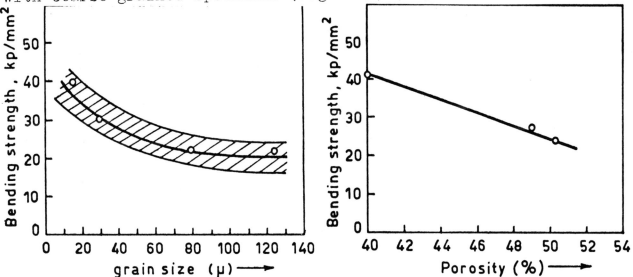

Fig.4: Infiltration of SiC-bodies (40% porosity) with aluminium. Dependence of bending strength on grain size of SiC

Fig.5: Infiltration of SiC-bodies (15 μ grain size) with aluminium. Influence of porosity on bending strength

References

1) R.Kieffer, F.Benesovsky: Hartmetalle, Springer Verlag Wien
        1965, 110,492

2) J.R.O'Connor and J.Smiltens: Silicon Carbide, a high tempera-
        ture semiconductor; Pergamon Press 1960, 235

3) J.R.Tinklepaugh, W.B.Crandall: Cermets; Reinhold Publishing
        Corp., 1960, 77

4) R.Kieffer, G.Jangg: Allg. u. prakt.Chemie; 1970, 153

5) R.Kieffer, E.Eipeltauer, E.Gugel: Ber.d.Dtsch.Keram.Ges.;
        46, 1969, 491

6) P.S.Kislyi, M.A.Kuzenkova, G.I.Shtainlyauf, M.A.Solovykh:
        Ogneupory, 30 (9) 1965, 36-39

7) L.B.Griffiths: J.Phys.Chem.Solids; 27 (8) 1966, 1368

8) T.C.Taylor: J.Appl.Phys. 29, 1958, 865-66

9) J.M.Guiot: Silicates Ind.; 31 (9), 1966, 363-7

10) G.A.Meerson, S.S.Kiparisov, M.A.Gurevich, Den Fen-San:
        Poroshkovaya Metallurgiya 26 (2), 1965, 15-21

11) T.A.Alexandrova, I.Ya.Prokhorova: Ogneupory 30 (9) 1965, 39-42

12) P.Romadin, E.N.Prosvirov, G.I.Pogodin-Alekseev: Metalloved.
        i Term. Obrabotka Metal (2) 1966, 46-48

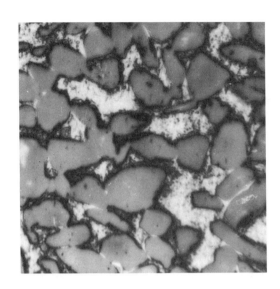

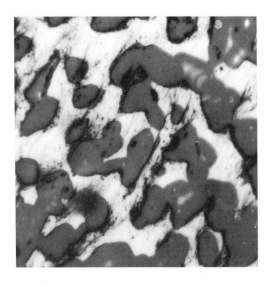

Plate No. 1: 700 x
SiC infiltrated with
Ni/Ti 99/1 ^{W}/o
grey: SiC, light: infiltrant

Plate No. 2: 700 x
SiC infiltrated with
Cu/V 97/3 ^{W}/o
grey: SiC, light: infiltrant

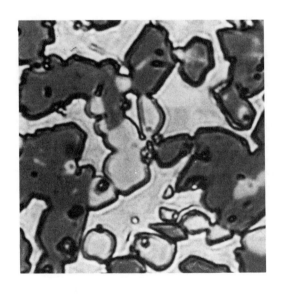

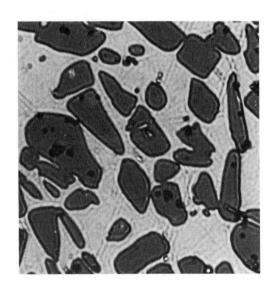

Plate No. 3: 1200 x
SiC + 20 % TiC infiltrated
with Co/Ti 99/1 ^{W}/o
dark grey: SiC, slight grey
TiC white: infiltrant

Plate No. 4: 600 x
SiC infiltrated with
Ni/Si 60/40 ^{W}/o
grey: SiC, light: infiltrant

Silicon Carbide Fibers with Improved Mechanical Properties: Study of Thermal Stability

J. L. Randon, G. Slama and A. Vignes

A continuous process was used to prepare silicon carbide fiber by chemical vapor deposition from methyltrichlorosilane and hydrogen upon resistance heated wire. The resulting fiber can be about 100 microns in diameter and its density 3.4. Rupture strength was statistically determined. The mean value is 3450 MN/m^2 (500000 PSI) with maximum value over 4700 MN/m^2 (680000 PSI). The rupture strength was found to decrease when tests were performed after high temperature treatments. In order to explain this strength decrease, the variations of SiC crystallites sizes, orientation of these crystallites, thickness of the interaction layer formed at W/SiC interface were examined as a function of temperature. It was found that the major cause for the loss of rupture strength is the growth of the interaction layer. This result is discussed.

INTRODUCTION

The feasibility of fibers vapor deposited silicon carbide on a tungsten substrate is well known {1} {2} {3}.

The present investigation was undertaken to develop this vapor deposition as an industrial fabrication method {4}. Another objective was the mechanical and thermal stability characterization of the fiber obtained. Silicon carbide has been deposited by thermal decomposition of methyltrichlorosilane at 1200 – 1300°C in a stream of hydrogen:

$$CH_3 SiCl_3 \xrightarrow{H_2} SiC + 3HCl$$

An outline of the apparatus is shown on Figure 1. The tungsten substrate is a 10 to 20μ diameter wire which is carefully cleaned. It enters the reaction chamber through mercury seals which permit electric heating of the wire. The gas mixture is recuperated after reaction.

The silicon carbide fiber obtained has a diameter of about 100μm and its density is about 3.4. Plate 1 presents a section of the fiber, with its tungsten core, a very thin interaction layer (0.3μm) and a coherent deposit of

silicon carbide. The surface of the fiber is very smooth (Plate 2). Continuous
lengths of many kilometers can be obtained depending on the length of the tungsten wire.

I. MECHANICAL CHARACTERIZATION

As silicon carbide is a brittle material, a statistical evaluation of its
rupture strength was undertaken. A continuous length of about 450 meters and
98μm diameter fiber was cut into 15cm lengths. Each sample was tested in an
Instron tensile machine fitted with pneumatic grips. The distance between the
grips was one inch. 2897 tests were carried out, 216 samples were broken in the
grips and were not taken into account for this study. The 2681 values of
rupture strength were divided into 2 types: high and low grade. These types
are presented later. The percentage of low grade fiber along the filament was
found to be very high for the first 75 meters which correspond to the time
necessary for the reactor to reach equilibrium. These values were therefore
disregarded. The histogram of the 2138 values of rupture strength is given in
Figure 2. The mean value is 3450 MN/m^2 (500,000 PSI) and maximum value
4700 MN/m^2 (680,000 PSI).

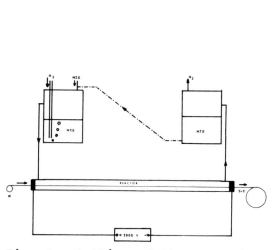

Fig. 1. Outline of the apparatus

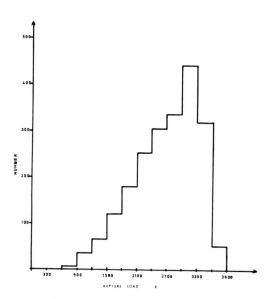

Fig. 2. Histogram for rupture strength

When studying the rupture of boron fibers Andre (5) distinguished between
different types of rupture (Figure 3) : 1-punctual rupture, 2-indefinite rupture
3-4, rupture with 1 or 2 remaining ligaments, 5-explosive rupture. These
different rupture types were also used for silicon carbide. As in the study of
boron fibers, the 3 first rupture types are associated with low grade fibers

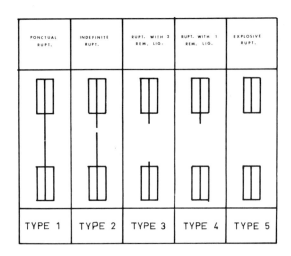

Fig. 3. Types of rupture

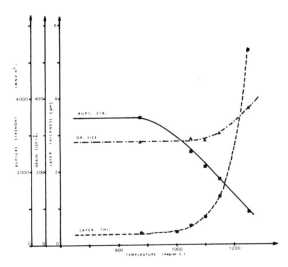

Fig. 4. Variation of rupture strength, grain size, and Interaction Layer as a function of temperature for a 2 hour treatment

N° of test	Temper-ature °C	Time (h)	Mean rupture strength σ MN/m^2	Shear Strength MN/m^2	Cristallite Size	Layer Thick-ness e,μ m	σ√e MN/m$^{3/2}$
0	–	–	3500	201.000	300	0,30	1,9
1 }	875°C	2	3500	201.000	280	0,30	1,9
2		100	2800	–	–	0,40	1,8
3 }	1010°C	15	2350	204.000	–	0,60	1,8
4		40	2150	–	–	0,95	2,1
5 }		2	2500	201.000	290	0,50	1,8
6 }	1050°C	155	1800	–	–	1,10	1,9
7		27	1600	–	–	1,50	2
8		2	2100	198.000	–	0,75	1,8
9	1100°C	10	1400	–	–	1,70	1,8
10		27	–	–	–	2,80	–
11 }	1150°C	2	1800	–	300	1,30	2,1
12		10	1000	–	–	3,05	1,7
13	1260°C	2	900	204.000	370	5,3	2

TABLE II
Results of the Heat Treatment experiments

and the two others with high grade fibers. This can be seen in Table I. The
more numerous fibers (85%) are the high grade fibers. An association of low
values with any particular defect or rupture pattern of the fiber could not be
established.

II. THERMAL STABILITY OF FIBERS

Thermal stability of fibers is a necessary quality for their use in composites for high temperature service. Cantagrel and Marchal {3} observed that the
rupture strength of SiC fibers is lowered after heating them for less than one
hour at temperatures above 1000°C.

Our aim was to understand why this decrease in rupture strength occurs and
to evaluate the ranges of time and temperature where silicon carbide fibers could
be used.

Type of rupture	Number of samples	Mean value of rupture strength MN/m^2
1	68	1670
2	60	2190
3	190	2950
4	877	3550
5	943	3670

TABLE I

Mean strength rupture for each rupture type

Samples of fibers of 3500 MN/m^2 rupture strength and 100µm diameter were heat
treated 1 to 100 hours under vacuum ($<10^{-5}$ Torr) at temperatures between 875°C
and 1260°C (Table II). Tensile tests were then performed on the samples and the
structure of the fibers was examined by X-Rays, optical or electronical microscopy.

Mechanical tests:

For each sample 30 tensile tests were performed and the mean value of
rupture strengths calculated (Table II). It decreases when the temperature or
the time of heat treatment increases. Figure 4 gives the variation of mean
rupture strength as a function of temperature for a 2 hour heat treatment.

A torsion pendulum was used to measure shear modulus , a value of
201,000 $\pm$ 3000 MN/m^2 was found for all the samples with or without heat treatment.

X Ray examination

Diffraction patterns show that the intensity of tungsten lines is reduced
when the temperature of heat treatment increases but no new phase can be detected;

the lines of β-SiC are unaffected by heat treatment. The size of cristallites
was also measured: fibers were stuck on a plexiglass plate to form a plane
surface of about 2 cm^2. X-ray curves were then analysed from Sherrer {6}. The
results given in Figure 4 were obtained for a heat treatment of 2 hours. The
cristalite size does not increase significantly for heat treatment under 1200°C.

     The texture of the fibers was also investigated for (111) lines using pole
figures: (111) planes of β-SiC are parallel to the axis of the fiber.

     No modification of the texture with heat treatment could be detected.

Microscopical examination

     Optical microscopy and scanning microscope were used on polished and etched
sections (Plate 3 and 4).

     The growth of the layer of interaction between tungsten and silicon carbide
was observed. Results are reported in Table II and on Figure 4.

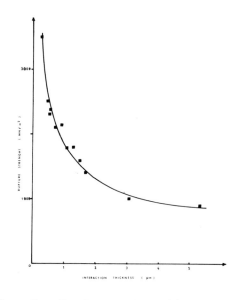

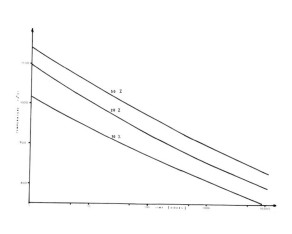

Fig. 5. Rupture strength as a
function of interaction layer
thickness

Fig. 6. % of rupture strength
loss as a function of time and
temperature heat treatment

     From these tests and examinations we could check that during heat treatment
only two characteristics of the fiber changed : the rupture strength and the
interaction layer. From Figure 5 it can be seen that for a given interaction
layer thickness there is one mean rupture strength of the fibers whatever the
time and temperature of the heat treatment.

DISCUSSION

The nature of the phase(s) which constitute the layer of interaction

between tungsten and silicon carbide is not well known. Adler {7} found
tungsten carbide WC after heat treatment of fibers at 2000°C. Burykina {8}
found WC and tungsten silicide W_5Si_3 when heating silicon carbide powder pressed
on tungsten at 1500°C. In any case this layer should be formed by diffusion.
To verify this we used Crank's {9} calculation for diffusion in a cylinder.
We found that the kinetics of growth of the interaction layer were diffusion
controlled. The energy of activation of the layer growth was 46000 Kcal. From
this result it is possible to extrapolate the kinetics of layer growth at any
temperature and therefore the mean value of rupture strength for a given heat
treatment. To explain the mechanism by which the growth of the layer of interaction decreases the mean rupture strength whilst the value of the shear modulus
is not significantly affected by the layer growth, we can suppose that this
growth promotes the formation of microcracks in the fiber which initiate rupture.

These microcracks can be caused by the growth stresses which appear when
compounds of volume ratios greater than unity such as WC, 1.3, or W_5Si_3, 1.7,
are formed or by the thermal stresses due to the cooling of the fiber after
heat treatment. This last hypotheses however seems less probable because the
coefficient of thermal expansion of W, SiC, WC are similar and the coefficient
of W_5Si_3 should not be very different, furthermore no difference in rupture
strength is observed for fibers of the same interaction layer thickness formed
at different temperatures and then cooled from these temperatures.

These microcracks are certainly too fine to be observed but we can suppose
that they are located in the interaction layer and that their length is of the
same order of magnitude as the layer thickness. This can be confirmed by
calculating the product of the rupture strength by the square root of the layer
thickness (Table II). (Griffith Criterion). This product is constant.

The results which were obtained allowed us to draw the curves of Figure 6.
We can see that silicon carbide can be used at 800 to 850°C for many thousands
of hours.

CONCLUSION

A statistical evaluation of silicon carbide fibers strength rupture was
made. The mean value found   was 3450 MN/m^2. Thermal stability was
investigated, it was shown that the decrease in rupture strength was due to the
growth of the interaction layer between SiC and W and the Griffith relationship,
$\sigma\sqrt{e}$ equals a constant, is verified. The growth of the interaction layer is

diffusion controlled. From these results it was possible to deduce that silicon carbide fibers on a tungsten substrate could be used from 800 to 850°C for many thousands of hours.

REFERENCES

1   F. GALASSO, M. BASCHE, and D. KUEHL. Applied Physics Letter, 9, 1, 37,
         1966

2   R.L. HOUGH. Journal of Polymerscience: Part C, 19, 183, 1967

3   M. CANTAGREL, M. MARCHAL. Rev. Int. Hautes Temperatures et Refractaires
         9, 93, 1972

4   M. TURPIN, D. MANDINEAU - French Patent N° 72/04355

5   H. ANDRE   These CNAM   1973

6   F. KAEBLE. Handbook of X-Rays. Mcgraw-Hill Book Company

7   R.P.I. ADLER, M.L. HAMMOND. Applied Physics Letters, 14, 11, 354, 1969

8   A.L. BURYKINA, L.V. STRACHINSKAIA, T.M. EVTUCHOK. Fiziko Chimicheskaia
         Mechanica Materialov, 1968, 4, 3, 301

9   J. CRANK. Mathematics of Diffusion. Oxford at the Clarendon Press, 1967

10   G.V. SAMSONOV, Y.S. UMANSKIY. Tverdyye Soyedineniya Tugoplavkikh Metallov.
         State Scientific-Technical Literature Publishing House (Moscow) 1957

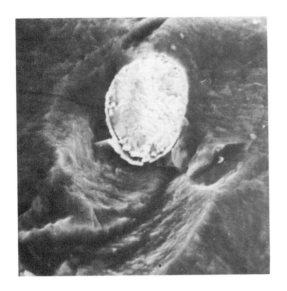

Plate 1. Section of a fiber
(scanning microscope) X2000

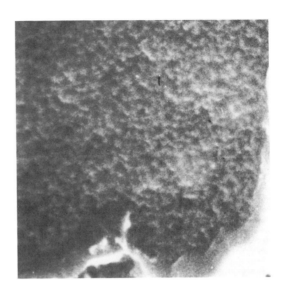

Plate 2. Surface of a fiber
(scanning microscope) X10000

Plate 3. Section of fiber heat
treated 2 hours at 1150°C X 1000

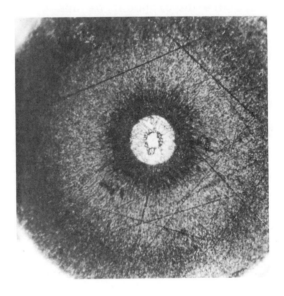

Plate 4. Section of a fiber heat
treated 2 hours at 1260°C X 1000

Abnormal Grain Growth in Polycrystalline SiC

Svante Prochazka

I. Introduction

Silicon carbide, prepared in polycrystalline form, is invariably a complex material since many crystalline polytypes of SiC exist and the polytypes that result in a given material are highly sensitive to impurities and processing conditions. Contrary to single crystals, SiC polycrystals have received little study apart from various SiC ceramics which are of a composite nature. The present work is concerned with observations of grain morphology, crystal type and grain growth in dense SiC prepared by hot pressing at high temperatures. While processing and microstructure development of dense, hot pressed silicon carbide has been previously reported (1-10) the present work is primarily concerned with the mechanisms and driving forces which lead to the development of a preferred SiC polytype and morphology in the final densified structure. Evaluations of mechanisms are made and the pertinent driving forces are estimated.

II. Materials and Procedures

Submicron size β-SiC powders, characterized in Table I, were pressure-sintered with an addition of 0.4 to 1.0% boron, which was introduced either during initial powder preparation or as an elemental amorphous boron powder additive to the base powder. Boron greatly enhances densification of SiC and thus a pressure of 0.35 to 0.7 Kb at 1900° to 1950°C yields densities above 99% of the theoretical compared to about 30 Kb of pressure required for pure SiC(6). Specimens (2" billets) were cut and sectioned for metallography and, after diamond polishing, were removed from the mount and thermally etched at 1550°C in an argon atmosphere. A 20 to 60 min hold was sufficient to reveal grain boundaries. Chemical etching by boiling Murakami's reagent was an alternate procedure occasionally employed. Some specimens were annealed at 2000°C for up to 8 hours prior to the etching

394

procedures. Annealing samples were generally 3 mm cubes in small carbon capsules
embedded in 200 mesh SiC grit and the annealing process was performed in a large
carbon crucible in argon atmosphere.

TEM studies were made on foils prepared by ion milling and these foils also
served for diffraction orientation studies. X-ray analyses were made by the
method suggested for SiC by Jagodzinski and Arnold (11), based on measurements of
relative intensities of specific peaks. Known mixtures of β and 6H-SiC powders
were used for calibration of these peaks. The procedure was relatively simple in
that only two SiC forms were observed in the majority of samples analyzed. Estimates of the fractions of the two polytypes could be made within an accuracy of
± 10%.

The distribution of the boron densification aid in the samples was evaluated
by neutron activation and autoradiography and also by microprobe scanning.

III. Results

The phase composition of the dense SiC bodies obtained by hot-pressing was,
with some exceptions, a mixture of β and 6H SiC polytypes. Minor amounts of other
phases (up to 1%) were detectable by microscopy and their presence was determined
to depend strongly on initial powder composition and stoichiometry. The results
of the investigation of these minor constituents and the conditions under which
they occurred are given in Table II. The most frequent minor constituent was a
phase rich in silicon metal which was typical for stoichiometric powders with oxygen content in the tenths of one percent. Its presence was always accompanied by
a high degree of transformation to the 6H polytype. It may have formed from a
residual oxide by the reaction:

$$SiO_2 + 2SiC = 3Si + 2CO$$

or from small amounts of unreacted elemental silicon in the initial powders which
are very difficult to detect. This phase could be eliminated by a stoichiometric
excess of carbon in the initial powder formulation. Nitrogen, when present at
levels comparable to the boron concentration was found to be a useful additive for
phase control and yielded pure β-SiC compositions. This additive probably acts as
a solute impurity for the stabilization of the cubic form (12)(13) of SiC.

The fraction of the β-SiC transformed on hot-pressing into the 6H polytype
was highly variable and very sensitivie to conditions of powder preparation, processing and densification. The results were not reproducible and showed degrees
of transformation ranging between barely detectable lines of the 6H form up to

almost full conversion to this form. The general trend to higher degrees of
transformation with increasing time and temperature was quite obvious but other
processing parameters, such as ball milling, acid leaching, applied atmosphere
composition and pressure also affected the degree of transformation.

In the early stages of transformation the 6H polytype always exhibited a
distinct morphology – thin plates up to several hundred microns long (Fig.1) with
a thickness which was approximately the matrix grain size (2-3μ). Sectioning,
therefore, rarely revealed the tabular morphology (e.g., Fig.2). The orientation
of the 6H crystals was determined by electron diffraction and by optical microscopy in thin sections (the large grains are frequently fully transparent). Without exception the growth direction was <nko>, i.e., the crystals developed basal
planes. While the growth rate could not be estimated with high accuracy, crystals
3 mm long were occasionally found after 30 min of heat treatment and thus indicated a linear advance rate of the growth front of about 200 micron/min.

In the latter stages one frequently observed interpenetration of the tabular
crystals. Such interpenetration results from the continuation of the two-
dimensional growth of two grains around each other after impingement (Fig.3).
Ultimately the whole matrix is consumed and Fig.4 shows a microstructure of a
specimen annealed for 1 hour at 2000°C which was transformed to 6H by more than
90%. At this stage the tabular morphology is no longer obvious as it was modified
by lateral growth and coalescence of the grains.

Besides the growth of the tabular grains normal grain growth of the matrix
grains was also observed. The resulting coarsening may be noted from the sequence of Fig.5a and b. A 30 min anneal brought about an increase of the average
matrix grain size by a factor of 1.6, i.e., from 2.6μ to 4.1μ. Thus one can view
the overall phenomenon as normal grain growth on which is superimposed the transformation and growth of another phase.

IV. Discussion

The distinct morphology and the unusually high growth rate of the 6H polytype
indicates that some extra driving force is involved in its formation. It is
tempting to attribute it to the free energy of the β to 6H transformation and
thereby classify the observed phenomenon as recrystallization. The tabular morphology could then be the result of the minimization of the strain energy (14)
induced by the change in lattice parameters during the transformation, i.e., due
to the increase in a c/a ratio. Simple calculations as shown below, cast doubt
on this possibility. Assuming that the growth mechanism does not change, the

velocity of the advancement of grain boundaries (of the same type) on crystal
growth is proportional to the driving force, i.e., to the energy expended to move
the grain boundary a unit distance. Hence, under the above assumption, the ratio
of the matrix growth rate to the growth rate of the idiomorphic grains will yield
an estimate of the energy involved in the growth of the 6H phase if the change in
energy content due to the matrix coarsening can be calculated. The latter can be
obtained as the product of the total matrix grain boundary area change and the
grain boundary energy.

An estimate of the grain boundary energy in SiC, 2500 erg/cm^2 was obtained
from the calculated data of SiC surface energy (15) and measurement of dihedral
angles at grain boundary-pore intersections in SiC which developed pores on annealing. The matrix coarsening (Fig.5a and b) from 2.6μ to 4.1μ involves a total
boundary area change 8460 cm^2/cm^3 which yields a total grain boundary energy drop
2.1×10^7 erg/cm^3 or 6.03 cal/mol.

The size of the largest grains in the microstructure shown in Fig.5b and the
annealing time permits an estimate of their growth rate, $3 \times 10^3 \mu/h$. The estimate
is probably low as essentially all of the grains impinged in this stage. The
ratio of the upper limit of this growth rate to the matrix grain growth rate cal-
culates to be 10^3 and thus in turn, one obtains 6 Kcal/mol as an estimate for the
energy driving the exaggerated growth of 6H crystals. Such a high energy would
probably bring about spontaneous transformation because the strain energy involved
is low and would not be, therefore, a serious barrier to the growth. It is well
known, however, that the transformation in single crystals is sluggish at temperatures near $2000^{\circ}C$ and that frequently both phases will coexist intergrown for an
appreciable period of time. We conclude, therefore, that it is likely that another growth mechanism is involved and the above assumption of constant growth
mechanisms, independent of polytype, does not apply.

To obtain further support for this view, pressure - sintering experiments
were conducted with a powder which was composed predominantly of the 6H polytype.
It was obtained by grinding up and jet milling selected single crystals, acid
leaching the powders with HCl and separating a submicron fraction by sedimenta-
tion. An x-ray pattern showed minor amounts of 4H, 15R and an unidentified SiC
polytype. However, at least 80% of the sample was 6H polytype. The powder was
mixed with 1% boron and sintered at $1980^{\circ}C$, 0.7 kb, for 10 min.

With the β-phase absent the fast-growth of the tabular crystals should have
not occurred were the energy of the β to 6H transformation the driving force since

the transformation energy would not be available. The fast growth, however, was observed to occur and did generate microstructures very similar to those observed in the fully transformed β-powders (Fig. 6). This result implies that the fast growth of the 6H phase is related to its crystalline structure and not to the free energy of the transformation, and that it is perhaps nucleation controlled. Similar type of grain growth designated as exaggerated or abnormal has been occasionally observed in ceramics such as impure α-Al_2O_3(16), β-Al_2O_3(17), ferrites(18) and, similarly, has not yet been satisfactorily explained(19).

V. Conclusions

On hot-pressing of boron doped β-SiC powders at 1950°C transformation to 6H-SiC occurs. It is always accompanied by rapid exaggerated growth of large tabular crystals. Observations of this exaggerated grain growth in the sintering of various SiC polytypes leads to the conclusion that the transformation energy of one polytype to another is not the origin of the driving force for abnormal grain growth.

REFERENCES

1. S. Prochazka, General Electric Rept. SRD-72-171 (Dec 1972).

2. S. Prochazka, R.J. Charles, Bull Am. Cer Soc (to be published).

3. R.A. Alliegro, L.B. Coffin, R.J. Tinklepough, J. Am. Cer. Soc., 39,386 (1956).

4. D. Kalish, E.V. Clougherty, J. Ryan, Final Rept. 1966, AD650883.

5. Y. Balloffet, E. Phillips, F. Hughes, AECL Final Rept. 3673, March 1971.

6. J. Nadeau, Bull. Am. Cer. Soc. 52, 170 (1973).

7. T.D. Gulden, J. Am. Cer. Soc. 52, 591 (1969).

8. F.F. Lange, Westinghouse Res. Lab. Final Rept. 73-9D4-SERAM-R1, 1973.

9. N.D. Antonova, et. al., Soviet Powder Met. Metal Ceram., No. 6, 444 (1962).

10. I.V. Dobromovo, A.A. Kalinina, V.I. Kudryavtsev, Soviet Powder Met. Metal Ceram. No. 2, 150 (1963).

11. H. Jagodzinski, H. Arnold in. Silicon Carbide, O'Connor, J. Smiltens Eds. 1960.

12. P.T.B. Shaffer, Mat. Res. Bull. 4, S13 (1969).

13. A.R. Kieffer et. al., Mat. Res. Bull. 4, S153 (1969).

14. J.W. Christian, The Theory of Transformations in Metals and Alloys, 1965, p. 415.

15. R.H. Bruce, in Science of Ceramics, G.H. Steward Ed., No. 2, 1965, p. 359.

16. I.B. Cutler in Kinetics of High Temperature Processes, W.D. Kingery Ed., 1959.

17. S.P. Mitoff, General Electric CR&D Schenectady, personal communication.

18. A.L. Stuijts, C. Kooy in Science of Ceramics, G.H. Steward, Ed., No. 2, 1965.

19. J.E. Burke in Ceramic Microstructures, R.M. Fulrath, J.A. Pask Eds., 1968.

TABLE I

Characteristics of SiC Powders

Characteristic	Made by SiO_2 Gel Reduction Unleached	Made by SiO_2 Gel Reduction HF Leached
Fe ppm	900	700
Al ppm	150	100
B ppm	4000	4600
Ca ppm	–	40
O_2%	4.9	0.59
Spec. Surface Area m^2/g	19.9	15.0
Mean Surface Average Crystallite Size, μm	0.1	0.13
X-ray Diffraction	β-SiC	β-SiC

TABLE II

Second Phases Occurring in Hot-pressed SiC

Phase Detected	Distribution	Conditions of Occurrence	Method of Detection
Silica rich melt	along three-grain intersections of the β phase	high oxygen, 2-5%, content in the starting powder	selective etching with HF
Silicon rich phase, liquid at the sint. temperature	three grain intersections; inclusions in the 6H SiC crystals	stoichiometric powders, 0.2-0.5% oxygen	microprobe, selective etching, melt. temp.
Carbon grains 1-5μ	random between grains	0.5-1.0% excess carbon introduced as an organic compound	microscopy, hardness
Boron rich crystalline phase, B_4C ?	random between β-grains engulfed into 6H-grains	high boron, >0.4%	Autoradiography after activation, microprobe

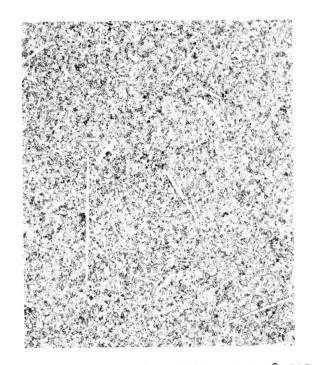

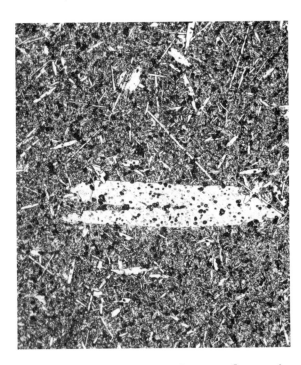

Fig. 1 Hot-pressed stoichiometric β-SiC showing an early stage of the growth of crystals of the 6H polytype. Thermal etch.   125X

Fig. 2 Early stage of transformation of polycrystalline β-SiC. Note tabular morphology of the 6H SiC. Thermal etch.   125X

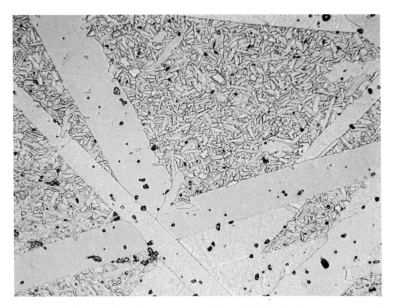

Figure 3.   Interpenetrating tabular crystals of 6H SiC grown on hot-pressing from
a β-matrix.   Thermal etch – 250X

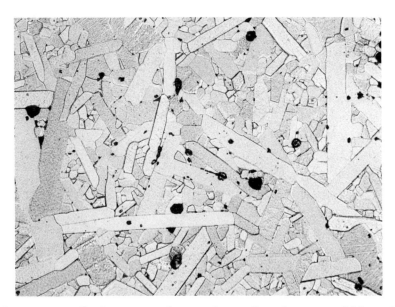

Figure 4.   Final stages of the transformation of polycrystalline β–SiC.   One hour
anneal at 2000°C.   Thermal etch – 125X

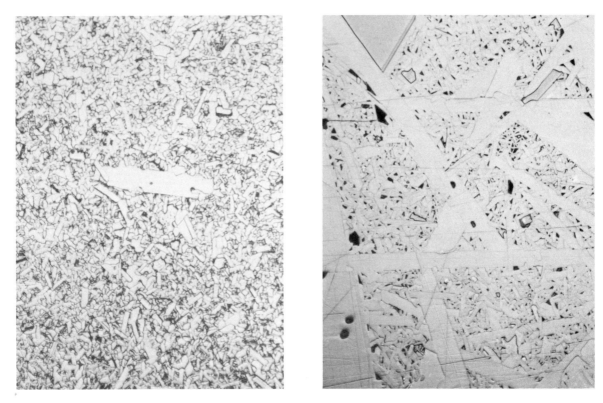

Figure 5.   Coarsening of the β–SiC matrix and exaggerated growth of the 6H grains
             due to annealing. (a) hot-pressed SiC at 2000°C.   (b) annealed 30 min.
             at 2000°C. Chemical etch – 250X

Figure 6.   Microstructure of dense polycrystalline SiC hot-pressed from a powder
             composed of 6H-SiC. Compare Figure 4.   Chemical etch – 150X

Striated Structures in Pyrolytic β–Silicon Carbide

D. E. Yeoman Walker

INTRODUCTION

Pyrolytic β-SiC produced by both static substrate or fluid bed techniques can contain striations, as reported by several authors,[1-15] who have offered reasons for them without realising that several types of striae exist. Consequently, various contradictory explanations have been proposed.

The present author has reported two distinctly different striae ("primary" and "secondary" bands) in fluid bed SiC.[9,11] Further work has revealed many more which merit detailed description.

EXPERIMENTAL

Pyrolytic SiC was made by cracking methyltrichlorosilane (MTS) vapour at $1500^{o}C$, using H_2 carrier gas. Two substrates were used: (a) uranium carbide granules (0.5 mm dia.) in a 1.5 in. dia. fluid bed;[10] (b) a heated graphite rod in a furnace. Ceramography and scanning electron microscope (SEM) etch fractography have been fully described elsewhere.[9,16]

RESULTS AND DISCUSSION

(i) Non-stoichiometry bands

It is believed that pyrolytic SiC formation involves a vapour-liquid-solid (VLS) mechanism[12,17]:-

$$CH_3SiCl_3 \longrightarrow CH_3\cdot + SiCl_3\cdot \qquad (1)$$

$$2\ SiCl_3\cdot \longrightarrow SiCl_2 + SiCl_4 \qquad (2)$$

$$SiCl_2 + SiCl_4 + H_2 \longrightarrow Si_{(liqu.\ film)} + HCl \qquad (3)$$

$$CH_3\cdot \quad\quad\quad\quad \longrightarrow \quad C \quad\quad + \; 3H. \quad (4)$$

$$Si_{(l.f.)} \quad + \quad C \quad \longrightarrow \quad SiC \quad\quad\quad\quad (5)$$

Reactions (3) and (4) occur at different rates, therefore when MTS/H_2 is passed over a heated carbon rod, silicon and carbon can deposit separately.[12] In an intermediary stage, a fluctuating reaction occurs (Pl.1) resulting in SiC layers interspersed with carbon (white bands) and grooves (black bands) containing traces of silicon. These "non-stoichiometry bands" rarely occur in a fluid bed, but these superficially resemble the "primary bands" discussed below. Non-stoichiometry bands can occur however if the MTS/H_2 ratio decreases, and are believed to be caused by either poor mixing in the gas or temperature fluctuation, which upsets the balance of equations (3) and (4).

(ii) Primary bands

The rapid fluid bed deposition of SiC ($\approx$ 100 g/h) usually produces striae. These appear as narrow white bands in dark field optical illumination (Pl.2), thick black bands in optical thin sections, and as lines of voids ($\approx$ 1 μm dia.) under Nomarski interference contrast. These were called "primary bands" because they occur in unetched, as-produced SiC sections. Their presence affects various properties, e.g. mode and strength of fracture, nuclear fission product retention and the β-SiC/UC reaction above 1600°C.[18] The primary band spacing varies in a single coated particle and also between coated particles in one coating run. The bands were thought initially to be gas bubbles,[14] but in fact they are vugs - i.e. polygonally shaped spaces resulting from crystal misfits. It was first thought from optical microscope etching studies that SiC deposited in this way occurred as laminas parallel to the substrate,[9,11,15] but later it was shown to consist of short radial columns with serrated sides.[11] (Pl.3). Vugs between neighbouring crystal sides are enlarged by etching, but those between the ends of neighbouring crystals become channels (Pl.3).

Primary bands probably result from irregularities or even stoppages in crystal growth, perhaps due to abrupt fluctuations in feed gas flow, and to the sheer speed of the deposition.

(iii) Secondary bands

Slow-coated fluid bed SiC ($\approx$ 15 g/h), when sectioned and etched, reveals myriads of striae, roughly concentric with the substrate (Pl.4). These were called "secondary bands" because they were not seen in the unetched state.[11]

Their width is ≥ 200 Å, which is consistent with the amount of SiC deposited per coating cycle, since the coating zone is only a small volume at the base of the fluid bed.[12]

Secondary bands appear to resemble the striae in Czochralski-grown single crystals,[19] but they are evidently much narrower spaced. Czochralski striae are due to minor variations in crystal growth caused by fluctuations in temperature and impurity deposition. Although this doubtless occurs in the present case, the SiC bands are thought to be polysynthetic twins, too small to register in X-ray diffraction analysis.[20] Indeed, infinite spinel twinning could permit the extension of a crystal in a roughly radial direction whilst maintaining both its girth and the normal crystallographic interfacial angles. The secondary bands would then be alternating {111} and {1$\bar{1}$1} strips, viewed in (say) a ⟨110⟩ direction. This theory agrees with the extensive micro-twinning observed in β-SiC whiskers.[21]

Secondary bands are not seen on the external surface of coated particles because they are parallel to that surface. Secondary bands are inside the crystals and are about forty times narrower than the primary bands, which are between the crystals. They are very rarely seen together,[11] as the narrowness of the crystals associated with primary bands renders the secondary bands generally unresolvable in the SEM. Therefore they should not be confused with over-etched primaries (c.f. Voice[8]).

(iv) Tertiary bands

When fluid bed pyrolytic SiC is fractured along a circumferential {111} cleavage plane, (which exposes only silicon atoms), and etched, most of the secondary bands lie parallel to it, and therefore do not appear. The few that do, produce circumferential lines of adjacent triangular etch pits (Pl.5). These are "tertiary bands", which are relatively rarely encountered.

(v) Pseudo primary bands

Pl.6 shows the fracture surface of a pyrolytic SiC (deposited on a static rod) after ion beam thinning. This has produced prominent, regularly spaced, bands wide apart (≈ 500 μm). Stereoscopy shows they are the top ends of rows of oblique crystals (lettered "B" in Pl.6 and Fig. 1). They are therefore an etching artefact, and are produced because the fracture does not coincide with the growth direction of the columnar crystals.

(vi)   Radial striae due to dendrites

    Slow-coated fluid bed SiC contains long (≤ 100 μm) radial dendrites with
striated spines (point "D", Pl.7), made up of twinned parallel rods or {111}
sheets. They grow in ⟨110⟩ and/or ⟨112⟩ directions and resemble the twinned cores
of germanium and silicon melt-grown single crystal dendrites.[22] The latter are
shaped, in section, like a letter H; this is also suspected to be true for this
SiC, but the minute crystal size has so far precluded proof of this.

(vii) Herringbone striae

    These are wide (up to 10 μm) "hill and valley" features (point "H", Pl.7),
and can be observed in unetched fracture surfaces. Generally they are perpendicular to a dendrite spine (revealed after etching). The "hills and valleys" are
thought to be twinned {1$\bar{1}$1} sheets with a common ⟨110⟩ axis, perpendicular to a
main, multiparallel spined ⟨110⟩ dendrite (Fig. 2).

(viii) Parallel growth

    Parallel multi-striae may be also due to parallel growth in a ⟨100⟩
direction (as distinct from {111} twins extending along ⟨111⟩) (Pl.8, Fig. 3).
This introduces a notable twofold appearance into faces with threefold symmetry,
and successive crystals grow progressively smaller.

(ix) Etch grooves in silicon faces

    β-SiC has no crystallographic centre of symmetry, and so two types of close-
packed faces appear on crystals - silicon atom faces {111} and carbon atom faces
{1$\bar{1}$1}. The threefold symmetry of the silicon faces is usually revealed by etching,
as triangular pits, but often the pits coalesce to form grooves parallel to two
of the three ⟨110⟩ directions (Pl.9, Fig. 4). The absence of the third is related
to the internal structure of such crystals.[23] These grooves differ from the
dendrite grooves (Section vi), which display twofold symmetry, and a much greater
length.

(x) Etch ridges on carbon faces

    The carbon atom faces {1$\bar{1}$1} of these crystals etch differently from the
silicon faces. Instead of pits and grooves, ⟨110⟩ ridges appear against a smooth
background surface (Pl.9, Fig. 4). The nature of the ridges and background is
imperfectly understood. The threefold disposition of the ridges, with one
missing, distinguishes them from those of multiparallel twin ridges, in similar

fashion to the case of silicon grooves discussed in Section (ix).

(xi) Vicinal faces

Sometimes, {113} faces occur as vicinal faces. These have the appearance of a narrow ledge adjacent to {100} (Pl.9, Fig. 4), and should not be confused with the nearby etch grooves on silicon {111} faces of Section (ix). {1$\bar{1}$3} faces are absent, as is required by the non-centrosymmetry of β-SiC.[24]

GENERAL REMARKS

Only three of the eleven types of striae discussed here (i.e. non-stoichiometry bands, primary bands and herringbone striae), are observed in unetched material. The most prevalent are the primary bands, which result from incomplete space-filling, and therefore from non-ideal coating conditions. Under ideal conditions, these bands, and also the non-stoichiometry bands, disappear. The other striae result from the crystallographic properties of SiC, and are only partially controllable. Secondary bands are rarely seen in fast-coated SiC, perhaps because the crystals are minute yet have prominent dendrite spines. Improved SEM resolution might reveal the bands in more coatings.

Several of the types of striae originate in the fact that β-SiC etches non-centrosymmetrically. Secondary bands, etch grooves, etch ridges and dendrite spines have in this sense a common denominator, although they are otherwise unrelated.

CONCLUSIONS

This paper shows that there are eleven distinct types of striae observable in pyrolytic β-SiC. Superficial similarities make some of these confusable with one another. Most of the striae are an inherent crystallographic property of β-SiC, so that only non-stoichiometry bands and primary bands can be eliminated, as they arise from a choice of poor crystal growth conditions.

ACKNOWLEDGEMENTS

The author thanks the Member for Reactors for permission to publish this paper, and Dr I F Ferguson for helpful discussion.

REFERENCES

(1) Popper P and Moyhuddin I. Special Ceramics 1964, ed. P Popper. Acad. Press,
     London & New York, 1965, p.65.

(2) Popper P and Riley F L. Proc. Brit. Ceram. Soc., 7, 99, (1967).

(3) Popper P and Cartwright B S. Science of Ceramics, 5, ed. C Brosser &
     E Knopp. Swedish Inst. Silicate Res., Gotenburg, 1970, p.473.

(4) Noone M J and Roberts J P. Nature, 212, 71 (1966).

(5) Huggins H W and Pitt C H. Bull. Amer. Ceram. Soc., 46, 266, (1967).

(6) Price R J. Bull. Amer. Ceram. Soc., 48, 859, (1969).

(7) Gulden T D. J. Amer. Ceram. Soc. 51, 424, (1968).

(8) Voice E H and Scott V C. Special Ceramics, 5, ed. P Popper. Brit. Ceram.
     Res. Assoc., Stoke-on-Trent, (1972), p.1.

(9) Walker D E Y. J Materials Sci., 2, 197, (1967).

(10) Hibbert N S, Ford L H, Ingleby B E and Walker D E Y. Special Ceramics, 4,
     ed. P Popper, BCRA, Stoke-on-Trent, (1968), p.121.

(11) Walker D E Y. Special Ceramics, 5, ed. P Popper, BCRA., (1972), p.33.

(12) Ford, L H, Walker D E Y and Ferguson I F. Ibid, p.49.

(13) Ferguson I F and Walker D E Y. Prakt. Metallographie, 6, 220, (1969).

(14) Walker D E Y, Prakt. Metallographie, 7, 445, (1970).

(15) Gyamarti E and Hoven H. Prakt. Metallographie, 7, 117, (1970).

(16) Walker D E Y. Micron, 1, 202, (1969).

(17) Ryan C E, Berman I, Marshall R C, Considine D P and Hawley J J. J.Crystal
     Growth, 1, 225, (1967).

(18) Walker D E Y, Bradshaw R P and Shaw E H. J Nucl. Matls, 40, 221, (1971).

(19) Witt A F and Gatos H C. Semiconductor Silicon, ed. R R Haberich and
     E Kern. Electrochem. Soc., New York, (1968), p.146.

(20) Walker D E Y. To be published.

(21) Van Thorne L I. J. Appl. Phys., 37, 1849, (1966).

(22) Faust Jr., J W and John H F. Metallurgy of elemental semiconductors,
     (Met. Soc. Conf. Vol. 12), ed. R O Grubel. Interscience, New York, 1960,
     p.127.

(23) Walker D E Y and Reid R D B. To be published.

(24) Wolff G A. Intermetallic Compounds, ed. J H Westbrook. Wiley, New York (1967), p.85.

(NB All micrographs enlarged 1.6 x for reproduction)

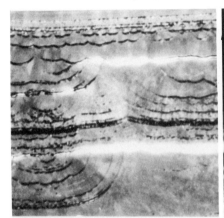

Non-stoichiometry bands (x100)

Pl.1

(x400

Primary bands (optical dark field)

Pl.2

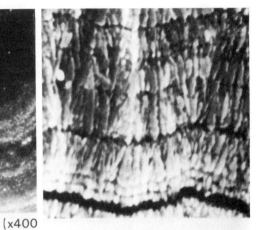

Primary bands, etched (x2270)

Pl.3

Secondary bands (x400)

Pl.4

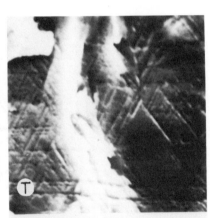

Tertiary bands (labelled 'T') (x2000

Pl.5

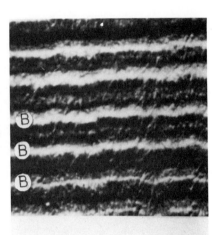

Pseudo-primary bands (x107)

Pl.6

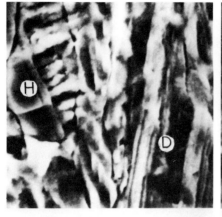

Dendrite 'D' and herringbone
striae 'H' (x900)

Pl. 7

Parallel growth in crystals (x430)

Pl. 8

Etch ridges and grooves

Pl. 9

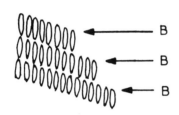

Vertical section through Pl 6

Fig. 1

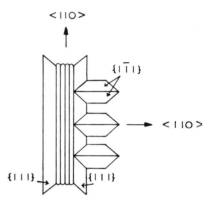

Herringbone striae—suggested structure

Fig. 2

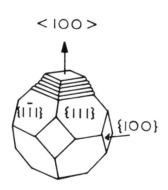

Basis of growth in Pl. 8

Fig. 3

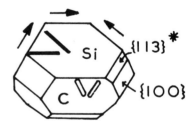

Crystal in Pl. 9

✳ Vicinal face.

⟨110⟩ directions arrowed

Fig. 4

Microstructural Effects in Silicon Carbide

C. H. McMurtry, M. R. Kasprzyk and R. G. Naum

INTRODUCTION

The properties of SiC are coming under close scrutiny as the demands for high performance ceramic materials are increased. Its high strength and excellent oxidation resistance make it one of the few possible materials capable of performing in many environments. Some knowledge of the properties of silicon carbide is essential to develop applications of this material. The fabrication technique used to produce high density SiC can result in variable microstructural features which effect the properties of the final product. It is the intent of this study to examine the effect of the microstructures resulting from varied fabrication techniques on properties such as strength, thermal conductivity, thermal expansion, Young's modulus and electrical resistivity.

Three fabrication techniques were used to produce nearly theoretically dense silicon carbide with extreme variations in microstructure, i.e., grain sizes ranging from 2000 microns to 0.05 microns. The coarse grain sized SiC was prepared by sublimation at high temperature. Hot pressing with a densification aid was used to produce a uniform microstructure with grains about 3 microns in diameter. Submicron grain size was prepared by chemical vapor deposition. In all cases densities greater than 99% of theoretical were achieved.

FABRICATION AND CHARACTERIZATION

Billets, 3-1/2" in diameter by 1" thick of coarse grain SiC were

produced by sublimation as described by Shaffer and Jun[1]. This sublimation was carried out at 2450 to 2550°C in an argon atmosphere in a
graphite tube furnace. The resulting SiC was found to be translucent
when sliced, with a light blue-green color. Average grain size was
about 2000 microns and the microstructure is shown in Plate 1. This
10X magnification shows highly twinned grains and very little evidence
of porosity, which was determined to be less than 0.1 volume percent by
quantimet measurement.

Four-inch diameter billets of SiC were hot pressed from Carborundum
1500RA silicon carbide powder. Two percent alumina was used as a
densification aid. Plate 2 shows the microstructure of the hot pressed
SiC as 1000X magnification. A density of 3.187 compared with the
3.186 gm/cc for the sublimed SiC although porosity was determined by
the quantimet to be about 1%.

Chemical vapor deposition was selected as a fabrication technique
to provide a material with extremely fine grain size. The pyrolysis
of methyltrichlorosilane at 1400°C in an argon/hydrogen atmosphere
resulted in the deposition of silicon carbide on an inductively heated
graphite substrate. Samples evaluated in this study were cut from 1 x
6 x 1/8" thick deposits after the graphite was removed by mechanical
means. The CVD material was predominately β-SiC with an average
crystallite size of 385Å as determined by x-ray diffraction. As shown
in Plate 3 a definite growth pattern can be seen at a magnification
of 200X although no grain boundaries are visible because of their small
size. Changes in deposition parameters result in a change of growth
habit as seen in the upper area of this plate. The density of 3.190
gm/cc and measured porosity of 0.1 volume percent make this material
comparable with the other SiC samples.

All three SiC materials were quite pure as seen in Table 1. Some
variation in the free silicon and carbon can be noted and of course, in
the aluminum used as a densification aid in the hot pressed materials.

Samples were cut from each of the materials and surface ground
using 200 grit diamond wheels. These samples are of variable size as
required for the individual tests as discussed in the following text.

PHYSICAL PROPERTIES

The elastic moduli were determined by the sonic resonance method from room temperature to 1400°C in argon atmosphere in a carbon resistance furnace described by Shaffer[2].

The sublimed SiC shows elastic modulus values somewhat higher than those of hot pressed silicon carbide. Measurements on the sublimed SiC with a density of 3.186 gm/cc yielded an elastic modulus of 68.3×10^6 psi at room temperature and if extrapolated to theoretical density from the data of Shaffer and Jun[1] would predict a modulus of 69×10^6 psi. Using the same extrapolation to theoretical density the 1500° elastic modulus would decrease to 66.6×10^6. The hot pressed SiC varied from 64.5×10^6 psi at room temperature to 60.5×10^6 psi at 1400°C. Comparable measurements were not made on the CVD SiC due to the maximum available thickness of 1/8. Attempts were made to make measurements on smaller bars and the results appear similar to the sublimed SiC.

Thermal expansion was determined using a technique described by Miccioli[3] which optically measures the elongation of a 1/8 x 1/8 x 3 bar heated in a graphite tube furnace. The coefficient of thermal expansion measured for the three SiC microstructures were similar. From room temperature to 1250°C a value of 4.78×10^{-6} cm/cm/°C was measured for the CVD material while the hot pressed material was measured to be 4.73×10^{-6} cm/cm/°C. A somewhat lower value of 4.5×10^{-6} cm/cm/°C was measured on the sublimed material with the large grain size possibly due to some perferred orientation in the material.

The thermal conductivity of the silicon carbide was determined from the thermal diffusivity measurements by the flash techniques originally developed by Parker[4]. The apparatus used for this work was described by Naum et al[5]. A pulse of energy is put into the sample with a laser and the thermal diffusivity determined from the time for the back surfaces to achieve half the maximum temperature risen. Thermal conductivity is then calculated from the relationship equating this quantity to the density, thermal diffusivity and the specific heat.

The thicknesses of samples varied from 50 to 1000 mils while a constant 1/2 inch diameter was employed. The half times of temperature were between 22 and 200 μ seconds.

The thermal conductivity determined on the CVD SiC is shown in Plate 4 and indicates a value of 0.137 watts/cm/°C at 100°C decreasing to about 0.1 at 1000°C. The specific heat used in calculating this conductivity was taken from the literature as 0.213 cal/gm/°C at 100°C. Hot pressed SiC exhibited a somewhat higher thermal conductivity ranging from 0.78 watts/cm/°K at 100°C to 0.36 watts/cm/°K at 1000°C. The results of three samples are shown in Plate 6 showing some scatter in the data with the scatter generally decreasing with increasing temperature. At 100°C this scatter amounted to a maximum of ±8% deviation from the mean.

The thermal conductivity of the sublimed SiC at 1.1 watts/cm/°K at room temperature is the highest measured on the three materials as shown in Plate 6. This decreases to about 0.57 watts/cm/°K at 700°C. Theoretically, nonmetallic crystals transport heat predominately by phonons and the phonon thermal conduction should be very high for the nonmetallic crystals which have (1) low atomic mass, (2) strong bonding, (3) simple crystal structures and (4) low anharmonicity. Therefore, SiC should have very high thermal conductivity at room temperature if it is very pure single crystal. The low thermal conductivities of CVD SiC in particular are due to phonon scattering caused by either impurities or grain boundaries. The small grain size of 0.05 microns makes a high concentration of grain boundaries in this material.

To put these measurements in perspective with other published data Plate 7 presents a summary along with values published[6] for single crystal SiC. Note that the coarsely crystalline sublimed SiC approaches the values for single crystal SiC.

Room temperature flexural strength was determined on an Instrom testing machine using three-point loading. Sample size was 1/8" x 1/8" x 1". Some scatter was seen in these values with the hot pressed materials being the strongest ranging from 80,000-125,000 psi with an average greater than 80,000. As expected the very coarse grained sublimed SiC was the lowest at an average of 35,500 psi. Somewhat lower results were observed with the CVD SiC than previously reported by other investigators[7] ranging from 37,700 to 61,400 and averaging slightly less than 50,000 psi. It is felt that the changes in growth pattern observed in the earlier microstructure resulted in internal strains which were responsible for the low strength values.

Electrical resistivity was measured on these three SiC materials using a four-point probe with a 0.25 span.   The resistivity values ranged from 5.9×10^3 ohm-cm for the sublimed SiC to 4.7×10^{-2} for the CVD SiC. This variation is thought to be more a function of impurities than micro-structure and the nitrogen contents are being determined to evaluate this dependence.

SUMMARY

Table 2 summarizes the data obtained to date in this study.   High densities, greater than 99% of theoretical were valuated in three SiC materials with grain sizes varying from 2000μ to 0.05μ.   Flexural strength was found to be highest with hot pressed SiC.   While this is no doubt effected by microstructure, it is felt that residual strain in the CVD material depressed the values measured in this material.

The coefficient at thermal expansion for the three materials was quite similar with the sublimed SiC somewhat lower due to orientation. Thermal conductivity decreased as the grain size decreased and the tem-perature increased.   Electrical resistivity varied in the three materials but it is thought to be a function of impurity rather than microstructure.

REFERENCES

(1) Shaffer, P. T. B. and Jun C. K., "The Elastic Modulus of Dense
        Polycrystalline Silicon Carbide", Mat. Res. Bul., 7, 63-70 (1972).

(2) Shaffer, P. T. B., Hasselman, D. P. H. and Chaberski, A. Z., "Factors
        Affecting Thermal Ceramic Bodies", WADD-TR-60-749, April 1961.

(3) Miccioli, B. R. and Shaffer, P. T. B., "High Temperature Thermal
        Expansion Behavior of Refractory Materials", J. Amer. Ceram. Soc.,
        47 (7) 352-6 (1964).

(4) Parker, W. J., "Flash Method of Determining Thermal Diffusivity, Heat
        Capacity and Thermal Conductivity", J. Applied Physics, 37
        1679-1684 (1961).

(5) Naum, R. G., Jun, C. K. and Shaffer, P. T. B., "Thermal Diffusivity
        and Thermal Conductivity of Carbon-Carbon Composites", XI Thermal
        Conductivity Conference, Albuquerque, New Mexico, September 28,
        1971.

(6) Slack, L., "Nonmetallic Crystals with High Thermal Conductivity",
        J. Phys. Chem. Solids, 34, 321-335 (1973).

(7) Price, J. R., "Structure and Properties of Pyrolytic Silicon Carbide".

Table 1

Characterization of SiC Solids

	Sublimed	Hot Pressed	CVD
Density			
g/cc	3.186	3.187	3.190
% Theoretical	99.31	99.35	99.44
Porosity, Vol.%	>0.1	1.0	>0.1
Grain Size, μ	2000	3	0.05
X-ray	α-SiC	α-SiC	β-SiC
Free Si, wt.%	0.05	0.35	<0.03*
Free C, wt.%	0.12	0.22	0.17
Spectrographic Analysis, wt.%			
Al	0.01	1-2	0.003
B	<0.001*	0.006	<0.001*
Mg	<0.0006	0.01	0.00
Ca	<0.001*	0.03	0.006
Fe	<0.0006	0.2	<0.003
Ni	<0.001*	0.02	<0.004
Co	<0.001*	<0.001*	<0.001*
Cu	<0.0006	0.01	<0.004
Cr	<0.001*	0.01	<0.001*
Ti	<0.001*	0.01	<0.001*

*Not detected but probably less than the indicated amount.

Table 2

Property Summary

	Sublimed	Hot Pressed	CVD
Density			
g/cc	3.186	3.187	3.190
% Theoretical	99.31	99.35	99.44
Grain Size, μ	2000	3	0.05
MOR, RT, psi x 10^{-3}	33-37,900	80-125,000	37-67,400
MOE, psi x 10^{-6}			
RT	69.0	64.5	~68
1400°C	66.6	60.5	~64
CTE, cm/cm x 10^6			
RT to 1250°C	4.51	4.73	4.78
Thermal Conductivity, watts/cm/°K			
100°C	1.1	0.78	0.137
1000°C	0.57	0.36	0.0995
Electrical Resistivity, ohm-cm	5.9 x 10^3	2.7	4.7 x 10^{-2}

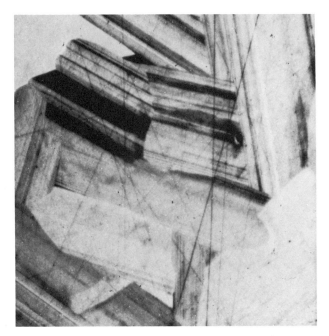

Plate 1. Microstructure Sublimed SiC

Plate 2. Microstructure Hot Pressed SiC

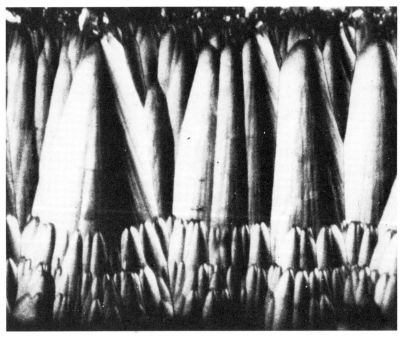

Plate 3. Microstructure of CVD SiC

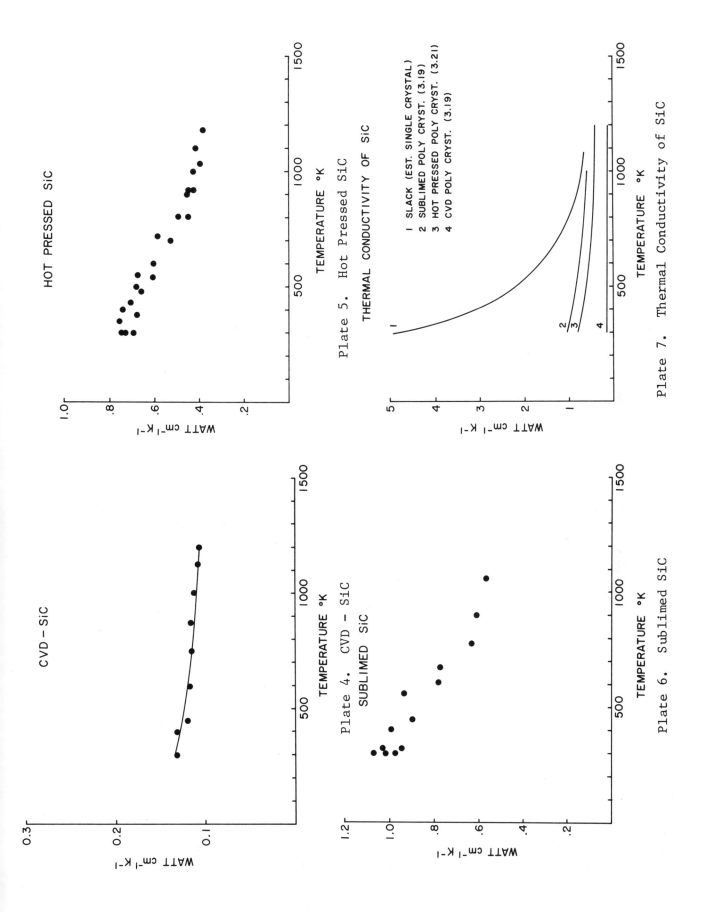

Plate 4. CVD - SiC

Plate 5. Hot Pressed SiC

Plate 6. Sublimed SiC

Plate 7. Thermal Conductivity of SiC

Fracture of Commercial Silicon Carbides

R. C. Bradt, S. D. Hartline, J. A. Coppola, G. Q. Weaver
and R. A. Alliegro

From a mechanical behavior viewpoint, commercial SiC refractory materials may be classified into three distinct groups. This classification may also be achieved by microstructural considerations and has a definite relationship to the processing. As is demonstrated in this paper, this classification occurs in a quite natural manner from a fracture mechanics treatment of the strength.

The first group consists of the SiC formed with the use of a bonding agent. Historically this bond has usually been some type of clay, which upon firing yields a composite body of SiC grains bonded by a silicate or glassy phase (1,2). A major advance in the technology of these refractories was replacing the glassy bond with silicon nitride or silicon oxynitride, both more refractory than the silicates or glass (3,4). Usually the processing of these types differ slightly, as the clays are generally added to the SiC grain during pressing, whereas the nitride and oxynitride bonds are developed by controlled atmosphere firing of pressed shapes. The mechanical properties of these bodies are frequently dominated by the characteristics of the bond phase.

The second group may be termed the direct, self-bonded bodies which structurally consist of direct bonding of the SiC grains. They are processed by several different techniques which might be broadly categorized into either a recrystallization process (5) or a reaction sintering (6,7). In the recrystallization process, the SiC grain is either pressed or slip cast to shape and then fired in a controlled atmosphere between 1900 and 2600°C. The reaction sintering process involves impregnating mixtures of silicon carbide and graphite with molten Si, and then heat treating to cause the Si and C to form additional SiC. This reaction thermal treatment is usually carried out above 2000°C and yields high density bodies, but with "free" Si present. Mechanically, these direct, self-bonded bodies are generally superior to the previously described bonded materials.

The third group of materials are those that are direct bonded and essentially 100% of theoretical density, including CVD (8,9) and dense hot pressed materials (10,11). The fabrication of dense hot pressed SiC was accomplished over a decade ago, but only recently have substantial quantities become commercially available. The processing of this form of SiC consists of hot pressing above 1500°C in the presence of a "pressing aid", usually containing iron, aluminum, or boron. It results in a body of very low porosity with mechanical properties vastly superior to the previously discussed types, and possessing excellent high temperature strength.

Plate 1 illustrates typical microstructures, representative of SiC formed by several of the different methods. The photomicrographs identify with Table I. The glassy bonded material (I) consists of a spectrum of sizes of fairly coarse SiC grains with substantial binder present. The recrystallized bodies (V) and (IV), unetched, illustrate the porosity of the former and the "free" silicon in the latter. Microstructurally the grain sizes of the self-bonded bodies are finer than the bonded materials, while the hot pressed structure (VI) is approximately an order of magnitude finer than the self-bonded materials. These four structures, as well as the others in Table I are commercially available and are reported in this study. For details concerning the other materials plotted on the graphs, please refer to their original references.

The elastic moduli reported here were measured by a resonance technique, employing an acoustic spectrometer (12,13). The strength measurements were made in air in three point bend on an Instron. Samples were approximately 1/4 x 1/4 x 2" and were broken over a 1-1/4" span with 1/16" radius of curvature alumina knife edges and a crosshead speed of 0.02" per min. The fracture surface energies were determined by the work of fracture method, (WOF), (14,15), except for the K_{Ic} calculations for the bonded bodies, which were determined by the notched beam test (15,16). All reported results were determined in air.

Elastic moduli of several forms of SiC are shown in Figure 1. It is apparent that Young's modulus is directly related to the processing behavior and the resulting microstructure. The moduli of both of the bonded bodies are only on the order of 10^7 psi. The glassy silicate bond exhibits a distinctive hysteresis associated with the α-β cristobalite transition (17), while the nitride bonded body exhibits a more uniform expansion with only a slight hysteresis, perhaps indicative of a minor concentration of silicates. The elastic modulus of the direct bonded and the hot pressed materials decrease only slightly through 1000°C, above which several of the bodies decrease more rapidly; however, the decreases between 1000 and 1400°C have been atrributed to the processing additions, rather than the SiC itself (18).

The fracture surface energies, Figure 2, vary widely between the different bodies. The bonded bodies, both silicate or glassy and the nitride have work of fractures in excess of 100,000 ergs/cm^2. These bodies fracture almost exclusively in an intergranular mode, that is the fracture path is restricted to the bonding phase. Fractography reveals little variation in this fracture mode, consequently the fracture surface energy reflects the character of the bond phase, the maxima corresponding to a viscous flow energy dissipation process within the bond (25).

The self-bonded bodies in Figure 2 are approximately an order of magnitude lower in work of fracture than the bonded bodies. There is general agreement of the trends reported here and by other investigators (22-24) of an increase in fracture surface energy with increasing temperature. Frequently, this is accompanied with a transition from a partially transgranular to a totally intergranular fracture mode, see Plate 2. This is probably also accompanied by extensive flow and dislocation activity, at, and in the vicinity of the grain boundaries for Stevens (22) has reported enhanced dislocation activity near fracture surfaces at temperatures as low as 300°C. Although the exact mechanisms have not yet been determined, it appears that the self-bonded SiC bodies increase in fracture surface energy at elevated temperatures.

TABLE I

Average Room Temperature Properties

Material Property	I*-CN-128 Glassy Bond	II*-CX-273 Nitride Bond	III* Crystar	IV** KT	V** Rex't'd	VI* Hot-Press.
ρ (gm/cc)	2.55	2.57	2.77	3.09	2.80	3.20
Porosity	10%	17%	16%	<1%	10%	<1%
E x 10^{-6}	7.0	15.0	36.9	53.9	44.5	63.7
Strength (psi)	3,600	5,300	18,600	23,100	17,000	105,000
Work of Fract. (ergs/cm^2)	132,000	102,000	11,100	23,500	14,600	24,800
Th. Exp. x 10^6	4.7	4.5	4.3	5.2	4.2	4.8

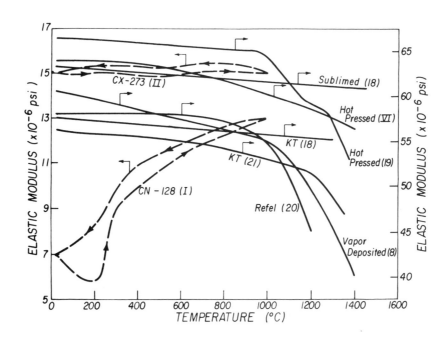

Figure 1. Elastic modulus vs. temperature. Note the two scales.

*Norton Co.; Worcester, Mass.
**Carborundum Co.; Niagara Falls, N.Y.

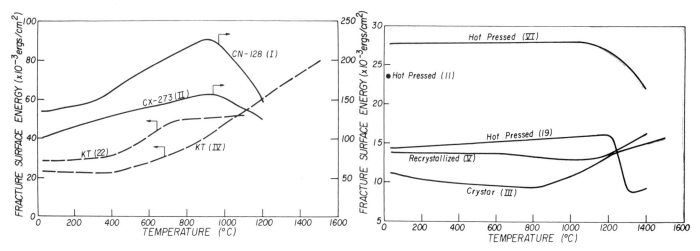

Figure 2   The fracture surface energy, note the two scales.

Available results on the dense hot pressed bodies are shown in Figure 2. These observations, and those of Lange (19), both indicate no change to approximately 1200°C, above which the hot pressed bodies appear to lose toughness. Fractography, shows only slight difference between the room and high temperature fractures; however, a transgranular to intergranular trend appears to exist and has been reported (26). The dense hot pressed bodies fracture in a predominantly elastic fashion to about 1200°C, beyond which they exhibit a decrease in toughness, probably the effect of the hot pressing additive.

Strengths are illustrated in Figure 3, demonstrating that the bonded bodies exhibit a strength trend analogous to that of the fracture surface energy, also possessing a maximum. As suggested by the refractory nature of the bond, the nitride bonded body is stronger and exhibits less temperature variation than the silicate or glassy bond. Generally the bonded bodies have strengths below 10,000 psi, and decrease rapidly above 1100°C. The direct or self-bonded bodies are considerably stronger than the bonded materials and maintain their strength above 1400°C. Occasional, minor minima and maxima are reported, but may not be significant. Not shown on the graphs is data of McLaren, et al. (20), for fracture strength determined in argon for a reaction sintered body containing "free" silicon. They reported a rapid decrease in strength above the melting point of silicon. Indications are that this decrease is not so severe in air, probably because of oxidation processes.

Results on hot pressed dense silicon carbide, Figure 3, yield very similar trends. The strength at room temperature is in the vicinity of 100,000 psi, and falls off very gradually with increasing temperature to about 1200°C, above which it decreases slightly more rapidly. At 1500°C, these hot pressed bodies retain a strength of approximately 50,000 psi. Clearly, these materials are superior to the others with regard to strength.

In comparing the strengths, fracture energies, moduli, etc., it appears that the bonded bodies and the hot pressed materials both exhibit analogous strength and fracture surface energy temperature trends, where the variation of strength is probably dominated by the fracture surface energy. The same can be said for the recrystallized and reaction sintered self-bonded bodies below 1000°C; however, above this temperature the fracture surface energy rises rapidly while the strength remains nearly constant.

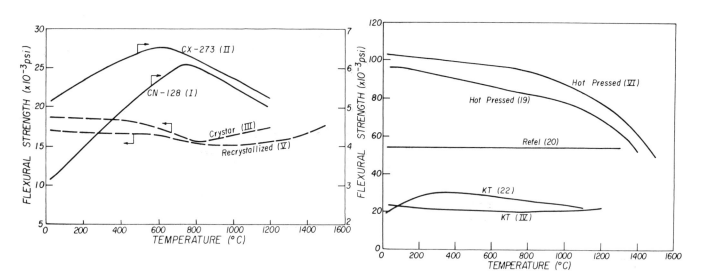

Figure 3.   Strength vs. temperature, note the two scales.

The application of fracture mechanics concepts to the room temperature strength of these commercial silicon carbides is illustrated in Figure 4, which plots K_{Ic} or $\sqrt{2E\gamma}$, the fracture toughness vs. the strength as measured by a transverse bend strength. It is apparent that the results fall into three distinct groupings previously noted, the bonded materials, the direct self-bonded bodies, and the dense hot pressed bodies. Note that the dense hot pressed materials correspond to the larger scale, and their slope is greater than the others. Applying the Griffith equation and assuming a zero intercept to yield these lines, results in flaw sizes of approximately 1400μm, 170μm, and 13μm respectively for the bonded, direct self-bonded, and hot pressed dense materials. With reference to Plate 1, it is apparent the magnitude of these flaw sizes is related, at least in a qualitative order, to the size of the largest micro-structural elements.

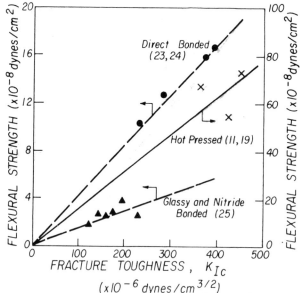

Figure 4.   A room temperature strength vs. fracture toughness plot for
            SiC. Again, note the two scales.

REFERENCES

1. Baumann, N.J., Jr., and Swenzel, J.P., Am. Cer. Soc. Bull., 16, (11), 419 (1937).

2. Kappmeyer, K. K., Hubble, D. H., and Powers, W. H., Am. Cer. Soc. Bull., 45, (12), 1060 (1966).

3. Butler, G. M., J. Electrochem. Soc., 104, (10), 640 (1957).

4. Washburn, Malcolm E. and Love, Robert W., Am. Cer. Soc. Bull., 41, (7), 447 (1962).

5. Billington, S. R., Chown, J., and White, A.E.S., Special Ceramics 1964, ed. P. Popper, Academic Press, London and New York, 19 (1965).

6. Popper, P., Special Ceramics, Heywood and Company Limited, London, 209 (1960).

7. Popper, P. and Davies, D.G.S., Powder Metallurgy, 8, 113 (1961).

8. Gulden, T. D., J. Amer. Cer. Soc. 52 (11), 585 (1969).

9. Hasselman, D.P.H., and Batha, H. D., Appl Phys. Lttrs. 2, 111 (1963).

10. Alliegro, R. A., Coffin, L. B., and Tinklepaugh, J. R., J. Am. Cer. Soc., 39, (11), 386 (1956).

11. Prochazka, S. and Charles, R. J., Final Rpt. of NASC Contract N00019-72-C-0129 (1972).

12. Spinner, S. and Teft, W. E., Proc. ASTM, 61, 1121 (1961).

13. Davis, W. R., Trans. Brit. Cer. Soc., 67, (9), 515 (1968).

14. Nakayama, J., J. Am. Cer. Soc., 48, (11), 583 (1965).

15. Davidge, R. W. and Tappin, G., J. Mat. Sci., 3, 164 (1968).

16. Brown, W. F. and Srawley, J. E., ASTM Special Technical Publication No. 410, 13 (1966).

17. Ault, N. N. and Ueltz, H.F.G., J. Am. Cer. Soc. 36, (6), 199 (1955).

18. Shaffer, P.T.B. and Jun, C. K., Mat. Res. Bull, 7, 63 (1972).

19. Lange, F. F., Personal Communication.

20. McLaren, J. R., Tappin, G., and Davidge, R. W., Proc. Brit. Cer. Soc., 20, 259 (1972).

21.   Wachtman, J. B., Jr., and Lam, D. G., Jr., J. Am. Cer. Soc., 41, (5),
      254 (1959).

22.   Stevens, R., J. Mat. Sci., 6, (4), 324 (1971).

23.   Coppola, J. A. and Bradt, R. C., J. Amer. Cer. Soc., 55, (9), 455 (1972).

24.   Coppola, J. A., Ph.D. Thesis, The Pennsylvania State University, (1971).

25.   Hartline, S. D., M.S. Thesis, The Pennsylvania State University, (1972).

26.   Kossosky, R., Personal Communication.

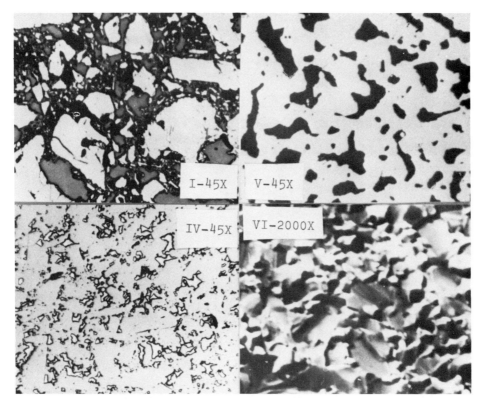

Plate 1. Microstructures of bonded silicon carbide (I),
          the recrystallized body (V), the KT material (IV),
          and the hot pressed body (VI).

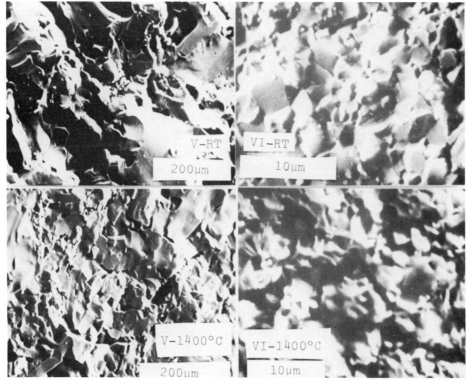

Plate 2. Room temperature and 1400°C fracture surfaces of
          recrystallized (V), and hot pressed (VI).

Corrosion Behaviour of Silicon Carbide Heating Elements

E. Buchner and O. Rubisch

The following report is given on corrosion studies carried out with SiC bodies of self-bonded material, especially considering their application as heating elements. Comparisons with respect to the metallic-conducting heating elements of $MoSi_2$ are made.

INTRODUCTION

The aim of the present studies is to clarify some of the appearances occurring during the application of SiC heating elements in different surroundings. $MoSi_2$ heating elements (1) are compared with those made from SiC.

CHEMICAL EFFECTS

Behaviour in Gases

The majority of SiC heating elements are used in an oxidizing atmosphere and only a small number in protective atmospheres. The SiC material is said to have less high temperature oxidation resistance than the $MoSi_2$ sintered material.

Studies on SiC bodies of different density (2) have clearly shown that such SiC bodies with raw densities exceeding 2.6 g/cm^3 cover themsleves in an oxidizing atmosphere, above 1350°C, with a coherent, transparent quartz glass layer. Free Si in the SiC favours the formation of the protective layer. Such self-coating SiC heating elements have the same high temperature oxidation resistance as $MoSi_2$ heating elements, i.e. they can be used at temperatures exceeding 1700°C.

Some remarkable observations have been made studying protective atmospheres. The diagram in figure 1 - determined on green SiC grains (size 0.1 - 0.15mm) with a thermoanalyzer - shows the following:

Pure nitrogen has the same effect as inert gas and results in an insignificant weight increase only above $1400^{\circ}C$. Repeated tests have confirmed that this increase was caused by free Si existing in the pores of the SiC test specimens. The pure gases CO ($>1250^{\circ}C$) and H_2 ($>1450^{\circ}C$) produced a larger weight change. All humid gases, such as N_2 ($>1450^{\circ}C$), CO ($>1200^{\circ}C$) and H_2 ($>1300^{\circ}C$), resulted in weight increases. In this case, water vapour alone was responsible for the oxidation. This is also shown by the results given in figure 2, from which follows that with increasing moisture content in the hydrogen gas the SiC specimens were increasingly oxidized at $1000^{\circ}C$. This result is of special interest for the application of SiC threephase heating elements used under $N_2 - H_2$ at about $1000^{\circ}C$ in float glass plants.

We also exposed the SiC specimens to dry and humid CO at $1000^{\circ}C$. The weight increase at this low temperature corresponds to the data given in figure 2. The weight increase of the SiC specimens due to pure CO gas was surprising and at first not understood. By chance we observed a deposition of carbon-black in the vicinity of the SiC specimens. In order to exclude any possibility of a carbon deposition caused by pyrolysis of hydro-carbons, the CO gas was piped through an Al_2O_3 tube filled with refractory clay and heated up to $1000^{\circ}C$. Nevertheless, carbon-black was, as before, deposited near the SiC specimens. This was caused by a shuttle containing a large amount of Fe. Due to the catalytic effect of the iron the Boudouard equilibrium of the CO was shifted to CO_2 + C. As a consequence, despite deposition of carbon-black and reducing surroundings, the SiC is already considerably oxidized by CO_2. The decomposition of CO by iron as well as cobalt was also ascertained in the temperature range of about 600 to $950^{\circ}C$. Similar observations were also made with refractory bricks (3).

Since in protective atmoshperes hydrogen sulphide again and again occurs as an impurity, the effect of H_2S on SiC was examined. Figure 3, shows a weight increase only from $600^{\circ}C$, and this increase only occurs in case of SiC containing Si.

On account of their protective layer of glass the $MoSi_2$ heating elements are comparatively insensitive to most gases. Only H_2 starts destroying the protective layer at approximately $1350^{\circ}C$.

Behaviour with Respect to Melts

Since SiC heating rods are used in the glass and ceramic industries, it is necessary to determine the resistance of SiC against different melts. The results have been summarized in figure 4. They show that the highly basic oxides, such as,

Na_2O and PbO, have a strongly corrosive effect at very low temperatures, about $500^{\circ}C$. Soda melts are less aggressive; they attack SiC bodies only above $800^{\circ}C$. V_2O_5 and its compounds as well as the alkali and earth alkali borates, silicates, phosphates and sulphates attack SiC to a considerably lesser degree. SiC can be brought into contact with these substances up to temperatures of $1200 - 1300^{\circ}C$ without significant corrosion. $MoSi_2$ heating elements respond in the same way as SiC heating elements (4). On account of the tightly closed quartz glass layer, a retardation of corrosion may occur.

Behaviour with Respect to Solid Substances

During furnace operation SiC heating elements again and again come into contact with dusts. It is known that small concentrations of aluminum oxide and corresponding Al silicates do only little damage. However, some oxides forming silicates, such as CaO, BaO, CuO, are dangerous in oxidizing atmosphere. In pure argon the stated oxides do not attack SiC. SiO_2 also reacts above $1700^{\circ}C$, this being a reaction which as a consequence of an effect mentioned below may have a quite considerable influence in practice. The transition metals of groups IV to VIII of the periodic system also have a destructive effect on the SiC. Thereby, the corresponding silicocarbides come into existence. The examination of the reactivity of ferrous metal oxides with SiC in argon atmosphere has shown that Fe_3O_4 starts to react at approximately $1150 - 1200^{\circ}C$, Co_3O_4 at $1300^{\circ}C$ and Ni_2O_3 at $1370^{\circ}C$.

$MoSi_2$ when in contact with the substances stated before responds in the same way as the SiC. It has to be especially mentioned that the $MoSi_2$ reacts in an exothermic way and more rapidly than SiC with all metals forming silicides.

PHYSICAL EFFECTS

Effect of the Formation of Cristobalite

With increasing period of use, the SiC heating elements are increasingly covered with SiO_2 in an oxidizing atmosphere. In case of stress by temperature change the known crystal structure change of the SiO_2 - the transformation of $\alpha-\beta$ cristobalite - results in forming microcracks in the structure of the SiC. As a consequence, the electrical resistance is increased and the useful life of the heating elements is reduced. Such an appearance is not known of $MoSi_2$ heating elements, since SiO_2 layers are present on the hot zone surface only.

Effects of Oxide Deposits

Tests have shown that due to the heat-insulating effect of oxide deposits, e.g. SiO_2, ZrO_2, Al_2O_3 or similar substances, local temperature increases occur resulting in destructive chemical reactions and a reduction of the cross-section.

OBSERVATIONS MADE IN APPLICATION PRACTICE

A comparison of the advantages and disadvantages of SiC and $MoSi_2$ heating elements shows the following:

Advantages of the SiC heating elements:

   a) high electrical resistance.

   b) good high-temperature strength.

   c) resistance to recrystallization.

   d) high resistance to thermal shock.

   e) great ruggedness.

   f) higher power density (referred to furnace volume) up to about $1350^{o}C$.

   g) unlimited installation possibility.

Disadvantages of the SiC heating elements:

   a) control problems caused by ageing (oxidation) and by the resistance-temperature characteristic.

   b) limited application of protective atmospheres.

It is possible to lessen the ageing disadvantge by using protective coatings for SiC heating elements. According to the application conditions this coating can increase the useful life to more than double. It also increases the application range of the SiC heating elements when used in protective atmospheres having a higher hydrogen content.

Advantages of the molybdenum disilicide heating elements:

   a) constancy of the electrical resistance, i.e. no change in the course of operation.

   b) better adjustibility of the temperature on account of the positive electrical resistance characteristic.

   c) higher resistance to corrosion than the SiC heating elements, since the heating conductor is protected by the quartz glass layer for a longer time and in a better way.

   d) higher power density (referred to the furnace volume) above about $1350^{o}C$.

   e) wide application of protective atmoshperes.

Disadvantages of the molybdenum disilicide heating elements:

    a) little high-temperature strength (deformation above 1250°C).

    b) little ruggedness.

    c) limited installation possibility.

    d) with rising temperature increasing recrystallization in the course of operation.

REFERENCES

(1) Fitzer E., 2. Plansee Seminar 19.-23.6.1955, Reutte/Tirol Wien: 1956 Springerverlag, S. 56/79.

(2) Rubisch O., and Schmitt R., Ber. Deutsche Keram. Ges. 43, 174/179 (1966).

(3) Rasch R., Ber. Deutsche Keram. Ges. 43, 81/82 (1966).

(4) Rubisch O., Werkstoffe and Korrosion 16, 467/472 (1965).

WEIGHT CHANGE OF SiC-POWDER IN GASES
(HEATING RATE 10°C / min)

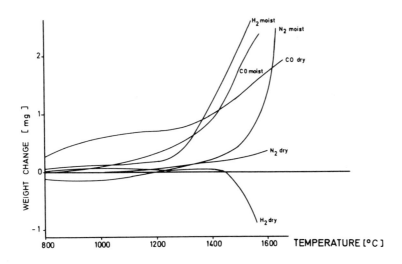

figure 1

CORROSION OF SiC - RODS IN MOIST HYDROGEN
AT 1000°C, 100 l/h H_2

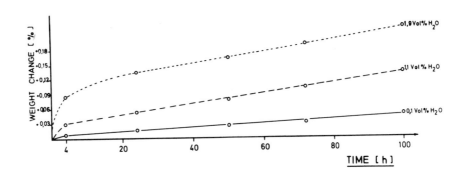

figure 2

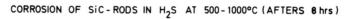

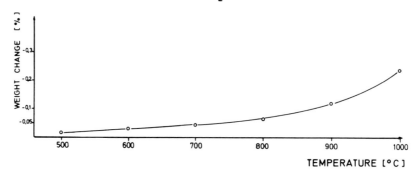

figure 3

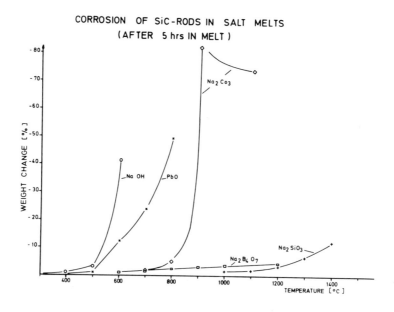

figure 4

Resistance Anomaly in "p" Type Silicon Carbide

H. D. Batha and L. H. Hardy

INTRODUCTION

Solid solutions of boron and boron carbide in silicon carbide have been studied for many years[1-7]. Early reports of resistivity variations as a function of boron content appear ambiguous[2,3] probably because the system was not adequately defined. A more definitive study reported a solid solution of up to 3 percent boron in SiC and on the other end, 13 percent silicon carbide in boron carbide[4]. Specific compositions corresponding to B_2SiC and $B_4Si_2C_3$ had lower resistance than SiC with the resistance decreasing with increased boron[5]. The effect of very small amounts of boron in pure SiC was to reduce the resistivity of SiC[6,7]. In this region, the boron seemed to replace carbon in the normal SiC lattice. The purpose of the present study was to resolve these apparent anomolies.

EXPERIMENTAL

Silicon Carbide Synthesis

Silicon carbide-boron solutions of high purity were grown in a high temperature vacuum furnace. Solar grade silicon from duPont, spectroscopic grade carbon from United Carbon and 99.6 percent boron from Cooper Metallurgical Associates were mixed in the proportions to obtain the desired boron content in silicon carbide and placed in crucibles made of spectroscopic grade carbon. The mixtures were heated in the furnace to 1900°C and held there until a vacuum of about 20 microns mercury was obtained. Argon was then introduced and the temperature raised to 2500°C for a period of 3 to 4 hours. After cooling, the crucibles were opened and the products crushed, oxidized, washed and analyzed. The samples were heated to 900°C in a stream of oxygen and washed with aquaregia, followed by a wash in hydrofluoric and nitric acids. After each acid immersion the silicon carbide was thoroughly cleansed with distilled water.

Emission spectrographic and chemical methods were used to analyze the samples. Electrical measurements on the silicon carbide grain used that size which passed through a 120 mesh screen and was retained on 140 mesh. The portion that passed through a 200 mesh screen was used in magnetic susceptibility measurements.

X-Ray Determinations

A Norelco® diffractometer was used to determine the shift in lattice parameters. Each time a series of samples was analyzed in the spectrograph, the unit was standardized by a single sample of silicon carbide known to be predominantly the 6H polytype. The scanning speed was 1 degree per minute. The angular precision was about 0.005 degrees.

ELECTRICAL MEASUREMENTS

Resistivity data were obtained from measurements on the silicon carbide single crystals and from measurements on the grain obtained from them.

The packed column technique, described by SCHWERTZ[8] was used to measure silicon carbide grain having a particle size distribution from 100 to 145 micrometers. The electrode area used was 0.25 sq. cm. and the height of the column under pressure about 6.35 mm. Contact pressure was maintained at 18.5 kg/cm^2. To avoid effects due to heating or anomalous field effects a low value of d.c. field was selected for the measurements. Data were obtained at room temperature and at elevated temperatures to +500°C.

Small disc shaped samples were ultrasonically cut from single crystals which had been ground and lapped to a uniform thickness. After they had been polished and cleaned in hydrofluoric and nitric acids, tantalum-gold contacts were alloyed onto the flat crystal surfaces. Resistance measurements were made at a frequency of one megahertz similar to the high frequency method proposed by CARROLL[9].

RESULTS AND DISCUSSION

X-Ray

X-ray examination of the boron-silicon material for evidence of solid solution revealed a drastic change in the c and a unit cell dimensions. Figure 1, showing a plot of the c and a lattice dimensions as a function of boron content, indicates that the solubility limit of boron in silicon carbide is near 0.1%. Using transmission microscopy SHAFFER[10] has, in fact, identified a second phase in crystals containing 0.2% boron. This would appear to be in agreement with VAN der BECK[3] who estimated on the basis of chemical analyses that the tolerance for boron in the silicon carbide lattice is in the neighborhood of 0.34% by weight.

To determine the effect of dissolved aluminum on the silicon carbide structure, samples of aluminum-silicon carbide powders prepared from carefully selected single crystals were also analyzed. Little change in the a and c unit cell dimensions was noted as a function of the aluminum concentration[11].

Resistivity Measurements

The conduction mechanism in packed columns of silicon carbide resistors has been investigated by HAGEN[12]. Grains with different impurity concentrations were studied and it was concluded that a semiconductor-metal-semiconductor model could be used for the conduction mechanism. In the low voltage region thermionic-field emission of carriers through the interfacial barrier between the grains dominates; in the higher voltage region, field emission. Surface states and the interfacial oxide layers are important, but current flow still depends upon the doping of the semiconductro.

Figure 2 shows a series of low field values of resistivity of a packed column of silicon carbide grain versus the boron content. The curve is characterized by a very sharp drop as the initial additions of boron are made. The minimum is followed by an abrupt increase to a second maximum and then a gradual decrease in resistivity as the boron content increases. This unusual curve is the result of our measurements as a function of boron content in the pseudobinary system of boron and silicon carbide. The curve bears out the literature discussed in the introductory remarks. The sharp drop as very small amounts of boron are added is in agreement with LELY and KROGER[7]. The increase in resistivisity as more boron is added is in agreement with VAN der BECK[3], and the subsequent decreasing values of resistivity are in agreement with RIDGWAY[2] and ALEKSANDROV, et. al.[4], who apparently worked in the two phase region of boron carbide and silicon carbide.

A similar characteristic is also shown in Figure 2 with data obtained from single crystals of silicon carbide. Both curves exhibit a minimum near 0.006% boron and a second maximum at 0.016% boron.

This anomaly in the resistivity curve is not present in the curve of aluminum-silicon carbide grain of Figure 3. The resistivity of the grain column changes with the amount of aluminum in the silicon carbide. A resistivity maximum is reached near 0.01% aluminum, after which a conductivity increase with aluminum concentration is seen. Conductivity-type of the single crystals crushed to obtain the data of Figure 3 was determined with a hot-probe. Those samples to the left of the resistivity maximum, i.e., below 0.004% aluminum, exhibit n type conductivity; those above 0.02% aluminum exhibit p type.

The aluminum-silicon carbide system is "well behaved". The first part of the curve showing the increase in resistance is caused by compensation of n type carriers by the additional aluminum. At approximately 0.01% aluminum complete compensation is obtained and increasing amounts of

aluminum result in increasing p type conductivity. This predictable behavior
is not totally unexpected in view of the constancy of the unit cell dimensions
with aluminum content.

When each system was investigated as a function of temperature up to
approximately $500^{\circ}C$, the magnitude of the resistivity decreased. However,
the basic shape of the resistivity as a function of impurity curve remained the
same. The minimum in the boron curve appeared at approximately the same
level of boron as at room temperature; no minimum appeared in the aluminum
curve.

The contract between the resistivity curves of the aluminum-silicon
carbide and the boron-silicon carbide clearly points to a basic difference in
the manner in which boron enters the silicon carbide lattice. Undoubtedly,
there are many interrated factors which contribute to this resistance anomaly.
Impurity band broadening would not be unexpected at the impurity levels
considered. Saturation by the boron impurity of the carbon lattice sites
resulting in boron atoms entering the lattice interstitially after approximately
0.006% boron may result in the well of the curve. Pressure effects have also
been known to shift energy levels within a lattice, at times even resulting in an
inversion of energy levels. The distortion of the unit cell, as shown in
Figure 1, could be sufficient to produce energy level shifts sufficient to cause
the unusual resistivity characteristic of Figure 2.

The first part of the resistivity curve showing the drop in resistivity may
be assumed to be due to the increase in carriers by the addition of boron.
Further additions of boron are reflected in a distortion of the silicon carbide
lattice, which undergoes increasing distortion with increasing boron concentration. Most rapid change is in the region 0.04% to 0.16%. In this region
there are shifts in energy levels within the silicon carbide which result in an
actually decrease in free carriers. This is the range in which a second
boron-silicon-carbon phase could be forming – a phase that SHAFFER[13]
observed. The decrease in resistivity with additional boron is then due to this
second, highly dispersed phase not detectable by our x-ray techniques. This

second phase could be boron carbide which would then decrease the resistivity in proportion to the amount of boron carbide as suggested by RIDGWAY[2]. Boron carbide was detected in the sample in which paramagnetic behavior was detected and in another sample in which about 13 percent boron was present. In both instances weak absorption lines attributed to boron carbide were detected.

An alternative explanation for the observed increase in resistivity beyond 0.004 weight percent of boron has been suggested. This level of impurity is quite high. Additional boron no longer enters the silicon carbide solely at the lattice sites by substitution for the carbon atoms. They may now enter interstitially thereby creating traps which result in a reduction of the mobility of the p type carriers. Over the range of from 0.004 to 0.02 weight percent boron this reduction in mobility may more than offset the increase in the number of carriers.

Beyond 0.02 weight percent boron the second phase becomes increasingly effective and the resistivity decreases with boron.

CONCLUSIONS

The resistivity of silicon carbide as a function of boron content has been studied and it appears that the apparent anomalous results in the literature are valid. The curve exhibits slope reversals at approximately 0.004 and 0.02 weight percent boron which are temperature independent and peculiar to boron. This resistance anomaly does follow the boron in solid solution and is not exhibited by aluminum or nitrogen impurities.

REFERENCES

1.   F. J. TONE, Ind. & Eng. Chem., 30, 231 (1938).

2.   R. R. RIDGWAY, USP 2,329,085, Sept. 7, 1943.

3.   R. R. VAN DER BECK, Jr., USP 2,916,460, Dec. 8, 1959.

4. A. A. KALININA and F. I. SHAMRAY, Doklady Instituta Metallurgii
    Imeni A. A. Baykova, Vol. 5, 151 (1960).

5. V. V. ALEKSANDROV, V. I. PRUZHININA, A. I. REKOV, T. S.
    TARAKANOVA, and A. E. TEPLOV, Fizika Tela, 1, 1587 (1959).

6. J. A. LELY, Ber. deut. keram. Ges., 32, 229 (1955).

7. J. A. LELY and F. A. KROGER, "Semiconductors and Phosphors,"
    Interscience Publishers, Inc., New York (1958) pp. 525-533.

8. F. A. SCHWERTZ and J. J. MAZENKO, Jour. App. Phys., Vol. 24,
    No. 8, August 1953, pp. 1015-1024.

9. P. E. CARROLL, Bull. Amer. Phys. Soc., 7, 432 (1962).

10. P. T. B. SHAFFER, Mat. Res. Bull., Vol. 4, pp. 13-20 (1969).

11. P. T. B. SHAFFER, Private Communication.

12. S. H. HAGEN, Phillips Res. Reports, 26, 486-518 (1971).

13. P. T. B. SHAFFER, Mat. Res. Bull., Vol. 5, pp. 519-522 (1970).

LATTICE PARAMETERS OF SILICON CARBIDE VS BORON CONTENT

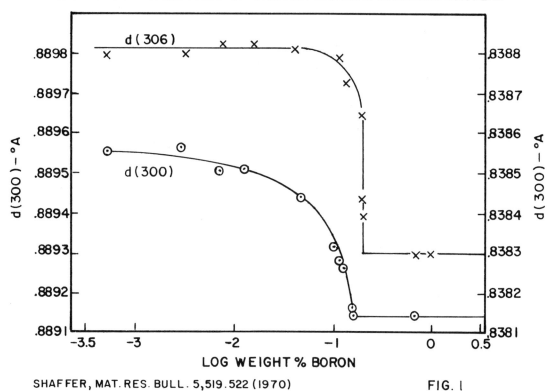

SHAFFER, MAT. RES. BULL. 5, 519.522 (1970)   FIG. I

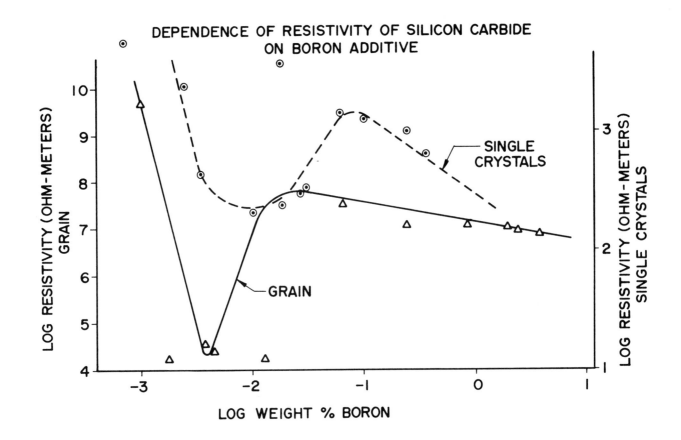

DEPENDENCE OF RESISTIVITY OF SILICON CARBIDE
ON BORON ADDITIVE

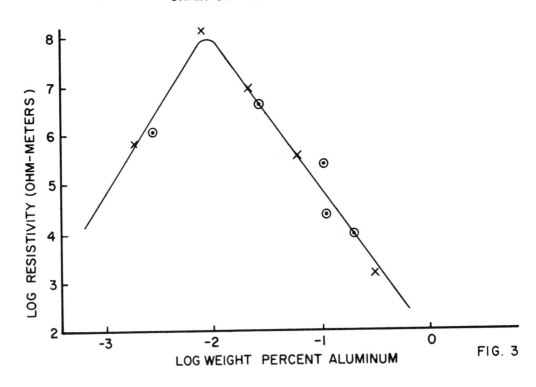

DEPENDENCE OF RESISTIVITY OF SILICON CARBIDE
GRAIN ON ALUMINUM ADDITIVE

FIG. 3

Non-Destructive Examination of Pyrolytic Silicon Carbide by Selenium Immersion

I. F. Ferguson

INTRODUCTION

Various nuclear fuels have been proposed for use in High Temperature Gas Cooled Reactor systems. An early fuel design for use in CO_2-cooled systems consisted of UC spheres coated with pyrolytic SiC and dispersed in a SiC matrix. Since the fission product retentivity and strength of such materials depends upon their detailed microstructure, methods have been developed to characterise them. One of these methods, developed to study non-destructively the interior of the coats on these particles, is described below.

After ceramographic preparation for examination under a microscope the SiC coatings often display cracks (Plates 1 and 2). Radial cracks represent a path for the escape of fission products, but they may result from the ceramographic preparation. A method was therefore required to examine the silicon carbide coatings for cracks without sectioning.

X-ray projection microscopy was first used (Plate 3)[1] but it has the following disadvantages:

(i)   Fine crack detection is difficult because of low contrast – generally a crack must lie parallel to the X-ray beam for it to be detected.

(ii) In a thick SiC coat the high X-ray absorption means that several photographs are required to reveal details of its inner and outer layers.

(iii) Only regions of very dissimilar X-ray absorptions can be distinguished; SiO cannot be detected in SiC.

Consequently, an optical method has been developed for the examination of SiC

443

coats by immersing the coated particles in selenium.

Now, the structure of spherical silicon carbide coat cannot be seen without sectioning because of its unusually high refractive index (2.63 in red light)[2] which has the following effects:

(i) The coat acts as a strongly convergent spherical lens. This generally allows only a magnified image of part of the kernel to be seen. The granules then have a dark metallic appearance although the separated coats are frequently pale yellow in colour.

(ii) The coat has a high reflectivity, calculated at 80%, and so it is difficult to illuminate the interior of a coat.

(iii) The critical angle, 22.35°, for silicon carbide is unusually high and imposes severe restrictions on any simple optical examination.

These difficulties vanish if the coated granule is immersed in a medium of the same refractive index, see for example Wells.[3]

Since the highest refractive index for a liquid at room temperature appeared to be 2.1 for selenium monobromide,[4] it was necessary either to find either a suitable glass or a solid which could be melted on a microscope stage. Selenium proved satisfactory but as it only transmits light poorly in the red, it was necessary to resort to photography to observe the fine details in the coats.

EXPERIMENTAL

Light's high-purity (99.9999%) selenium was melted on a glass microscope slide, heated to 350°C to remove any SeO_2 which would cause it to become opaque on cooling. A few coated granules were stirred into the selenium. The mixture was allowed to stand to allow any trapped air to escape. A cover slip was pressed into the molten selenium until it touched the granules, and the selenium was allowed to solidify. The cover slip was then generally removed. These operations were carried out in a glove-box because of the toxicity of selenium.

The cooled slide was viewed through a Reichert MeF microscope with transmitted light from a zirconium arc lamp fitted with an infra-red filter. Photographs were then taken using infra-red extra rapid plates, which have an exceptionally high sensitivity between 6700 and 8800 Å. Exposure times were in the range $\frac{1}{8}$ to $\frac{1}{2}$ sec.

RESULTS AND DISCUSSION

Observations of cracks in coats

Various granules examined by this method clearly showed circumferential cracks as in ceramographic sections (Plate 4). Now the method has the advantage that the whole three dimensional nature of the crack is visible rather than its projection (c.f. Plate 3). Radial cracks are not found and so they must represent a preparative artefact.

Coating thickness measurements

Although the method appears to be ideal for coating thickness measurement it is shown, practically, to be unsuitable. Thus accurate measurements require the refractive index of selenium which varies appreciably with wavelength. An attempt was therefore made to assess the effective wavelength at which the visual observations were made. This was decided to be 7000 $\overset{o}{A}$, which corresponds to a refractive index of 2.73 and ray tracings were carried out to assess the relation between the apparent size of a kernel as determined by this method and its true size, and so the relation between apparent and real coating thickness. It was demonstrated that more than a hemisphere of the kernel is illuminated by a microscope condenser while the kernel should appear diminished in size (Plate 5). In practice the kernels generally appear larger than expected on the basis of either sectioning or X-ray projection microscopy. This is ascribed to the presence of an opaque zone of unpredictable thickness adjacent to the kernel. The method is thus unsuitable for measuring the thickness of silicon carbide coats. The method cannot, moreover, be used to examine any gap between kernel and coat for the space between kernel and coat can be detected only if the kernel has a radius of less than half that of the central hole and lies at the bottom of this hole (Plate 6).

Optically opaque zones in coats

The method has found that SiC coats are wholly or partially opaque (Plates 4, 7, and 8). The origin of this opacity is discussed as follows:-

The first possibility, impurities, is unlikely because:

(i)   The impurity levels in these coats are low and cannot be correlated with their optical density (Tables I and II). Moreover, the absorption spectra of optically clear and opaque SiC coatings do not differ significantly in the range 0.5 to 15 μm.[5]

(ii)   X-ray diffraction reveals no phases other than β-SiC in these coats, and this is confirmed by electron microscopy.

(iii) The lattice parameter of the coats equals that of β-SiC to within 1 part in 10,000.

TABLE I Typical analytical and density data on SiC coats

Analytical data		Density (g/cm^3)	Porosity (%)	Optical appearance
Oxygen (wt.%)	Uranium (p.p.m.)			
0.28	< 50	3.20	0.3	Clear
0.52	2100	V. probably > 3.18	V. probably < 1.7	very clear
0.10	350	3.19	1.3	Fairly clear
1.50	475	3.18	1.7	Fairly clear

TABLE II Relation between coating conditions, analytical data, densities and optical appearance

Code	Analytical data				Coating conditions	Density (g/cm^3)	Porosity (%)	Optical appearance
	O_2 (wt.%)	U (ppm)	N_2 (ppm)	H_2 (ppm)				
50	0.15	815	25	20	$H_2/Ar/CH_3SiCl_3$	3.06	5	Opaque
58	0.13	775	25	13	H_2/CH_3SiCl_3	3.22	0	Clear

The second possibility, α-SiC, is unlikely for the amounts indicated optically disagree with the levels indicated by X-ray powder diffraction and the optical examination of thin sections of the coats. The refractive indices of α-SiC do, however, differ from those of β-SiC, vis: in sodium light $\omega = 2.655$, $\epsilon = 2.69$ (c.f. β-SiC: 2.63 in red light).

The third possibility, faulty wetting of the granules by the selenium, which would lead to total internal reflexion phenomena, is unreasonable.

The fourth possibility, porosity, has been shown to be correct by electron microscopy which reveals circumferential zones of pores in certain samples (Plate 9). The number of these pores can be linked with the opacity of the coats. Although the space introduced by this porosity is frequently too low to be detected by density measurements with a pyknometric fluid, the pores can be seen using dark field microscopy because their presence is exaggerated optically since the coat is seen in projection while each pore acts as an optical scattering centre which is considerably larger than its real size (Plates 10 and 11). The fine structure of the opaque zones indicates a radial structure which may be related to the grain distribution observed ceramographically (Plate 4).

Polarisation effects in coats

When the coats are examined in selenium between crossed Nicols, a Maltese cross pattern results (Plate 12). This is typical of a spherulite where many crystals radiate from a point in this case presumably as though from the granule centre.[6]

CONCLUSIONS

A non-destructive method has been developed for the optical examination of pyrolytic SiC. This method reveals circumferential cracks and also circumferential zones of high optical absorption which are due to microporosity in the coat. It has been shown that radial cracks in the coatings are artefacts. The technique cannot be used to measure coating thickness because of the opaque zones and it cannot be used to detect spaces between kernel and coat.

REFERENCES

1.   SHARPE R S. X-ray microscopy applied to the examination of coated nuclear fuel particles. 1963. AERE-R.4252

2.   THIBAULT N. Am. Mineralogist, 1944, vol. 29, pp 357-358

3.   WELLS H G. 1897. 'The Invisible Man', Chapter 19

4.   'Handbook of Chemistry and Physics'. The Chemical Rubber Publishing Company, 1959

5.   KNIPPENBERG W F. Philips Res. Repts., 1963, vol. 18 p.161

6.   BUNN C W. 1961. 'Chemical Crystallography', Chapter III, p.94

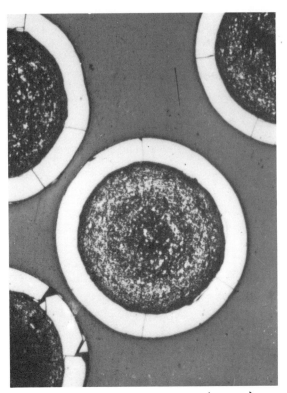

Plate 1   SiC coated UC (x 100)
standard ceramographic section,
showing radial cracks.

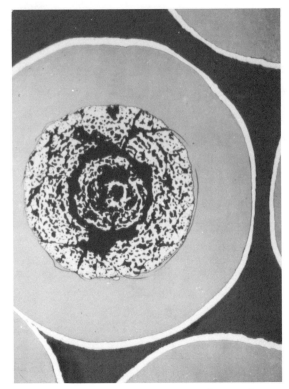

Plate 2   SiC coated UC (x 150)
standard ceramographic section,
showing circumferential cracks.

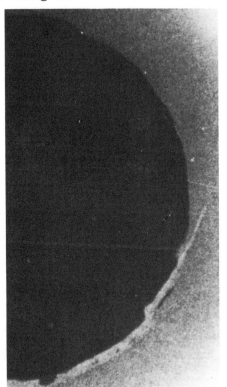

Plate 3   SiC coated UC (x 400)
X-ray projection micrograph.

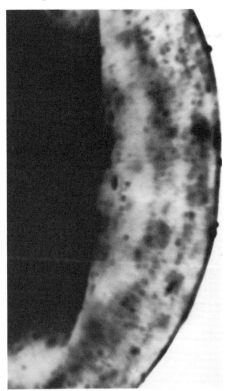

Plate 4   SiC coated UC (x 400)
viewed in selenium.

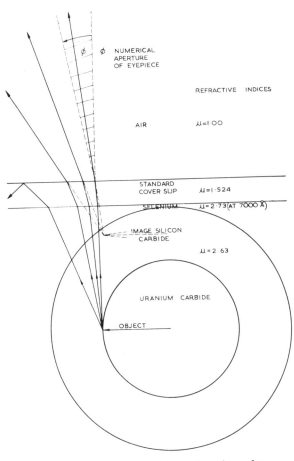

Plate 5 Ray tracing showing how a kernel should appear reduced in size when viewed in selenium.

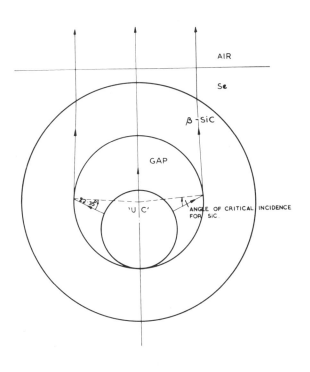

Plate 6 Ray tracing showing why it is difficult to examine any gap between kernel and coat using the selenium immersion method.

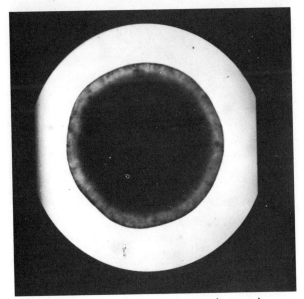

Plate 7 SiC coated UC (x 100) showing 'clear' SiC coat, as viewed in selenium.

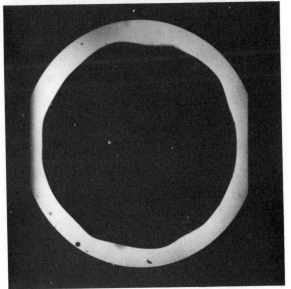

Plate 8 SiC coated UC (x 100) showing 'opaque' SiC coat, as viewed in selenium.

Plate 9 Electron micrograph of a
shadowed replica of a fractured
SiC coat showing a line of
porosity (x 11, 500).

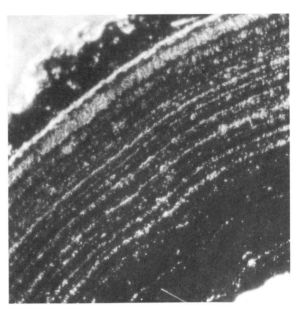

Plate 10 SiC coated UC (x 400)
dark field optical photograph
showing circumferential light
scattering regions in a SiC coat
which is opaque when viewed in Se.

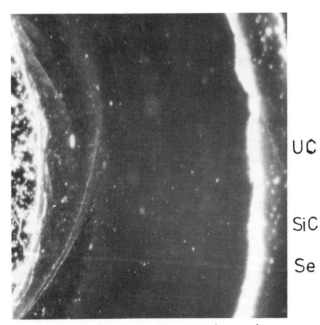

Plate 11 SiC coated UC (x 400)
dark field optical photograph
showing 'clear' SiC which is
transparent when viewed in
selenium.

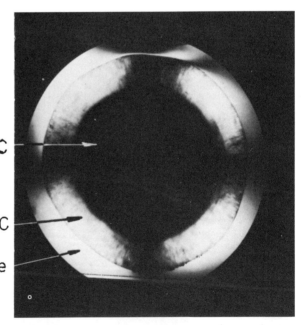

Plate 12 SiC coated UC (x 400)
examined in transmission between
crossed nicols showing Maltese
cross effect.

Anomalous Young's Modulus Behaviour of SiC at Elevated Temperatures

J. D. B. Veldkamp and W. F. Knippenberg

1. Introduction

The stiffness and strength of selfbonded dense sinte-red poly-crystalline α-SiC at elevated temperature depends on the grain size and on the impurity content. The stiffness of such a material containing 5% aluminium was measured by Alliegro et al (1956) from room temperature up to 1370°C showing a gradual de-crease. Polycrystalline material containing several percents of free silicon loses in tensile strength from about 500°C (Stevens 1971).

The strength of aluminium bonded material generally also drops with temperature but at low aluminium contents an increase in strength with temperature up to 1370°C was observed (Alliegro et al 1956).

Kern et al (1969) determined the bending strength of poly-crystalline β -SiC obtained by chemical vapour deposition. To about 500°C a small decrease was found but from 500 to 1100°C there was an increase to a maximum.

Hasselman and Batha (1963) measured the bending strength of pure α SiC single crystal 6H (33) platelets at room temperature and at about 1750°C. They observed at 1750°C a ben-ding strength, on the average, three times higher than at room temperature.

From the literature quoted above it appears that the increase in strength at elevated temperatures is the higher the lower the impurity content and might be due to an increase of the theoretical tensile strength, thus to an increase of some elastic moduli (Kelly 1969).

By the techniques of growing SiC crystals in the form of whiskers as developed by Knippenberg and Verspui (1969) α SiC whiskers grown parallel and perpendicular to the hexagonal axis became available.

In this communication[*] the determination of static Young's moduli from stress-strain measurements on α SiC whiskers is reported from which the temperature behaviour of two maxima of the elasticity body of α SiC viz. perpendicular and parallel to the hexagonal axis is obtained. It is to be expected that this property is only slightly influenced by the particular polytype and will show a temperature behaviour characteristic for all types (Patrick 1969).

2. Measurements

The measurements were performed in a tensile testing machine for thin filaments of high strength and stiffness described previously (Bouma et al 1970). The machine was provided with a high temperature tube furnace (see fig. 1), in which temperatures up to 2000°C can be maintained.

The furnace used showed temperature intervals between the sides of about 100° around 850° and 50° around 1400°C respectively, due to differences in contact resistance.

From the force-displacement curves the relative changes in modulus with temperature can be obtained.

The sensitivity of the modulus determination was limited by the axial gradient of the furnace. Thus the Young's moduli have to be considered as a kind of average moduli in the temperature intervals.

The measurements were performed on 4H(22) whiskers grown in a [00·1] direction having a circular cross-section and on 6H(33) whiskers grown in a [10·0] direction with a rectangular cross-section. For whiskers which were tapered an effective cross-section was calculated by the method given by Op het Veld et al

---

[*] To be reported in the J. of Applied Physics D.

(1970) enabling also the calculation of absolute modulus values. A representative example of the modulus determination is given in fig. 2. In this figure the increase of the moduli, amounting at room temperature to 52.5 GN/m^2 in [00·1] and to 36 8 GN/m^2 in [10·0] is indicated. The latter value was calculated for an irregularly tapered whisker and must be most probably looked upon as lower than the true absolute value.

The evidence resulting from a series of measurements indicate that the elasticity body shows a fluctuating expansion for elevated temperatures. In the temperature interval from 700° to 1400°C in both directions three maxima were observed almost coinciding at about 850, 1080 and 1300°C respectively, of which in any case the first two are followed by pronounced minima at about 950 and 1200°C respectively.

The effects were reversible and time-independent as far as could be measured within the limitations of the experiment. High temperature X-ray analysis confirmed the absence of crystallographic transformations in the time-temperature interval of the measurements in agreement with Bootsma et al (1971).

In the stress-strain plots no plastic deformation could be observed.

3. Discussion

The measurements indicated in fig. 2 show an increase of the Young's modulus from room temperature to about 1300°C by about 14% for the [00·1] and about 34% for the [10·0] direction.

As in the temperature range of investigation plastic deformation was not observed the strength of the material is proportional to the theoretical tensile strength and thus according to Orowan's expression proportional to Youngs modulus (Kelly 1966). The temperature behaviour found for the Young's moduli of α-SiC is rather uncommon. Increases of Young's modulus with temperature coupled with defect movements or reversible allotropic transitions have been reported for other substances (Westbrook 1960). The whiskers used, however, were practically perfect and in the time-temperature interval of the measurements no phase transitions have

been observed. The thermal expansion data of Taylor and Jones
(1960) for 6H SiC, representative in this respect for α-SiC in
general do not at first sight give a clue to a solution either, the
lattice expands isotropically up to about 100°C. From 100 to 700°C
there is a preferential expansion along the c-axis, leading to a
maximum in the $^c/a$ axis ratio at about 700°C (see fig.2). There is,
however, an unexplained large "scatter" in the data above 700°C
which practically seems to coincide with the fluctuations in the
increase of the Young's moduli with temperature (fig.2) and which
possibly therefore possesses a real meaning.

Where two of the three Youngs moduli give a fluctuating
increase with temperature the compression modulus can be expected
to show an influence of this behaviour. The compression modulus
(K) is related to the specific heat at constant pressure (C_p) and
the coefficient of cubic expansion (α_v) by the Grüneisen relation

$$K = \frac{\gamma C_p}{\alpha_v V (1 + \gamma \alpha_v T)}$$

in which V is the molar volume, γ the Grüneisen constant and T the
absolute temperature (Dekker 1969). In the temperature range
studied the second term within the brackets is small in comparison
with the first term. The temperature dependence of K is thus largely given by the temperature dependence of C_p/α_v. C_p measurements
on β SiC by Kern et al (1969) show a continuous increase from 0°
to about 800°C. From 800-1400°C only one measurement is reported.
It is clear, however, that anomalous behaviour is present in that
temperature range. Above 1400°C, C_p is constant. Values of α_v for
α SiC from Taylor and Jones (1960) show a continuous increase from
0° to 300°C. From 300 to 730°C α_v is constant and from 730° α_v
measured to about 1230°C, increases continuously. The quotient
C_p/α_v and thus K from 0° to 300°C decreases, then increases to a
maximum at about 730°C after which there is a new decrease (see
fig. 2). Due to the lack of measurements from 800° to 1200°C and
from 1200°C to 1400°C the behaviour of K in these regions cannot
be predicted: An uncommon behaviour of K is in any case evident.

From the relation between K and the Young's moduli one would
expect that at least one of the Young's moduli will show a minimum in the temperature range between 0° and 700°C and that C_p will
show a fluctuating behaviour between 700° and 1400°C.

References

Alliegro, R.A., Coffin, L.B. and Tinklepaugh, J.R., 1956,
J.Am. Ceram. Soc. 39, 386-389.

Bootsma, G.A., Knippenberg, W.F. and Verspui, G. 1971,
J. Cryst. Growth 8, 341-353.

Bouma, J., Francken A.J.J. and Veldkamp J.D.B., 1970,
J. of Physics, E., 3, 1006-1008.

Dekker, A.J., 1969, Solid State Physics, Mc Millan, London

Hasselman, D.P.H. and Batha, A.D. 1963,
Appl. Phys. Letters, 2, 111-112.

Kelly, A., 1966, Strong Solids, Clarendon, Oxford.

Kern, E.L., Hamill, D.W., Deem H.W. and Sheets, H.D.,
1969, Mat. Res. Bull., 4, S25-S32.

Knippenberg, W.F. and Verspui G., 1969,
Mat. Res. Bull., 4, S33-S44.

Knippenberg, W.F. and Verspui G., 1969, Mat. Res. Bull.,4,S45-S56.

Op het Veld, A.J.G. and Veldkamp J.D.B., 1970,
Fibre Sc. Techn. 2, 269-281.

Patrick L, 1969, Mat. Res. Bull., 4, S129-S139.

Stevens, R.J.J. 1971, J. Mat. Sc. 6, 324-331.

Taylor, A. and Jones, R.M., 1960,
Silicon Carbide, O'Connor, J.R. and Smiltens, J. Eds
Pergamon Press Beford, 147-154.

Westbrook, J.H.
1960, Mechanical Properties of intermetallic compounds,
Westbrook J.H. Ed., Wiley, New York.

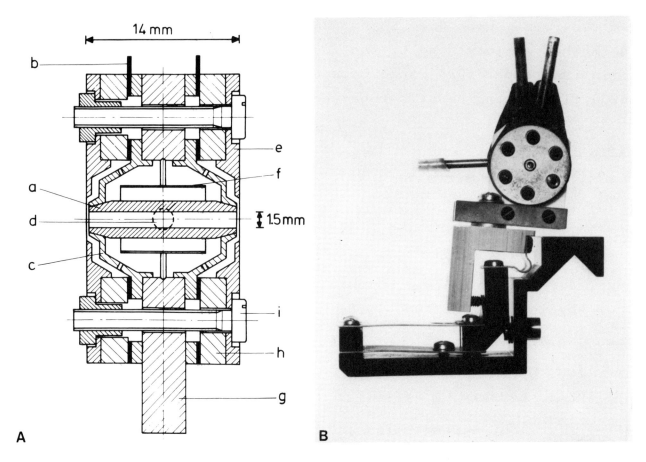

A

B

Fig. 1. Microfurnace for measurements on whiskers. A. Longitudinal cross-section: porous graphite heating tube (a) fed by copper contacts (b) on perforated tantalum electrodes (c). An inert gas enters the space between (a) and (c) via gas inlet (d), passes through (a) and (c), leaving the furnace via the inside of the heating tube (a) and via the space between the electrodes (c) and outer tantalum lids (e). For further protection of the heating element and temperature equalization a tantalum shield (f) is provided. The furnace is mounted in an alundum plate (g) with the aid of alundum spacer rings (h) and screws (i). B. Photograph showing end view and adjustable spring mounting on optical rail sliding (See Bouma et al 1960).

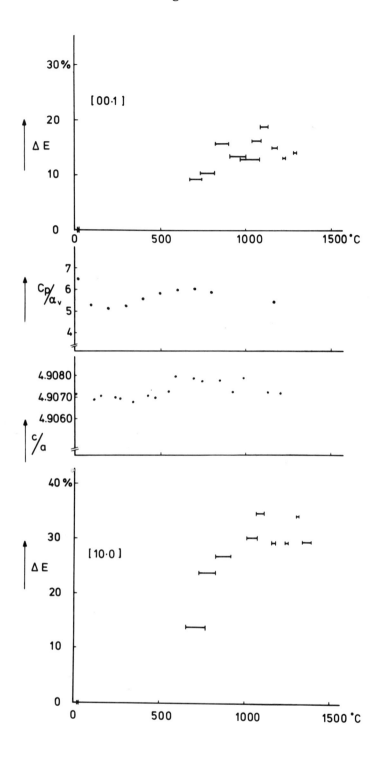

Fig. 2. Young's modulus increase of α-SiC versus temperature. The increase (ΔE) is plotted as a percentage increase of the room temperature value. The axial ratio of the unit cell of 6H and the normalized C_p/α_v values of α-SiC are also indicated (Data from Taylor and Jones 1960 and Kern et al 1969).

Photoelectrical Properties of Polycrystalline Silicon Carbide Layers on Spinel at High Temperatures

E. Wagner and J. Graul

Polycrystalline silicon carbide layers were prepared by means of chemical conversion of 0.4 µm thick silicon films on spinel. As investigation of optical absorption and cathodoluminescence indicate, both α- and ß-phase SiC are grown on spinel substrate. Photoconductivity shows a particular variation with operation temperature in the range between room temperature and 600^{0}C. These silicon carbide layers may find application in high radiation flux measurements in the temperature range considered, but would need an appropriate compensation arrangement.

Photoresistors operating at high temperatures are of special interest for a wide range of applications. In particular this is the case for incore measurements in nuclear reactors. But for this application not only operation at high temperatures is required but also extremely high radiation resistance, due to the high radiation fluxes to be measured. Radiation resitance with respect to photoresistors means insensitivity of the output signal and hence of the electrical parameters to radiation damage. First experiments with silicon layers of different crystallinity showed promising results in respect of radiation resistance of polycrystalline material. However, as application of silicon is rather limited as regards operation temperature, we studied some properties of polycrystalline silicon carbide layers concerning their applicability as photoconducting material.

A disadvantage of polycrystalline semiconductors is the fact, that the quantum efficiency is appreciably lower than that of monocrystals. So, polycrystalline material can only be applied for measurements of high flux densities, and the crystallinity is to be dimensioned in an appropriate way to yield a measurable output signal on the one side and a practicable operation time of the sample on the other.

The polycrystalline silicon carbide layers were prepared by means of chemical conversion of silicon (1,2,3). In a first step silicon films of a thickness of about 0.4 μm were deposited upon electrically insulating substrates by cathode sputtering technique. As substrates we used quartz and spinel platelets as well as oxidized silicon, whereby better photoelectrical properties of the SiC layers were achieved with monocrystalline spinel platelets than with other substrates. The silicon films deposited by sputtering on polished and etched spinel surfaces were undoped and showed amorphous structure.

The silicon layers were converted to silicon carbide by a carbidization procedure, performed in an open-tube set-up, at temperatures above 1050 ^{o}C. Methane, the purest available, was used as hydrocarbon gas in our experiments, with impurities of less than 3 ppm, the main impurity being nitrogen. Hydrogen, which had passed a palladium diffuser, was used as carrier gas. The methane concentration was adjusted between 0.5 and 5 vol%. The samples were positioned on a silicon support, that had been coated with silicon carbide and acted as susceptor for rf-heating.

Care must be taken the sample temperature not to exceed 1200 ^{o}C approximately. Above this temperature chemical reaction between silicon and spinel took place, which caused rapid deterioration of electrical properties. On the other hand, at temperatures below 1000 ^{o}C no formation of SiC could be observed. We therefore prepared the silicon carbide layers at temperatures between 1100 and 1150 ^{o}C.

No essential dependence of growth rate or crystallinity on methane concentration was found, which is consistent with previous experimental results, that growth of SiC layers by carbidization is mainly determined by diffusion of silicon in the layer (1). The fact, that SiC-on-spinel layers show higher photoelectric efficiency than those prepared on more or less amorphous substrates, such as fused quartz or oxide films, is corresponding with the observation, that methods for improving the crystallinity of SiC layers increase the photoelectric yield : e.g. annealing of the amorphous silicon films on monocrystalline spinel in inert gas atmosphere before the carbidization procedure increased the ratio of photo- to ohmic conductivity of the carbide layers. Thus, the following results refer to layers annealed at

1050 °C in argon for 30 min before carbidization.

According to our growth rate measurements at surfaces of silicon platelets (1)
a carbidization procedure lasting for half an hour is sufficient for the thickness of the silicon layers of 0.4 μm. The layers were converted to greenish-
yellowish opaque silicon carbide films. As electron micrographs showed, the
average grain size is smaller than 1000 Å in the plane of the surface.

Spectral transmissivity of the layers in the optical range has been measured
and is shown in Fig.1 . According to the dependence of the transmissivity on
wavelength it is likely, that a mixture of α- and β-phase silicon carbide
material has grown on the monocrystalline spinel substrate, whereas in
SiC layers grown on silicon substrate or on sapphire only the cubic β-phase
has been identified (2-5).

The presence of the hexagonal configuration in SiC-on-spinel layers is further
indicated by cathodoluminescence measurements shown in Fig.2 . Two representative intensity curves are plotted against wavelength resp. photon energy, as

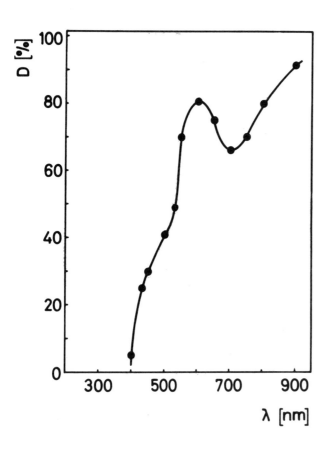

Fig. 1

Spectral transmissivity D of
a SiC-on-spinel layer
prepared by means of chemical
conversion of Si at 1130 °C.

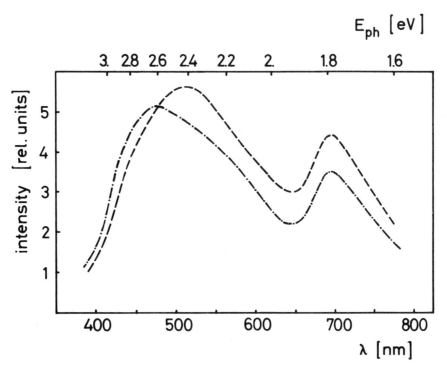

Fig. 2   Cathodoluminescence spectra of a SiC-on-spinel
layer, recorded at different points of the same sample.

measured at room temperature at different points of the same sample. It is
interesting to note that the long wavelength maximum is at the same wavelength
for all points examined on the sample, while the position of the left hand
maximum varies from point to point. This variation might be a result of the
different phases of silicon carbide grown in the layer.

Finally, the electrical properties, which are important for applicability,
were examined as functions of operation temperature. Ohmic conductivity
shows a temperature dependence with an activation energy of 0.45 eV between
room temperature and 600 ^{o}C. Photoconductivity, however, displays a rather
special variation with temperature, as is shown in Fig.3 . The samples were
irradiated with ultraviolet light and photocurrent was measured as a function
of temperature by means of a lock-in amplifier, the applied voltage being
constant. Despite the special behaviour of photoconductivity at variation of
temperature there is to be emphasized, that the photoresponse at elevated
temperatures is in any case higher than at room temperature.

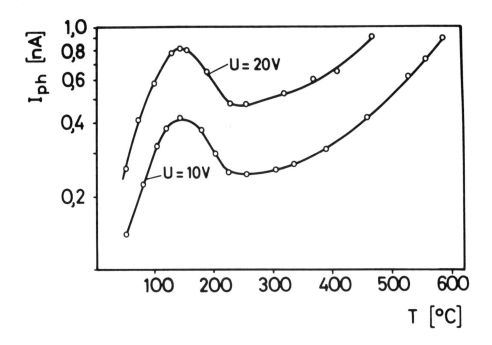

Fig. 3   Photocurrent of a SiC-on-spinel layer as
           function of operation temperature under uv irradiation.

The results show, that polycrystalline silicon carbide layers – prepared by
means of chemical conversion of silicon – may find application as photoconducting material at elevated temperatures up to about 600 °C. Due to the low
quantum efficiency they are limited to measurements of rather high radiation
flux densities. As the ratio of photo- to ohmic conductivity is a function of
operation temperature, it is further necessary to use an appropriate compensation arrangement. Such an arrangement might consist, for instance, of two
identical detector areas, the one being irradiated, the other being shielded.
These could be prepared on the same platelet simultaneously and, according to
our experience, with sufficient reproducibility.

This work was supported by the Bundesministerium für Bildung und Wissenschaft.
Appreciation is expressed to Mr. W. Just of the Forschungslaboratorium für
Festkörperchemie of the University of Munich, head Dr. J. Nickl, for perfor-
mance of cathodoluminescence measurements.

References:

1   J. Graul, E. Wagner, Appl. Phys. Lett. $\underline{21}$ (1972) 67

2   K.E. Haq, I.H. Khan, J. Vac. Sc. Techn. $\underline{7}$ (1970) 490

3   P. Rai-Choudhury, N.P. Formigoni, J. Elchem. Soc. $\underline{116}$ (1969) 1440

4   G.O. Krause, phys. stat. sol. (a) $\underline{3}$ (1970) 899

5   I.H. Khan, A.J. Learn, Appl. Phys. Lett. $\underline{15}$ (1969) 410

Light Emission during the Fracture of Silicon Carbide Filaments

F. E. Wawner, Jr., H. E. DeBolt and V. J. Krukonis

ABSTRACT

Silicon carbide filaments give off light in the form of a spark when broken in tension. The spark occurs at each point where there is simultaneous fracture. Preliminary experiments show that factors such as strain rate variation, testing in an argon atmosphere, annealing to $1100^{\circ}C$ in air and in vacuum, and testing at low temperature do not appear to have any influence on the observed phenomenon. It is felt that the light emission is due to triboluminescence.

Much interest has been generated in silicon carbide (SiC) as an engineering material during the past several years. Not only is the material a semiconductor showing potential for electronic devices, but when produced in filamentary form SiC displays an average tensile strength of 400-500,000 psi and a Young's modulus of 65,000,000 psi. These excellent mechanical properties coupled with the material's relatively low density and chemical inertness make it quite attractive for use as a reinforcement in composite structures and especially in metal matrix applications.

Continuous silicon carbide filaments are produced by chemical vapor deposition from a silane gas, such as methyldichlorosilane or methyltrichlorosilane. A tungsten substrate typically 0.0005 inch in diameter is heated in a continuous reactor to about $1300^{\circ}C$. The resulting filament, 0.004 or 0.0056 inch in diameter consists primarily of a sheath on the tungsten core of β-SiC with some α-SiC.

During a recent study of the mechanical properties of the filaments an unusual and interesting phenomenon was observed. It was seen that during tensile
testing of the filaments, flashes of light in the form of sparks were emitted
from the fracture site in the filament. Plate 1 is a magnified photograph of a
spark that occurred during fracture of the filament. Plate 1 was experimentally
determined by tensile testing SiC filaments on an Instron tester in complete
darkness using a camera, with open shutter, focused on the filament to record
the flash.

Further attempts to capture evidence of the phenomenon were made by double-
exposure whereby a photograph was taken of the filament in light, then the room
darkened and the shutter re-opened while the filament was being strained to
fracture. When the spark occurred it was recorded on the same film with the
filament which produced it. Plate 2 is a representative example of this experiment. In Plate 2, the positive photograph is the double exposure of the filament and sparks and the negatives are enlargements of the localized region of
light to show more detail. The filament in this figure produced two sparks as
it broke simultaneously in two places. These can be seen in the enlargements.
The spark and filament are displaced because tension was not applied to the filament during the initial exposure and hence it shifted during actual pulling. As
many as four sparks have been visually observed during the testing of a single
filament implying that light is emitted from each fracture surface as the filament breaks. During the tests the filaments were completely insulated from the
Instron tester to insure no electrical contribution from the machine.

Several experiments were also performed to determine what factors influence
the spark. It was advanced that one contributory factor to the formation of the
spark was a rapid oxidation of free silicon at the fracture surfaces which mani-
fested itself as a heat release. Tensile tests were made in an inert argon

atmosphere devoid of air to determine if either atmospheric oxygen or nitrogen contributed to the spark. The filaments tested in an argon enclosure still sparked when broken. Filaments were heated in air to $1100^{\circ}C$ to form an insulating SiO_2 layer, and these samples also produced sparks. Similarly, filaments heated in vacuum to remove physically occluded gases that could produce luminescence produced the same effect. It was further determined that the sparking phenomenon is not strain rate dependent for cross-head speeds in the range of .005 cm/min. to 5 cm/min. Experiments were made to cool the filaments by blowing liquid nitrogen directly on them, withdrawing and immediately testing. These tests showed that the sparking occurred at low temperatures (although the exact temperature was not known, it probably was not as low as the temperature of liquid nitrogen).

It is felt that this effect is due to triboluminescence (triboelectric effect) which is defined as luminescence or light emission resulting from mechanical deformation and subsequent fracture of a material. A qualitative mechanism that has been suggested to explain this phenomenon is that as crystal planes separate during the fracture process there is an unbalance of charge on the two faces. The high electric field gradient causes electrons to leave the negative face and bombard the positive face. As a result luminescence occurs in the material or an intervening gas (1).

Silicon carbide has been reported to exhibit several forms of luminescence such as photoluminescence, electroluminescence and cathodoluminescence. A recent report also claims that the material exhibits triboluminescence when tested by the sandblasting technique (2). The present observations verify that triboluminescence is exhibited by SiC.

REFERENCES

1. H. Leverenz, <u>Luminescence of Solids</u>, John Wiley & Sons, New York, 1950.

2. L. Sodomka, Phys. Stat. Sol. (a), <u>7</u>, K65 (1971).

Plate 1. Light Emission During SiC Fracture

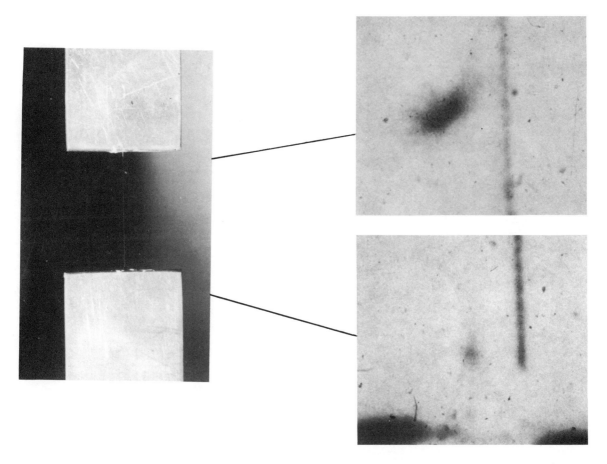

Plate 2. Double Exposure of SiC Filament and Light Emission During Fracture

PART V
DEVICES AND DEVICE TECHNIQUES

Ion Implantation in SiC

O. J. Marsh

INTRODUCTION

With increased requirements for electronic equipment operating at elevated temperatures, there has been considerable interest in wide bandgap semiconductors such as gallium arsenide and silicon carbide. The hexagonal, or alpha, form of SiC has many inherent advantages. The material is chemically stable and mechanically rugged, and its bandgap of 2.9 eV is more suitable for high-temperature operation than silicon's 1.1 eV or gallium arsenide's 1.4 eV.

The fabrication of semiconductor devices in SiC using diffusion requires process temperatures of $1800^{\circ}C$ or greater, and long process times.[1] It is not possible to use passivating oxide layers at these temperatures to mask against diffusion, making precision fabrication of SiC planar devices very difficult.[2] p-n Junctions have been fabricated by growth processes, but these generally involve high temperatures and long times.[3] Recently, p-n junctions have been produced by epitaxial solution growth techniques at process temperatures of $1650^{\circ}C$,[4] but long process times still are required, and techniques for precise geometry control have not yet been developed.

The difficulties experienced in controllably doping SiC by the standard processes led us to look at ion implantation as a doping process in SiC.[5,6,7,8] In this paper I will discuss some of the major investigative areas that must be covered in order to apply implantation to the doping of SiC. I will show also that implantation can be used to form useful devices in this material. Finally, some of the problem areas experienced in attempting to apply ion implantation to doping of SiC will be discussed, and suggestions for possible reasons for these difficulties will be made.

It was hoped in this review to include work from many investigators, but because of lack of time I have had to rely mostly on the work performed by the

471

group at our laboratories. This work is the result of many investigators, and proper recognition should go to my co-workers, H. L. Dunlap and R. R. Hart. We also must thank R. Campbell of Westinghouse, J. Blank of G. E., and W. F. Knippenberg of the Philips Research Laboratories for their most generaous contribution of the SiC crystals used in our work.

FORMATION AND ANNEAL OF DISORDER

The major energy loss mechanisms for the dopant ions and energies used in these investigations are to the electrons in solid and to collisions of the ions with host atoms.[5] It is the latter mechanism that is responsible for disorder in the host lattice. The formation and annealing of this disorder is of primary interest to the investigator interested in applying ion implantation to the doping of semiconductors. We have used the ion backscattering technique[7] to analyze atom displacement in crystals and lattice location of implanted elements.

Backscattered energy spectra for both random and aligned orientation after implantation at room temperature to various doses of 40 keV Sb^+ are shown in Fig. 1.[7] The pronounced peaks near the Si surface edge at $\sim$ 160 keV are caused primarily by backscattering of the aligned He^{++} from disordered Si atoms, i.e., Si atoms displaced greater than approximately the Thomas-Fermi radius, a $\sim$ 0.2 $\overset{o}{A}$, from normal lattice sites.[7,10] It can be seen that the peak height increases with increasing dose and coincides with the random level at a dose of $\sim$ 9 x 10^{13} Sb^+/cm^2, indicating a near-saturation level of disorder, an amorphous layer, as measured by backscattering. The peak near 240 keV represents He^{++} backscattered from the implanted Sb atoms after a dose of 9 x 10^{13} Sb^+/cm^2. Measurements[7] after 30 keV N^+ implantations show a similar buildup of disorder only at greater depth and the formation of an amorphous layer after 9 x 10^{14} N^+/cm^2.

The backscattered energy spectra for an amorphous layer formed by such an implantation is shown in Fig. 2, indicated as "Not Annealed," along with spectra for increasing anneal temperatures.[7] In the "Not Annealed" case the depth distribution of disordered Si atoms is approximately three times deeper and twice as wide as that obtained with the Sb^+ implantations. Further, the major part of the disorder is clearly submerged below the target surface; some crystallinity remains near the target surface, since the aligned yield is lower than the random yield near the surface.

The study of the annealing behavior of disorder formed by ion implantation

is important to proper interpretation of observed electrical phenomena. Annealing normally was carried out in nitrogen atmosphere for periods of 15 min. at
temperatures to 1200°C.

In Fig. 2 we can see that regrowth of the disorder first proceeds from both
the underlying substrate and the target surface, as observed after the 500°C
anneal.[*] This result suggests that lightly disordered SiC anneals significantly
by 500°C. The disorder peak then decreases in height at higher anneal temperatures. Although not presented, the backscattered spectra for the 9×10^{13}
Sb^{+}/cm^{2} implanted layer shows regrowth first from only the underlying substrate
and then a disorder peak decreasing with increasing anneal temperature.

To obtain quantitative anneal data, the disorder was analyzed after each
anneal in the manner discussed in Ref. 11. Figure 3 gives the relative disorder remaining for both implants as a function of anneal temperature. Both
implants behave similarly with an annealing stage centered around $\sim 750^{o}$C and
then leveling off to a small amount of residual disorder even after the highest
temperature anneals studied here. The dashed line indicates surface decomposition. The residual disorder observed after anneals above 1200°C is more likely
due to defects such as dislocation loops in the implanted layer, as has been
observed in silicon,[12,13] than to local amorphous regions. It is also interesting to note that the first strong indications of a p-n junction formed by
nitrogen implants into p-type SiC are found after 750°C anneal.[6] Although there
appears to be little measurable annealing taking place at temperatures above
1200°C, as shown in Fig. 3, the measured electron Hall mobility of nitrogen-
implanted layers in SiC[5,6], which will be described later, continues to increase significantly with annealing to 1700°C, indicating a strong reduction in
the density of electron scattering centers.

The formation of conducting layers and p-n junction devices by ion implantation and the influence of annealing on these layers will be discussed next.

FORMATION OF p-n JUNCTIONS BY ION IMPLANTATION

Ions implanted at constant energy result in a Gaussian distribution if
channeling is neglected. This requires multiple implants to approximate a
linear distribution with depth which is desirable for device studies. For
example, Fig. 4 shows the predicted profile[14] for a double nitrogen implant of

---

[*]Further studies on the annealing of amorphous layers formed by implantation of
N in SiC are presented in Ref. 19.

$10^{15}/cm^2$ at 84 keV and $10^{15}/cm^2$ at 25 keV. The profile is far from linear, but can be considered as approximately $10^{20}/cm^3$ throughout the region of peak concentration. Substrates customarily were held at room temperature during implantation, although some have been maintained at 500°C.

As discussed above, annealing is required to recover the crystalline quality of the lattice and the attendant semiconducting properties; temperatures as high as 1800°C have been used. In order to prevent oxidation, we have used nitrogen atmospheres at temperatures below 1300°C, and vacua for temperatures to 1800°C.

In our early work on junction formation studies with ion implantation we felt that mesa-type junctions would allow for a more straightforward interpretation of diode behavior than would planar devices. Mesa diodes were formed by anodization[15] and the procedures for etching and ohmic contacting are described more fully in Refs. 5 and 6. The mesas were used for capacitance-voltage and current-voltage measurements.

Current-voltage and capacitance-voltage measurements made on devices annealed at temperatures of less than 1500°C indicate a p-i-n device behavior.[6] The thickness of the " i" region decreases with increasing anneal temperature with apparent normal p-n junction behavior observed after a 1500°C, 2-min. anneal.

The forward and reverse current-voltage behavior of a diode after anneal at 1500°C is shown in Fig. 5 as a function of operating temperature following surface cleaning. A major portion of each curve contains a current proportional to $\exp(eV/2kT)$. A plot of J_s vs. $1/T$ at $V=0$ (not shown) shows $J_s \propto \exp(-2.9/2KT)$, indicating that the forward current-voltage behavior of these diodes at low currents appears to be explainable in terms of recombination of injected carriers in the space charge region.[16] A component of the current at 300° and 400°C near 1 V is proportional to $\exp(eV/kT)$ which may be due to diffusion current. At higher values of forward current, the I-V characteristics are controlled by a series resistance attributable to the substrate resistivity, although a contribution from a thin i-region cannot be entirely ruled out.

The reverse current-voltage behavior of these diodes is shown in Fig. 5 for operating temperatures of 23° and 400°C and are of excellent quality; they are similar to those found by Brander and Sutton on epitaxial grown junctions. The onset of breakdown between 45 and 50 V is correlated with the appearance of small blue microplasmas[4,17,18] at visible defects in the original surface of the device.

Devices fabricated from antimony-implanted junctions exhibit essentially the same characteristics as those described for the nitrogen-implanted devices.

ELECTRICAL PROPERTIES OF IMPLANTED LAYERS

In our investigations[5,6,7,8] of ion implantation doping of SiC we have attempted to produce n- and p-type layers using the Column III and V elements , as well as other elements that have been reported to produce dopant action. We have used van der Pauw-Hall effect and sheet resistivity measurements to characterize these layers where possible. We will discuss first the donor doping behavior of Column V elements, and then describe some of the efforts to attain p-type behavior with Column III elements, as well as others. Finally, we will summarize the present state of understanding of implanted layers in SiC.

A. n-Type Dopants

A p-type SiC sample doped with Al with a concentration of $10^{18}/cm^3$ implanted with nitrogen at a dose of 10^{15} N^+/cm^2 at both 25 and 85 keV (see Fig. 4 for the predicted impurity distribution) showed a photoresponse to ultraviolet light after a $500^{\circ}C$, 15-min. anneal in a nitrogen atmosphere, suggesting the presence of a p-n junction. Annealing at $750^{\circ}C$ in nitrogen for 15 min. produced positive indications of an n-type layer when tested with a thermal probe and uv photo-response measurements. After annealing at $1100^{\circ}C$ for 2 min., ohmic contacts to the implanted layer could be made and Hall measurements performed. Current-voltage measurements from the implanted layer to the substrate confirmed the presence of a p-n junction.

The results of Hall effect and sheet resistivity measurements on this implant, as a function of annealing temperature for 2-min. anneals in vacuum, are shown in Fig. 6(a). The notable features are that the number of donors measured is essentially constant with a value of one-half the implanted nitrogen dose of $2 \times 10^{15}/cm^2$ for anneal temperatures up to $1700^{\circ}C$, and that the decrease in sheet resistivity with increasing anneal temperature results entirely from an increase in the effective carrier mobility. The increase in carrier mobility with anneal temperature probably is associated with annealing of disorder in the implanted layer. The decrease in donor concentration after anneal at $1800^{\circ}C$ and the corresponding increase in sheet resistivity result from dissociation of SiC at that temperature. No further increase in mobility is observed above $1700^{\circ}C$, indicating that the annealing of disorder is complete.

The annealing results of Hall effect and sheet resistivity measurements on a phosphorus-implanted layer in p-type α-SiC are shown in Fig. 6(b). The implant was performed at room temperature with a dose-energy series of 1 x 10^{14} and 5 x 10^{14} p^+/cm^2 at 50 and 145 keV, respectively. In contrast to the annealing behavior of nitrogen-implanted layers, no measurable type conversion was observed in the phosphorus-implanted layers for anneal temperatures as high as $1300^{\circ}C$. Annealing in vacuum at $1500^{\circ}C$ for 2 min. produced a measurable n-type layer and p-n junction. The carrier concentration is constant over the anneal range investigated, as was observed with the nitrogen implants. The measured donor concentration is approximately one-third of the implanted phosphorus dose. The decrease of sheet resistivity with increasing anneal temperatures is again associated with the increase in effective carrier mobility as the disorder anneals.

The mobility values obtained in the implanted regions after anneal at $1700^{\circ}C$ appear to be consistent with bulk values obtained in SiC highly doped with nitrogen.[20] This conclusion is discussed more fully in Ref. 6.

The results of our investigations of various ion species and their dopant behavior in SiC after annealing from 1000 to $1700^{\circ}C$ for 2 to 10 min. are summarized in Table I. The Column V elements are expected to be donor impurities in SiC, and donor-type behavior has been observed for implanted N, P, Sb, and Bi in p-type α-SiC. Hall effect measurements have been made in N, P, and Sb layers, confirming n-type behavior as indicated by the thermal probe and the presence of p-n junction behavior. The Bi-implanted layers showed n-type behavior with thermal probing and the presence of a broad area p-n junction. Sb implants have been used to form n^+ contacts to n-type α-SiC after annealing at $1300^{\circ}C$.

TABLE I

SUMMARY OF ION SPECIES IMPLANTED INTO SILICON CARBIDE

SUBSTRATE	IMPLANTED ION[*] AND EFFECT					
	COLUMN V	ELECTRICAL BEHAVIOR	COLUMN III	ELECTRICAL BEHAVIOR	OTHER	ELECTRICAL BEHAVIOR
α-SiC (N- OR P-TYPE)	N	N	B	HIGH ρ	BE	HIGH ρ
	P	N	AL	HIGH ρ	H	HIGH ρ
	SB	N	GA[†]	P, HIGH ρ [X]	HE	HIGH ρ
			IN[†]	P, HIGH ρ [X]	A	HIGH ρ[Δ]
	BI	N	TL	--		
β-SiC (N-TYPE)	SB	N	AL	P (?)		
			IN[†]	HIGH ρ		

[*] IMPLANTED AT ROOM TEMPERATURE; ANNEALED AT $1000^{\circ}C$ TO $1700^{\circ}C$ FOR 2 TO 10 MINUTES.

[X] HIGH ρ AFTER ANNEALS ABOVE $1200^{\circ}C$.

[†] IMPLANT $350^{\circ}C$.

[Δ] AFTER IMPLANT AT ROOM TEMPERATURE AND AFTER ANNEAL AT $1400^{\circ}C$.

B.   p-Type Dopants

The Column III elements are expected to behave as acceptors in SiC. Aluminum and boron have been used extensively in both growth and diffusion to form p-type SiC.[4,21] We have studied B, Al, Ga, In, and Tl implants[5,6,7,8] made with energies from 10 to 50 keV into n-type SiC held at room temperature at dose levels calculated to form a p-type layer doped at $10^{19}/cm^3$ or above.

The implanted layers appeared to be of high resistivity, and except in a very few cases, either measured n-type by a thermal probe, or gave no response to this instrument after the 500 or $1000^\circ C$ anneals. Continued annealing to as high as $1700^\circ C$ did not result conclusively in a p-type layer for any of the Column III elements tried. The implanted surface was either high resistivity, or weakly n-type.

Kalnin, et al.,[22] and Maslakovits and coworkers[23] have reported that Be in SiC forms deep acceptor levels and is an activator for red luminescence. We attempted to dope n-type SiC with Be with doses as high as $10^{15}/cm^2$ at energies from 5 to 60 keV. Annealing these layers to $1700^\circ C$ in a manner similar to that described for the Column III and V dopants did not produce a p-type layer.

Possible reasons for this lack of p-type conduction include: (1) out-diffusion of the implanted species during anneal, (2) decomposition of the implanted layer during anneal, (3) nonsubstitutional site location of the implanted atoms, or (4) the presence of electrically compensating defects. We have used backscattering analysis[8] in conjunction with electrical studies of SiC implanted with the Column III element, In, to determine which of the possible reasons listed above may be responsible for this lack of p-type doping.

Backscattering analysis on the room temperature In implants (1×10^{14}, 40 keV In^+/cm^2, annealed at $1700^\circ C$) showed that at least 40% of the In atoms were not located on substitutional sites.[8]

In an attempt to obtain a greater substitutional fraction of the implanted In, as well as to be able to measure this fraction in a simpler lattice, 3×10^{14}, 40 keV In^+ was implanted at $450^\circ C$ into β-SiC.[8] The backscattered spectra are shown in Fig. 7.

The strong reduction of the In peak in both the $\begin{bmatrix}111\end{bmatrix}$ and $\begin{bmatrix}110\end{bmatrix}$ spectra indicates that ∿90% of the In atoms are on substitutional sites after the $1200^\circ C$ anneal. A $350^\circ C$ implant of α-SiC with 1×10^{14} In^+/cm^2 showed that after $1200^\circ C$ anneal ∿ 95% of the In atoms were located along the $\begin{bmatrix}001\end{bmatrix}$; based on the results of the implant in β-SiC, we believe that these also are substitutional. It should be noted also that In is highly soluble in SiC, since the peak concentration of substitutional In measured in Fig. 7 exceeds 10^{20} In/cm^3.

Electrical evaluation of the hot (450°C) In implant into β-SiC did not produce a clear indication of p-type behavior either immediately after implant, or after 1200°C anneal, even though the substitutional fraction determined from the backscattering in Fig. 7 was ∿ 90%. However, the small size of the sample made effective electrical evaluation difficult.

Electrical probing of the implanted layer of an α-SiC sample implanted at 350°C with 1 x 10^{14}, 40 keV In$^+$/cm^2 indicated weak p-type behavior by both thermal probe and p-n junction response. The response obtained after anneals at 600° 800°, and 1000°C continued to indicate the presence of a weak p-type layer, although the sheet resistivity of the layer was greater than 10^5 Ω/□ . However, a clear indication of the type behavior of the implanted layer could not be obtained after a 1200°C anneal. We also have obtained results very similar to the lattice location and electrical behavior for 350°C In implants in α-SiC and for 350°C* Ga implants.

Implanted layers with In concentrations of less than 10^{20}/cc in n-type, 10^{18}/cc α-SiC result in very high resistivity layers, "i" regions, at anneal temperatures from 500° to 1800°C with depths equivalent to the depth of the In implants indicated by capacitance-voltage measurements on Au surface barriers.[8]

ELECTRICAL EFFECT OF OTHER IONS

We have investigated the electrical behavior of elements that presumably would have no doping effect in SiC. Helium implanted at room temperature into n-type (∿ 10^{18} N/cc) and p-type (∿ 10^{18} Al/cc) produces a high resistivity layer. Au surface barriers have been used to study the width of these i-regions through C-V measurements for various anneal temperatures.[†]

Calculation of the carrier concentration vs. depth from C-V measurements on Au surface barriers are shown in Fig. 8 for the unimplanted material. The Au barriers were removed and two separate areas implanted with He, one with 10^{13}, 40 keV He$^+$/cm^2, and the other with 10^{14}, 100 keV He$^+$/cm^2, into both the n- and p-type samples. After evaporating Au barriers upon the samples, the C-V measurements indicated the presence of an i-region of the thickness indicated as

---

[*]Not published.

[†]H. L. Dunlap, C. L. Anderson, R. R. Hart, and O. J. Marsh, to be published.

"No anneal" with measured carrier concentration beyond the i-region of the concentration of the substrate. The I-V characteristics of the barriers on the implanted areas also indicate the presence of a high resistivity layer. The deeper half of the gaussian He ion profile was theoretically calculated for each implant and is shown for comparison purposes. In this case the He profile would represent the approximate distribution for disorder in the SiC lattice generated by the high energy He atoms. The similarity between the i-region depth and the He penetration is obvious. A low mass ion such as He at the concentrations used here would not form appreciable disorder, but has generated a sufficient number of defects to completely compensate the 10^{18}/cc n- and p-type material.

Removing the Au barriers before annealing and re-applying them after annealing, we find little change in the i-region thickness after a 600°C anneal, whereas an anneal at 1500°C has restored the original dopant profile and eliminated the i-region. No surface decomposition was observed after the 1500°C anneal.

Argon implants into p-type SiC to a dose sufficient to form an amorphous layer and subsequently annealed at 1400°C for 2 min. in vacuum result in high resistivity layers.

SUMMARY AND CONCLUSIONS

Our observations on the electrical properties of various dopants implanted into SiC are summarized in Table I. N-type layers have been formed in both n- and p-type SiC through implantation of the Column V dopants, N, P, Sb, and Bi. Rectifying junctions of excellent quality have been formed with N implants into p-type SiC after anneals at 1500°C for 2 min. Hall-effect analyses on N-and P-implanted layers indicate donor concentrations of 50% and 30%, respectively, of the implanted dose. Campbell, et al.,[19] have found from their work that $\sim$ 50% of implanted nitrogen (for similar implant and anneal conditions) is located along lattice rows, indicating possible substitutional behavior of 50%, and in good agreement with our Hall measurements on N implants. We have observed very similar behavior for Sb implants from backscattering analysis.[7] After implantations into SiC at room temperature, all ions we have tried result in high resistivity layers after implant.

Presumably neutral elements such as He and A implanted in SiC indicate either no effect on the electrical conduction in the substrate, or produce a high resistivity layer after annealing at 1400°C. In neither case do the layers result in low resistivity n-type layers, indicating that the doping behavior observed for Column V elements is a chemical rather than a disorder effect.

Implantation at room temperature of the Column III elements B, Al, Ga, In, and Tl to concentrations of 10^{19}/cc or greater result only in high resistivity layers at all anneal temperatures to as high as 1800°C. Implantation of In and Ga at temperatures of 350° to 500°C result in weakly doped p-type layers that convert to high resistivity after 1200°C anneals or higher. The backscattering data show that the lack of p-type behavior of layers implanted with In and annealed at 1200° to 1700°C is not caused by outdiffusion of the implanted impurity as has been observed for Al implants in SiC[24], and suggested from the work on B implants.[25] Neither can decomposition of the implanted layer, nor nonsubstitutional site location of the implanted atoms be a cause.

Since only high resistivity layers are observed electrically after 1200° to 1800°C anneals of the SiC layers which were all implanted with In^{+} at concentrations greater than the substrate donor concentration, (whereas similar implantations of Column V elements result in highly conducting n-type layers after 1200° to 1800°C anneals) the presence of deep-donor-type defects are suggested. These defects may consist of an impurity-vacancy complex, as suggested by the luminescence work of Choyke and Patrick.[26] The p-type layers observed in the sample implanted at 350°C after anneals of 1000°C or less may be due to acceptor behavior of the implanted In and Ga. However, it is possible that some as yet unidentified acceptor-type defect exists at these temperatures.

REFERENCES

1) L. J. Kroko and A. G. Milnes, Solid State Electron. 9, 1125 (1966).

2) R. B. Campbell and H. S. Berman, Mat. Res. Bull. 4, S211 (1969).

3) C. Goldberg and J. W. Ostroski, "Silicon Carbide - A High Temperature Conductor," p. 453, Pergamon Press, New York (1960).

4) R. W. Brander and R. P. Sutton, Brit. J. Appl. Phys. 2, 309 (1969).

5) H. L. Dunlap and O. J. Marsh, Appl. Phys. Letters 15, 311 (1969).

6) O. J. Marsh and H. L. Dunlap, Radiation Effects 6, 301 (1970).

7) R. R. Hart, H. L. Dunlap, O. J. Marsh, Radiation Effects 9, 262 (1971)

8) R. R. Hart, H. L. Dunlap and O. J. Marsh, "Backscattering Analysis and Electrical Behavior of SiC Implanted with 40 keV Indium," II Internat'l Conf. on Ion Implantation in Semiconductors, p. 134, Garmisch-Partenkirchen, Germany, May, 1971.

9) J. W. Mayer, L. Eriksson and J. A. Davies, "Ion Implantation in Semiconductors," Academic Press, N.Y. (1970).

10) J. A. Davies, J. Denhartog, L. Eriksson, and J. W. Mayer, Can. J. Phys. 45, 4054 (1967).

11) R. R. Hart, Radiation Effects 6, 51 (1970).

12) L. N. Large and R. W. Bicknell, J. Mater. Sci. 2, 589 (1967).

13) R. W. Bicknell, Proc. Roy. Soc. (London), Ser A311, 75, (1969).

14) J. Lindhard, M. Scharff and H. E. Schiøtt, KgL. Dansk Videnskab, Selskab, Mat. Fys. Medd. 33, No. 14 (1963).

15) V. J. Jennings, Mat. Res. Bull. 4, S199 (1969).

16) C. T. Sah, R. N. Noyce and W. Schockley, Proc. IRE 45, 1228 (1957).

17) C. A. A. J. Greebe, Phil. Res. Rpts., Suppl. No. 1 (1963).

18) L. Patrick, J. Appl. Phys. 28, 765 (1957).

19) A. B. Campbell, J. B. Mitchell, J. Shewchun and D. A. Thompson, "Use of the Channeling Effect to Study Damage and Nitrogen Atom-Site-Location in Nitrogen Implanted α-SiC," Third International Conference on Silicon Carbide, Miami, September, 1973.

20) G. S. Kamath, Mat. Res. Bull. 4, S57 (1969).

21) J. M. Blank, Mat. Res. Bull. 4, S179 (1969).

22) A. A. Kal'nin, Yu. M. Tairov and D. A. Yas'kov, Soviet Phys.-Solid State 8, 755 (1966).

23) Yu. P. Maslakovets, E. N. Mokhov, Yu. A. Vadakov, and G. A. Lomakina, Fiz. Tverd. Tela. 10, 809 (1968).

24) J. Comas, W. Lucke and A. Addamiano, Bull. Am. Phys. Soc. 18, 606 (1973).

25) A. Addamiano, G. W. Anderson, J. Comas, H. L. Hughes, W. Lucke, J. Electrochem. Soc. 119, 1355 (1972).

26) L. Patrick and W. J. Choyke, Phys. Rev. 5, 3253 (1972).

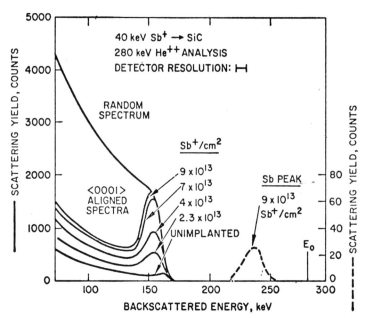

Figure 1.   Backscattered energy spectra as a function of dose for
E_o = 280 keV He^{++} incident on α-SiC after implantation of 40 keV Sb$^+$
A random-equivalent spectrum and an aligned spectrum from unimplanted
crystal are included for comparison.

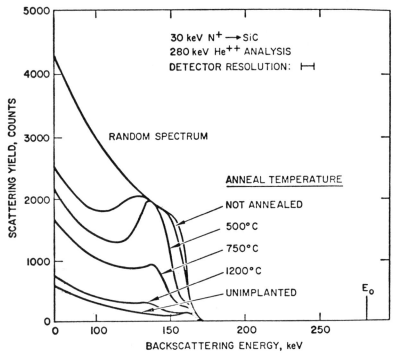

Figure 2.   Backscattered energy spectra of SiC as a function of anneal temperature for a 9 x 10^{14} N$^+$/cm^2 implant.

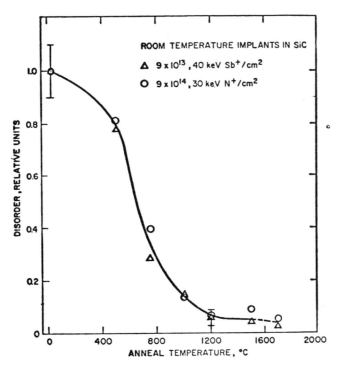

Figure 3. The annealing of disorder in SiC introduced by 9×10^{13} Sb^+/cm^2 or 9×10^{14} N^+/cm^2 implantations.

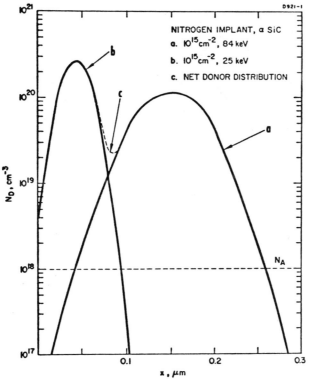

Figure 4. Theoretical distribution of nitrogen ions implanted in SiC. Junction depth seen to be 0.26 μm with presumed bulk doping of 10^{18} cm^{-3} acceptors. (Calculation courtesy H. Schiøtt.)

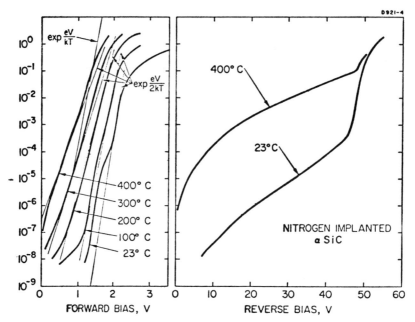

Figure 5.   Log J vs. forward and reverse bias for diode of well-annealed
            nitrogen-implanted α-SiC at various operating temperatures.

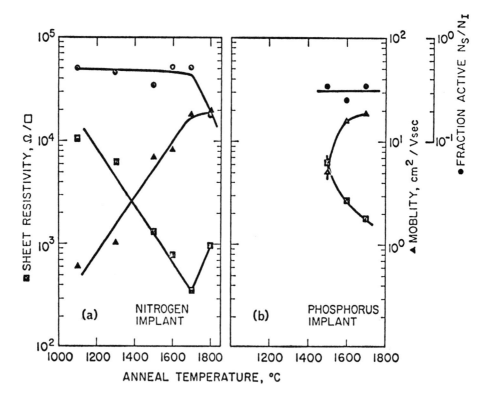

Figure 6.   Electrical carrier parameters of implanted ions in α-SiC vs. annealing
            temperature. (a) With a nitrogen fluence of 10^{15}cm^{-2} at 84 keV and
            10^{15}cm^{-2} at 25 keV; (b) With a phosphorus fluence of 5 x 10^{14}cm^{-2}
            at 145 keV and 10^{14}cm^{-2} at 50 keV.

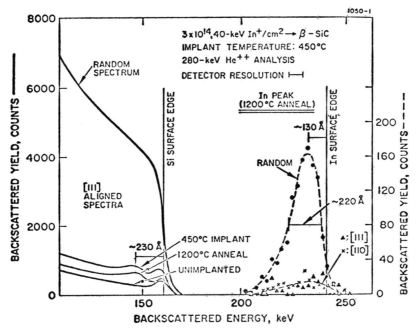

Figure 7.   Backscattered energy spectra from β-SiC implanted at 450°C with
3×10^{14}, 40 keV In$^+$/cm^2.   For clarity, a small background correction
was applied to the In peak.

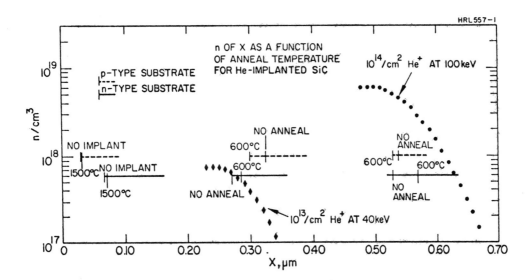

Figure 8.   Carrier concentration vs. depth as calculated from C-V measurements
made on Au surface barriers on n- and p-type SiC.   Curves are shown
before and after bombardment with He$^+$ ions and as a function of
anneal temperature.

Use of the Channeling Effect to Study Damage and Nitrogen Atom-Site-Location in Nitrogen Implanted α-SiC

A. B. Campbell, J. B. Mitchell, J. Shewchun, D. A. Thompson and J. A. Davies

INTRODUCTION

Silicon carbide has long been considered an ideal material from which to form high temperature semiconductor devices, but the formation of these devices has been hampered by the high diffusion temperatures required ($\sim$2200°C). Consequently, ion implantation is an attractive alternate technique, provided the associated radiation damage anneals out at reasonable temperatures and the implanted atom becomes electrically active.

Previous work[1,2] on the nitrogen/silicon carbide system has established that an n-type layer can be formed by nitrogen implantation at room temperature and a subsequent 750°C, 15 minute anneal. The resistivity of this layer following the 750°C anneal was too high to allow Hall measurements. However, Hall effect and sheet resistivity measurements performed as a function of annealing temperature in the range 1100-1700°C have confirmed the n-type region and have shown that 50% of the implanted nitrogen has become electrically active. Although the donor concentration remains relatively constant over this anneal temperature range, the mobility increases with increasing anneal temperature, suggesting that the density of scattering centers is decreasing.

The most extensive damage study on the nitrogen/silicon carbide system[3] has used 280-keV He^{++} backscattering to study the damage following implantation and subsequent anneal to 1500°C. Although this damage study confirmed the annealing stage at $\sim$750°C found in the electrical measurements mentioned above, and determined that the damage regrowth during anneal was occurring from both the crystal surface and the underlying substrate, the depth resolution was too poor to describe the nature of the annealing.

This paper reports a further investigation of the nitrogen/silicon carbide

486

system in which we use channeling effect techniques to study not only the radiation damage problem but also the lattice location of the implanted nitrogen atoms. The damage following implantation and subsequent anneal treatments was studied using a 2 MeV He^+ ion beam and standard channeling techniques[4]. Once it was established that the crystal had reordered, the lattice location of the implanted nitrogen was investigated using the $^{15}N(p,\alpha_o)$ nuclear reaction together with channeling techniques.

EXPERIMENTAL CONDITIONS

The silicon carbide samples used in this study were α-type, aluminum doped with a net acceptor concentration of $10^{18} cm^{-3}$, kindly provided by R.B. Campbell of the Westinghouse Astronuclear Laboratory.

For the damage study, implants with ^{14}N were performed at 80 keV with doses of approximately 10^{15} ions/cm^2, using the McMaster 150 keV ion implantation accelerator. Implants of ^{15}N (for the (p,α_o) study) were performed using the Chalk River isotope separator; these were double implants of 5×10^{14} $^{15}N/cm^2$ at 45 keV and 20 keV in order to provide a relatively uniform implanted region. In all cases, the nitrogen dose was sufficiently high to overcome the compensating effects of the high concentration of acceptor impurities. Implants were performed at room temperature. All anneals were of 3 minute duration and were performed in vacuum. The maximum anneal temperature available was $1450°C$.

The techniques and general principle of channeling (combined with backscattering and/or nuclear reaction measurements) to study lattice damage and foreign-atom location in crystals has been described in detail in the literature[4]. Basically, a channeled ion can interact only with those atoms that are displaced more than ~0.2 Å (the Thomas-Fermi screening radius) from a lattice site. Hence, the interaction yield provides a direct measure of the displaced lattice atoms (ie. lattice disorder) and of the non-substitutional foreign atoms in the crystal.

For the lattice disorder studies, He^+ backscattering measurements were conducted with the McMaster and Chalk River van de Graaff accelerators using 10 nA beams of well collimated 2 MeV He^+. A silicon surface barrier detector with 15-keV resolution located at an angle of $150°$ with respect to the incoming beam was used to provide energy spectra of backscattered He^+ ions. Comparison of the spectra obtained when the sample is in an aligned direction (ie. with the c-axis parallel to the incoming beam) to that obtained when the sample is in an non-aligned direction provides information on the amount of radiation damage.

In order to determine the atom-site-location of the implanted nitrogen, a nuclear reaction is needed which is specific for nitrogen and is sensitive enough to analyse nitrogen at the 10^{15} atoms/cm^2 level. Two possible reactions were investigated: $^{14}N(d,\alpha)^{12}C^{(5,6)}$ and $^{15}N(p,\alpha_o)^{12}C^{(7,8)}$. The first reaction turned out to be unacceptable because of background effects due to surface nitrogen and $^{12}C(d,p)$ reactions. The second reaction proved acceptable with a background equivalent to 3×10^{14} ^{15}N atoms/cm^2.

A silicon sample implanted with 5×10^{15} ^{15}N/cm^2 was used to calibrate the energy of the α_o-peak. We chose 800 keV as the optimum proton energy for the (p,α_o) studies, because at this energy a sufficiently thick absorber ($\sim$8.5μm aluminized mylar) to stop most of the backscattered protons is still thin enough to allow the α_o particles to reach the detector.

RESULTS AND DISCUSSION

Figure 1 shows the backscattering spectra of 2 MeV He$^+$ from a SiC sample before implantation and after an implant of 10^{15} N ions/cm^2 at 80 keV and

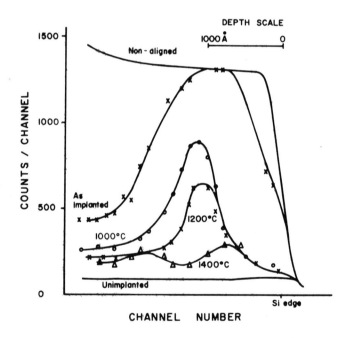

Figure 1:   Aligned backscattering spectra for 2 MeV He$^+$ on nitrogen implanted SiC as a function of anneal temperature (10^{15} ions/cm^2, 80 keV implant)

subsequent annealing up to 1400°C. The channel number is proportional to the energy of the backscattered He$^+$ and can be converted to a depth scale by calculating the energy loss of He$^+$ in SiC. This depth scale indicates that the maximum of the as-implanted damage peak is 1000Å below the surface. This value agrees reasonably well with the theoretical damage distribution[9]; it is also consistent with the projected range values measured by Addamiano et al[10] and inferred from the junction depths measured by Marsh & Dunlap[1].

Note that the damage anneals predominantly from the surface inward, thus producing a buried damage peak which becomes progressively narrower with increasing anneal temperature. This anneal behavior correlates well with the conclusions of Marsh and Dunlap[1] based on their study of the electrical properties of implanted diodes. They found evidence of a semi-insulating buried layer (probably due to the presence of defects) below the n-type surface region and the thickness of this insulating layer decreased with increasing anneal temperatures.

The area of the damage peaks in figure 1 was measured as a function of anneal temperature. Figure 2 shows the results of this measurement along with the earlier results of Hart, Dunlap & Marsh[3]. They measured the damage in

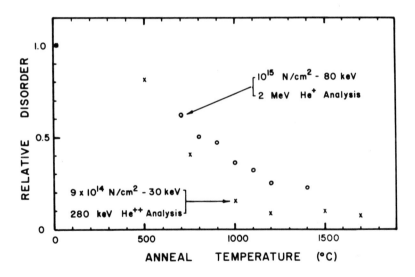

Figure 2: Comparison of relative disorder vs. anneal temperature for this work (o) with the previous study of Hart, Dunlap, and Marsh[3] (x)

a sample implanted with 9×10^{14} ions/cm^2 at 30 keV, using a 280-keV He^{++} beam for the backscattering analysis. Hart et al used the height of the aligned spectrum behind the damage peak as a measure of the amount of lattice disorder, whereas we use the area of the damage peak (after subtracting a linear dechanneling correction). Both results are in fair agreement and indicate that most of the disorder has annealed by 1200°C. However, as shown in figure 1, an anneal of 1400°C is required to reduce the buried damage peak to a sufficiently low level for the nitrogen lattice-location study.

Figure 3 shows the energy spectra of the α-particles produced in the $^{15}N(p, \alpha_o)^{12}C$ reaction for a ^{15}N standard and the implanted sample after a 1450°C

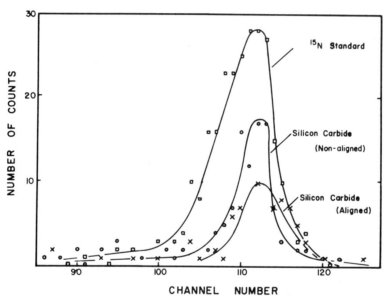

Figure 3: α_o-spectra from the $^{15}N(p, \alpha_o)^{12}C$ reaction on implanted and annealed SiC (10^{15} ions/cm^2, 45 keV and 20 keV, anneal at 1450°C) after standard proton doses of 160μC; a spectrum from a ^{15}N standard is shown for comparison.

anneal. The area of the aligned peak is 62% of the non-aligned peak.

The substitutional percentage of the implanted nitrogen can be calculated from

$$S = \frac{1 - N_A/N_R}{1 - \chi^*_{MIN}} \times 100 \qquad (1)$$

where N_A and N_R are the areas of the α-peaks in the aligned and non-aligned directions and χ^*_{MIN} is the average value of the ratio of the aligned and non-aligned backscattered yields in the implanted region (cf. fig. 1).

To get reasonable statistics, proton doses of the order of 1000μC are necessary but attempts to improve the statistics over those shown in Figure 3 (obtained with a total dose of 320μC) were unsuccessful due to excessive damage from the proton beam. Even after the 320μC dose of fig.3, subsequent backscattering analysis (using 2 MeV He^+) indicated that χ_{MIN}^* has increased to 36%, compared to a χ_{MIN}^* value of 15% at the beginning of its analysis.

Substituting into equation (1) the above area ratio of 62% and a χ_{MIN}^* value of 25% (a reasonable estimate for the average χ_{MIN}^* value during the analysis), we obtain a value of 51% for S, indicating that ∿51% of the ^{15}N atoms are shadowed by the atomic rows along the c-axis and therefore are probably on substitutional sites. This is in good agreement with the substitutional component at 50% deduced by Marsh and Dunlap[1] from their electrical measurements.

Samples implanted to a dose of 10^{15} $^{14}N/cm^2$ at 45 keV and 20 keV have been made into planar diodes and a preliminary study has been made of their I-V characteristics. We observed diode action after an 1100°C anneal, with improved performance after a 1400°C anneal. The characteristics are comparable to those measured by Dunlap and Marsh[2].

These experiments have provided results in excellent agreement with previous work[1-3] and have extended the study of nitrogen-implanted silicon carbide to include nitrogen-atom-location experiments which correlate well with electrical activity measurements.

REFERENCES

1) O.J. Marsh and H.L. Dunlap, Rad, Effects <u>6</u>, 301 (1970)

2) H.L. Dunlap and O.J. Marsh, Appl. Phys, Letters <u>15</u>, 311 (1969)

3) R.R. Hart, H.L. Dunlap, and O.J. Marsh, Rad. Effects <u>9</u>, 261 (1971)

4) J.W. Mayer, L. Eriksson, and J.A. Davies, "Ion Implantation in
    Semiconductors," Academic Press, New York (1970), Chapters 3 and 4
    and reference therein

5) G. Amsel and D. David, Rev. Phys. Appl. <u>4</u>, 383 (1969)

6) G. Amsel, J.P. Nadai, E. D'Artemare, D. David, E. Girard, and J. Moulin
    Nucl. Inst and Meth. <u>92</u>, 481 (1971)

7) F.B. Hagedorn and J.B. Marion, Phys. Rev. <u>108</u>, 1015 (1957)

8) A.V. Cohen and A.P. French, Phil Mag, <u>44</u>, 1259 (1953)

9) J. Lindhard, V. Nielsen, M. Scharff and P.V. Thomsen, Kgl. Danske
    Videnskab. Selskab, Mat. Fys. Medd. <u>33</u>, No. 10 (1963)

10) A. Addamiano, G.W. Anderson, J. Comas, H.L. Hughes, and W. Lucke,
    J. Electrochem. Soc. <u>10</u>, 1355 (1972)

Resistivity and Hall Coefficient Measurements on SiC

H. Kang and R. B. Hilborn, Jr.

ABSTRACT

This paper is a report of a Van der Pauw experiment set up for the routine measurement over a temperature range of $77^{\circ}K$ to $1000^{\circ}K$ of resistivity and Hall coefficient of crystals of SiC furnished from a variety of sources. A description of the facility, results of measurements on a number of different polytypes and recommendations for the proper employment of this experiment on SiC are given.

INTRODUCTION

The work reported in this paper has resulted from an effort on the part of the Air Force Cambridge Research Laboratory to establish a center at the University of South Carolina for the routine characterization of crystals furnished from a variety of sources. The main goal towards which this effort is directed is the determination of the growth parameters most important for producing single crystals of SiC of commercial quality. This goal must be attained if SiC is to play a significant role in the semiconductor device industry in the future.

DESCRIPTION OF FACILITY

The experiment used for measuring the resistivity and Hall mobility of our SiC samples is that commonly referred to as the Van der Pauw method[1]. Using this technique in an environmental chamber similar to the one described by Van Daal[2] we are able to perform our measurements on two samples simultaneously as a function of temperature from $77^{\circ}K$ to $1000^{\circ}K$.

As the typical morphology of samples received is in the form of flat platelets,

our electrical contacts are attached to the periphery of the basal plane of the
platelet; and the sample is situated with the basal plan perpendicular to the
applied magnetic field. Our data is therefore only representative of the magneto-
electric transport properties in the basal plane of the sample. Electrical connection to the sample is made through four Au-5%Ta contacts alloyed in vacuum.
This type of connection has been found by us and others[3] to form adequate low
resistance ohmic contacts over the temperature range of interest.

The environmental chamber used to vary the sample temperature from 77°K to 1000°K is as shown in figure 1. It is constructed of non-magnetic stainless steel tubing so as to minimize any perturbation of the uniformity and magnitude of the externally applied magnetic field. The chamber essentially con-sists of two vacuum

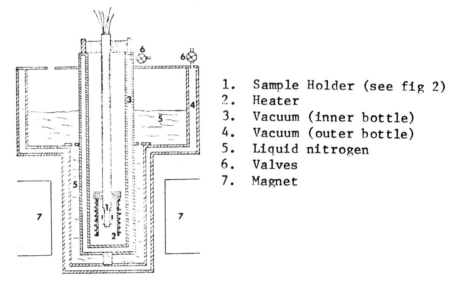

1. Sample Holder (see fig 2)
2. Heater
3. Vacuum (inner bottle)
4. Vacuum (outer bottle)
5. Liquid nitrogen
6. Valves
7. Magnet

Figure 1: Environmental chamber

bottles, one inside the other, separated by a chamber filled with liquid nitrogen.
The outer vacuum bottle serves to prevent the absorption of heat from the ambient
room temperature atmosphere by the liquid nitrogen, thus conserving the amount of
nitrogen needed for the experiment. The inner vacuum bottle serves as a variable
heat exchanger between the sample, inside the inner vacuum bottle, and the liquid
nitrogen reservoir. The rate of heat exchange is made variable by varying the
degree of vacuum within the walls. With the chamber evacuated ($\sim 10^{-5}$ torr), the
heat transfer between the sample and the liquid nitrogen reservoir is at a minimum. In this state, with the sample initially at 300°K, the rate of decrease
of the sample temperature is approximately 15°K/hour. The rate of decrease of
the temperature of the sample decreases as the sample temperature is lowered.
Thus at a temperatue of 230°K, the rate of temperature decrease drops to about
10°K/hour. It is necessary at a sample temperature of 190°K to leak air into the
walls of the inner vacuum bottle in order to further decrease the sample temperature in a realistic time.

Even though the cooling rate of the sample is kept low, it is necessary to stabilize the sample temperature at the points where the Van der Pauw measurements are to be made as they take several minutes to carry out completely. The temperature is stabilized by a non-inductively wire wound electric heater concentrically surrounding the sample. Power to the heater is controlled by a chromel-alumel thermocouple connected between the sample and a temperature controller. By using this technique we are able to stabilize the sample temperature to within $\pm 0.5^{\circ}$K at any temperature between 77°K and 300°K.

The same wire heater is used to raise the temperature of the sample from 300°K to 1000°K. During the measurement over this temperature range, the liquid nitrogen is removed and the walls of the inner vacuum bottle are evacuated to prevent excessive heat loss from the heater to the ambient atmosphere. We are able to stabilize the sample temperature while we take our measurement at any desired value within this range to within $\pm 1^{\circ}$K.

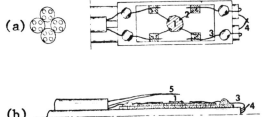

1. Sample
2. Gold bonding wire
3. Platinum lead
4. Thermocouple for controller
5. Thermocouple for temperature
       measurement

Figure 2: Sample holder; (a) top view, (b) side view (upper half only).

The samples themselves are attached to two alumina plates oriented back to back in a sample holder as indicated in figure 2. In this arrangement the basal planes of the two samples are parallel to each other while being on opposite sides of the sample holder.

The magnetic field is furnished by a six inch electromagnet whose pole faces are butted up against the environmental chamber as indicated in figure 1. With this magnet the magnetic field can be varied in polarity and in intensity from 0 to 8 kilogauss, (with a uniformity of ± 0.1 gauss over the area occupied by the samples).

The electric current is applied to the sample by a well regulated, constant current power supply. The current is measured by voltage drop across a standard resistor. All voltage measurements are made by potentiometric methods with an accuracy of $\pm 0.03\%$.

FIELD STRENGTH RESTRICTIONS ON MEASUREMENTS

In the course of setting up and calibrating our facility we made Van der Pauw measurements on a dozen or more single crystal samples furnished us by the Westinghouse and General Electric Corporations. We were thus able to compare the results of our measurements with the results of two other independent laboratories on the same samples. The electrical contacts, in all cases, had been formed by the supplier of the crystal. They were all alloyed Au-5%Ta with the exception of two p-type samples which had Si-Au-B contacts.

Upon comparing the results of our measurements with those of the suppliers of the samples we found agreement in some cases and disagreement in others. Further investigation into the matter revealed that in the cases where we had used the same value of current as the suppliers of the crystal our results were in agreement with their's to within a few per cent. However, for those cases where the same magnitude of current was not used we had discrepancies between our results and their's of as much as 25%.

This bit of information then precipitated a rather thorough investigation on our part of the dependence of the results of the Van der Pauw measurements in SiC on the magnitude of the electric current and magnetic fields applied. The result of this study is that we found the measurement to be independent of magnetic field, at least up to a magnitude of 8 kilogauss which was the limit of our magnet. We did, however, find a strong dependence of the results of our measurements on the magnitude of the electric current being supplied to our samples. The resistivity value with a current of 20 mA was approximately 5% lower than that measured when using a current of 10 mA. The difference got larger for greater currents. For currents below 10 mA, however, we found no current dependence for the measured value of resistivity.

We do not know, at this time, if the dependence of the resistivity on current, for values greater than 10 mA, is due to non-ohmicity of the contacts or due to local heating from the I^2R losses, or both. In any event it is apparent that caution should be exercised in using currents greater than a few milliamperes when making Van der Pauw measurements on SiC.

MEASUREMENT RESULTS ON VARIOUS n-TYPE POLYTYPES

The results of our Van der Pauw measurements on six samples over a temperature range of 77°K to 1000°K are shown in figures 3, 4, and 5. A summary of the

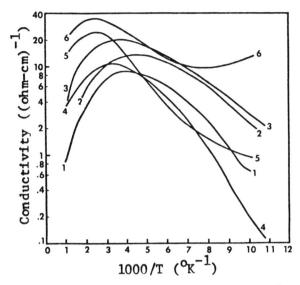

Figure 3: Conductivity vs 1000/T.

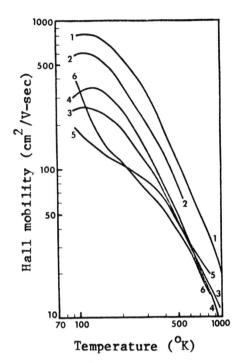

Figure 4: Hall mobility as a function
of temperature.

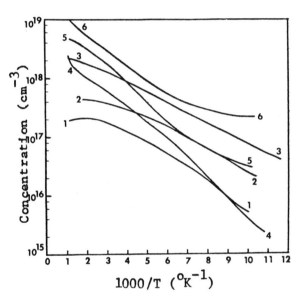

Figure 5: Carrier concentration
vs 1000/T.

room temperature properties as well as polytype identification of these samples is given in table 1. The samples numbered 1 through 4 were furnished from the Westinghouse Corporation, whereas, samples number 5 and 6 were supplied by the Air Force Cambridge Research Laboratory. All of the samples are nitrogen doped and compensated n-type.

From the plots of conductivity (reciprocal of resistivity) vs 1/T shown in figure 3, it can be seen that all of the samples go from a low temperature region having a negative temperature coefficient of resistivity to a higher temperature region with a positive temperature coefficient of resistivity. It would appear also upon comparing these curves with the data in table 1 that the changeover point increases in temperature as the ionization energy of the donor impurity increases. This would be indicative that the changeover point is due to the sample going from a region of extrinsic conduction at the lower temperature to a region of saturated extrinsic conduction in the higher temperature region.

Table 1

Room Temperature Electrical Properties

Sample No.	Polytype	Mobility (cm^2/V-sec)	Carrier Concentration (cm^{-3})	Ionization Energy (eV)
1	4H	332	1.6×10^{17}	0.04
2	15R	220	3.5×10^{17}	0.04
3	15R	110	1.2×10^{18}	0.04
4	6H	136	5.2×10^{17}	0.07
5	6H	84	1.5×10^{18}	0.07*
6	6H	72	2.5×10^{18}	0.05*

*From straight line portion of curve at high temperatures.

Looking at the plot of Hall mobility vs. temperature in figure 4, we note that the slope of these curves for temperature above 500°K appears to vary with impurity concentration independently of polytype in agreement with similar information published by Patrick[4]. The temperature dependence of μ_H in this region is seen to vary as $T^{-2.5}$ for the purer sample, no. 1; to $T^{-1.5}$ for the more highly doped sample, no. 6. It would appear from the room temperature values given in table 1, that the mobility for a single polytype would tend to decrease with increasing carrier concentration, this can be seen not to hold true in general, however, from the curves in figure 4, by noting the mobilities of the 6H samples, 4, 5, and 6 at temperatures below 200°K. In this region the less pure sample, 6, has a higher mobility than the purer sample, 5. This inconsistancy is believed to be the result of the presence of compensating impurities. We show in another paper, to be presented at this meeting[5], that the mobility does decrease monotonically with increasing carrier concentration in uncompensated n-type 6H polytypes.

Carrier concentrations obtained from the Van der Pauw measurements using the relation, $n = 1/(e R_H)$, where e is the electronic charge and R_H is the Hall coefficient, are shown in figure 5. The ionization energies given in table 1 are obtained from the slopes of these curves taken, with the exception of sample 6, in the low temperature region. It was necessary to take the slope of sample 6 in the high temperature region of the curve due to its peculiar behavior at low temperatures. This behavior though not presently understood is similar to that we have seen in other samples and it has been reported by other researchers[2,6,7].

CONCLUSION

In conclusion we wish to say that the Van der Pauw measurement, as described herein, utilized over a broad range of temperature, represents a powerful tool for the characterization of SiC single crystal polytypes. Caution is advised, however, in keeping the current levels low while making the measurements.

ACKNOWLEDGMENTS

We wish to thank the Westinghouse and General Electric Corporation for furnishing us some samples. We would also like to thank Mr. Y. Tung for his identification of the polytype by transmission Laue X-ray techniques.

This work was partially supported by the Air Force Cambridge Research Laboratories under contract #F19628-72-C0136.

(1) L. J. Van der Pauw, Philips Res. Repts., 13, 1 (1958).

(2) H. J. van Daal, Philips Res. Repts., Supp. 3 (1965).

(3) D. L. Barrett and R. B. Campbell, J. Appl. Phys., 38, 53 (1967).

(4) L. Patrick, J. Appl. Phys., 38, 50 (1967).

(5) R. B. Hilborn, Jr. and H. Kang, recent news paper presented this conference.

(6) S. H. Hagen and C. J. Kapteyns, Philips Res. Repts., 25, 1 (1970).

(7) G. A. Lomakina, Sov. Phys. Solid State, 8, 1038 (1966).

Carrier Velocity Measurements on Silicon Carbide

H. S. Berman, T. M. Heng, H. C. Nathenson and R. B. Campbell

The scattering limited velocity, V_{SAT}, was measured on several n-type and p-type sublimation grown SiC crystals. The sample geometrics, which were of small length and cross section were prepared by an oxidation-photoresist-chlorine etching techique. The n-type samples were measured up to a velocity of 8.2×10^6 cm/sec with the curve still unsaturated. Silicon, by comparison begins to saturate at 4×10^6 cm/sec.

INTRODUCTION

A charge carrier in the space charge region will have its velocity increased by an applied electric field. At a certain value of field, however, the carrier velocity no longer increases with increasing field, but saturates and thereafter remains essentially constant. This saturation occurring at the threshold field E_s, is due to scattering, and the scattering-limited velocity, V_{SAT}, is an important parameter in high frequency devices. The intrinsic rise time of a device is inversely proportional to V_{SAT}, and in addition, velocity saturation is required for linear operation of the high frequency device since the carrier transport is relatively insensitive to minor electric field variations.

V_{SAT} appears in two figures of merit relating to the ultimate performance of semiconductor devices.

The Johnson[1] figure of merit is:

$$P_M F_M^2 Z_M = \frac{E_c^2 V_{SAT}^2}{4\pi}$$

where: E_c is the critical field and P_M, F_M, and Z_M are the maximum power output, maximum operation frequency and maximum load impedance respectively.

The Keyes[2] figure of merit is:

$$\sigma_t (C\, V_{SAT}/4\pi\epsilon)^{1/2}$$

where: C   = velocity of light

   σ_t = thermal conductivity

   ϵ = dielectric constant

The various parameters, E_c, σ_t, ϵ , etc., are fairly well known for SiC.   If V_{SAT} for SiC is near 10^7 cm/sec, then SiC high frequency devices could compare quite favorably with those fabricated from Si, Ge or the III-V's.

CARRIER SATURATION VELOCITY

The carrier saturation velocity for SiC has not been previously determined. The results of the Johnson and Keyes figures of merit point out the need for this quantity in the evaluation of the true device potential.   Assuming the same scattering mechanism as we find in silicon where the hot electrons are relaxed by optical phonons we have the relationship[3]:

$$V_{SAT} \approx [8E_{Lo}/3M_o]^{1/2} \text{ (cm/sec)}$$

where: E_{Lo} = longitudinal optical phonon energy

   M_o = free electron mass

The longitudinal optical phonon energy in SiC has been found[4] to be about 0.107 eV which gives a predicted value of about 1.3×10^7 cm/sec for V_{SAT}.   There is uncertainty in this prediction in that the bulk mobility for electrons in SiC tends to be lower than that for Si. Lower bulk mobility (which is determined by impurity scattering and acoustic phonon scattering) might lead to a lower V_{SAT} if the high field relaxation mechanism is not optical phonon dominated.   Thus, it was necessary to measure the carrier velocity as a function of electric field to determine if the V_{SAT} of SiC was lower or higher than that of silicon.

Measurement Technique

Of the various techniques available for generating carrier velocity versus electric field strength curves, the most applicable is a Ryder[5] type conductivity measurement since our relatively high doping levels do not experimentally lend themselves to time-of-flight measurements.

Using the Ryder technique, if one has a material of known carrier concentration (N_o) and known dimensions, then a sample current (I) versus applied voltage (V) plot can be manipulated to give current density (J), electric field (E), drift field mobility (μ) and carrier velocity (V).

We have: $E = V/\text{length}$ (volts/cm)

$J = I/\text{Area}$ (amperes/cm^2)

$J = N_o e \mu E = N_o eV$

Thus $V = J/N_o e$ (cm/sec)

and $\mu = dV/dE$ (cm^2/volt-sec)

The power densities involved in such a measurement are quite high, and to avoid heating effects, a pulsed arrangement is necessary.

The block diagram of Figure 1 shows the experimental arrangement. Typically, the pulse width was near 0.2 microseconds with a 200 Hz repetition rate.

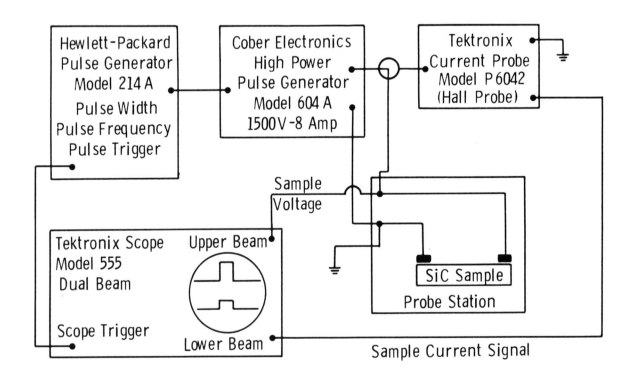

Figure 1.   Experimental Arrangement for Obtaining Pulsed I-V Data

A 50 ohm impedance matching termination and a dropping resistor were also required in the experimental arrangement. This is shown schematically in Figure 2 where we have:

I = measured current

V_o = voltage measured at oscilloscope

V_s = voltage across SiC sample

I_s = current through the sample

R_s = sample resistance.

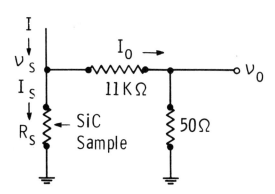

Figure 2. Schematic of Current-Voltage Sampling Circuit

Thus, measuring I and V_o we can calculate I_s and V_s (the quantities of interest) where:

where: $\quad I_s = I - (V_o/50)$

$\qquad\quad V_s = 221 \, V_o$

and $\qquad R_s = V_s/I_s$

Sample Preparation

Conductivity samples were made from SiC crystals which had been characterized with Hall measurements in previous work. The n-type samples were typical SiC material with carrier concentrations near 10^{17} and Hall mobilities near 250 $(cm^2/volt\text{-}sec)$. The p-type samples were more of a problem since such crystals are generally compensated. Thus there are few characterized p-type crystals and these tend to be highly compensated and of low mobility. Additionally, ohmic contacts to these highly compensated p-type crystals has so far proved impossible to fabricate. Ohmic contact to the n-type samples was made by heating them in contact with tungsten discs up to 1900°C to form a metallurgical band with no intervening alloy.

Originally, the V_{SAT} samples were in the form of thin rectangular bars made by lapping and diamond cutting before contacting. However, since the power supply was rated at 1500 volts and 8 amperes (with overruns to 2000 volts and 16 amperes by "ringing" the line) and field strengths in the range of 10^5 volts/cm were needed at current densities of the order of 1.8×10^5 amps/cm^2, it became obvious that thermal and power supply limitations required samples of very small length and cross sectional area. The only convenient method of producing such a sample is an oxidation–photoresist–chlorination technique as shown in Figure 3. Final trimming of the channel width was accomplished with an abrasive technique.

This procedure is very difficult since one has to mechanically handle an unsupported crystal with a channel of about 50 micron thickness. Since the tungsten contacts are attacked by the oxidation and chlorination procedures, they must be the last step. As a result, yield of samples is very low and limited by the number of characterized crystals available.

V_{SAT} Results

The data generated from these V_{SAT} measurements are shown in the curves of Figure 4. As can be seen, both GaAs and Si samples were also measured as an experimental check. Good agreement with published data is demonstrated. Additionally, the slope of the SiC V–E curves (plotted on linear paper) gave excellent agreement with previously obtained Hall mobility ($\mu_{Hall} = \mu_{drift} = dV/dE$).

The n–type samples have been measured up to a velocity of 8.2×10^6 cm/sec. The curve is still unsaturated (Ohm's Law region) up to this value. We were not able to measure to higher value due to a combination of instrumentation and sample geometry constraints. Si, by comparison, starts to show saturation at about 4×10^6 cm/sec. Thus it is presumed that the V_{SAT} for SiC will be at least as high as that of Si and will in fact probably reach between 1.3 to 2×10^7 cm/sec.

Contact problems (carrier injection) with available p–type material (sample D–49–4 shown on Figure 4) frustrated attempts to obtain hole data.

Dwg. 6162A96

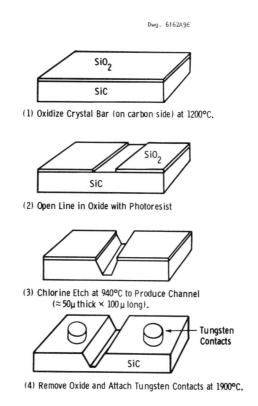

(1) Oxidize Crystal Bar (on carbon side) at 1200°C.

(2) Open Line in Oxide with Photoresist

(3) Chlorine Etch at 940°C to Produce Channel
(≈ 50μ thick × 100μ long).

(4) Remove Oxide and Attach Tungsten Contacts at 1900°C.

Figure 3.   Oxidation-Photoresist-Chlorine Etching Technique

Curve 645831-A

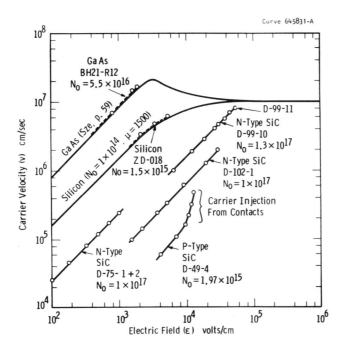

Figure 4.   Results of Saturation Velocity Measurements

FIGURES OF MERIT

Using what might be considered a minimum value 1×10^7 cm/sec of V_{SAT} for SiC and other known materials parameters, we can compute the Johnson and Keyes figures of merit for SiC, Si and GaAs. The following tables give these values.

Table 1.   Johnson Figure of Merit

Material	E_c (V/cm)	V_{SAT} (cm/sec)	$E_c^2 V_{SAT}^2$ (V^2/sec^2)
Si	3×10^5	10^7	9×10^{24}
SiC	$2\text{-}4 \times 10^6$	1×10^7	9×10^{26}

Table 2.   Keyes Figure of Merit

Material	Dielectric Constant	Conductivity (watts/cm °C)	V_{SAT} (cm/sec)	$\sigma_t (C V_{SAT}/4)^{1/2}$ (watts/deg-sec)
Si	12	1.5	1.0×10^7	6.7×10^7
GaAs	12	0.5	1.5×10^7	2.7×10^7
SiC	7	5.0	1.0×10^7	29.0×10^7

CONCLUSIONS

The data given here indicates that the measured saturation velocity of SiC largely confirms theoretical prediction. When this value of V_{SAT} is taken into account in the Johnson and Keyes figures of merit it is seen SiC has a great potential for high frequency, high power devices.

ACKNOWLEDGMENT

This work was supported by the Office of Naval Research under contract N00014-71-C-0405.

REFERENCES

1.   A. Johnson, "Physical Limitations on Frequency and Power Parameters of Transistors," RCA Review, Vol. 26, June 1965, p. 163.

2.   R. W. Keyes, "Figure of Merit for Semiconductors for High-Speed Switches," Proc. IEEE, p. 225, Feb. 1972.

3. Sze, Physics of Semiconductor Devices, John Wiley and Son, p. 57 (1969).

4. W. J. Choyke and L. C. Patrick, "Exciton Recombination Radiation and Phonon Spectrum of 6H SiC, " Phys. Rev. 127 (6): 1868-1877, Sept. 1962.

5. E. J. Ryder, "Mobility of Holes and Electrons in High Electric Fields, " Phys. Rev. 1953, Vol. 90, p. 766.

Diffusion and Solubility of Impurities in Silicon Carbide

Yu. A. Vodakov and E. N. Mokhov

ABSTRACT

The data on the solubility and diffusion of B, Al, Be, Ga, Se, N in pure and doped SiC are given for a wide range of experimental conditions. Possible diffusion mechanisms are discussed.

INTRODUCTION

The diffusion of impurities in SiC is now one of the principal methods of creating various kinds of structures including p-n junctions for the purposes of solid-state electronics. The diffusion as applied to SiC, which hardly lends itself to direct refining and doping, provides an additional method for controlled doping with a material intended for the study of the parameters of introduced impurities (1 to 4).

This paper summarizes our studies on the diffusion of a number of impurities in SiC and gives a brief account of the chief results of other workers. We should naturally focus our attention on impurities that affect appreciably the properties of SiC. The donors N, P, O, and the acceptors B, Al, Ga, In, Be as well as Sc and Cr are known to be such impurities. As will be seen from the review of the data on solubility, the study of the diffusion in SiC cannot unfortunately be done except for a few impurities.

THE SOLUBILITY OF IMPURITIES IN SiC

So far, the solubility of impurities in SiC has not been systematically studied. The data on the solubility of some impurities obtained by various workers are listed in Table 1.

Table I

Impurity	Temperature (°C)	Solubility (cm^{-3})	Test method	Particulars of impurity introduction
N	2550	3×10^{19} /5/	electrical measurements	when growing SiC by sublimation from N$_2$
N	2450	2.6×10^{20} /6/	mass-spectrometry	when growing in excessive N$_2$ (pN$_2$=35 atm gauge pressure)^2
N	1750	$\geqslant 10^{20}$	electrical measurements	upon doping from acetonitrile /3/
Al	-	10^{21} /7/	emission spectroscopy	when growing SiC by sublimation
Al	to 2000 2400	10^{21} /8/	electrical measurements	upon diffusion saturation of the surface region
B	2500	3.6×10^{20} /9/	emission spectroscopy	when growing SiC
B	to 1600 2550	1.2×10^{19} to 3×10^{20} /10/	neutron activation analysis	during growth and upon diffusion saturation by certain methods
Ga	2200	2.8×10^{18} /12/	neutron activation analysis	when growing SiC by sublimation
Be	2500	7×10^{18} /11/	emission spectroscopy	when growing SiC by sublimation
Be	to 1600 2300	1.5×10^{17} to 2×10^{19} /13/	electrical measurements	Upon diffusion saturation C$_{Be}$=1/20A, since Be is a double-level acceptor
Sc	to 1700 2500	3×10^{17} /12/	neutron activation analysis	when growing SiC
Lantanoids	to 1600 2200	10^{16} /12/	neutron activation analysis	when growing SiC

Besides, there is some information about the low solubility ($\leq 5 \times 10^{17}$ cm^{-3}) of In and Mg in SiC. The solubility of oxygen may probably be rather high (4), although according to the results of electrical measurements it is not higher than 10^{18} cm^{-3}. However, reliable methods for the identification of oxygen are lacking (4). The solubility of P is unlikely to be over 10^{18} cm^{-3}. Figure 1 shows the temperature dependencies of the solubility of B and Be in SiC (it is not clear, whether or not the solubility limit is achieved at low temperatures).

Thus, such impurities as In, Mg, P, Sc and lantanoids are only slightly soluble in SiC. Consequently, the study of the diffusion of these impurities appears to be impracticable in the present state of the art. Yet we attempted to estimate the diffusion parameters of Sc because of its important luminescent properties (14). The chief data obtained by us refer to the diffusion in SiC of B, Be, Al, Ga, and N.

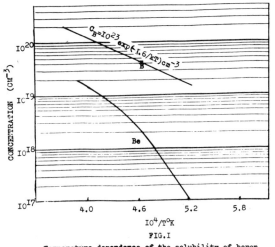

FIG.I
Temperature dependence of the solubility of boron and beryllium.

TECHNIQUES FOR PERFORMING AND STUDYING DIFFUSION IN SiC

For successful study of the diffusion in SiC, it is necessary to solve a number of serious problems due to the particularities of this compound, i.e.:

1. The temperature in the diffusion chamber must be above 1500°C.

2. It is necessary to suppress SiC dissociation and vapor transfer processes, especially at temperatures above 1800°C.

3. A means must be provided for varying the conditions of the diffusion annealing, i.e. diffusant surface-concentration, external-phase composition, temperatures, etc.

4. A suitable technique must be found for studying the diffusion of impurities in SiC.

The last problem is complicated by the fact that most of the above-mentioned impurities do not have appropriate radioisotopes and the use of such classical semiconductor techniques as the electrical conductivity method present difficulties due to high activation energies of the impurities (which are only partially ionized at room temperatures) and to the presence of surface inversion layers with an energy barrier higher than 1 eV.

We carried out diffusion in a quasiclosed system made from vacuum-tight graphite of spectroscopic purity under conditions that exclude SiC evaporation and epitaxial growth. The diffusant was either applied to the surface of the sample or introduced from a vapor phase. In the latter case, the use of two temperature containers permitted the vapor-phase composition to be controlled over a wide range.

The diffusion duration was varied from 5 minutes to 50 hours. The diffusion depths were found to increase as the square of the diffusion duration, except for the Al distribution branch corresponding to a region near the surface. To determine the diffusion parameters, p-n junction and electrical conductivity methods (9, 16) were used as well as our technique for studying diffusion in heavily doped semiconductors suggested in (17). The capacitance characteristics of p-n junctions and the luminescent parameters of diffusion layers were studied in a number of cases.

Since all of the above methods for studying diffusion are not direct so that the results obtained thereby depend also on the parameters of the initial crystals, much consideration was given to the choice of uniform crystals differently doped with nitrogen and a special preliminary study was made of the effect of high-temperature annealing without diffusant on the electrical parameters of samples (18). Most of the information on the diffusion in SiC was obtained on samples of 6H polytype.

DIFFUSION OF NITROGEN AND OXYGEN

The diffusion of nitrogen had been studied in (5, 6, and 8). Our experiments indicate that at N_2 excess pressures of 10^{-3} to 2 atm the data of (5) are the most correct, according to which N is characterized by an extremely low diffusivity ($D_N < 10^{12}$ cm^2/s at 2550°C) (Figure 7, curve 1). Therefore, the migration of N may be neglected, when studying the diffusion of acceptor impurities.

n-SiC layers on p-SiC(Al) samples are also formed upon diffusion annealing in an oxygen containing atmosphere, their thickness at diffusion temperatures below

2200°C being even somewhat higher than with the diffusion of N. In these layers, a characteristic luminescence is observed which is different from N-activated luminescence (4). However, the difficulties in quantitative identification of 0 did not allow determining its diffusion parameters in SiC.

DIFFUSION OF ALUMINUM AND GALLIUM

The diffusion of Al in SiC was first studied by Chang (19) in the temperature range 1800 to 2150°C and also by Griffits (20). An extremely low mobility of Al when diffused from a solid phase as compared to the diffusion from vapor is reported in (21). The nature of this effect is not yet known.

Figure 2 shows the characteristic diffusion profiles obtained by the p-n junction method at various temperatures (1900 to 2400°C) (22). It is seen from Figure 2 that the bulk ("volume") distribution may be described by a complementary error function with the effective boundary concentration C_{Ieff}^{Al} equal to (1 - 1.5) x 10^{18} cm^{-3}.

The values of D^{Al} obtained by us are lower than those reported in the earlier works (19, 20) and the value of the impurity activation energy is 1.2 eV higher. It should be noted that the character of the bulk distribution of Al much depends on the conditions of the diffusion annealing. Thus, if the diffusion of Al is accompanied by the formation of epitaxial layers heavily doped with it, then a sharp decrease in the thickness of the p-n junctions is observed, as in (21), especially on samples with $N_d - N_a > 3$x 10^{17} cm^{-3}. However, unlike (21), the

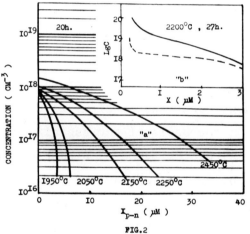

FIG.2
Distribution of diffused Al for various annealing temperatures a)"volume" parts of the distribution
b)"surface" parts of the distribution
-- p-n junction method
— electrical conductivity method
X_{p-n} - depth of the p-n junction
X - depth of the diffused layer

bulk branch still shows up on samples with $N_d - N_a < 1$ x 10^{17} cm^{-3}. On the other hand the diffusion parameters of Al depend on the value of the surface concentration C^{Al}. Thus, lowering C_s^{Al} down to 10^{19} cm^{-3} results even in an increase in the depth of the p-n junction, mainly due to an increase in C_{Ieff}^{Al}. Unfortunately, the diffusion parameters in the region near the surface could not be determined by the p-n junction method, since the depth of the p-n junctions formed on the samples with $N_d-N_a > 1.5 \times 10^{18}$ cm^{-3} did not exceed 1 μm even at the maximum temperature and diffusion duration. Much more information for this distribution region was obtained by the electrical conductivity method (9). Thus, at 2200°C the diffusivity of Al was found to be 5.6×10^{-14} cm^2/s with $c_s^{Al} \simeq 10^{21}$ cm^{-3}.

Discrepancy between the results obtained by various methods (Figure 2-b) for the region near the surface seems to be associated with the decelerating action of nitrogen in SiC samples.

The diffusion of Ga has not been studied so far, although it has long been known that Ga is a luminescence activator (23). Using the p-n junction method, we have established that under the assumption of a normal distribution the diffusivity of Ga varies from 2.8×10^{-13} cm^2/s to 2.9×10^{-12} cm^2/s in the temperature range 2050 to 2300°C. The value of C_{eff}^{Ga} does not exceed $(3-5) \times 10^{17}$ cm^{-3}, which is much below the solubility limit of Ga (2.8×10^{18} cm^{-3}), as determined by the neutron activation method.

The character of the temperature dependence of Ga diffusion and the diffusion parameters are given in Figure 7.

DIFFUSION OF BORON

Despite the wide use of the diffusion method for doping SiC with B, little is known about the particulars of the diffusion of this impurity.

On the basis of the capacitance characteristics of diffused p-n junction, an abnormal distribution of B characterized, among other things, by a steep section near the surface was stated (24). Data on the anisotropy of diffusion in the polar directions <0001> were also reported.

We gave much attention to the study of diffusion of B (16, 25).

The diffusion was carried out at temperatures of 1600 to 2550°C and the process duration varied between 0.5 hour and 50 hours.

The chief results of our study may be summarized as follows.

1. With small c_s^B ($\leq 3 \times 10^{18}$ cm^{-3}) the distribution is of simple form and

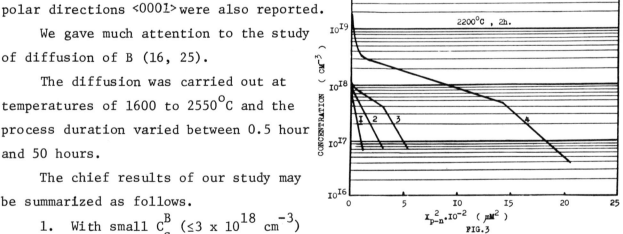

Distribution of the diffused boron for various surface concentrations c_s^B (cm^{-3}): I - 5.10^{17}, 2 - 8.10^{17}, 3 - 5.10^{18}, 4 - 7.10^{19}.

FIG.3

can be described by one diffusion coefficient whose temperature dependence
within the range 2000 to 2400°C is shown
in Figure 7.

With increasing C_s^B, the character of the diffusion profile is found
to become more complicated (Figure 3).

Table II

$^{\circ}$C	D_N^B/D_i^B		D_{Al}^B/D_i^E	
	$c^N=3\times10^{19}$cm^{-3}	$c^N=5\times10^{18}$cm^{-3}	$c^{Al}=5\times10^{19}$cm^{-3}	$c^{Al}=1\times10^{20}$cm^{-3}
2300	0.3	0.9	1.0	4.0
2150	0.2	0.6	3.0	10.0
1850	0.08	0.5	10.0	18.0

D_N^B, D_{Al}^B, and D_i^B are the effective diffusivities of B in SiC doped with N, Al, and in pure material, respectively.

A steep section S appears near the surface, the diffusivity in the bulk region
increases considerably, the latter region taking a stepped shape above 2100°C and
breaking into I and T sections, which is particularly evident in the representa-
tion $lgC = f(x^2)$ (Figure 3). The temperature dependence of the I section at 1600
to 2550°C is given in Figure 7 (curve 5), C_{Ieff}^B changing from 2×10^{18} to 6×10^{18}
cm^{-3}. T section shows up only on the distribution tails at B concentrations below
1×10^{18} cm^{-3}. It should be noted that the values of D_{Teff}^B calculated for this
section fall well on the straight line $D = f(\frac{1}{T})$ with small C_s^B (Figure 7, curve 4).

3. The enrichment of the vapor phase in Si results in a marked increase in
the depth of the distribution region near the surface (Figure 5). In the bulk
distribution branch this effect is less pronounced, although the whole distribution
shifts toward higher concentrations. The same is observed also in the presence of
Al in the vapor phase.

4. The diffusion of B is affected by doping SiC crystals, nitrogen decelerating and aluminum accelerating the diffusion (see Table II).

Note that the decelerating effect of N on the diffusion of B manifest itself
also in the comparative distribution study by the electrical conductibility and
p-n junction methods (Figure 4).

5. The diffusion of B from a previously grown epilayer is characterized by
low diffusivity, especially with not too high values of C_s^B (Figure 5). However,
with C_s^B approaching the solubility limit of B, the diffusivity is found to decrease
only in the region near the surface (Figure 5).

6. The study of the diffusion of B in different polytypes (6H, 15R, 3C, 21R)
has not revealed any considerable difference in the diffusion parameters. Somewhat smaller depths and luminescent layer thicknesses have been observed in
4H-SiC.

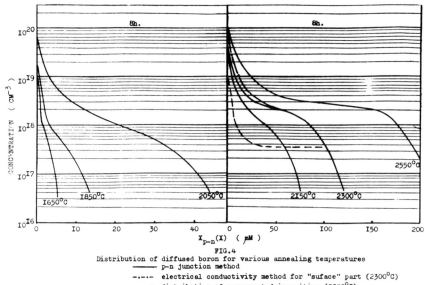

FIG.4
Distribution of diffused boron for various annealing temperatures
——— p-n junction method
-.-.- electrical conductivity method for "suface" part (2300°C)
- - - distribution of compensated impurities (2300°C).

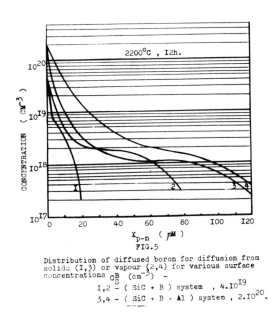

FIG.5
Distribution of diffused boron for diffusion from
solids (I,3) or vapour (2,4) for various surface
concentrations c_S^B (cm^{-3}) —
I,2 — (SiC + B) system , 4.10^{19}
3,4 — (SiC + B + Al) system , 2.10^{20}.

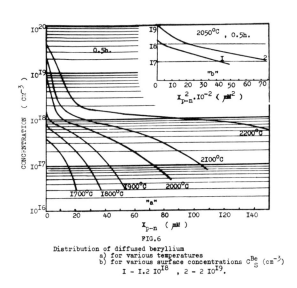

FIG.6
Distribution of diffused beryllium
a) for various temperatures
b) for various surface concentrations c_S^{Be} (cm^{-3})
I - $1.2 \cdot 10^{18}$, 2 - $2 \cdot 10^{19}$.

DIFFUSION OF BERYLLIUM

Beryllium is an acceptor impurity in SiC and a luminescence activator (2). However, but for (13), data on the diffusion of Be in SiC are lacking in the literature.

We studied the diffusion of Be at 1600 to 2300°C, the diffusion duration varying from 10 minutes to 12 hours. The results are given below.

1. Figure 6-a shows the concentration distributions of Be at various temperatures. The distribution of Be at temperatures above 1800°C is seen to have two

distinct sections, one in a region near the surface and another in the bulk of the material. Each of these sections can be described by a complementary error function with the following boundary conditions: $c_s^{Be} = 6 \times 10^{18}$ to 4×10^{19} cm^{-3} at 1900 to 2200°C, respectively; $c_{Ieff}^{Be} = 3 \times 10^{17}$ to 2×10^{18} cm^{-3} at 1600 to 2200°C respectively.

2. The temperature dependence of diffusivity for both branches is given in Figure 7 (curves 6, 7). With D_s^{Be} below 1×10^{18} cm^{-3}, no special region near the surface is seen to form. However, unlike the diffusion of B, the value of D_{eff}^{Be} for the bulk branch (Figure 6-b) is little affected by varying c_s^{Be}.

3. When studying the diffusion distributions of B and especially Be by the electrical conductivity method, abnormal temperature behavior of the Hall curves is observed in a thin (up to 10 μm) intermediate layer between S and I sections. This behavior seems to be associated with the non-monotonical concentration change (the presence of a maximum), whose nature is not clear.

4. The presence of N in SiC samples somewhat decelerates the volume diffusion of Be, while the presence of Al and B results in a substantial change in D_{eff}^{Be} (Table III).

5. Interesting features of the behavior of Be in SiC have been revealed, when studying the effect of re-annealing on the diffusion distribution for samples previously doped with Be by diffusion. Thus, upon low-temperature annealing (below 1800°C), no appreciable change in the diffusion profile of Be occurs, whereas annealing at higher temperatures (2000 to 2200°C) results in an increase in the depth of the p-n junctions formed on samples only slightly doped with nitrogen ($N_d - N_a < 1 \times 10^{18}$ cm^{-3}). Besides, an n-type surface layer forms on such samples at the cost of Be leaving the sample. By contrast, the changes in the thickness of the p-type layer of the samples with $N_d - N_a > 1.5 \times 10^{18}$ cm^{-3} have been insignificant even after prolonged annealing at the above-mentioned temperatures.

SUMMARY AND DISCUSSION

Analysis of the above experimental results shows that the diffusion of N, Al, Ga, B is characterized by rather high activation energies. It may be thought that the diffusivity of these impurities is limited by displacement though the vacancies of the corresponding sublattices, i.e., N and B though carbon vacancies (V_C), while Al and Ga though silicon vacancies (V_{Si}). Within the framework of the vacancy model, it seems plausible to associate the experimentally observed anomalies in the distribution of Al and B with the fact that at high concentrations of acceptor

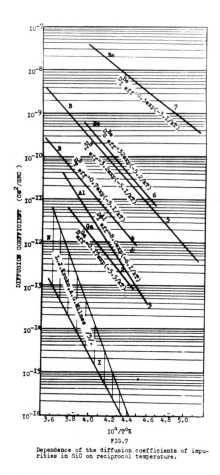

FIG.7
Dependence of the diffusion coefficients of impurities in SiC on reciprocal temperature.

impurities, doping may be accompanied by the generation of compensating defects such as V_C (26), i.e., with high C_s of these impurities the region near the surface will be found to be V_C-enriched. The propagation of these vacancies deep into the crystal may be the cause of the acceleration of the diffusion of B and the occurrence of the I section of the diffusion curve. By contrast, in the case of the diffusion of Al, a deceleration of the diffusion should occur, since the concentration of V_{Si} responsible for the diffusion of Al would decrease. Of course, with the present-day knowledge, it can be hardly imagined that in the bulk of the crystal $V_C \gg V_c^o$, where V_c^o is the equilibrium concentration of vacancies (27). It seems necessary to take into account the recently established vacancy stabilization by impurities in a number of semiconductors (28). Thus, it is quite realistic to suppose the entrapment of a vacancy by a diffused atom giving rise to diffusion-active complexes, specifically in $SiC(B-V_c)$. With decreasing stability of such complexes, as the temperature increases, the probability of vacancy entrapment by B decreases, especially at a depth where C^B is small. Therefore, D_{eff}^B (T region) should decrease on the tails of the distribution of B.

The elevated concentration of compensating donors in the heavily acceptor-doped layers near the surface of samples (Figures 2 and 4) as well as the appearance after the diffusion of B of low-temperature line luminescence close in spectral composition to the defect-activated luminescence in neutron-irradiated crystals (25) may be well associated with impurity-generated vacancies. We also observed the effect of stabilization by B of the above-mentioned luminescence (25).

The same processes seem to take place also after the diffusion of other

Table III

$^\circ C$	D_N^{Be}/D_1^{Be} $c^N = 4 \times 10^{19} cm^{-3}$	D_{Al}^{Be}/D_1^{Be} $c^{Al} = 2 \times 10^{20} cm^{-3}$	D_B^{Be}/D_1^{Be} $c^B = 1.5 \times 10^{20} cm^{-3}$
2050	0.6	40	10
1850	0.4	50	15

impurities, in particular N and Be.

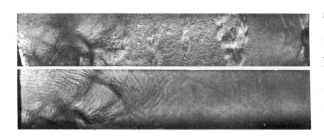

FIG.8

a) Disturbances near the surface of a SiC sample after Al diffusion ($c_s^{Al} \approx 10^{20}$ cm^{-3}).

b) Photograph of the same crystal before Al diffusion. Berman's method, 7.5X.

It could be supposed that the decrease in the diffusivity in a region near the surface is due to the formation of slow diffusing associates. However, we do not have sufficient evidence to accept this assumption. The investigation of various electrical and physical parameters of diffused layers, the study of the EPR spectra of SiC(B) and comparison of the distribution of paramagnetic B to the previously obtained diffusion profiles revealed no differences in the nature of the impurities in the region near the crystal surface, on one hand, and in the bulk of the crystal, on the other hand. The abrupt decrease in the diffusivity of B and Al in a layer near the surface seems to have a common nature with disturbances of the strain type arising in this layer at $c_s^B > 3 \times 10^{18}$ cm^{-3} and $c_s^{Al} > 10^{19}$ cm^{-3} (30) as found by X-ray topography techniques (Figure 8).

In the case of the diffusion of Be, when treating the distribution near the surface, the interaction of Be with N must be also considered which manifest itself upon annealing samples with Be diffusion (refer to paragraph 5 of the foregoing section).

Unlike other impurities, the bulk diffusion of Be seems to be effected by an interstitial or dissociative mechanism, as evidenced first of all by large diffusivities and low (as for SiC) activation energy.

In conclusion, it should be noted that most impurities in SiC, such as N, Al, Ga, Sc, and probably O, have extremely low diffusion mobilities. Therefore, high temperatures (above 2200°C) are required for the introduction of these impurities by diffusion. Besides, at high surface concentrations of impurities, the diffusion may be accompanied by disturbances in the crystalline structure (Figure 8) (3). However, the equipment currently available, in particular that used by us, offers strong possibilities of obtaining by the diffusion or epitaxial-diffusion method absolutely different structure types of p-n junctions on the basis of SiC doped with one or more specified impurities.

The authors express their gratitude to G. F. Kholuyanov, Yu. I. Kozlov, G. A. Lomakina, V. G. Oding, I. L. Shulpina, S. I. Taits, A. S. Tregubova, M. M. Usmanova, G. F. Yuldashev, and B. S. Zverev for technical assistance and valuable discussions.

REFERENCES

(1) Yu. A. Vodakov, G. F. Kholuyanov, and E. N. Mokhov, Fix. Techn. Polupr., 5 (8), 1615 (1971).

(2) V. I. Sokolov, V. V. Makarov, E. N. Mokhov, G. A. Lomakina, and Yu. A. Vodakov, Fiz. Tverd. Tela, 10 (10), 3022 (1968).

(3) V. I. Sokolov, V. V. Makarov, and E. N. Mokhov, Fiz. Tverd. Tela, 12 (1), 285 (1970).

(4) G. F. Kholuyanov, Yu. A. Vodakov, E. E. Violin, G. A. Lomakina, and E. N. Mokhov, Fiz. Tekhn. Polupr., 5 (1), 39 (1971).

(5) L. T. Kroko and A. G. Milnes, Sol. St. Electr., 9 (11/12), 1129 (1966).

(6) R. I. Scace and G. A. Slack, J. Chem. Phys., 42, 805 (1965).

(7) P. Carrol, in: Silicon Carbide, Pergamon Press, N. Y., p. 341 (1960).

(8) G. A. Lomakina, Yu. A. Vodakov, E. N. Mokhov, V. G. Oding, and G. F. Kholuyanov, Fiz. Tverd. Tela, 12 (10), 2918 (1970).

(9) P. T. B. Shaffer, Mat. Res. Bull., 5, 519 (1970).

(10) B. P. Zverev, Yu. F. Simakhin, and M. M. Usmanova, Atomn. Energ., 33 (2)(1972).

(11) I. G. Pichugin and N. A. Smirnova, Neorg. Mater., 5 (2), 231 (1969).

(12) G. F. Yuldashev, M. M. Usmanova, Yu. A. Vodakov, Atomn. Energ., 32, 592 (1972).

(13) Yu. P. Maslakovets, E. N. Mokhov, Yu. A. Vodakov, and G. A. Lomakina, Fiz. Tverd. Tela, 10 (3), 809 (1968).

(14) Kh. Vakhner and Yu. V. Tairov, Fiz. Tverd. Tela, 11 (9), 2440 (1969).

(15) B. I. Boltaks. Diffuziya v Poluprovodnikakh (Diffusion in Semiconductors), Fizmatgiz Publishers, Moscow 1961.

(16) E. N. Mokhov, Yu. A. Vodakov, G. A. Lomakina, V. G. Oding, G. F. Kholuyanov, and V. V. Semyonov, Fiz. Tekhn. Polupr., 6 (3), 482 (1972).

(17) E. N. Mokhov, S. Kh. Koprov, and Yu. A. Vodakov, Fiz. Tverd. Tela, 13, 3695 (1971).

(18) G. A. Lomakina, G. F. Kholuyanov, R. G. Verenchikova, E. N. Mokhov, and Yu. A. Vodakov, Fiz. Tekhn. Polupr., 14, 1133 (1972).

(19) H. C. Chang, in: Silicon Carbide, Pergamon Press, New York, 496, 1972.

(20) L. G. Griffits, J. Appl. Phys., 36, 571 (1965).

(21) G. Van Opdorp, Sol. St. Electr., 14 (7), 613 (1971).

(22) E. N. Mokhov, Yu. A. Vodakhov, and G. A. Lomakina, Fiz. Tverd. Tela, 11, 519 (1969).

(23) G. F. Kholuyanov, Fiz. Tverd. Tela, 7 (11), 3241 (1965).

(24) R. M. Potter, J. M. Blank, and A. A. Addamiano, J. Appl. Phys., 40 (5), 2253 (1969).

(25) E. N. Mokhov, Yu. A. Vodakov, G. A. Lomakina, M. B. Reyfman, V. I. Popov, and O. A. Koloso in: Trans. III All-Union Conference on SiC, Moscow (1970), p. 93.

(26) F. Kroger, Chemistry of Imperfect Crystals, 1964.

(27) S. M. Hu and M. S. Mock, Phys. Rev. B., 1 (6), 2582 (1970).

(28) R. N. Choshtayore, Phys. Rev. B., 3 (2), 397 (1971).

(29) V. A. Uskov and V. V. Vaskin, Neorg. Mater., 7, 1843 (1972).

(30) A. S. Tregubova, E. N. Mokhov, and I. L. Shulpina, Fiz. Tverd. Tela, 14, 3655 (1972).

(31) A. A. Pletyushkin and L. M. Ivanova, in: Trans. III All-Union Conference on SiC, Moscow, (1970), p. 43.

Electrical Properties of Various Polytypes of Silicon Carbide

G. A. Lomakina

ABSTRACT

A comparative study is presented of the transport phenomena in various poly-types of both n- and p-SiC. It is shown that various polytypes of p-SiC are very similar in electronic properties, while n-SiC polytypes differ substantially. Analysis is performed of the conductivity anisotropy in various polytypes as a function of the length of their unit cell. The results obtained are discussed from the point of view of the formation of minizones.

INTRODUCTION

It is well known that various polytypes of SiC appreciably differ in semi-conductive properties such as indirect band gap, electron mobility, luminescence and absorption spectra, etc.

This paper is concerned with the comparison of the electrical properties of 4H, 15R, 6H, 10H, 21R, 27R, 33R, 330R polytypes of the electron-conduction type and 4H, 15R, 6H polytypes of the hole-conduction type.

SILICON CARBIDE OF THE HOLE-CONDUCTION TYPE

To obtain hole crystals of 4H, 15R, 6H polytypes with equal concentration of acceptor and compensating donor impurities the diffusion of Al was carried out for 24 hours at 2200°K in previously studied electronic crystals of these polytypes with an equal nitrogen concentration of $\sim 6 \times 10^{16}$ cm^{-3} (1). The conductivity of hole layers at the same depth from the surface, the hole mobility in them as well as the temperature dependence of conductivity and mobility were found to coincide for all the three polytypes within the experimental error (Figure 1).

From this it follows that the activation energy of Al impurity as well as the hole mobility are the same in all equally doped crystals of the polytypes studied. Similar results were obtained on samples doped with B by the diffusion method.

Thus, in the polytypes studied, the effective mass of a hole changed little, if at all, from polytype to polytype, nor does the structure of the valence band near its maximum.

SILICON CARBIDE OF THE ELECTRON-CONDUCTION TYPE

The study of the electrical properties of electronic crystals revealed a substantial dependence of these properties on the polytype. It was noted in a number of papers (2,3) that the electron mobility in 4H crystals was higher than in 15R; while in 15R, in turn, it was higher than in 6H. In the purest samples with a nitrogen concentration of $\sim 6 \times 10^{16}$ cm^{-3}, the Hall mobility at right angles to the C-axis (μ_{nh}) at 300°K was equal to 330, 500, 700 cm^2/Vs in 6H, 15R, 4H crystals, respectively (3).

Measurements of the temperature dependence of the Hall effect made on these crystals showed that the activation energy of nitrogen impurity was also different in different polytypes, equaling 0.095, 0.047, 0.033 eV, respectively (1,4).

Unfortunately, there were no crystals at our disposal of other polytypes with low concentrations of nitrogen impurity. Therefore, in an effort to compare the values of the thermal ionization energies of various polytypes, we used specially chosen homogeneous crystals (containing no interlayers of foreign polytypes) grown under equal conditions with close concentrations of nitrogen impurity of $\sim 3 \times 10^{18}$ cm^{-3}. The activation energy of nitrogen impurity in all the polytypes studied was found to be higher than that in 4H and lower than in 6H, equaling 0.03 eV for 27R, 0.035 eV for 10H and 33R, 0.04 eV for 21R, 0.06 eV for 330R (Figure 2) (with a concentration of nitrogen of $\sim 3 \times 10^{18}$ cm^{-3}, its ionization energy equals 0.066 eV for 6H, 0.032 eV for 15R, 0.029 eV for 4H (5).

Because of possible small differences in the concentration of compensating impurities, an accurate comparison of the electron mobilities in these polytypes can hardly be made. However, it is safe to say that the temperature dependence of the mobility at right angles to the C-axis has the same appearance as that in the simplest polytypes (6H, 15R, 4H) and the previously established correlation

holds, i.e. the higher the activation energy of nitrogen impurity, the lower the
electron mobility (Figure 3).

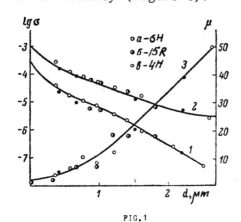

FIG.1
Dependence of the conductivity (1, 2) and hole mobility
(3) of the remaining diffused layer on the depth of the
etched-off layer. 1-at 250°K, 2-at 400°K, 3-at 300°K.

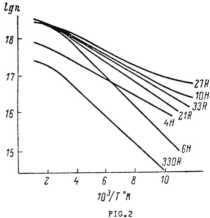

FIG.2
Temperature dependence of the concentration of free
electrons in the polytypes 4H, 6H, 10H, 21R, 27R, 330R.

CONDUCTIVITY ANISOTROPY IN
ELECTRON AND HOLE
SILICON CARBIDE

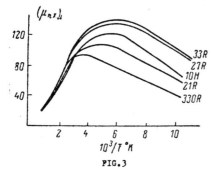

FIG.3
Temperature dependence of the Hall mobility of electrons
at right angles to the C-axis.

Since the difference in the crystalline structure between various polytypes lies in the alternation, along the C-axis, of cubic or hexagonal close-packed layers, special attention was paid to the comparative study of the anisotropy of the electrical conductivity and its temperature dependence. The conductivity anisotropy was measured by Schnabel's method (6).

The very first measurements showed that at room temperature and below, for the vast majority of crystals, the resistivity along the C-axis was much higher than in the perpendicular direction, the corresponding ratio reaching 10^5. Measurements of the potential distribution along the C-axis, using a metal probe moving from one end to the other, showed that in the presence of even one very thin interlayer of a foreign polytype, at 300°K the whole potential dropped on layers of several µm thickness at the polytype-polytype interface, thus resulting in large values of the conductivity anisotropy observed in most crystals.

In this connection, crystals were carefully chosen on the homogeneity principle to attain reliable results reproducible from crystal to crystal. The absence of thin interlayers of foreign polytypes was checked by inspection of polished cross-sections under polarized light. Then, the conduction homogeneity throughout the crystal thickness was checked by measuring the potential distribution in a cross-section along the C-axis. The polytype was identified by X-ray diffraction analysis. The temperature dependence of conductivity anisotropy was measured only on samples which had been chosen in the described manner.

The investigation of the chosen homogeneous crystals revealed the following facts.

(1) The ratio of the resistivity in the parallel and the perpendicular direction to the C-axis $\rho_\parallel/\rho_\perp$ varies only slightly with temperature for the n-type 4H, 15R, 6H, 33R polytypes (Figure 4) and is equal to 0.9, 1.5, 3.7, 2.0, respectively, at $1000^\circ K$ (5), which agrees well with the data of (7) for 6H polytype.

(2) For 27R, 10H polytypes, $\rho_\parallel/\rho_\perp$ varies appreciably with temperature, increasing from ~1, 2 at $100^\circ K$ to ~5, 6 at $1000^\circ K$, respectively (Figure 5).

(3) The resistivity anisotropy of 21R is large, amounting to ~25 at $300^\circ K$, and is characterized by a complex temperature dependence, peaking at $250^\circ K$.

In 27R, 10H, 21R polytypes, the dependence $\rho_\parallel/\rho_\perp = f(T)$ is of the form $\rho_\parallel/\rho_\perp = CT$, where C is 8.3×10^{-3} (for 27R), 2×10^{-2} (for 10H), 1×10^{-1} $^\circ K$ (for 21R) at $T < 300^\circ K$.

(4) The resistivity anisotropy reaches its highest value in 330R polytype, i.e. 2×10^3 in the temperature range $100^\circ K$ to $300^\circ K$. At $T > 400^\circ K$ the value of $\rho_\parallel/\rho_\perp$ rapidly decreases, dropping to ~10 at $1000^\circ K$ (Figure 6).

It should be noted that the temperature dependence of $\rho_\parallel/\rho_\perp$ in crystals containing interlayers of foreign polytypes is very close to that observed in 330R polytype. This may well be associated with the fact that in many cases, near interfaces between different polytypes -there are thin interlayers of multi-layer or disordered modification (7,8) with a high resistivity along the C-axis.

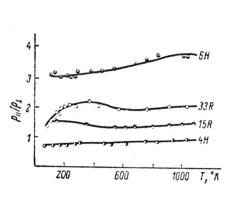

FIG.4

Temperature dependence of resistivity anisotropy for 6H, 33R, 15R, 4H.

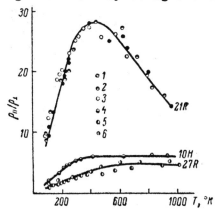

FIG.5

Temperature dependence of resistivity anisotropy for 21R, 10, 27R.

Measurements of the conductivity anisotropy in p-type crystals containing a great number of interlayers of multilayer modifications showed that in these crystals $\rho_\parallel / \rho_\perp = 1$, just as in the case of the homogeneous crystals of only one polytype (5).

SUMMARY AND DISCUSSION

Thus, the electrical properties of hole crystals practically do not differ in various polytypes, while those of electron crystals are markedly polytype-dependent.

A possible explanation of the polytype dependence of free-electron mobility, donor-impurity activation energy, and conduction anisotropy may be the following. As the length of a unit cell along the C-axis increases, becoming a multiple of itself, energy discontinuity planes appear in the large Brillouin zone which serve as new boundaries of Brillouin zones (9,10). The number of possible energy discontinuities should increase with increasing layer number, their value depending on the structure of a particular polytype.

To a first approximation, as the layer number increases, the width of the first allowed Brillouin zone should decrease, while the component along the C-axis of the tensor of the effective mass of an electron should increase. The series 4H, 15R, 6H furnishes an example of such a sequence.

On the assumption of an isotropic relaxation time valid for the valley-to-valley scattering mechanism, the resistivity ratio is equal to the ratio of the components of the effective mass tensor. Assuming that the effective mass approximation holds true for nitrogen and knowing its activation energy and $m_\parallel^* / m_\perp^*$ ratio, the values of $m_\parallel^*$ and $m_\perp^*$ can be estimated. The calculation results are summarized in Table 1.

The value of $m_\parallel^*$ is found to depend on the polytype much more strongly than $m_\perp^*$ does. The weak polytype dependence of $m_\perp^*$ speaks in favor of the assumption that the location of the minimum of the conduction band does not vary from polytype to polytype, while the variation of the electrical properties is associated with energy discontinuities. At the point K = 0, the influence of these discontinuities is small (for p-SiC and n-ZnS), which is probably related to the large wavelength of free carriers at the point K = 0.

TABLE 1

Polytype	$m_{\parallel}^{*} / m_{0}$	$m_{\perp}^{*} / m_{0}$	E_{D} , (eV)
4H	0,19	0.21	0.033
15R	0.37	0.25	0.047
6H	1.3	0.35	0.095

In the polytypes with higher layer
number (10H, 27R, 21R), the first allowed
conduction minizone gets still narrower,
so that higher minizones begin contributing more substantially to the activation energy of nitrogen impurity, which
results in the fact that a further
increase in the activation energy is no

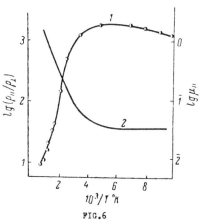

FIG.6

Temperature dependence of the resistivity anisotropy (1)
and of the mobility along the C-axis for 330R (2).

longer observed with increasing layer number. At the same time the resistivity
anisotropy increases and follows a more complex temperature dependence. In the
case where the width of the first allowed band becomes comparable to kT (10H, 27R,
21R, at T 100°K) and the discontinuity to the next band is larger than kT, the
conduction along the C-axis decreases with increasing temperature because of a
more uniform distribution of electrons throughout the band. With further increasing temperature (above 500°K for 21R), electrons begin penetrating into higher
bands, so that the conduction in this direction increases with temperature again.
For the multilayer polytype 330R, the allowed bands get so narrow that the value
of $\rho_{\parallel} / \rho_{\perp}$ is high even at temperatures as low as 100°K. Conduction along the
C-axis is effected by tunneling transitions and the mobility $(\mu_{n})_{\parallel}$ does not depend on temperature in the range 100 to 300°K. At higher temperatures, the
energy of an electron becomes higher than the height of the superperiodic poten-
tial and $\rho_{\parallel} / \rho_{\perp}$ falls off, while $(\mu_{n})_{\parallel}$ increases, following the law $(\mu_{n})_{\parallel}$
$\sim \exp (\Delta E/kT)$, where $\Delta E \simeq 0.1$ eV.

If the above interpretation of the experimental facts is correct, then silicon
carbide does offer a means, still rare, of investigating the transport phenomena
in narrow (of the order of hundredths of the electron-volt) bands due to the
existence of one-dimensional superperiodicity in SiC crystals.

The author is grateful to G. B. Dubrovsky, G. F. Kholuyanov, E. N. Mokhov,
V. G. Oding, S. I. Taits, and Yu. A. Vodakov for their valuable experimental
assistance and suggestions.

REFERENCES

(1) G. A. Lomakina, Yu. A. Vodakov, E. N. Mokhov, V. G. Oding, and
      G. F. Kholuyanov, Fiz. Tverd. Tela, 12, 2918 (1970).

(2) D. L. Barrett and R. B. Campbell, J. Appl. Phys., 38, 53 (1967).

(3) G. A. Lomakina, G. F. Kholuyanov, R. G. Verenchikova, E. N. Mokhov, and
      Yu. A. Vodakov, Fiz. Tekhn. Polupr., 6, 113 (1972).

(4) S. H. Hagen and C. J. Kapteyns, Philips Res. Repts., 25, I (1970).

(5) G. A. Lomakina and Yu. A. Vodakov, Fiz. Tverd. Tela, 15, 123 (1973).

(6) P. Schnabel, Philips Res. Repts., 19, 43 (1964).

(7) G. Bench, J. Phys. Chem. Sol., 27, 795 (1964).

(8) J. P. Golightly and L. J. Beaudin, Mat. Res. Bull., 4S, 119 (1969).

(9) D. R. Hamilton, L. Patrick, and W. J. Choyke, Phys. Rev., 138, 1472 (1965).

(10) G. Iaamarchi, Philips Res. Repts., 20, 213 (1965).

A Comparison of SiC with Related Diamond-Like Crystals

Glen A. Slack

Silicon carbide is formed from elements of the first and second rows of the periodic table. In searching for other elements or compounds that resemble SiC with respect to bonding and crystal structure the same section of the periodic table will be considered. This gives us the elements C (diamond structure), and Si; and the binary compounds AlN, AlP, BN (cubic), BP, BeO, and BeS. The compounds LiF, NaCl etc. are excluded from consideration because they do not possess a diamond-like or adamantine crystal structure, and are too ionic. The crystals such as graphite, hexagonal BN, $B_{12}C_3$, $B_{12}P_2$, $B_{12}Si_2$' etc. are also excluded because they have greatly different crystal structures.

MICROHARDNESS AND ELECTRONEGATIVITY

One of the important parameters in discussing the adamantine compounds is their electronegativity difference, ΔX. This concept, due to Pauling,[1] has been discussed in several reviews[2-5]. The values used here are taken from Gordy[5]. The electronegativity differences, on an arbitrary scale, between the elements of the binary compounds vary from 0.28 for BP to 2.09 for BeO. The elemental crystals C, Si, and Ge are zero on this scale. Another important parameter is the interatomic distance, r, between nearest neighbor atoms. With these two parameters we can plot the identation micro-hardness, H, versus r at values of constant ΔX as in Fig. 1. This plot is similar to that given by Wolff et al[6] with the addition of ΔX as a variable. The effect of ΔX on H has been discussed by Goryunova et al[7]. The line in Fig. 1 for ΔX = 0 is well defined. The lines for ΔX = 1.0 and 2.0 are only approximate, but clearly indicate the trends. The larger the value of ΔX for a crystal, the more ionic the bonding, and the lower the hardness, H. The H values have been selected, where possible, for the (111) or (0001)

crystal faces measured with zero atmospheric moisture(8,9) at room temperature.(10) Where such results were not available, H values have been taken
from other literature sources(6,7,11). The range of values for BP has been
taken from four sources(12-15), while the values for AlN are from two sources.
(11-16) The BN value of DeVries(17) is 4600 ± 100 kgm/mm^2, and is preferred
to the less certain values of Filonenko et al(18).

From Fig. 1 one can clearly see that, for the two compounds AlN and BP
that are the closest relatives of SiC, BP has about the same hardness as SiC,
and both have small ΔX values. The hardness of AlN is considerably less than
that of SiC because ΔX is larger, i.e. its bonding is more ionic than that
of SiC.

BAND GAP

The valence band to conduction band gap is important for characterizing
the optical and electrical properties of these crystals. These room temperature
band gap values E_g, have been collected from the literature for C(19), BN(20),
BeO,(21) BP,(22-24), SiC,(25) BeS,(26) AlN,(27) MgO,(28) Si,(29) AlP,(30),
BeSe(26), BAs(31), GaN(32), and ZnO(33). Note that the general increase of
Eg with increasing ΔX. The three compounds BP, SiC, and AlN have increasing
Eg and increasing ΔX in the series. Again ΔX is a significant factor in
determining the differences between these various adamantine compounds.

MOBILITY

The measured electron and hole Hall mobilities, u_N and u_P, of the
various crystals are difficult to attach much significance to unless the
mechanisms limiting the mobilities are understood. For many of the values in
Table I the limitation is probably impurity scattering, but for some it is the
interaction with the lattice. Roughly speaking u decreases as ΔX increases
for a given row of the periodic table. The values of u at 300K in Table I
are taken from the literature for C(34,35), BN(36), BP(12,22), SiC(37-39),
AlN(40,41), MgO(42), Si(43), AlP(44,45), BAs(31), GaN(46,47), and ZnO(48).
In general the highest mobility values measured are the ones used in Table I.
Where only n-type or p-type conductivity has been seen but no u values are
available, it is stated only that $u > 0$. Note that for SiC the value of u
depends on the polytype.(49) The mobility trend in Table I indicates that
pure BP will have higher mobility than cubic SiC, and that AlN will have
lower mobility.

OTHER CRYSTALS

Other crystal compounds made from elements of the first two rows of the periodic table that may possess some properties similar to those of SiC are numerous. Some binary compounds such as Be_3P_2, Be_2C, Al_4C_3, Be_3N_2, Mg_3N_2, and Si_3N_4, arranged in order of increasing ΔX, can be mentioned. None of these have the adamantine crystal structure, but many are tetrahedrally bonded. Their "average" electronegativity differences, ΔX, range from 0.10 for Be_3P_2 to 1.88 for Si_3N_4. Very little is known about their optical or electrical properties, and much work remains to be done.

Of the ternary compounds the most promising are the adamantine structure ones $MgSiP_2(50)$, $BeSiN_2(51)$, $MgSiN_2(52)$, and $Al_2OC(53)$. The first is an analog of AP, while the next three are analogs of AlN. From what little is known about their properties they appear to be interesting high temperature materials, with many similarities to AlP and AlN.

CONCLUSIONS

Even though SiC is a very versatile and interesting material, there are a number of other crystals which may be able to compete with or surpass SiC in particular applications. These should not be overlooked, and perhaps some time should be spent in further study on a few selected ones. An important parameter in determining the properties of these compounds is the electronegativity difference between the elements in it.

Table I: Properties of Adamantine Crystals at 300K (a)

Crystal	ΔX	H kg/mm^2	Eg eV	u, cm^2/volt sec N	P
C	0.00	13200	5.6	1800	1550
BN	1.10	4600	6.5	>0	>0
BeO	2.09	1400	10.5	—	—
BP	0.28	~3600	2.0	>0	500
SiC (3C)	0.70	3720	2.35	1000	20
SiC (6H)	"	"	3.00	300	55
SiC (2H)	"	"	3.30	—	—
BeS	1.30	—	~5.5	—	—
AlN	1.53	1200	6.2	>0	14
MgO(b)	2.31	845	7.77	10	—
Si	0.00	1120	1.12	1880	400
AlP	0.71	—	2.41	80	>0
BeSe	0.97	—	4.3	—	—
BAs	1.13	—	1.46	—	~200
GaN	1.53	—	3.4	300	—
ZnO	2.26	490	3.3	200	—

(a) For the hexagonal crystals the values of H, Eg, and u, are slightly anisotropic. The value quoted is an average over the anisotropy.

(b) MgO is included for comparison, because of its large ΔX, although it has the rocksalt structure.

REFERENCES

1. L. Pauling, "Nature of the Chemical Bond", 3rd Edition, Cornell Univ. Press, Ithaca (1960).

2. J. C. Phillips, Rev. Mod. Phys. 42, 317 (1970).

3. J. A. Van Vechten, Phys. Rev. 182, 891 (1969).

4. H. O. Pritchard and H. A. Skinner, Chem. Rev. 55, 745 (1955).

5. W. Gordy, Phys. Rev. 69, 604 (1946).

6. G. A. Wolff, L. Toman Jr., N. J. Field, and J. C. Clark, in "Semiconductors and Phosphors,   ed. by M. Schon and H. Welker, Interscience, New York, 1958, p. 463.

7. N. A. Goryunova, A. S. Borshchevskii, and D. N. Tretiakov, in "Semiconductors and Semimetals", ed. by R. K. Willardson and A. C. Beer, Academic Press, New York, 1968, Vol. 4, p. 3.

8. R. E. Hanneman and J. H. Westbrook, Phil Mag. 18, 73 (1968).

9. J. H. Westbrook and P. J. Jorgensen, Am. Mineral. 53, 1899 (1968).

10. J. H. Westbrook, Rev. Hautes Temp. Refract. 3, 47 (1966).

11. C. F. Kline and J. S. Kahn, J. Electrochem. Soc. 110, 773 (1963).

12. B. Stone and D. Hill, Phys. Rev. Letters 4, 282 (1960).

13. Ya. Kh. Grinberg, Z. S. Medvedeva, A. A. Eliseev, and E. G. Zhukov, Dokl. Akad. Nauk SSSR 160, 337 (1965).

14. Ya. Kh. Grinberg, Z. S. Medvedeva, and L. A. Klinkova, Neorg. Mater. 1, 478 (1965).

15. B. V. Baranov, V. D. Prochukhan, and N. A. Goryunova, Neorg. Mater. 3, 1691 (1972).

16. K. M. Taylor and C. Lenie, J. Electrochem. Soc. 107, 308 (1960).

17. R. C. DeVries, General Electric Report 72CRD178, June 1972, unpublished.

18. N. E. Filonenko, V. I. Ivanov, and L. I. Feldgun, Dokl. Akad. Nauk SSSR 164, 1286 (1965).

19. J. F. H. Custers and F. A. Raal, Nature 179, 268 (1956).

20. R. Chrenko, General Electric Report 73CRD103, March 1973, unpublished.

21. D. M. Roessler, W. C. Walker, and E. Loh, J. Phys. Chem. Solids 30, 157 (1969).

22. C. C. Wang, M. Cardona, and A. G. Fischer, RCA Review 25, 159 (1964).

23. R. J. Archer, R. Y. Koyama, E. E. Loebner, and R. C. Lucas, Phys. Rev. Letters 12, 538 (1964).

24. T. L. Chu, J. M. Jackson, A. E. Hyslop, and S. C. Chu, J. Appl. Phys. 42, 420 (1971).

25. W. J. Choyke, Mat. Res. Bull. 4, S141 (1969).

26. W. M. Yim, J. P. Dismukes, E. J. Stofko, and R. J. Pfaff, J. Phys. Chem. Solids 33, 501 (1972); see also D. J. Stukel, Phys. Rev. B2, 1852 (1970).

27. W.M. Yim, E. J. Stofko, P. J. Zanzucchi, J. I. Pankove, M. Ettenberg, and S. L. Gilbert, J. Appl. Phys. 44, 292 (1973).

28. D. M. Roessler and W. C. Walker, Phys. Rev. 159, 733 (1967).

29. N. Hannay, "Semiconductors

30. B. Monemar, Solid State Commun.. 8, 1295 (1970).

31. T. L. Chu and A. E. Hyslop, J. Appl. Phys. 43, 276 (1972).

32. D. L. Camphausen and G. A. N. Connell, J. Appl. Phys. 42, 4438 (1971).

33. R. E. Dietz, J. J. Hopfield, and D. G. Thomas, J. Appl. Phys. 32, S2282 (1961).

34. A. G. Redfield,,Phys. Rev. 94, 526 (1954).

35. I. G. Austin and R. Wolfe, Proc. Phys. Soc. (London) B69, 329 (1956).

36. R. H. Wentorf Jr., J. Chem. Phys. 36, 1990 (1962).

37. A. Rosengreen, Mat. Res. Bull. 4, S355 (1969).

38. H. J. van Daal, Philips Res. Rept. Suppl. 3, 1 (1965).

39. D. L. Barrett, J. Electrochem. Soc. 113, 1215 (1966).

40. K. Kawabe, R. H. Tredgold, and Y. Inuishi, Electr. Engr. Japan 87, 62 (1967).

41. T. L. Chu, D. W. Ing, and A. J. Noreika, Solid State Electronics 10, 1023 (1967).

42. J. H. Pollard, J. Phys. Chem. Solids 26, 1325 (1965).

43. J. Messier and J. M. Flores, J. Phys. Chem. Solids 24, 1539 (1963).

44. H. G. Grimmeiss, W. Koschio, and A. Rabenau, J. Phys. Chem. Solids 16, 302 (1960).

45. D. Richman, J. Electrochem. Soc. 115, 945 (1968).

46. R. Dingle, D. D. Sell, S. E. Stokowski, and M. Ilegems, Phys. Rev. 4B, 1211 (1971).

47. J. I. Pankove, et al,RCA Review 32, 383 (1971).

48. A. R. Hutson, J. Phys. Chem. Solids 8, 467 (1959).

49. L. Patrick, J. Appl. Phys. 37, 4911 (1966).

50. A. J. Springthorpe and J. G. Harrison, Nature 222, 977 (1969).

51. P. Eckerlin, et al, Z. anorg. allgem. Chem. 353, 113 (1967).

52. J. David, et al,Bull. Soc. Fr. Mineral. Cristallog. 93, 153 (1970).

53. L. M. Foster, et al, J. Am. Ceramic Soc. 39, 1 (1956).

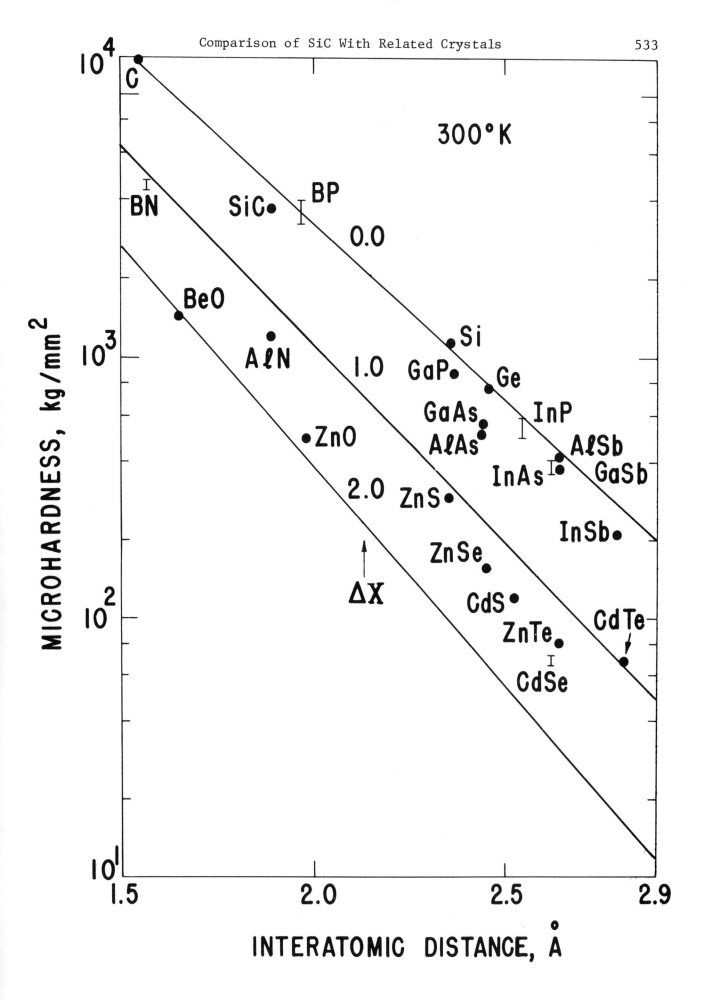

Silicon Carbide from the Perspective of Physical Limits on Semiconductor Devices

Robert W. Keyes

The properties that can be anticipated for devices made in SiC are compared with those of devices made in Si. The principal advantage of SiC is found to be its high breakdown voltage, which should permit very small bipolar transistors to be fabricated and also is favorable to the use of SiC in high power transistors.

INTRODUCTION

There are two aspects to semiconductor device technology: The semiconductor must have physical properties that allow a device to perform some useful function under a given set of conditions and a processing technology that permits devices to be made economically must be possible. Naturally, these two kinds of considerations are not entirely independent. It is difficult, however, to make generalizations about process technology because of the very many specific chemical considerations that are involved.

On the other hand, the theory of the physical properties of semiconductors and of the relation of these physical properties to the operation of devices is rather well understood. It is possible to make some generalizations about semiconductor devices from the point of view of physical theory. These generalizations are useful and interesting: Semiconductor device technology has advanced so rapidly over the last two decades that one would like to ask about what kinds of factors will eventually limit the progress toward such things as higher packing densities and higher speeds of operation, and higher power handling capability. Further, the generalizations from the physical theory of devices can be applied to any semiconductor and provide a basis for comparison of the device potential of different materials. The results of generalizations from theory can often be expressed as "figures of merit" or as "physical limits".

The advantages of silicon carbide as a material for light-emitting diodes and for devices intended for operation at high temperatures are well known and will not be mentioned further here. Instead, the object of this work is to investigate the suitability of SiC as a material in which to pursue those objectives towards which development work in silicon devices is now oriented. Thus, Table 1 presents values of the relevant physical properties of SiC and silicon. In the early stages of transistor technology, germanium was also a promising candidate for a wide variety of semiconductor devices. Therefore, a comparison of silicon with germanium, illustrating why (and in spite of what unfavorable properties) silicon rather than germanium has become the basis of so much modern technology, helps to illuminate the relative merits of SiC and silicon, and properties of germanium are also presented in Table 1. It is indeed seen that the properties of SiC are in many respects extrapolations of the sequence Ge-Si. Of course, a major reason that most semiconductor devices are now made of silicon does not appear in the table, namely the success of silicon processing technology.

FIGURES OF MERIT

As a first example of a figure of merit, consider the problem of speed or high frequency response of devices. The ease with which charge can move through a semiconductor is measured, at least to first order, by mobility, the velocity that a carrier acquires per unit of applied electric field. A figure of merit for transistors, in which both holes and electrons must move, is [Giacoletto 1955]

$$Q \equiv \mu_p \mu_n / K^{1/2} \tag{1}$$

Here μ_p and μ_n are, respectively, the hole and electron mobilities and K is the dielectric constant, which enters in the denominator because the amount of change stored in the device decreases with decreasing dielectric constant. The values of Q calculated from the information in Table 1 are presented as the first line of Table 2. It is seen that Q is about an order of magnitude less for Si than for Ge and is another order of magnitude smaller for SiC. The speed of a transistor is, of course, determined by many factors beside mobility, and, in particular, high speed or high frequency transistors must

be very small. Nevertheless, as suggested by the figure of merit, transistors
made to the same dimensions and with similar technology in silicon and in
germanium showed the germanium transistors to be faster by a factor of three
[Dill, Farber, and Yu 1968]. This advantage of germanium has not proved
decisive, however, and silicon has become the material of choice even for high
frequency devices. The lower mobilities in SiC may not be as important a
disadvantage as would appear at first sight.

A somewhat different view of high speed switching transistors for computer
logic observes that increase of speed is obtained by decrease of size and
that electric fields are very high in modern small transistors. The electron
velocity is then determined by the "limiting velocity" rather than the low
field mobility. The limiting velocity, v_L, is essentially the velocity at
which an electron has enough energy to emit an optical phonon [Sze 1969].
Apparently a high limiting velocity permits high speed devices. Another
problem of small devices is the very high density of power dissipation, or
production of heat. The heat flows away from the device by conduction
through semiconductor material, encountering thereby a thermal resistance
that is inversely proportional to the size of the device. A lower limit on
size is set by the maximum permissible thermal resistance, so that high
thermal conductivity is desirable. A figure of merit that encompasses these
small size effects is [Keyes 1972]

$$Q_2 = \lambda(cv_L/K)^{1/2} \qquad (2)$$

Here λ is the thermal conductivity of the semiconductor and a factor c, the
velocity of light, has been included for dimensional convenience. The values
of the parameters in Eq. (2) are given in Table 1, with v_L for SiC being
estimated from the phonon energy, and the figure of merit calculated from
Eq. (2) is given as the second line of Table 2. The high thermal conductivity
of SiC may be an advantage where removal of heat from a small high power device
is a problem.

E. O. Johnson [1965] considered the high frequency--high power capability
of materials. He argued that the basic limitation on various transistor
characteristics is set by the parameter $F_B v_L$, the product of the breakdown
electric field and the limiting velocity. Breakdown of a semiconductor occurs
when carriers acquire enough energy to excite an electron across the gap,
creating an electon-hole pair. The high energy gap of SiC gives it a high

breakdown field. Johnson's figure of merit, $F_B v_L / 2\pi$, is given in Table 2.
The high breakdown field causes SiC to stand out by an order of magnitude
in this voltage-frequency product.

An example of the way in which $F_B v_L$ limits transistor characteristics is

$$V_m \, f_T \, < \, F_B v_L / 2\pi$$

Here V_m is the maximum allowable applied voltage and f_T is the transit time
cutoff frequency. Another form is

$$(I_m V_m X_c)^{1/2} \, f_T \, < \, F_B v_L / 2\pi$$

X_c is the impedance of the collector-base capacitance at frequency f_T and I_m
is the maximum current that can be passed through the transistor without
producing significant base widening. Johnson found that the characteristics
of silicon and germanium transistors indeed obey the above relations.

DEVICE DENSITY

Higher density, that is, more devices per unit area, is perhaps the most
prominent objective of development efforts in silicon technology at present.
Higher densities of devices are attained by making the devices smaller, so an
inquiry into the lower limits on device sizes has interest. Hoeneisen and
Mead [1972] have used the following chain of reasoning to find a lower limit
on the size of metal-oxide semiconductor field effect devices. The p and n
regions of a semiconductor device are separated by a space charge or depletion
region, and devices must have linear dimensions at least as great as the width of
the space charge region. Now, the space charge width can be reduced by heavier
doping. The doping level is limited, however, in a field effect device by the
requirement that the field applied through the insulating oxide must be able to
invert a layer. The narrower the space charge region, the higher the field
that is required in it to make up the necessary potential difference. If the
depletion region is made too narrow, the field required in the oxide will cause
it to break down. This minimum depletion layer width sets a minimum size on
the field effect transistor.

It can be seen rather directly that the relation between the width of a
depletion region, w, the potential drop across it, ξ, and the concentration
of charged impurities in it, N, is

$$\xi = 4\pi Nqw^2 / 2K$$

and that the maximum electric field is $2\xi/w$. When the electrtric displacement
in the oxide is matched to that in the semiconductor and required to be less
than that corresponding to the breakdown field F_{ox}, it is found that

$$w < 2K \, \xi/K_{ox} \, F_{ox}$$

Here K_{ox} is the dielectric constant of the gate oxide. The potential drop,
ξ, must be of the order of the energy gap. Thus, a figure of merit for
"smallness" might be

$$w = 2K \, E_g/K_{ox} \, F_{ox} \tag{4}$$

Values of w are given in Table 2. It is seen that from the point of view of
making very small field effect devices SiC is at a disadvantage as compared to
silicon. The principal reason is that larger potential drops would have to be
supported by the depletion layer because of the larger energy gap of SiC.

The limit on the width of the depletion region in bipolar transistors and,
thus, on the doping level, is set by the condition that a junction must not
break down under the voltage applied to it [Hoeneisen and Mead 1972]. The
applied voltage must be something like the energy gap voltage, so a measure of
width would be $w = E_g/F_B$, values of which are given in Table 2. The doping
levels needed to attain the allowable small junction widths are very high, so
that the simplified notion of breakdown field used to characterize materials in
Table 1 does not provide an accurate comparison. Breakdown by internal field
emission becomes important at the high doping levels needed for very small
transistors. Thus, a favorable aspect of SiC with respect to breakdown is that
the degeneracy concentrations, estimated in Table 1, are higher in SiC, so that
field emission breakdown will not play a role until higher concentrations in SiC.

The minimum depletion layer width and, thus, transistor size, also depends
on the voltage that is applied to the transistor. For any applied voltage it is
possible to determine the maximum doping level that a semiconductor material may
have without breakdown occurring in a junction in the material. The depletion
widths for junctions with this doping level can then be calculated and a minimum
base width can be found. The base must be wider than the junction depletion
layers to prevent the base material from being entirely depleted (punch-through).
A curve of base width as a function of voltage can thus be constructed. Figure 1
shows such curves for silicon (after Hoeneisen and Mead [1972]) and SiC, the

latter using the breakdown data of van Opdorp and Vrakking [1969] and estimates for higher concentrations. Considerably higher doping levels and
smaller base widths can be used with SiC than with silicon. The other linear
dimensions of the transistor may be expected to scale with the base width.
The use of higher voltages in SiC circuits must be anticipated, but it appears
that smaller bipolar transistors can be used in SiC than in silicon.

Random fluctuations in the number of dopant atoms contained in a very small
volume can degrade performance in the devices being considered [Hoeneisen and
Mead 1972]. Note in Fig. 1 that the doping levels are higher for a transistor
of given base width in SiC, so that statistical fluctuations will be smaller
than in silicon.

A caution must be added to this glowing picture of SiC, however. The
acceptor and donor levels of SiC are far from the band edges, and some
impurities will not be ionized at 300°K. This trapping of carriers will
undoubtedly have an adverse effect on device operation [Dumke 1970].

CONCLUSIONS

The property of SiC that appears to have the most significance for electrical devices operated in the vicinity of 300°K is its high breakdown field.
Although breakdown in SiC has not been very thoroughly studied, we estimate that
the breakdown field is three to five times that of silicon junctions with comparable concentrations of impurities. The high breakdown fields imply that
planar bipolar transistors can be made very small in the light of the considerations by Hoeneisen and Mead and that SiC has a high power handling capability
in terms of the criteria of Johnson. The large energy gap is a disadvantage
with respect to small devices, as it increases the width of depletion regions, a
disadvantage that shows up in Hoeneisen and Mead's estimate of limiting sizes
of field-effect transistors. The low carrier mobilities of SiC may also be a
problem in some applications, but mobility does not appear to be an important
factor in most of the "figures of merit" and "physical limits" that we have
studied.

REFERENCES

Dill, F. H., Farber, A.S., and Yu, H. N.: 1968, IEEE J. Solid State Circuits, SC-3, 160.

Giacoletto, L. J.: 1955, RCA Review 16, 34.

Hoeneisen, B., and Mead, C. A.: 1972, Solid State Electronics 15, 819, 897.

Johnson, E. O.: 1965, RCA Review 26, 163.

Keyes, R. W.: 1972, Proc. IEEE 60, 225.

Sze, S. M.: 1969, Physics of Semiconductors Devices (John Wiley and Sons, New York).

van Opdorp, C., and Vrakking, J.: 1969, J. Appl. Phys. 40, 2320.

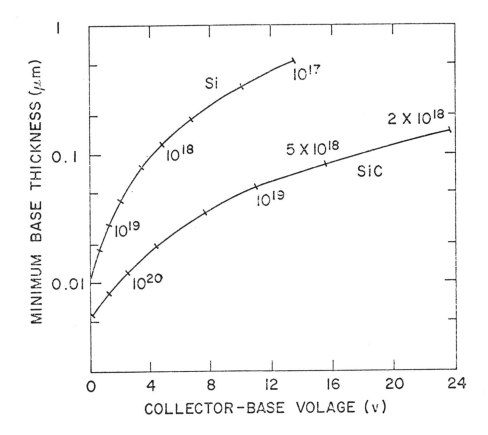

Fig. 1.   Minimum base thickness of a planar bipolar transistor for a given collector-base voltage determined by collector junction breakdown and base punch through for silicon and SiC, calculated according to the method of Hoeneisen and Mead [1972].

TABLE 1.   PROPERTIES OF GROUP IV SEMICONDUCTORS

Property	SiC[a,b]	Si[c]	Ge[c]
Electron mobility[d], μ_n (cm^2/Vsec)	500	1500	3900
Hole mobility[d], μ_p	50	600	1900
Dielectric constant, K	10	12	16
Energy gap, (ev)	3.0	1.2	0.7
Breakdown field, F_B (10^5 V/cm)	25[e]	3	1
Thermal conductivity (w/cm°K)	5[f]	1.5	0.5
Optical phonon energy (ev)	0.098	0.063	0.037
Limiting velocity (10^6 cm/sec)	10	10	6
Degeneracy N (10^{18} cm^{-3})	14[g]	2	0.15

(a) 6H polytype; (b) Silicon Carbide-1968 (published as Volume 4 of Materials Research Bulletin), pp. 365-370, summarizes the physical properties of SiC; (c) Sze [1971], pp. 57-8, tabulates many properties of Si and Ge; (d) Highest mobility at 300°K; (e) van Opdorp and Vakrking [1969]; (f) G. A. Slack, J. Appl. Phys. 35, 3460 (1969); and (g) H. J. van Daal, W. F. Knippenberg, and J. D. Wasscher, J. Phys. Chem. Solids 24, 109 (1963).

TABLE 2.   CERTAIN QUANTITIES DERIVED FROM TABLE 1

	SiC	Si	Ge
Q, (Eq. 1), (10^{6}cm^{4}V^2/s^2)	0.01	0.26	1.85
Q_2, (Eq. 2), (10^{7}w/s°K)	85	24	5
$F_B v_L / 2\pi$ (10^{11}V/s)	40	5	1
w, (Eq. 4) (10^{-8}cm)[a]	300	132	112
E_g/F_B (10^{-8}cm)	100	300	700

(a) Using $K_{ox} = 4$, $F_{ox} = 5 \times 10^6$

Optical Properties of Amorphous Silicon Carbide Films

E. A. Fagen

INTRODUCTION

Silicon carbide belongs to the large family of tetrahedrally coordinated semi-conducting binary compounds. Only a few members of this family have been investigated in the amorphous state, and among these silicon carbide is unique in that both of its constituents belong to the fourth column of the periodic table. Thus it should provide significant tests of current thinking about tetrahedrally coordinated amorphous semiconductors in general, and amorphous germanium and silicon in particular. An advantage from the investigator's point of view is that the bonding is partially ionic. Hence, in contrast to germanium and silicon, the vibrational modes are inherently infrared-active, and spectroscopic techniques may be brought to bear on problems of structural analysis and lattice dynamics. To these theoretical considerations may be added potential technological usefulness, inasmuch as the hardness, thermal stability, and other desirable properties of crystalline silicon carbide appear to be largely preserved in the amorphous state.

Despite these attractions, amorphous silicon carbide seems to have received scant attention. A literature search reveals only two studies of its electrical properties (1,2) and none whatever of its optical properties. We report here initial measurements of the optical constants of sputtered thin films over the range 0.08 to 4.0 eV, embracing both the interband absorption edge and the principal reststrahl. Special emphasis is placed on comparison with optical properties of the crystal. We conclude with a brief discussion of the significance of these results with respect to short range order in the amorphous phase.

SAMPLE PREPARATION AND CHARACTERIZATION

Thin films of amorphous silicon carbide were prepared by r.f. sputtering in a conventional oil-pumped vacuum system. The system was throttled before argon admission to prevent saturation of the diffusion pump, resulting in a residual pressure ca. 3×10^{-5} torr. The target was a hot pressed polycrystalline disc 89 mm in diameter and 7 mm thick, of alleged 99.5% purity, furnished by Semi-Elements, Inc., Saxonburg, Penna. Substrates of mica, alumina, fused quartz, Corning 7059 glass, or polished silicon were selected in accord with experimental needs. Virgin films exhibited severe compressive stress and adhered poorly to mica, alumina, or molybdenum-overcoated substrates, but adhered strongly to glass and silicon.

Electron microprobe analysis of the deposited films revealed an average Al content ca. 0.30% and an average Fe content ca. 0.31%. These metallic impurities apparently arise from the sputtering cathode itself. No analysis was made for reactive gas incorporation or departures from stoichiometry. These findings cast doubt on the validity of the experimental studies then in progress. More recently, susceptors from the silicon wafer processing industry (graphite discs coated with ultrahigh purity SiC by chemical vapor deposition) have been employed as sputtering cathodes. Preliminary measurements on these newer films substantially confirm those reported below, indicating that metallic impurities at least are not the source of the unusual features we observe.

EXPERIMENTAL RESULTS AND DISCUSSION

Optical transmission and reflection were measured in the neighborhood of the principal interband absorption edge with Cary 11 and Beckman DU spectrophotometers. Films of several thicknesses were employed, and measurements were made before and after annealing in high vacuum at 600°C for one hour. Reflectivity increased slowly over the range 1 to 4 eV, averaging about 3% less than that of the crystal (3). The absorption coefficient is displayed in Fig. 1. Error bars show the effect of ±2% experimental uncertainty in reflectivity, becoming significant only at small values of absorptance. Discrepancies among nominally similar specimens lie outside this uncertainty and have not been smoothed over; their origin is presently unclear. Other traces have been included in the figure to facilitate comparison. The solid lines, taken from Philipp and Taft (3) represent high purity cubic (β) and hexagonal (α) monocrystals. The lower dashed line represents an "electrical grade" crystal of ca. 1 Ω-cm resistivity, containing approximately 330 ppm Al and 30 ppm Fe, and showing the resultant free carrier absorption. (Note that this impurity content is less than a tenth that of our films.) The upper dashed line represents amorphous silicon films, a composite of the results of Brodsky et al. (4), and Fischer and Donovan (5), differences between which are scarcely apparent on this scale of presentation. The striking features of the new data are the red shift with respect to the crystalline state and the extreme shallowness of the edge, both of which exceed that of any amorphous semiconductor known to us.

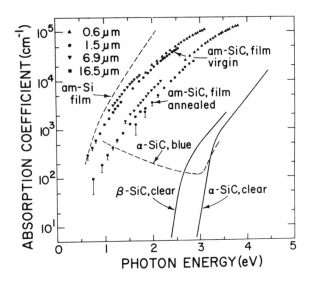

FIG. 1. Room temperature absorption coefficient of virgin and annealed amorphous SiC films of the indicated thickness, near the interband edge. See text for explanation of additional traces.

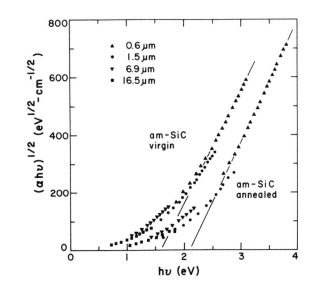

FIG. 2. Room temperature absorption coefficient of virgin and annealed amorphous SiC films, replotted from Fig. 1 to test for quadratic dependence on photon energy.

In Fig. 2 the data of the previous figure have been replotted to test for a relation of the form $\alpha h\nu = B(h\nu - E_o)^2$. The straight lines shown yield E_o = 1.6 and 2.1 eV in the virgin and annealed states respectively, and B = 1.6 x 10^5 and 1.7 x 10^5 $eV^{-1} cm^{-1}$ correspondingly. (These values clearly derive from the thinnest specimen only.) By comparison, the coefficient B in more typical amorphous semiconductors ranges from 2 to 9 x 10^5 eV^{-1} cm^{-1}. Under certain assumptions regarding the densities of states and matrix elements of interband transitions, Davis and Mott (6) have derived the expression $B \simeq 4\pi\sigma_o/ncE^*$, where σ_o is the preexponential factor in the electrical conductivity, n the refractive index, and E^* the width of the region of localized states. Insofar as these assumptions are applicable to the present case, the shallowness of the absorption edge signifies very extensive tailing of localized states into the pseudogap.

Fig. 3 shows the shift of the interband absorption edge with temperature. Inasmuch as this shift would be indiscernibly small if presented in the manner of Fig. 1, it has been expressed with respect to the position of the edge at room temperature, arbitrarily defined as zero. The large relative scatter of the data precludes any conclusion regarding obedience to Urbach's rule. If, as in other amorphous semiconductors, we ascribe the temperature shift of the edge primarily to the electron-phonon interaction rather than lattice dilation (7), this shift is proportional to the average phonon population (8),

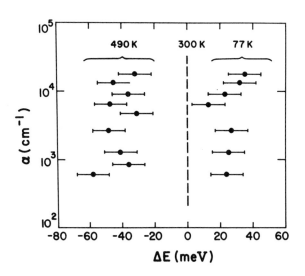

FIG. 3. Temperature shift of interband absorption edge of annealed amorphous silicon carbide films, expressed relative to position of edge at room temperature, arbitrarily defined as zero.

$$\langle n \rangle \sim [\exp(h\nu_o/kT) - 1]^{-1} \simeq \exp(-h\nu_o/kT)$$

wherein the low temperature approximation is justified if $\nu_o \simeq 2.38 \times 10^{13}$ sec^{-1}, the reststrahl frequency. A crude fitting of the data of Fig. 3 to a relation of the above form yields $(\partial E/\partial T)_V \simeq -1.7 \times 10^{-4}$ eV deg^{-1} at room temperature. Thus the smallness of the thermoöptical coefficient of amorphous silicon carbide appears to be directly related to its high Debye temperature.

Optical transmission and reflection were measured in the neighborhood of
the principal reststrahl with a Perkin-Elmer 700 spectrophotometer. Data from
specimens of various thicknesses were combined in reduction to achieve a reasonable degree of internal consistency. Results are shown in Figs. 4 and 5. The

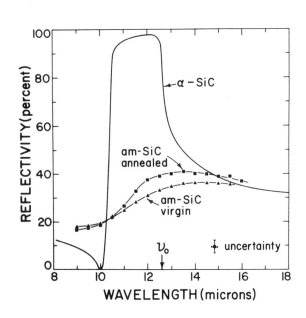

FIG. 4. Room temperature reflectivity of thick (16.5 μm) virgin and annealed amorphous SiC films near the
principal reststrahl. Solid line is
theoretical fit to ordinary ray reflectivity of α-SiC from Ref. (9), using
Lorentzian model at $\nu_o = 2.38 \times 10^{13}$
sec^{-1}.

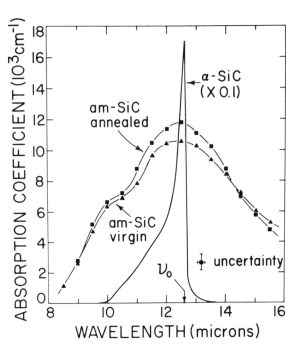

FIG. 5. Room temperature absorption
coefficient of thin virgin and annealed
amorphous SiC films near the principal
reststrahl. Solid line is replotted from
Ref. (9). Note change of scale.

solid line in Fig. 4 is a one-oscillator fit at the frequency ν_o via classical dispersion theory to the ordinary ray reflectivity of hexagonal SiC, replotted from
Ref. (9). The solid line in Fig. 5 is computed from the theoretically determined extinction coefficient in the same work, and diminished tenfold for convenience of
presentation. The significant feature is that the resonance is broadened but not
shifted in the amorphous state. The resonance appears to sharpen somewhat with
annealing at 600°C for 1 hour, but the increment cannot with justice be said to
lie outside the experimental uncertainty. Further annealing at 1000°C appears to
sharpen the resonance still more, but results are presently incomplete. On the

assumption that the reststrahl in the amorphous state is a single, highly damped Lorentzian line (10), one can extract from Figs. 4 and 5 an estimate of its oscillator strength relative to that of the crystal. We make use of the sum rule in the form

$$N_{eff} \sim \int \omega \epsilon_2(\omega)d\omega \sim \gamma n_o \alpha_o$$

where N_{eff} is the effective density of ion pairs, γ is the dimensionless damping factor, n_o and α_o are the refractive index and absorption coefficient evaluated at their respective maxima, and the constants of proportionality contain only factors such as reduced ionic masses, presumably common to both amorphous and crystalline phases. γ can be evaluated from the maximum reflectivity (11) or the half-width of the absorption band (12); it is ca. 0.50 in the virgin state and ca. 0.42 in the annealed state. We thus obtain $\gamma n_o \alpha_o \simeq 2.0 \times 10^4$ cm^{-1} for the amorphous phase in either state of anneal, comparable to the value 1.8×10^4 cm^{-1} taken from the computations of Spitzer et al. (9) for the crystalline phase. We conclude that within the accuracy of these estimates there is no diminution of oscillator strength in the amorphous phase, and hence no significant fraction of "wrong" (i.e., homonuclear) bonds.

This conclusion places powerful constraints on structural models of amorphous silicon carbide. It clearly excludes phase-separated models, which would contain chiefly homonuclear bonds. It likewise excludes any random pairing or diatomic models, which would contain 50% homonuclear bonds. Aggregation of random pairs into tetrahedra (13,14), or clusters of any size, does not alter the argument. Apart from microcrystalline models, which fail to explain the shift of the fundamental edge, there remain only ordered binary versions of tetrahedral random network models, of which the Polk model (15) is the prototype. Because of the occurrence of five- and seven-membered rings, however, an adaptation of the Polk model to a stoichiometric binary compound must contain not less than 12% homonuclear bonds (16). The accuracy of our data is not sufficient to test this crucial point, although a more refined dispersion analysis would certainly do so.

CONCLUSIONS

The optical properties of amorphous silicon carbide appear to be broadly consistent with those of other tetrahedrally coordinated amorphous semiconductors, and particularly germanium and silicon. The generic features of the latter are substantially reproduced: relative insensitivity to the incorporation of impurities, a shallow optical absorption edge at higher energies, and large annealing effects. Dispersion analysis of the strong reststrahl can yield useful information regarding short range order of the amorphous phase. It would now seem appropriate to apply additional analytic techniques to this unique material, and especially to determine the radial distribution function and the ultraviolet reflectivity.

ACKNOWLEDGEMENTS

We are indebted to R. S. Nowicki for sample preparation, to R. W. Seguin for assistance in data taking, to L. Allard of the Scanning Electron Microscopy Laboratory, University of Michigan, for electron microprobe analysis, and to H. Fritzsche for numerous helpful discussions.

REFERENCES

1. C. J. Mogab and W. E. Kingery, Journal Appl. Phys. 39, 3640 (1968).

2. T. E. Hartman, J. C. Blair and C. A. Mead, Thin Solid Films 2, 79 (1968).

3. H. R. Philipp and E. A. Taft, Silicon Carbide, J. R. O'Connor and J. Smiltens, eds., (Pergamon, Oxford, 1960) p. 366.

4. M. H. Brodsky, R. S. Title, K. Weiser and G. D. Pettit, Phys. Rev. B1, 2632 (1970).

5. J. E. Fischer and T. M. Donovan, Journal Non-Crystalline Solids 8-10, 202 (1972).

6. E. A. Davis and N. F. Mott, Phil. Mag. 22, 903 (1970).

7. E. A. Fagen, S. H. Holmberg, R. W. Seguin, J. C. Thompson and H. Fritzsche, Proceedings Tenth International Conference on the Physics of Semiconductors, (USAEC Division of Technical Information, 1970) p. 672.

8. H. Y. Fan, Phys. Rev. 82, 900 (1951).

9. W. G. Spitzer, D. A. Kleinman, C. J. Frosch, and D. J. Walsh, Silicon Carbide, J. R. O'Connor and J. Smiltens, eds., (Pergamon, Oxford, 1960) p. 347.

10. D. L. Mitchell, S. G. Bishop and P. C. Taylor, Journal Non-Crystalline Solids 8-10, 231 (1972).

11. S. S. Mitra, Crystallography and Crystal Perfection, G. N. Ramachandran, ed., (Academic Press, London, 1963) p. 347.

12. C. Haas and J. A. A. Ketelaar, Phys. Rev. 103, 564 (1956).

13. H. R. Philipp, Journal Non-Crystalline Solids 8-10, 627 (1972).

14. H. R. Philipp, Journal Electrochemical Society 120, 295 (1973).

15. D. E. Polk, Journal Non-Crystalline Solids 5, 365 (1971).

16. N. J. Shevchik, Technical Report Nos. HP-29, ARPA-44: Structure of Tetradrally Coordinated Amorphous Semiconductors, (Harvard University, Cambridge, Mass., 1972) p. 1-53.

Luminescent Devices

J. M. Blank

Luminescent devices encompass a very broad field of scientific endeavor. Injection electroluminescence (EL) alone is an area of considerable scope. Excellent general reviews of EL were published in 1972 by Bergh and Dean (1) and by Brander (2) which are recommended reading for anyone seriously interested in light-emitting diodes (LED's). This paper will limit itself to SiC and will concentrate on its performance in blue LED's because the blue LED's appear to be the most attractive area for continuing SiC research.

SiC is beset by many problems in crystal growth, efficient hole injection, and efficient radiative recombination at high injection levels. But SiC offers some unique advantages too in the wide band gaps of the 2H, 4H, and 6H polytypes, easy doping for both p- and n-type conductivity, relatively good transparency to its own light, and superb chemical stability. Thus, the feasibility of a blue light-emitting SiC diode is mainly a question of efficiency.

In the sections which follow we shall consider: (1) Factors that determine internal quantum efficiency, (2) Injection efficiency, (3) Phosphor properties and efficiency, (4) Alternative approaches to blue diode construction.

Factors that Affect Internal Quantum Efficiency

The internal quantum efficiency, η, of a LED is the product of the injection efficiency, η_I and the radiative recombination efficiency η_R.

$$\eta = \eta_I \cdot \eta_R \qquad (1)$$

Injection efficiency is defined as the ratio of the current of minority carriers (electrons or holes) injected into the light-producing region of a diode to the total current through the diode. At useful current levels the total current is composed of diffusion currents $I_{D,N}$ and $I_{D,P}$ corresponding to recombination in the N- and P-regions respectively and space charge recombination current, I_{SC}.

The spacial relationships of the several regions in a diode are shown in Fig. 1 for a diode in which the light generation takes place in the N-region. For this case the light generation is associated with $I_{D,N}$ so that

$$\eta_{I,N} = \frac{I_{D,N}}{I_{D,N} + I_{SC} + I_{D,P}} \qquad (2)$$

SiC diodes with light generation in the P-region are also readily achieved, in which case the light generation is associated with $I_{D,P}$ and the injection efficiency is

$$\eta_{I,P} = \frac{I_{D,P}}{I_{D,N} + I_{SC} + I_{D,P}} \qquad (3)$$

Diodes with light generation in the space charge region are not unknown and diodes with light generation taking place simultaneously in two or all three regions are not impossible although reproducible emission spectra from such diodes might be hard to maintain. Factors which influence η_I will be discussed later.

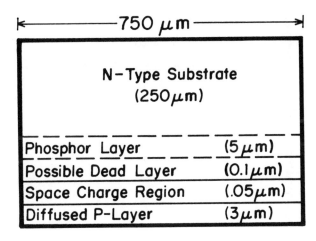

FIG. 1

Diagram of a LED made by diffusion. Layer thicknesses are shown in parentheses.

Let us limit the following discussion to $I_{D,N}$ keeping in mind that a completely analogous situation exists for $I_{D,P}$.

In SiC all of the injected holes associated with $I_{D,N}$ become bound to trapping centers or experience recombination in about 10^{-8} sec. Most of the recombination is nonradiative. This recombination behavior leads to a characteristic density of injected holes which decreases rapidly with distance, x, into the n-type layer.

Expressed quantitatively, the injected hole concentration, p, depends upon x as

$$p \propto \exp(-x/L_p) \tag{4}$$

where L_p is the hole diffusion length and

$$L_p \propto (\mu_p \tau_p)^{\frac{1}{2}} \tag{5}$$

In Eq. (5) μ_p is the mobility of holes in the N-region and τ_p is the lifetime of holes in the N-region. L_p is less than 1 μm in SiC diodes.

Thus, an abundance of holes available for recombination (radiative and non-radiative) is confined to a layer about 1 μm thick in the region adjacent to the space charge layer. This layer in which radiative recombination can be generated and in which radiative recombination centers must be provided will hereafter be called the phosphor layer. The dependence of the p(x) profile upon x and its relationship to L_p is illustrated in Fig. 2.

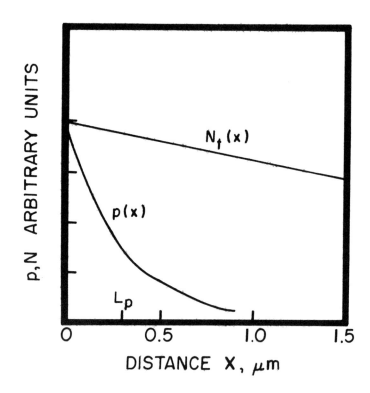

FIG. 2

Hypothetical profile of concentrations of injected holes p(x) and radiative recombination centers $N_t(x)$ versus distance into the phosphor layer, x, measured from the junction.

Also shown in Fig. 2 is the concentration of radiative recombination centers, N_t. In general, N_t is not uniform throughout the phosphor layer and the change of N_t with distance from the SC layer depends upon the method used to fabricate the junction and phosphor layer. In diffused diodes N_t will probably have a complimentary error function dependence on x.

$$N_t \propto \text{erfc}(-x/L_D) \tag{6}$$

where L_D is a characteristic length. L_D depends upon the type of the diffusing specie and upon the time and temperature used in the diffusion process. The case for $L_D \gg L_p$ is illustrated in Fig. 2 where N_t is essentially constant over the thickness L_p. While a uniform distribution of N_t is desirable, it is by no means assured. Especially detrimental to LED efficiency is the occurrence of a "dead layer" in the phosphor which can be introduced by such things as unwanted doping during meltback in liquid epitaxy or formation of dislocations or precipitates along a diffusion front or a growth front.

Dead layers and low efficiency phosphors suffer from unfavorable competition between radiative and nonradiative recombination processes. It is customary to characterize radiative recombination processes by their associated lifetime τ_R and nonradiative processes by the total lifetime τ_p (in n-type material) then the radiative recombination efficiency becomes

$$\eta_R \propto \tau_p/\tau_R \tag{7}$$

The radiative lifetime of the centers in phosphors are a natural property of materials and LED improvement in terms of these centers requires the discovery and recognition of the best that nature can provide. This subject will be discussed in a later section. The nonradiative lifetime also has a natural limitation but this limit is seldom reached because methods of material preparation, crystal growth and device fabrication introduce multitudes of nonradiative centers. These man-made non-radiative centers are associated with impurities and imperfections whose reduction or elimination constitute another opportunity for LED improvement. How this can be done is intimately related to the chemical and physical structure of the LED and will be discussed later.

Injection Efficiency

There is no way to measure directly the electron and hole diffusion currents in a LED. In fact it is possible to distinguish between I_{SC} and the sum $I_{D,P} + I_{D,N}$ only in special cases. Ralston (3) has recently performed such an analysis on GaP red and green emitting diodes and has pointed out the many

special qualities that must be possessed by diodes to qualify for analysis by
his methods.

On the other hand, $\frac{I_{D,N}}{I_{D,P}}$ can be calculated if the doping profile in the
neighborhood of the junction is known using

$$\frac{I_{D,N}}{I_{D,P}} = \frac{L_e}{L_h} \cdot \frac{\mu_h}{\mu_e} \cdot \frac{p}{n} \tag{8}$$

Even this information is practically unobtainable because the pertinent values
arise from dopant concentrations within one micron on either side of the
junction. Besides L_e and L_h are not well known at any doping concentration.

However, the ratio p/n is in most cases the controlling factor in Eq. (8)
and probably

$$\frac{L_e \mu_h}{L_h \mu_e} \geq 0.1 \tag{9}$$

Therefore a rough approximation to the ratio $\frac{I_{D,N}}{I_{D,P}}$ can be calculated for a
symmetrical linearly graded junction of the type shown in Fig. 3. In this case
(N_A-N_D) at the P-side boundary of the space charge layer is equal to (N_D-N_A) at
the N-side boundary. N_D and N_A are the donor and acceptor concentration
respectively. Both the N- and P-regions will be strongly compensated near the
space charge region so that the free electron and hole concentrations in the
N and P regions can be calculated from Ref. (5).

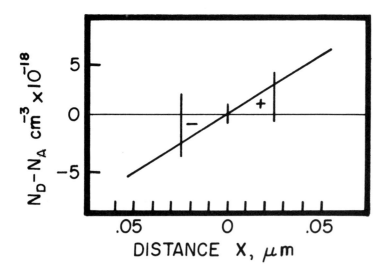

FIG. 3

Simplified model of diffused
junction. In reality N_D-N_A
is much distorted by ion drift
effect during diffusion. See
Ref. (4).

$$n = \frac{N_D - N_A}{N_A} \frac{N_c}{g} \exp\left(-\frac{E_D}{kT}\right) \tag{10}$$

$$p = \frac{N_A - N_D}{N_D} \frac{N_v}{g} \exp\left(-\frac{E_A}{kT}\right) \tag{11}$$

where N_c and N_v are the effective density of states in the conduction and valence bands and g is the multiplicity factor. In the rough approximation that

$$\frac{N_c}{g} = \frac{N_v}{g}$$

$$\frac{p}{n} = \frac{\exp(-E_A/kT)}{\exp(-E_D/kT)} \tag{12}$$

Taking the Lely and Kroger (6) values for $E_A = .28$ eV and $E_D = .08$ eV, then $\frac{p}{n} = 3.3 \times 10^{-4}$. Using Eqs. (8) and (9)

$$\frac{I_{D,N}}{I_{D,P}} \geq 3 \times 10^{-5} \tag{13}$$

Such a situation would make injection of holes into an n-type phosphor very inefficient. However, at practical current densities ($5A/cm^2$) conductivity modulation effects can greatly increase the hole concentration on the P-side and by injection on the N-side as well. Such action must occur in order to maintain electrical neutrality on the P-side when the injected electron concentration far exceeds the majority (hole) concentration. The extra holes must come from the P-side further from the junction or they must be generated at the P-side contact. Brander (2) has pointed out that in the limit of very high currents ($10A/cm^2$ for SiC) the ratio

$$\frac{I_{D,N}}{I_{D,P}} = \frac{\mu_h}{\mu_e} \tag{14}$$

in which case the ratio might be .03 - .20.

Potter (7) measured the electroluminescent efficiencies at the same current densities on the same diode prepared by simultaneous diffusion of boron and aluminum into an n-doped SiC crystal at 1900°C. This allowed him to separate the η_I and η_R factors in Eq. (1) and yielded the value $\eta_I = .03$. One must conclude that either Potter's diode was operating in the high current regime at 10^{-2} A/cm^2 or that the model built around Fig. 3 does not apply to his diode. In any event over 90% of the total diode current remains to be accounted for in I_{SC} and $I_{D,P}$.

Since SiC diffused diodes rarely exhibit I-V curves with $\frac{eV}{nkT}$ where n = 1 being mostly n $\geq$ 2 throughout the current range, it is not possible to separate I_{SC} from the diffusion current by measurements of I-V. In principle, electron diffusion currents could be estimated from measurements on diodes with p-type phosphors using Potter's method (7). It will be seen in a later section that red emitting SiC diodes in which electrons are injected into a p-type phosphor enjoy relatively high quantum efficiency suggesting that electrons might be injected with higher efficiency than holes.

Other interesting results from Potter (7) are $L_h = \frac{1}{3}$ μm, $\eta_R = 0.4$ at low excitation levels decreasing to .02 at 5A/cm^2. Thus, he attributes saturation of the light output to the phosphor, not to injection.

Phosphor Properties and Efficiency

The energy levels and spectra associated with the most useful activators of luminescence are shown in Fig. 4. If the thesis of this review is correct that blue emitting LED's offer the best possibilities for SiC lamps, then phosphors containing nitrogen and aluminum either alone or in combinations hold the greatest interest.

Ref. (8) shows the performance of LED's made by deposition of n-type epitaxial layers on p-type substrates doped with aluminum. When the doping in the n-type layer was principally nitrogen, emission peaks at 2.75 eV and 2.92 eV were observed. Author attributes the former to transitions between a nitrogen donor level and the valence band, the latter to a band to band transition. The 2.75 eV peak was just detectable above the background of a continuous spectrum at 30 ma/mm^2 and 213°K but dominated the spectrum at higher currents. However, when the temperature was raised, the 2.75 eV peak dropped relative to the 2.92 eV peak. From room temperature to 100°C diodes of this type would emit light of a deep blue color. Author did not state the efficiency but it must have been small for band to band radiation.

When the nitrogen doped epitaxy layer was partially compensated with aluminum Ref. (8) found emission with peaks at 2.41, 2.51 and 2.61 eV. Author attributed the 2.61 eV peak to a transition between the conduction band and aluminum, the other two peaks to transitions between nitrogen donor and aluminum acceptor levels. At room temperature the 2.51 and 2.61 eV peaks dominated the spectrum. A diode of this type would emit a convincing blue light. No statement of efficiency was given.

Ref. (14) described similar diodes made on high resistivity n-type crystals doped with nitrogen and nearly compensated with aluminum. LED's were made with alloy junctions and with diffused (aluminum) junctions. Both types of diodes produced spectra with peaks at 2.32, 2.42, and 2.52 eV at low temperatures. Authors attributed the peaks to transitions between N-Al donor-acceptor pairs. At room temperature and 1.2 A/cm^2 the 2.52 eV peak dominated the spectrum. The spectrum is shown in Fig. 4. It would appear blue-green. Authors mention that many diodes of this type did not show saturation of the EL with increasing current. Authors also observed photoluminescence(PL) at 77°K in p-type layers made by diffusing aluminum into n-type crystals. The p-type phosphor favored transitions from the conduction band to the aluminum level with peaks at 2.57, 2.63 and 2.74 eV. The 2.63 eV peak was strongest.

The performance of blue, green, yellow and red silicon carbide LED's is given in Ref. (10) which represent state of the art in the U.S.S.R. in 1968. Data are shown in Table I. Several relationships stand out: (1) Blue and green LED's made by diffusion of aluminum have the lowest brightness. (2) Green LED's made by alloying are better than ones made by diffusion. Perhaps alloyed junctions enjoy more efficient injection. (3) The highest quantum efficiency is attained by the Be doped LED's with alloy junctions. These LED's have p-type phosphors and might profit from relatively efficient injection. (4) The

FIG. 4

Energy level diagram for donor and acceptor centers in 6H SiC and EL (if not PL) spectra from diodes which contain these centers. All of these diodes may be presumed to be doped with nitrogen and oxygen to some degree. Energy values which are shown in the figure without parentheses have been specified by the referenced authors. Tie lines have been drawn between energy levels where electronic transitions have been identified. Values shown in parentheses are simply arithmetic consequences of the other data and are supplied to assist in further interpretation. Two different spectra which arise from two different fabrication techniques are given for aluminum doped junctions because they peak in or near the blue part of the spectrum. The spectra of diodes doped with other species also depend upon fabrication details. Data for this figure were obtained from the following sources. Nitrogen: Refs. (8) and (9). Aluminum: Refs. (8), (10), (11), (12), (13), (14), (15). Boron: Refs. (10), (16), (19). Beryllium: Refs. (10), (18), (19), (20), (21), and (22).

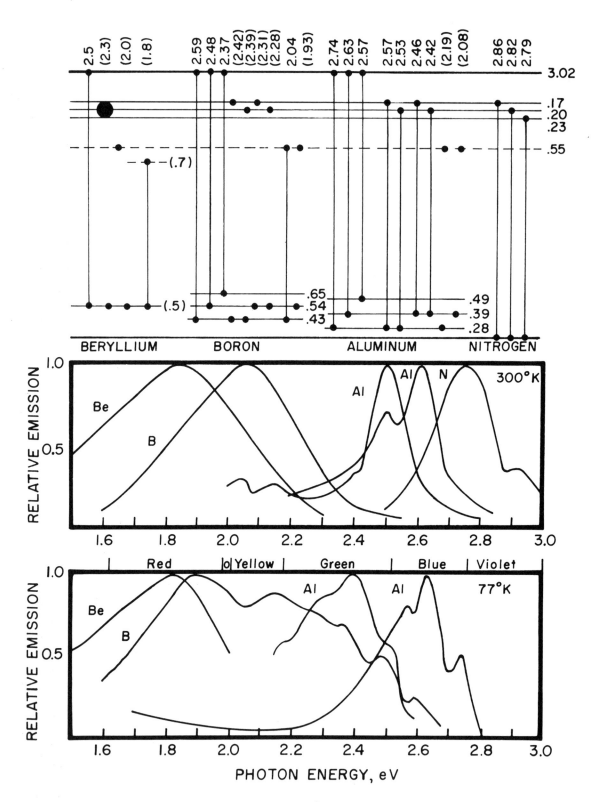

FIG. 4

TABLE I

Performance of Various SiC LED's
USSR - 1968 Ref. (8)

Acceptor	Color	Method	Brightness	Current	Ext. Q.E.
Aluminum	Blue	Diffusion	10-20 nits	30 ma	
Aluminum	Green	Diffusion	10-20 nits	30 ma	
Aluminum	Green	Alloy	10-100 nits	10-30 ma	.0001
Boron	Yellow	Diffusion	10-100 nits	10 ma	
Beryllium	Red	Alloy	10-100 nits	10-30 ma	.005

aluminum diffused blue emitting LED's have the same brightness as the green which indicates that the blue lamp has higher quantum efficiency.

The data of Table I are obviously out of date. Just how far out of date can be judged from the following performance quoted by Violin and Tairov in Ref. (23). For an LED made by diffusion of boron into an n-type crystal doped with nitrogen the brightness was 50-100 nits at 1 ma/mm^2 and 2.5 - 3.0 volts. This represents a tenfold increase in efficiency over the value given in Table I. Nevertheless, Table I provides us with an evaluation of several types of dopants at the same point in time by authors who had direct experience with the devices.

We have already noted the possible advantages in injection efficiency in alloy junction diodes over diffused junction diodes and similar advantages in diodes with p-type phosphors over diodes with n-type phosphors. In addition the two best acceptors, beryllium and boron, appear to have luminescence associated with a deep donor-like center which is not nearly so effective in phosphors doped with nitrogen and aluminum. Practically nothing is known about the nature of these deep centers or the luminescence mechanisms associated with them. Ref. (17) suggested that the deep donor level might be due to nitrogen atoms on silicon sites. Support for this idea was drawn from the necessary condition that the nitrogen concentration must exceed 10^{18} cm^{-3} to make an efficient boron doped phosphor. More recently Ref. (16) suggested that the recombination center for the yellow boron radiation is due to boron-nitrogen donor-acceptor pairs or to a more complex center involving a pair of nitrogen centers bound to a boron atom. Of course these speculations were made before the role of oxygen was recognized.

Phosphors doped with beryllium and nitrogen behave very differently from those doped with boron and nitrogen. Ref. (19) gives details relating the doping levels of beryllium and nitrogen to the temperature ranges for observation of bright line blue PL, as well as a green band, two yellow bands, and the well-known red band. The latter was generated in p-type phosphors with nitrogen concentration less than $7 \cdot 10^{17}$ cm^{-3}. Ref. (20) states that efficient EL was obtained in p-type crystals with alloy junctions and net acceptor concentration of 2-5 $\times$ 10^{17} cm^{-3}. Also the 1.85 eV PL increased with increasing resistivity of the host crystal up to 10^2 ohm on the highest resistivity evaluated. Thus, the beryllium-doped crystals do not appear to require as large amounts of nitrogen as boron-doped crystals to make efficient phosphors. Ref, (22) showed that at room temperature red (1.85 eV) PL is generated in p-type phosphors and yellow (2.10 eV) PL is generated in n-type phosphors. Ref. (22) also suggested that beryllium-coactivator nearest neighbor complexes might be the luminescence centers but also advanced the possibility that donor-acceptor triplets could be formed. Ref. (18) mentioned that the red EL had superlinear dependence on current up to 0.1 A/cm^2.

Recent investigations reported in Ref. (24) show that oxygen plays an important role in the blue PL of n-type crystals at 77°K. Oxygen diffusion annealing of nitrogen-doped crystals with weak blue PL at 77°K increased the blue PL by several factors of 10. The source of oxygen was variously CO_2 or SiO_2. Evidently oxygen is a donor since diffusion of oxygen into n-type crystals produced a layer of higher conductivity than the crystal. Oxygen diffusion annealing a p-type aluminum-doped crystal for 3 hours at 2200°C produced an n-type layer 1 μm thick. Cathodoluminescence spectra (CL) at 77°K of diffusion annealed and epitaxial layers doped with nitrogen and oxygen taken from Ref. (24) are shown in Fig. 5. Authors attribute the differences in the two spectra to the greater amounts of oxygen in the epitaxial layer not to differences in nitrogen which had little effect.

Since most SiC crystals used in research to date contain oxygen in unknown amounts, it is impossible to tell how much oxygen has influenced the properties of SiC phosphors previously attributed to nitrogen, aluminum, boron and beryllium.

Oxygen doping has been pursued in Refs. (25) and (26) where the goal was to improve the efficiency of phosphors doped with nitrogen and boron. Up to three times improvement was found in the yellow EL of treated versus untreated crystals.

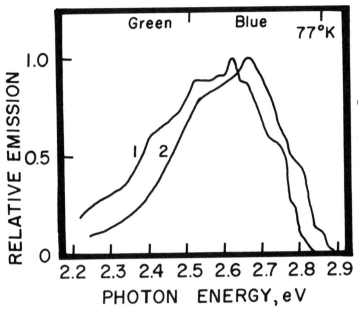

FIG. 5

Cathodoluminescence spectra of 6H SiC crystals doped with nitrogen and oxygen. Sample (1) was diffusion annealed. Sample (2) is an epitaxy layer.

Specifically Ref. (25) states 80-100 nits at 0.2 A/cm^2. Ref. (26) reports that oxygen promotes PL in p-type phosphors doped with nitrogen, aluminum and boron and further asserts that strong EL could be obtained from such p-type layers. The authors also suggest that aluminum increases the solubility of oxygen in SiC.

Alternative Approaches to Blue Diode Construction

We have seen that in 6H SiC blue light is generated from radiative recombinations in phosphors containing nitrogen, aluminum and their combinations. Also the light-producing efficiency is increased by doping with oxygen. The green light-emitting phosphors could be promoted to blue light-emitting phosphors if the same dopants were employed in the 4H polytype, for example, the N-Al donor acceptor pair spectrum, the nitrogen and gallium doped phosphors Ref. (17), and the nitrogen and scandium doped phosphors Ref. (27). The yellow boron luminescence of 6H is shifted only into the green in 4H Ref. (28). It is by no means certain that a phosphor system when employed in a 4H crystal and which emitted bright green light in 6H will outperform existing blue emitting 6H phosphors. The spectral luminous efficiency (SLE) of blue light compared to green light works against it. For light in the green part of the spectrum beyond 2.4 eV an increase in photon energy of 0.2 eV results in a 75 to 80% loss of SLE. Because of SLE considerations it will be important to choose phosphor compositions that emit as much as possible in the low energy side of the blue

region of the spectrum. The low energy boundary of the blue region at 2.52 eV
is rated 190 lumens per watt while the high energy boundary at 2.76 eV is rated
30 lumens per watt. Compositional variations of the Al-N-O doping in 6H SiC
produce a variety of phosphors with substantial radiation in the range from
2.30 to 2.85 eV but the phosphor and injection efficiencies that might be
attainable with these compositions in LED structures are not at all well known.
Consequently optimization of structures would have to be mostly empirical.

While other more efficient radiative recombination centers might be found
in SiC, the shallow isoelectronic traps that have been so beneficial to gallium
phosphide and gallium arsenide phosphide are not expected to perform well in
SiC. See Ref. (1).

Crystal perfection and purity considerations would seem to strongly favor
epitaxial deposition of the phosphor layers. Both vapor and liquid epitaxy
methods have been successfully applied to SiC and are known to be feasible at
temperatures 800°C below those used in the Lely process. Ref. (29). At these
lower temperatures growth occurs with lower dislocation densities and diffusion
is effectively stopped for the essential impurities. Consequently compositions
of p- and n-sides of junctions can be independently controlled to optimize
phosphor and injection efficiency.

Summary and Conclusions

We have reviewed the factors that determine the quantum efficiency of LED's
and searched the literature for quantitative information about the performance
of silicon carbide phosphors and LED's. Feasibility of a blue light-emitting
SiC diode was continually explored because SiC appears to be well suited for
blue LED's.

It is concluded that there are sufficient undeveloped potentialities for
SiC in blue LED's to warrant further research in that area.

Acknowledgment

The author wishes to thank R. M. Potter for his council and help in the
preparation of this paper.

References

1. A. A. Bergh and P. J. Dean, Proc. IEEE 60, 156, Feb. (1972).

2. R. W. Brander, Rev. of Phys. in Tech. 3, 145 (1972).

3. J. M. Ralston, J. Appl. Phys. 44, 2635 (1973).

4. R. M. Potter, J. M. Blank, and A. Addamiano, J. Appl. Phys. 40, 2253 (1969).

5. J. H. DeBoer and W. C. van Geel, Physica 2, 186 (1935).

6. J. A. Lely and F. A. Kroger, "Semiconductors and Phosphors," Proceedings of International Colloquium at Garmisch-Partenkirchen, p. 525. Interscience Publishers, Inc., New York (1958).

7. R. M. Potter, J. Appl. Phys. 43, 721 (1972).

8. R. W. Brander, Brit. J. Appl. Phys. 2, 309 (1969).

9. D. R. Hamilton, W. J. Choyke, and L. Patrick, Phys. Rev. 131, 127 (1963).

10. A. A. Vasenkov, I. I. Kruglov, V. I. Pavlichenko, I. V. Ryzhikov, V. P. Sushkov, IEEE Journal of Solid State Circuits, SC-4(6) December (1969) p. 421.

11. V. I. Pavlichenko, I. V. Ryzhikov, Yu. M. Suleimanov, and Yu. M. Shvaidak, Fiz. Tver. Tela 10, 2801 (1968) ⌈Soviet Phys.-Solid State 10, 2205 (1969)⌋.

12. I. S. Gorban, Yu. A. Marazuev, and Yu. M. Suleimanov, Fiz. Tekh. Poluprov. 1, 612 (1967) [Soviet Phys.-Semiconductors 1, 514 (1967)].

13. I. S. Gorban, G. N. Mishinova, and Yu. M. Suleimanov, Fiz. Tver. Tela 7, 3694 (1965) [Soviet Phys.-Solid State 7, 2991 (1966)].

14. Yu. S. Krasnov, T. G. Kmita, I. V. Ryzhikov, V. I. Pavlichenko, O. T. Sergeev, and Yu. M. Suleimanov, Fiz. Tver. Tela 10, 1140 (1968). [Soviet Phys.-Solid State 10, 905 (1968)].

15. V. I. Pavlichenko, I. V. Ryzhikov, Yu. M. Suleimanov, and Yu. M. Shvaidak, Fiz. Tver. Tela 10, 2801 (1968) [Soviet Phys.-Solid State 10, 2205 (1969)].

16. V. I. Pavlichenko and I. V. Ryzhikov, Fiz. Tver. Tela 10, 3737 (1968) [Soviet Phys.-Solid State 10, 2977 (1969)].

17. G. F. Kholuyanov, Fiz. Tver. Tela 7, 3241 (1965) [Soviet Phys.-Solid State 7, 2620 (1966)]

18. A. A. Kal'nin, V. V. Pasynkov, Yu. M. Tairov, and D. A. Yas'kov, Fiz. Elektron-Dyrochnykh Perekhodov Poluprov. Prib. p. 75 (1969).

19. V. I. Sokolov, V. V. Makarov, E. N. Makhov, G. A. Lomakina, and Yu. A. Vodakov, Fiz. Tver. Tela 10, 3022 (1968) [Soviet Phys.-Solid State 10, 2383 (1968)].

20. A. A. Kal'nin, V. V. Pasynkov, Yu. M. Tairov, and D. A. Yas'kov, Fiz. Tekh. Poluprov. 1, 484 (1967) [Soviet Phys.-Semiconductors 1, 401 (1967)].

21. A. A. Kal'nin, Yu. M. Tairov, and D. A. Yas'kov, Fiz. Tver. Tela 8, 948 (1966) [Soviet Phys.-Solid State 8, 755 (1966)].

22. A. A. Kal'nin, V. V. Pasynkov, Yu. M. Tairov, and D. A. Yas'kov, Fiz. Tver. Tela 8, 2982 (1966) [Soviet Phys.-Solid State 8, 2381 (1967)].

23. E. Ye. Violin and Yu. M. Tairov, Abstract No. 62, This Conference.

24. G. F. Kholuyanov, Yu. A. Vodakov, E. E. Violin, G. A. Lomakina, and E. N. Mokhov, Fiz. Tekh. Poluprov. 5, 39 (1971) [Soviet Phys.- Semiconductors 5, 32 (1971)].

25. E. E. Violin, Yu. M. Tairov, and O. A. Fayans, Fiz. Tekh. Poluprov. 6, 2301 (1972) [Soviet Phys.-Semiconductors 6, 1941 (1972)].

26. Yu. A. Vodakov, G. F. Kholuyanov and E. N. Mokhov, Fiz. Tekh. Poluprov. 5, 1615 (1971) [Soviet Phys.-Semiconductors 5, 1409 (1972)].

27. Kh. Vakher and Yu. M. Tairov, Fiz. Tver. Tela 11, 2440 (1969) [Soviet Phys.-Solid State 11, 1972 (1970)].

28. A. Addamiano, J. Electrochem. Soc. 111, 1294 (1964).

29. R. W. Brander, Abstract No. 1, This Conference.

Light-Emitting Devices Based on Silicon Carbide

E. E. Violin and Yu. M. Tairov

Silicon carbide electroluminescent (EL) devices are of great practical
interest owing to a number of specific features of SiC. First of all, on the
basis of this material it is in principal possible to obtain EL devices with any
colour of luminescence in the visible region, up to violet. (The energy of violet
radiation quantum amounts to $\sim$3 eV. This exactly corresponds to the energy gap
of the most common silicon carbide polytype - 6H). On the other hand, the extremely
high physical-chemical stability of silicon carbide is universally known. It may
be recalled that it is practically unaffected by either hot acids (with the
exception of orthophosphoric) or by alkaline water solutions. It can be decomposed
only by molten salts and certain metals (1,2). A noticeable interaction with oxygen
begins only at $\sim$900°C. It is known, as well (3) that impurity diffusion coefficients
in silicon carbide are relatively small. All this provides good grounds for constructing highly-stable semiconducting devices based on silicon carbide. In
particular, the practical absence of degradation in electroluminescent SiC devices
should not be surprising.

A well-known phenomenon of the existence of polytypes in SiC permits, in principle, the control of the luminescence colour of SiC EL devices. The differences in
the energy gap of polytypes permit a change in the energy of radiated light quanta
by 0.8 eV without changing the technique of dense production. The success achieved
in recent years in controlled growth of silicon carbide of different polytypes
(4,5) makes it really possible to control, in this way, the colour of luminescence
of SiC EL sources of light.

Moreover, the insertion of different luminescence activators into silicon
carbide of one polytype permits obtaining devices with different colour of

luminescence. Silicon carbide electroluminescent devices with red, yellow, green and blue luminescence can be produced using one polytype by doping with beryllium (6), boron (7), scandium (8), and aluminum (9), respectively (Figure 1).

Silicon carbide EL devices can be produced by different semiconductor techniques i.e. by alloying epitaxy, during the process of growing, and by diffusion. Beryllium can be introduced into silicon carbide by different methods. The greatest content of Be in silicon carbide is obtained when crystals are grown from Be-SiC or Be-Cr-SiC solutions (10) by increasing the temperature of growth from 1700°C to 2000°C, the solubility of Be in SiC is greatly increased when large concentrations of beryllium are introduced, the crystal acquires a dark-blue tint. Liquid-phase epitaxy SiC (Be) from a solution of SiC (Be) on the n-type substrate results in the formation of EL-p-n junction, which are located in the substrate due to the rapid diffusion of Be. These are uniformly luminescent in the plane parallel to the face $(000\bar{1})$. p-n junctions are formed even when growing p-SiC (Be) crystals by sublimation, if during the final stage of the process a small amount of nitrogen is added into the furnace. At present, using p-SiC (Be), alloyed light-emitting diodes with the red radiation can be produced.

Doping of silicon carbide with scandium can be accomplished during crystal growth by the traveling solvent method from solutions using a temperature gradient, and from the gas phase by sublimation. When growing from the solution, scandium is introduced into a molten zone either from the vapour or by placing a thin platelet of pure scandium between two crystals of silicon carbide. Different materials, among them Cr, Si, Dy and others, are used as solvents. SiC grown from the solution possesses n-type conductance. This evidently is caused by the high solubility of nitrogen, a small amount of which is always present in the atmosphere of the growth furnace, as well as in starting SiC charge. When growing SiC (Sc) by sublimation, one succeeds in obtaining both n-type and p-type crystal. SiC (Sc) EL p-n junctions are obtained mainly by epitaxial deposition of n-SiC (Sc) layers on the p-SiC substrate doped with B or Al.

Doping of SiC with aluminum is carried out predominantly in the process of forming p-n junctions by the method of alloying diffusion or epitaxial growth (9). Diffused SiC (Al) EL p-n junctions are produced by diffusing pure Al from the vapour phase at 2000 to 2100°C for 4 to 10 hours. These relatively strict conditions of diffusion are due to a small value of the diffusion coefficient of Al into SiC. In the process of epitaxial growth on n-SiC substrates one may succeed in producing p-SiC layers, 100 to 200μ thick, singly doped with Al.

Boron may be introduced into silicon carbide in the process of crystal growth, by epitaxial deposition of layers from the vapour phase and from solutions, as well as by diffusion. However, because of a number of advantages EL SiC (B) p-n junctions are principally formed by diffusion. Boron diffusion into n-SiC crystals is carried out at 1800 to 2000°C for 30 to 60 minutes. These conditions were selected due to a relatively large diffusion coefficient of boron into SiC, as compared to Al. A relatively low solubility of boron in the SiC lattice and high activation energy result in the formation of a sufficiently high resistance p-region of p-n junction (the resistivity of p-SiC layer amounts to about 10^3 ohm cm). This property of p-layer combined with its thinness in microns, ensures the isolation of luminescent regions formed in a single p-n junction. The simplicity of production of SiC (B) EL devices with various configurations of luminescence is responsible for their more extensive application at present as compared to other SiC EL devices.

Two groups of electro-luminescent devices based on silicon carbide are of the greatest interest: a) for the visual display of data output from small-size computers, measuring instruments and other electronic devices; b) for data recording on photo-sensitive materials. The first group of devices includes single light-emitting diodes, digital and character indicators, devices with controlled geometry of light field, EL dials for analog computers, and so on. The principal advantage of the above devices, in addition to their high stability, is the small value of operating currents. This permits their use in microelectronic circuits. Devices built on one single crystal may possess luminescent areas of up to 50 to 80 mm^2. The brightness of luminescence amounts of 50 to 100 nits, with the current density of 1 ma/mm^2 and operating voltage 2.5 to 3 volts. For example, digital seven-segment indicators with digit dimensions of 5 x 3 mm^2 and luminescent segment 0.4 mm thick, consume no more than 2 mA for all seven segments, with luminescence brightness of 20 nits, and 7 to 10 ma, with luminescence brightness of 100 nits. The large dimensions of the luminescent area guarantee a reliable signal vizualization even with a relatively low brightness of 20 nits. A device for coordinate determination according to "Loran C" system, having such indicators, is shown in Figure 2.

The devices with the controlled geometry of the light field are described in (11). The simplest model of the device is represented by the EL p-n junction, in which the n-region is equipotential, a relatively high-resistance p-region supplied with two electrodes acts as a voltage divider (Figure 3). When the current

passes through contacts 1 and 2, the p-n jucntion has uniform luminescent over
all its area. When the current passes through contacts 2 and 3 simultaneously,
there appears a voltage drop on the voltage divider. In this case a part of p-n
junction may be under a voltage which is smaller then required for passing the
current through the p-n junction. A smooth variation of current through the
divider (a control current) results in a corresponding displacement of the
boundary between luminescent and non-luminescent parts of the crystal. The current sensitivity of such a device is 1.5 mm/mA with 4 x 1 mm^2 dimensions. The
maximum control current is 3 mA. The contrast of the crystal light and dark
field boundary is sufficient for the amplitude resolution of the signal with
the accuracy of 10 per cent.

Such an EL p-n junction with controlled luminescence boundary can also be
used for recording signals on photographic film (12). In this case the film
blackening density is uniform and the signal recorded will be an envelope of the
exposed region.

For data recording on light-sensitive materials in a digital form, a digital
EL indicator may be constructed which allows one to synthesize any three-digit
numbers. The dimensions of each digit presented are 1.25 x 0.75 mm^2. The clearance
between adjacent elements of the digit is 50µm. In order to increase the density
of data recorded it is desirable to continue decreasing the dimensions of a
luminescent digit (or character). This in turn requires the computation of a
minimum allowable clearance between adjacent elements (x_o). The minimum value
of the clearance must satisfy the condition that the voltage at the boundary of
the element adjoining the luminescent one should not exceed 2 volts (i.e. the
voltage at which no radiation is yet observed from SiC p-n junctions). In
Figure 4 one may see a cross-section, in a diagrammatic form, of a p-n junction
passing through two luminescent elements and the voltage distribution over the
p-region when one of the elements is switched on. Assuming the resistivity of a
p-layer equal to ρ we may write down:

$$du = -\rho \, \frac{dx}{\delta b} \cdot i(x) \qquad (1)$$

$$di = -b \cdot dx \, j(u) \qquad (2)$$

where j(u) is the current density through the p-n- junction
(voltage-current characteristic).

By differentiating (1) and then substituting (2) we shall receive:

$$\frac{d^2 u}{dx^2} = \frac{\rho}{\delta} \cdot j(u) \qquad (3)$$

The voltage-current characteristic of silicon carbide p-n junctions formed by
boron diffusion into n-SiC, in a wide range of current densities, may be represented
with a function of the type (13): $j = A(u - u_1)^4$. Substituting the expression
for the current density into (3) and integrating within the limits when the voltage
varies from the operating voltage on the element (U_{oper}) to the voltage at which
no luminescence of p-n junction is observed ($u_o = 2V$), we receive the expression
for the minimum allowable distance between the elements (x_o):

$$x_o = \frac{\frac{2}{3}\left(\sqrt{\frac{1}{(u_o - u_1)^3}} - \sqrt{\frac{1}{(u_{oper} - u_1)^3}}\right)}{\sqrt{\frac{0.4\,\rho\,A}{\delta}}} \qquad (4)$$

Substituting into (4) the values for the p-region resistivity ($\rho = 10^3$ ohm cm),
p-layer thickness ($\delta \approx 1\mu m$) as well as the values computed for real light-emitting
diodes $u_1 = 1.43$ V and $A = 0.215$ A/cm^2 V^4, one can obtain the values of x_o for
different operating voltages. Since when recording the data on the photographic
film relatively high operating voltages are required to increase the brightness
and therefore the speed of recording, expression (4) can be simplified by neglecting the second term in the numerator. In this case the minimum allowable value
of the clearance between the elements is 17μm. Thus, the exciting technique of
manufacturing silicon carbide character indicators allows one to further diminish
the character dimensions by the factor of 2-3 without changing the relation between the elements and the clearance. This permits an increase in the photographic
recording density.

For the photographic recording of coded data, for example, in binary system,
silicon carbide light-emitting diodes with the light output from the edge of p-n
junction are used. The construction of such a device is shown diagrammatically
in Figure 3. Due to a short diffusion length of current carriers in silicon
carbide, the width of the luminescent strip measured in the direction normal to
the plane of the p-n junction is about two microns. A light-emitting diode line
of such a construction makes it possible to perform recording on a photographic
film, possessing the sensitivity of 250 A.S.A., in 2mK sec time with 1A current.
In this case the optical density of the film blackening exceeds unity.

When constructing similar light-emitting diodes based on silicon carbide, a sublinear dependence of light flux (ϕ) on current density should be taken into consideration with sufficiently great current densities. It is known that such a dependence is described by the expression of the form.

$$\phi = C \times j^n$$

where C is the proportionality coefficient

  n – the exponent equal to about 0.5 for the majority of specimens.

   The existence of such a dependence leads to the necessity of selecting the optimal contact length (1 in Figure 5). Thus when the contact length is too long, a part of the light generated under the contact is dissipated in the crystal through absorption (the absorption coefficient (1) of the light of corresponding wavelength is about 20 cm^{-1}). With the decrease in the contact length the loss of light through absorption is decreased. However, the current density through the p-n junction is increased (if the value of the current is maintained constant) and, in accordance with the above expression, the quantum output of p-n junction radiation drops. The computations show that the maximum light flux from the crystal with the specified current is obtained if the contact length satisfies the equation:

$$\exp{(-1k)} \; (1k + 0.5) - 0.5 = 0$$

The solution of this equation by numerical method gives the value of the product 1k equal to 1.26. If k = 20 cm^{-1} the optimal contact length should be 0.63 mm.

   One of the most important characteristics of the device for data recording on photosensitive materials consists in the rate of EL rise and decay. The relaxation times for SiC(E) EL sources is considered in detail in (14). The rise time of EL for such sources is 10 to 30μ sec, and the decay time – 0.1μ sec. The EL rise time is decreased to a few μ sec, when the current density is increased up to 100 a/cm^2. The EL rise and decay time for SiC (Be) fused sources in $\sim$0.25μ sec, and for SiC (Se) p-n junction by one order of magnitude greater (15). The relaxation times for SiC (Al) fused sources amount to a few μ sec which permits one to make high-speed EL radiation sources based on them.

   A brief discussion of SiC EL devices and their properties in this paper shows that various types of such devices with parameters enabling their practical applications have already been developed and are being manufactured. Further studies of the electroluminescence of silicon carbide doped with different impurities will permit both the improvement of parameters of already developed SiC electro luminscent devices and the creation of new types of them.

REFERENCES

(1) J. W. Faust, Jr. in "Silicon Carbide", (ed. J. R. O'Connor and J. Smiltens),
       New York; Pergamon Press, 1960, pg. 403.

(2) V. J. Jennings, Mat. Res. Bull., v. 4, S199 (1969).

(3) Ye. N. Mokhov, Yu. A. Vodakov, G. A. Lomakina, Fiz. Tverd Tela, $\underline{11}$, 519 (1969).

(4) Kh. Vakhner, Yu. M. Tairov, Fiz. Tverd Tela, $\underline{12}$, 1543 (1970).

(5) W. F. Knippenberg, G. Verspui, Phil Res. Repts., $\underline{21}$, 113 (1964).

(6) A. A. Kal'nin, Yu. M. Tairov, D. A. Yas'kov, Fiz. Tverd. Tela, $\underline{8}$, 948 (1966).

(7) E. E. Violin, G. F. Kholuyanov, Fiz. Tverd Tela, $\underline{6}$, 1696 (1964).

(8) Kh. Vakhner, Yu. M. Tairov, Fiz. Tverd. Tela, $\underline{11}$, 2440 (1969).

(9) V. A. Migunov, G. F. Kholuyanov, Izvestijy LETI, $\underline{80}$, 14 (1969).

(10) A. A. Kal'nin, V. V. Pasynkov, Yu. M. Tairov, D. A. Yas'kov, Fiz. Tekh.
       Polupr., $\underline{1}$, 484 (1967).

(11) A. A. Kal'nin, Yu. M. Tairov, D. A. Yas'kov, Pribory Tekh. Eksp, No. 2,
       233 (1969).

(12) V. A. Barkov, A. A. Kal'nin, L. M. Kogel, Yu. M. Tairov, Prib. Tekh, Ekspet.,
       No. 5, 229 (1970).

(13) I. V. Ryjikov, V. I. Pavlichenko, T. G. Kmita, P. M. Karageorgi-Alkalayev,
       Radiotekhn, and Electronics (USSR), $\underline{12}$, 842 (1967).

(14) E. E. Violin, G. F. Kholuyanov, Fiz. Tverd. Tela, $\underline{8}$, 3395 (1966).

(15) A. A. Kal'nin, V. V. Pasynkov, Yu. M. Tairov, D. A. Yas'kov, Fiz. Tekh,
       Polupr., $\underline{4}$, 1467 (1970).

(16) V. I. Pavlichenko, I. V. Ryjikov, "Phys. of p-n Junctions and Semicond,
       Devices", USSR, "Nauka", Leningrad, p. 326 (1969).

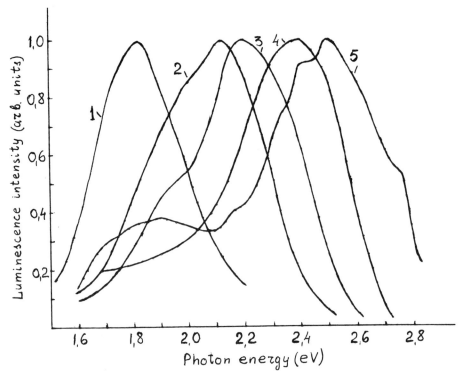

Figure 1.   Luminescence spectra of silicon carbide crystals activated with various impurities.

1.   6H – SiC(Be),   2.   6H – SiC(B),   3.   6H – SiC(Sc),
4.   4H – SiC(Sc),   5.   6H – SiC(Al).

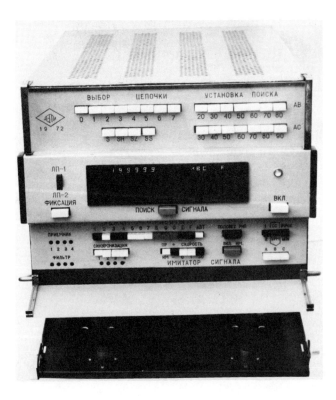

Figure 2.   Appearance of the device having silicon carbide indicators for the determination of the object coordinates by the "Loran C" system.

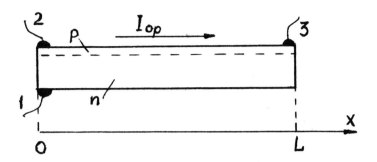

Figure 3.   Model of the device with the displaceable boundary of the luminescent
            part.

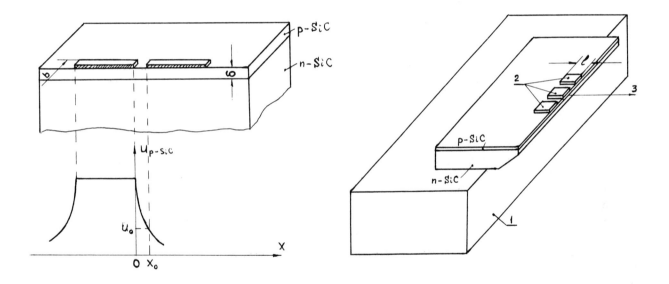

Figure 4.   Electron-hole junction
with two radiating elements and voltage
distribution over p-region when one of
the elements is switched on.

Figure 5.   Light-emitting diodes with
the light output from the p-n junction
edge.

1.   contact to n-region;

2.   contact to p-region;

3.   light output.

Semiconductor Light Sources from Silicon Carbide

G. F. Kholuyanov and Yu. A. Vodakov

The promise offered by a number of luminescence activating dopants in SiC is discussed for the fabrication of injection light sources. The luminescence in B-doped SiC is considered in the most detail in conjunction with the preparation on its basis of electroluminescent B-activated p-n junctions. Comparison is given between SiC-LED's and GaP-LED's.

Inspite of great advances achieved in the development of light-emitting diodes (LED's) on the basis of III-V compounds, silicon carbide still remains the only material from which LED's can be made that have outstanding long-time stability of radiation as well as current-surge and thermal stability.

By reason of its structure, the possibility is exluded for SiC for obtaining efficient radiations due to band-to-band recombinations or to annihilation of free excitons. Even at high excitation levels the efficiency of radiation identified for SiC as being due to excitons localized at neutral or charged impurities falls to $\leq 10^{-4}$% upon approaching room temperatures. The same limitation is also inherent in those kinds of luminescence in SiC that are due to defects arising, say, upon irradiating SiC with high-energy nuclear particles or implanting into it ions of a number of light elements.

Of much greater interest from the view-point of LED application are luminescences occurring upon doping SiC with such acceptors as Al, Ga, Be, Sc, and B. Table 1 shows some physicochemical features of these impurities.

It is seen from the table that of all the known acceptors Al has the lowest thermal activation energy in SiC, combined with high solubility limit. These properties are favorable to the formation of low-resistivity p-type layers in LED's. However, very low diffusion coefficients of this impurity involve

difficulties in the preparation of
electroluminescent (EL) p-n junctions by
the diffusion method. This disadvantage
is characteristic of Ga and Sc to a still
greater extent. Relatively high solubilities and diffusion coefficients are
best combined in B and Be. The diffusion coefficients of these impurities

Table I

	Solubility limit (cm^{-3})	Diffusivity (cm^2/s)		Activation energy at low concentrations (eV)	Surface solubility (cm^{-3})
		at 1700°C	at 2300°C		
Al	10^{21}	8×10^{-15}	1×10^{-11}	0.27	1.5×10^{18}
Ga	3×10^{18}	6×10^{-15}	3×10^{-12}	0.25	5×10^{17}
Be	$(5-7) \times 10^{17}$ (at 1700°C)* $(5-7) \times 10^{19}$ (at 2300°C)	8×10^{-10}	1×10^{-7}	0.43, 0.62	$(2-3) \times 10^{18}$ at 2300°C
Sc	$3 \cdot 10^{17}$		10^{11}	0.5	
B	3×10^{19} (at 1700°C) 1.5×10^{20} (at 2300°C)	1.5×10^{-12}	5×10^{-10}	0.4	$(2-6) \times 10^{18}$

* With impurities introduced by diffusion.

in SiC are substantially higher than those of other acceptors, which fact facili-
tates doping SiC with these acceptors by diffusion. At the same time, with
decreasing temperatures the diffusion coefficients decrease to the extent that
they allow us to get rid of instability effects in LED's caused by the migration
of these impurities in the p-n junction, such effects being observed for some
acceptors in III-V compounds. Limitations of B and Be used as acceptors are their
high activation energies, which will be discussed further. By suitably doping the
most common 6H-SiC polytype with the above acceptors, luminescence bands can be
obtained which cover practically the whole visible spectral range, thus providing
raditors ranging from the blue to the red light.

Al and Ga in 6H-SiC activate luminescence bands in the blue and green spectral
ranges, the radiation maxima lying in the range 2.4 to 2.67 eV. Unfortunately,
these kinds of luminescence have their high efficiencies, reaching 15 to 20% in
SiC-LED's, at low temperatures only. Above 77°K the luminescence efficiency drastically falls off with increasing temperature and at ∿300°K is situated at a level
of <10^{-3}%. Such a low-temperature extinction of Al- and Ga-activated luminescence
is accounted for by donors with relatively low activation energies present at the
luminescent centers.

Much more efficient luminescence at room and higher temperatures occurs in
SiC doped with Be, Sc, or B. Figure 1 shows the typical luminescence spectra
(for 6H-SiC at 20°C) of these impurities. The radiation light ranges from the
red-orange for p-SiC(Be) to the yellow green for n-SiC(Sc).

Several authors studied the possibilities of using Be, Sc, and B for the
fabrication of SiC-LED's. Using Be-doped p-SiC grown by the Lely method, LED's
were manufactured by alloying Si in an atmosphere containing N$_2$ to obtain n-type
layers (1). The LED's radiated in the red-orange spectral range due to the injection of electrons into the luminescent p-type region. Beryllium-activated
SiC-LED's provided no gain in radiation efficiency compared to LED's based on

B-doped SiC. Practical realization of Be-activated SiC LED's presents difficulties
due to poor reproducibility of the properties of the alloyed p-n junctions as well
as to failure to obtain large and shaped radiating areas by alloying. These LED's
also feature large voltage drops due to relatively high resistivities of the
starting p-type material and large compensated regions near the p-n junction.
Also, the n-type layers are very difficult to realize by diffusion because of very
low diffusion coefficients of the known donors in SiC.

Beryllium introduced into nitrogen-doped ($\sim 10^{18}$ cm^{-3}) 6H-SiC crystals gives rise to yellow radiation (2) which is very similar in spectrum and behavior to B-activated luminescence (BL) (Figure 1). At 20°C the efficiency of the corresponding type of Be-activated luminescence may be only slightly inferior to BL and its special maximum is found to be displaced several hundredths of the electron-volt toward shorter wavelengths with respect to that of BL. All this together with its acceptable diffusion coefficient at

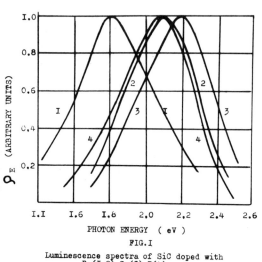

FIG.I

Luminescence spectra of SiC doped with
Be(I,2),Sc(3),B(4).

temperatures 200 to 300°C less than those used in the manufacture of B-activated
LED's makes Be attractive for obtaining a new type of SiC-LED. Unfortunately,
practical realization of this type of LED using as simple a process as that
developed for B-activated LED's got into difficulties for the following reasons.
Beryllium, when introduced by diffusion into SiC, exhibits rather a strong temperature dependence of surface solubility (Table 1), so that at temperatures of 1400
to 1700°C it is difficult to form by diffusion a p-type layer on SiC doped to the
required level. At higher temperatures the distribution of Be is such that introducing it into SiC gives rise to rather extensive compensated regions near the
p-n junction, these LED's having as a result elevated working voltages. All the
above does not allow p-n junctions with structures best suited to LED's to be
formed by the diffusion introduction of beryllium alone. These difficulties may
be bypassed through the use of multistaged epitaxial diffusion techniques as well
as probably ion implantation followed by short-time dispersion of Be upon diffu-
sion-annealing.

Sc, as a luminescence activator (3), is attractive from the following view-points. The spectral band of Sc-activated luminescence is in best agreement with the sensitivity curve of the human eye. The efficiencies of photo- or cathodoluminescence (PL and CL, respectively) achieved on best samples are not inferior to those of BL. Also, despite the relatively low solubility limit of Sc in SiC, the dependence of radiation I_{CL} on excitation level I, e.g. when exciting luminescence by an electron beam, remains linear in those current ranges where the brightness of B- or Be-activated luminescence becomes a sublinear function of excitation. However, realization of LED's using Sc-activated luminescence is rather difficult. The relatively low solubility and slow diffusion rate of Sc does not allow a p-n junction to be formed by the diffusion of Sc alone. Besides, it has been possible to obtain Sc-doped SiC with good luminescence parameters only through the melt growth of SiC layers (in this way luminescent n-SiC partially compensated with Sc is obtained). Unfortunately, such layers generally contain inclusions of another phase in the form of aggregates of scandium silicides and carbides (8). Creation of p-n junctions with reproducible properties and small leakage currents in these layers, the efficient Sc-activated luminescence being preserved, is still an unresolved problem.

Thus, B, as a luminescence activator (4), still retains its leading role in the fabrication of SiC-LED's. Therefore, some factors affecting the efficiency of LED's using BL are below discussed in more detail.

We studied two kinds of samples. One kind was made up of single crystals grown by the Lely method (L samples). The Hall concentration of uncompensated donors N_d-N_a of these crystals lay within the range 10^{17} to 7×10^{18} cm^{-3}, which was ensured by different levels of doping with nitrogen. It follows from the temperature dependencies of the Hall effect and electrical conductivity that these samples contained uncontrolled acceptors amounting to $(2-5) \times 10^{17}$ cm^{-3}. The other kind of test samples are constituted by 15 to 50 μm thick 6H-SiC single-crystal epitaxial layers (E samples) prepared on 6H-SiC substrates at substantially lower temperatures (1700 to 2000°C) than in the Lely method. The concentration of N_d-N_a in the E samples lay within the range 4×10^{17} to 8×10^{18} cm^{-3}. Luminescent layers were created by the usual B diffusion. B was introduced into some epitaxial layers also when growing them. Also samples were annealed in gaseous media containing N and 0 donors. Luminescence was studied upon exciting it with an electron beam having relatively small penetration depths (with energies < 10 to 15 keV). This type of excitation made it possible a) to study radiation in thin layers, b) to obtain sufficient resolution when studying microsections

of the p-n junctions, c) to easily vary the excitation level over a wide range,
providing also excitation levels comparable to those of EL. The main results of
our experiments may be summarized as follows.

1. B introduced (e.g. by diffusion) into initially non-luminescent crystals
participate simultaneously in the formation of two types of luminescent centers,
that of low-temperature orange luminescence with a maximum at ~1.7 eV (LTBL) and
that of high-temperature yellow B-activated luminescence (BL). From this it follows that a portion of the B atoms introduced into a crystal is unfortunately spent
to form LTBL centers which are of no interest for the use in LED's operating at
room temperatures. The minimum B concentration N_B, starting from which luminescence may be identified as LTBL is at least by an order of magnitude lower than
that for BL.

Analysis of a great number of B-doped samples with different impurity concentrations shows that within the range of N_B from ~10^{17} to ~10^{18} cm^{-3} a superlinear
increase of BL occurs with increasing N_B. This may be associated, among other
things, with that portion of the B atoms which is spent to form LTBL centers.

2. The effect of N_B on the dependence $I_{CL} = f(I)$ was studied on microsections made at 0.5° to the p-n junctions. Figure 2 shows the curves $I_{CL} = f(I)$
taken at 20°C with the same electron-beam area in increasing order of N_B : 1 - near
the boundary of the luminescent layer where $N_B \simeq 10^{17}$ cm^{-2}; 2 - at the boundary
of the p-n junction where $N_B \simeq 2 \times 10^{18}$ cm^{-3}; 3 - in the intermediate region.
For clarity, all curves have been normalized to the minimum I. With increasing
N_B, the region of transition to the sublinear section of $I_{CL} = f(I)$ is clearly
seen to shift toward higher I's. Because of these dependencies of $I_{CL} = f(I)$,
an increase in the BL brightness with increasing N_B from ~10^{18} cm^{-3} shows up more
vividly at relatively high I's.

3. The B diffusion was carried out simultaneously for L and E crystals with
different values of N_d-N_a, so as to preserve the n-type conduction and the value
of N_B in the samples compared. With increasing N_d-N_a, the slope of the curves
$I_{CL} = f(I)$ increased, the efficiency of BL tending to decrease with decreasing I
for N_d-N_a comprised between 10^{18} and 10^{19} cm^{-3}. Regularity with which this occurred was particularly evident, when comparing BL in E samples grown in one experiment on the silicon and the carbon face of a 6H-SiC substrate. Due to the
polar character of SiC, the donor solubility appears to be higher on the carbon
face and so does N_d-N_a as compared to the silicon face. The curves $I_{CL} = f(I)$
for the BL of one of the pairs of the epitaxial layers being compared are shown

in Figure 3. The initial values of N_d-N_a of the layers on the carbon and the silicon face are 3.5×10^{18} cm^{-3} (curve 1) and 1.5×10^{18} cm^{-3} (curve 2) respectively.

4. No p-type B-doped L or E sample showed BL comparable in brightness to the BL in the n-type crystals. The same is characteristic of the p-type layers formed in n-SiC by the diffusion of B at 1500 to 2000°C. Exceptions were the p-type layers adjacent to the p-n junction which were obtained by diffusion at 2300 to 2400°C.

5. The observed regularities given in 1 to 4 for BL show up also in LED's obtained by the diffusion of B. A decrease in N_d-N_a in crystals which is always accompanied by a decrease in N_B

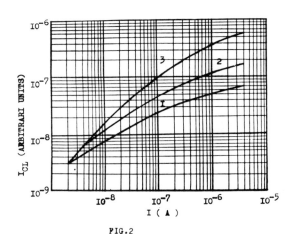

FIG.2

Intensity of B-activated CL vs.excitation level for vatious B concentrations (1,2,3).

in the vicinity of the p-n junction results in a decrease in current densities at which transition from the linear to the sublinear section of the dependence $I_{CL} = f(I)$ occurs. This means that to provide higher radiation brightnesses with larger current densities samples with as large values of N_d-N_a as possible should be used

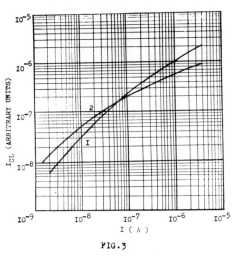

FIG.3

Intensity of BL vs. excitation level for different concentrations of uncompensated donors.

1 - 3.5 10^{18} cm^{-3}
2 - 1.5 10^{18} cm^{-3}

under otherwise equal conditions. However, raising N_d-N_a up to 10^{19} cm^{-3} proves to be unreasonable because of the difficulties involved in obtaining p-n junctions with satisfactory electrical properties. Besides the regularities depicted by Fig. 3 are also characteristic of LED's with relatively high N_d-N_a, so that these devices would be less efficient at relatively small current densities. Thus, the starting n-SiC material intended for the fabrication of LED's should be differently doped, depending on the required brightness and working current densities. At working current densities of ~ 0.1 A/cm^2, the optimum impurity concentration is situated about 10^{18} cm^{-3}.

6. The comparison of the BL of the p-n junctions in L and E crystals has shown: a) E crystals provide, on the average, higher BL brightnesses, b) the BL of E crystals is better reproducible from sample to sample in both specific brightness and working voltage. This suggests that crystals show promises for LED's, the more so as E crystals are substantially easier to obtain than L crystals. Nevertheless, the use of E crystals in LED's does not eliminate the sublinearity of the curves $I_{CL} = f(I)$ at relatively large current densities, which is a drawback of BL.

7. Apart from the sublinearity, a serious limitation of B-activated p-n junctions is that B forming the p-type region has a high activation energy (see Table I). As a result, the p-type layer has a high resistivity in the working temperature range (5) and does not provide satisfactory injection of holes into the luminescent n-type layer. This situation is generally typical of those kinds of SiC-LED's in which the luminescent material is of the n-type. It is therewith difficult, if not impossible, to provide effective injection by simple means. The activation energies of the known acceptors in SiC are several times as high as that of N, the principal donor impurity. Even Al, that is rather a shallow acceptor, cannot provide acceptable injection of holes into a n-type layer except that one succeeds in so introducing it that the Al concentration in the p-type layer exceeds that of excessive nitrogen N_d-N_a in the n-type region of the p-n junction by at least 2-3 orders of magnitude. With a doping level of N_d-N_a of 10^{18} cm^{-3}, which is typical of SiC(B)-LED's, doping with Al should be above 10^{20} cm^{-3}. Such a high doping level in the immediate vicinity of the p-n junction (this is required by small diffusion lengths) easily gives rise to various kinds of defects, including current leakages through tunneling, thus deteriorating the quality of LED's. The other known acceptor impurities, including B (see Table I), can provide no acceptable injection at all into n-SiC at the required temperatures.

Therefore, attention has been focused on finding an acceptable method of introducing Al into the p-type layer of B-activated p-n junctions. So far, the method of separate diffusion (6) has been proved to be the most efficient, enabling a sufficiently doped p-type layer to be obtained by the diffusion of Al alone and a luminescent n-type layer near the p-n junction to be created by the diffusion of B in n-SiC. The potentialities of the separate diffusion method may be extended by carrying out the diffusion of Al in an atmosphere containing a donor, oxygen (7). As this takes place, an activated region under the heavily doped p-type layer is formed, in which the BL is found to be 2-3 times more efficient than in LED's not subjected to such a process. The diffusion conditions must be so chosen that the p-n junction remains in the activated region and is located as close to the heavily doped p-type layer as possible. Also, the luminescent n-type region near the p-n junction should not have too low a resistivity.

Figure 4 shows the specific nit-ampere characteristics of SiC(B)-LED's obtained using this process (see region between curves 1 and 2) and those of one of the best samples (curve 3) compared to high-quality GaP(ZnO)-LED's (curve 4) and GaP(N)-LED's (curve 5) with efficiencies of 3% and 0.1%, respectively, while Table II shows some data of SiC(B)-LED's. All data refer to flat GaP- and SiC-LED's which are neither covered with special antireflection coating nor sealed.

The above data show that with working current densities below 0.1 A/cm² SiC(B)-LED's are superior to the comprehensively studied and well developed GaP-LED's for brightnesses up to 100 nits. Further improvement in the properties of SiC(B)-LED's especially in the range of large working current densities and brightnesses will be achieved by increasing the injection efficiency and decreasing the sublinearity of nit-ampere characteristics, since according to our findings and to (9) the efficiency of PL or CL with low excitation levels seems to be sufficiently high, reaching tens of %.

Unfortunately, the nature of the centers of BL has not yet been studied to the extent that the cause of the observed sublinearity may be convincingly indicated. However, if BL is compared to the behavior of Sc-activated luminescence with different excitation levels (taking into account their close internal efficiencies with low excitation levels), the two following facts may be concluded to be responsible for the observed sublinearity:

Table II

Working area	Working voltage	Maximum wavelength	Current density through p-n junction (A/cm²)									
			10^{-3}		10^{-2}		10^{-1}		1		10	
			Specific brightness	Efficiency	Specific brightness	Efficiency	Specific brightness	Efficiency	Specific brightness	Efficiency	Specific brightness	Efficiency
m²	V	nm	$\frac{nit \cdot cm^2}{A}$	%	$\frac{nit \cdot cm^2}{A}$	%	$\frac{nit \cdot cm^2}{A}$	%	$\frac{nit \cdot cm^2}{A}$	%	$\frac{nit \cdot cm^2}{A}$	%
3×10^{-8} to 5×10^{-5}	2.3 to 3	585	-	-	1000	0.04	500	0.02	200	0.01	80	-
up to 2×10^{-5}	2.2 to 2.6	580	4000	0.16	4000	0.16	1600	0.07	430	0.02	110	

Upper line: typical SiC(B)-LED, lower line: best SiC(B)-LED.

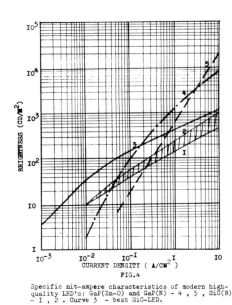

Specific nit-ampere characteristics of modern high-
quality LED's: GaP(Zn-O) and GaP(N) - 4 , 5 , SiC(B)
- I , 2 . Curve 3 - best SiC-LED.

a) the concentration of the centers of BL is much lower than 10^{17} cm^{-3}, the concentration of the centers of Sc-activated luminescence being not higher than 10^{17} cm^{-3}, since the solubility of Sc is equal to 3×10^{17} cm^{-3}, b) the centers of BL are such that the probability of nonradiative recombination through the boron center itself drastically increases with higher excitation levels. If the former cause is valid, it will be necessary to find out ways and means for increasing the concentration of BL centers. The latter cause is unlikely to be eliminated by technological methods.

The efficiency of the injection of impurities into SiC may be raised in various ways. The potentialities have not yet been exhaused for designing better p-n junction structures, with special reference to the use of epitaxial diffusion techniques and ion implantation. However, the creation of efficient luminescence in p-SiC seems to have the greatest promise. Following this way, hopes are pinned on rather efficient BL which has been initiated in the p-type material. Also promising are Sc-, Be-, and Ga-activated luminescences whose efficiency in p-SiC may be rather high. Mastering the controlled growth of 4H-SiC and developing Sc-activated LED's on its basis will enable the maximum of the EL of SiC to be shifted far in the green-and-blue spectral range.

The authors would like to thank G. A. Lomakina, E. N. Mokhov, V. I. Sokolova, and S. I. Taits for their laboratory help and valuable suggestions.

REFERENCES

(1)   A. A. Kalnin, V. V. Passynkov, Yu. M. Tairov, and D. A. Yas'kov, Fizika p-n-Perekhodov i Poluprovodnikovykh Priborov (The Physics of p-n Junctions and of Semiconductor Devices), Nauka Publishers, Leningrad (1969).

(2)   V. I. Sokolov, V. V. Makarov, E. N. Mokhov, G. A. Lomakina, and Yu. A. Vodakov, Fiz. Tverd. Tela, 10, 3022 (1968).

(3)   Kh. Vakhner (H. Wachner) and Yu. M. Tairov, Fiz. Tverd. Tela, 11, 2440 (1969).

(4)   A. Addamiano, R. M. Potter, and V. Osarov, J. Electrochem. Soc., 110, 517 (1963).

(5) E. E. Violin and G. F. Kholuyanov, Fiz. Tverd. Tela, <u>6</u>, 1969 (1964).

(6) V. I. Pavlichenko and I. V. Ryzhikov, Fiz. Tekhn. Polupr., <u>4</u>, 2409 (1970).

(7) Yu. A.Vodakov, G. F. Kholuyanov, and E. N. Mokhov, Fiz. Tekhn. Polupr., <u>5</u>, 1615 (1971).

(8) G. F. Yuldashev, M. M. Usmanova, and Yu. A. Vodkov, Atomn, Energ., <u>32</u>, 592 (1972).

(9) R. M. Potter, J. Appl. Phys., <u>43</u>, 721 (1972).

Silicon Carbide Cold Cathodes

R. W. Brander

The electron emission from SiC p-n junctions is reviewed and its merits contrasted with other cold cathodes. The potential areas of application of the different devices is indicated.

For many years, almost since the invention of the valve, there has been a desire to replace the thermionic cathode with an electron emitter capable of operating at room temperature. This desire was unlikely to be fulfilled as long as the hot cathode was capable of meeting the vast majority of device needs and as long as our understanding of emission processes was relatively elementary. In recent years the limitations of thermionic emitters have begun to retard advancement in a number of fields and this had lead to a general increase in interest in producing new cathodes, not necessarily with any hope of replacing the thermionic emitter but primarily intended to compliment it in more specialist applications for, with all its drawbacks, the thermionic cathode is at least inexpensive. Some of the potential advantages offered by a cold cathode are simplified valve and equipment construction, robustness, long life, contamination resistance, low power consumption, high current densities, fast response, narrow energy distribution and reduced noise. Not all cold cathodes will fulfil all requirements and the device application will determine which cathode is most suited.

Extracting electrons from unheated solids can be accomplished either by activating the electrons to high energies compared with their electron affinities, by reducing the surface work function, or both. The former effect can be achieved by avalanching, tunnelling, injection or optical excitation while the latter is produced by coating the surface of the material with BaO,

Cs or CsO or by field lowering. Cathodes based on all these effects have been produced.

REVERSE BIASED P-N JUNCTION EMITTERS

Avalanche multiplication can be used to generate hot electrons by applying a reverse bias to a p-n junction. Minority carriers generated within a diffusion length of the high field region being accelerated until they gain sufficient energy to cause ionization of electrons and holes. The threshold for ionization, E_i, is very approximately 1.5 E_g and, hence, if the band gap, Eg, is sufficiently large, electrons with energies greater than the surface work function, ϕ, will be produced. If these electrons are generated within a diffusion length of the surface it will be possible for them to escape from the solid. Silicon, with its well developed planar technology, would be the ideal choice of material for this cathode, however, with E_i < 2eV and ϕ > 3.5eV the best emission obtained has been 10^{-12} Acm^{-2} although by lowering the work function with a BaO coating 10^{-6} Acm^{-2} were obtained[1]. α-SiC with its higher energy gap and better thermal conductivity and stability has proved more successful, but the shorter diffusion length of hot carriers has limited the emission to regions where the junction intersects the surface. The most successful emitters have been prepared by ultrasonically cutting mesas in planar p-n junctions produced by epitaxial growth, emission then occurs from points distributed round the circumference of the island[2]. The shape of the emitting area can be selected to suit requirements and areas of the periphery can be prevented from emitting with thin layers of silicon oxide. Multiple emitting regions can also be constructed to give larger areas of emission or to provide a number of independently modulatable emitters suitable for multi-beam devices[3].

Before emission can be obtained from these cathodes an activation procedure has to be carried out in vacuum[4]. This consists of either warming the crystal to $500/600^0$C for a few seconds or alternatively applying a reverse bias to the device for a longer period. It is likely that the avalanche current (a few mAs) which flows during this period causes local heating and that the activation mechanisms are similar. During this treatment the reverse current increases and the emission current begins to flow (Fig 1). Emission levels of 50μA per millimeter of periphery have been obtained, giving beam currents in excess of 50μA into a 0.5μm diameter spot on a CRT. On re-admission of ambient air to the

system the characteristic reverts back to its original form, while further
pumping and activation will regain the emission. If the ambient introduced to
the vacuum system after activation is a dry inert gas, emission is regained
immediately on subsequent evacuation without reactivation being necessary and
it would appear that moisture is the prime cause of loss of emission. Operation in water vapour pressures of $>10^{-6}$ torr result in the complete suppression
of emission. After activation the devices can be stored for many months in dry
gas and emission immediately gained on pumping down. In addition devices can
be stored in organic vapours, eg hexane, without deterioration and even operation in these vapours is possible without adverse effects (D R Lamb, private
communication). Devices operated in vacuum have given stable emission

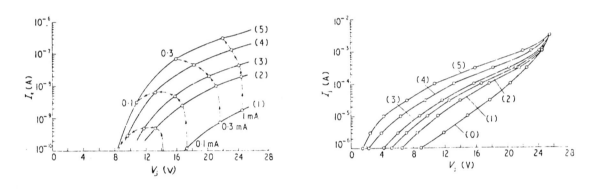

Fig 1   Characteristics of a SiC cathode after successive stages
of activation. I_j, junction current; I_e, emission current;
V_j, junction voltage. (After Bellau, Chanter and Dargan, Ref 3)

characteristics for periods exceeding two years, the limit of these tests.
Retarding field measurements of the energy distribution of the emitted electrons
although showing a larger spread of energies than from a thermionic emitter
indicate that the SiC cathode has a much sharper high energy cut-off. This and
the partial collimation of the electrons by the peripheral junction field,
resulting in a small transverse electron velocity, allows the electrons to be
readily focussed to a sharp point. The presence of the junction also causes a
reduction in emission current by directing some of the emitted electrons back
into the crystal; hence a strong extraction field is desirable for maximum
efficiency. The best values so far reported have been 1mA/watt corresponding
to 100μA emission for a 5mA, 20V input. By isolating individual spots,
emission current densities greater than 1A cm^{-2} can be achieved.

Two possible mechanisms for electron emission from SiC junctions have been proposed but no experimental evidence is yet available to indicate conclusively which mechanism predominates. Many of the observed results can be explained on the basis of electron acceleration and multiplication in the reverse biased field and the subsequent emission of the hot electrons over the surface work function. In particular, the electron emission originating from spots at the junction periphery resembles the blue light-emitting microplasma regions known to be due to avalanche breakdown and sometimes, but not always, correlates with these spots. In addition, the desirability of a damaged (ultrasonically cut) surface for optimum emission could indicate that surface breakdown was necessary; etched surfaces giving much poorer results. The relative emission capabilities

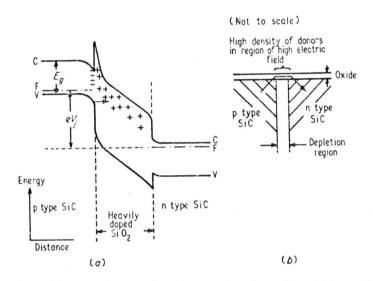

Fig 2   An alternative emission mechanism for SiC cathodes; (a) the energy diagram through the current path shown in (b). (After Widdowson and Rose, Ref 5)

of silicon and silicon carbide, which correspond to their relative energy gaps and electron affinities, also tend to support this theory. An alternative mechanism has recently been proposed which could explain many of the phenomena which are incompatible with the above mechanism particularly the remarkable stability of the emission in ambients which would not be expected to give stable and reproducable surfaces[5]. This mechanism postulates the presence of a thin carbon-doped silicon oxide layer as the emitting region, this covering the junction and being formed at near room temperature so that the carbon is not oxidised. Activation at 500/600°C causes the carbon to become electrically active providing donor electrons and giving rise to the excess conductivity

observed in Fig 1. Electrons are injected into the layer either from the
microplasmas in the SiC or by tunnelling as shown in Fig 2. Since SiO_2 has a
relatively low work function, 0.9eV, emission is readily obtained by acceleration of the electrons in the film. This theory explains the substantially
different shapes of the bulk (unactivated) and surface (activated) currents in
Fig 1; the remarkable stability of the emission in vacua as low as 10^{-5} torr
and the poisoning effect of water vapour which can be attributed to the
oxidation and subsequent removal of the carbon donors.

OTHER COLD CATHODES

Field Emitters[6]

When electric fields in excess of $10^6 Vcm^{-1}$ are applied to a surface the
vacuum energy level is reduced by bending and electrons can tunnel through the
surface barrier into vacuum. In order to achieve the high fields, the voltage
must be applied at a sharp point, and whiskers, needles or electrolytically
sharpened wires with μm diameter tips are employed. Emissions of $100\mu A$ per point
can be achieved, giving current densities of greater than $10^7 Acm^{-2}$, higher than
that available from a thermionic cathode. The electrons have a large angular
spread and the beam has to be trimmed to a few μAs for most device applications,
enabling focussed spots of $0.1\mu m$ to be achieved. The device has to be operated
in ultrahigh vacuum, less than 10^{-10} torr, in order to achieve reasonable
stability and operating life, although SiC and carbon tips have been operated
satisfactorily at 10^{-7} torr[7]. Degradation occurs due to heating, particle
bombardment, chemical reaction, evaporation and adsorbtion; all affecting either
the shape of the tip or the work function. Large total currents can be obtained
using matrices of tips, eg chemically deposited or eutectically grown needles,
the main problem being to obtain a uniform field at each tip.

Metal-Insulator-Metal Emitters[8]

These simple thin film sandwich structures offered the greatest hope of low
cost cold cathodes, but unfortunately have been the least successful of the
devices discussed. Two modes of operation are possible dependent on the
thickness of the film; tunnelling for films of $50\overset{o}{A}$ or less, and field excitation
and hopping in thicker films. Tunnelling films are extremely difficult to produce without flaws and most work has been undertaken on thicker structures.
Emission currents of $200\mu A$ cm^{-2} with efficiencies of $25\mu A/watt$ for $200\overset{o}{A}$ films
to $500\mu A/watt$ for $1000\overset{o}{A}$ films have been achieved but uniformity in the latter

case was very poor. As in the SiC device, a forming process is required and is
believed to be related to the diffusion of electrode atoms into the insulator
to form donor levels. Electrons are emitted through the top electrode which is
thin and discontinuous, resulting in easy damage. Operational life is in fact
very short due to the burning out of the electrode and destructive breakdown in
the insulator. The emitted electron beam exhibits considerable noise and fluctuation as well as having a large energy and directional spread, the latter due
to scattering at the exit surface.

Negative Electron Affinity Emitters

When a p-type semiconductor is coated with a thin layer of material with a
low work function the energy bands at the surface are bent and electrons,
injected into the conduction band, will have an energy above that of the vacuum
level, escaping from the solid if within a diffusion length of the surface.
Photocathodes and secondary electron emitters have been constructed on this
principle. For direct electrical excitation p-n junctions are used to inject
electrons into the p-type material. The p-layer must be thin enough to minimise
recombination but thick enough to give conductivity over the surface of the
emitter. In addition the band gap must be sufficiently large to separate the
conduction band from the vacuum level by an amount which is insensitive to
fluctuations in vacuum level due to contamination. Silicon[9] and 3-5 compounds
have been used successfully for these devices with the 3-5 ternaries, GaAsP[10]
and GaAlAs[11], giving the more reproducable and practical results. The prime
advantage of the device is the low energy spread of the emitted electrons
(160meV compared with 230meV for a thermionic cathode) and in particular the
sharpness of the high energy cut-off. Emission currents of 1A cm^{-2} with
efficiencies of 5mA $watt^{-1}$ can be readily obtained, the main problem being
emission stability during life in ambients other than UHV, owing to the sensitive nature of the surface. Metal-Schottky barrier contacts can be used in a
similar manner but the yield of devices with the very thin metal layers
essential for electron emission is likely to be low.

The excitation of photocathodes, either of the conventional or negative
electron affinity type, with a suitable independent source of light has been used
to provide cold cathodes. The main problem is the efficient matching of the
system and the bulky nature of the device. With GaAlAs technology the light
emitter and the photocathode can be produced in one block and efficient coupling
achieved[11]. The emission mechanism is similar to that for other negative

electron affinity cathodes but frabication is simplified by eliminating current spreading problems.

CONCLUSIONS

Cold cathodes can now be fabricated commercially by a number of different techniques and it is likely that substantial improvements in performance will be obtained from all these structures with further development. The merits of each particular device governs its field of application and Table 1 summarises the properties of each device in accordance with the advantages predicted for cold cathodes at the beginning of this article. All the cathodes are more difficult to construct than thin metal filament emitters but very little more difficult than the more specialised oxide cathodes. The cost difference will be negligible when the devices are assembled in a system since certain features necessary for thermionic emitters, such as thermal and optical shielding and heater transformers, can be eliminated. When more than one electron beam is required in a system, cold emitters, since they can be fabricated in a matrix with identical emission characteristics, offer considerable advantages. For high power requirements the thermionic cathode is still supreme; the advantage of the cold cathode being its high efficiency at modest emission currents which render it particularly suitable for portable battery operated equipment. In all other areas the cold cathode shows the substantial benefits predicted, particularly in robustness, life, fast response (10ns or less), sharper energy cut-off and lower noise in the case of the negative electron affinity devices.

With these features in mind the application of the different devices can be discussed. Field emitters are ideal where small focussed spots and relatively low emission levels and high current densities are required. Their surface sensitivity, however, restricts their use to UHV ambients and therefore continuously pumped systems such as electron microscopes. The short life and instability of thin film cathodes renders them unsuitable at present for any application. The negative electron affinity cathodes and the silicon carbide cathode would appear to have the greatest potential. They both have many common advantages but it is the few features of difference which indicate the areas of application. The low energy spread of the negative electron affinity devices make them ideal for camera tubes where they should give a substantial reduction in beam discharge lag. They may also find application in other devices where low noise beams are required and good vacuums can be provided. Silicon carbide's main potential is its extreme stability in contaminated ambients which render it suitable for use in ionization gauges[12] and mass spectrometers as well as

POTENTIAL ADVANTAGES	FIELD EMITTER	REV. BIAS JUNCTION	THIN FILM	NEGATIVE AFFINITY
SIMPLER CONSTRUCTION - cathode	-	X	-	X
- no thermal shielding	✓	✓	✓	✓
- no radiation shielding	✓	✓	✓	X
DEMOUNTABLE	✓	✓	✓	X
RESISTANCE TO CONTAMINATION	X	✓	✓	X
ROBUSTNESS	-	✓	✓	✓
LONG LIFE	X	✓	X	?
LOW POWER CONSUMPTION	X	✓	✓	✓
HIGH EMISSION CURRENT DENSITIES	✓	X	X	X
FAST RESPONSE	✓	✓	✓	✓
NARROW ENERGY DISTRIBUTION	✓	X	X	✓
LOW NOISE	X	-	X	✓

Table 1 The potential advantages of cold cathodes compared with the properties of a thermionic emitter. ✓ better, X worse, and - similar

cathode ray tubes and demountable systems. SiC emitters have been shown capable of emitting electrons into hexane and other organic vapours without deterioration and have even been found to inject substantial currents into polar liquids such as nitrobenzene[13] indicating their potential in previously unconsidered areas.

ACKNOWLEDGMENT

The author would like to thank R V Bellau, C L Dargan and A E Widdowson, G.E.C. Hirst Research Centre, and A W Bright and D R Lamb, Southampton University for allowing him to quote unpublished information.

592   R. W. Brander

REFERENCES

(1) R J Hodgkinson; Solid State Electronics, 5, p269-272, (1962)

(2) R W Brander, A Todkill; Materials Research Bull. 4, p5303-10 (1969)

(3) R V Bellau, R A Chanter, C L Dargan; J. Phys. D. 4, p2022-30 (1971)

(4) R V Bellau, A E Widdowson; J. Phys. D. 5, p656-666, (1972)

(5) A E Widdowson, F W G Rose; J. Phys. C. 6 p437-449 (1973)

(6) R Gomer; Field Emission and Field Ionization, Oxford Univ. Press (1961)

(7) F Baker; Nature 225, p539-40 (1970)

(8) G Dittmer; Thin Solid Films 9, p141-172 (1972)

(9) E S Kohn; IEEE Trans, ED-20, p321-9 (1973)

(10) P J Deasley, K R Faulkner; Proc. 5th Int. Photoelectric Imaging Symp.,
     London, p459-67 (1971)

(11) H Kressel, H Schade, H Nelson; J. Lumin. 7, p141-161 (1973)

(12) E E Windsor; Vacuum 20, p7-9 (1970)

(13) A W Bright, J King; Advances in Static Electricity, Vol. 1, p276-287 (1970)

Thin Films of α and β Silicon Carbide Prepared by Liquid Epitaxy and by Sputtering

C. E. Ryan, I. Berman and R. C. Marshall

Abstract

The preparation of thin films of alpha silicon carbide by a liquid epitaxy process is discussed. Silicon is used as a solvent at 2250°C. The silicon carbide substrate is wetted by molten silicon saturated with carbon. As the silicon evaporates, the liquid layer supersaturates precipitating carbon to form an epitaxed SiC layer on the substrate. Experiments on the wettability of silicon carbide by molten silicon vs. temperature and the time and pressure dependence of the process are discussed. The layers formed by liquid epitaxy are used as substrates for sputter deposited beta layers which are briefly described.

I. Introduction: One of the good features of silicon carbide is that junctions can be prepared by a wide variety of techniques. In this conference we have heard about junctions and thin films prepared by vapor growth, C.V.D., ion implantation, solution, diffusion and liquid epitaxy techniques. In this paper I will be talking primarily about film preparation by one form of liquid epitaxy.

The preparation of thin films of semiconductors by liquid epitaxy methods has been widely and successfully developed during the past decade. In most instances one of the constituents of the compound, like gallium in GaAs, GaP and $Al_xGa_{1-x}As$, is used as the solvent. Since gallium has a low melting point, a high boiling point and does not appreciably react with quartz or carbon, it makes tipping, dipping and sliding techniques quite convenient for the preparation of epitaxial and heteroepitaxial layers with well controlled properties[1,2,3].

To apply liquid epitaxy techniques to silicon carbide presents some difficulties as can be seen from the Plate One which is the well known solubility

curve of Scace and Slack presented long ago at the first Silicon Carbide Conference[4]. In order to get a solubility of about one atomic percent of carbon in
silicon, it is necessary to raise the temperature above 2100°C. The higher
temperature required to increase the solubility causes the vapor pressure to become quite excessive leading to rapid loss of the solvent. One approach to overcome this difficulty has been the use of additives in the melt to increase the
carbon solubility at lower temperatures[5,6].

In this investigation silicon has been used as the solvent at temperatures
above 2000°C. The procedure and apparatus which were used are closely patterned
after the patent disclosure by Sanjiv Kamath[7]. We are also indebted to Dr.
Kamath and to Dr. Al Miller for discussions of their experiments. We believe,
however, that our growth mechanism is somewaht different than Kamath's and that
it may have some implications on silicon carbide growth mechanisms.

II. Experimental Procedure:

Plate 2 is a schematic of the construction of the furnace showing the R.F.
Coil driven by a 30KW 450KC generator, the heater shields, the susceptor, and
the crucible whose volume is 2.5cm^3. The pedestal on which the substrate is
placed is shown at the center of the crucible. The silicon charge is placed
around the pedestal.

Plate 3 shows the individual graphite components of the furnace assembly.

Plate 4 shows the fully assembled furnace.

Plate 5 shows the furnace in operation.

A typical deposit was made as follows: an alpha substrate of proper doping was
lapped on the rougher surface to 1/4 micron. The substrate was then etched in
KOH/Na$_2$O$_2$ at 450°- 500°C for one minute. After being thoroughly washed and
dried, the substrate was placed on the pedestal in the crucible. Two hundred
milligrams of pure silicon chips were put into the crucible surrounding the
pedestal. The pedestal was high enough so that the silicon was not in contact
with the substrate.

The furnace was assembled and positioned in a quartz jacket. The system
was evacuated to the 10^{-3} torr and then backfilled with helium. The helium was
allowed to flow through the furnace assembly continuously at the rate of one
liter per minute. After 15 minutes of flushing with helium the furnace temperature was brought to 2250°C for 5 minutes. The temperatures cited are corrected
pyrometer readings. A 10°C temperature differential existed between the top and

bottom of the crucible with the bottom being hotter.

As the temperature of the crucible increased the silicon melted, wetted and climbed the pedestal and covered the silicon carbide substrate with a molten silicon layer which was saturated with carbon at the temperature of the furnace, i.e., approximately 1% carbon in solution. As the silicon evaporated, the molten layer became supersaturated and alpha silicon carbide was deposited and epitaxed on the surface. After 5 minutes the excess silicon had all evaporated, reacted with the crucible and substrate to form SiC, or had escaped from the crucible around the lid. The growth cycle was then complete. After disassembling the crucible, the alpha SiC sample was readily lifted from the graphite pedestal. This was very convenient because not only was the substrate with film readily available without further processing, but the crucible assembly could be reused.

The distinction between this technique and the normal Kamath technique was simply as follows:

In the normal Kamath technique the SiC layer was grown on the bottom side of the crystal from a saturated layer of carbon in molten silicon in a temperature gradient. Hence it was essentially a traveling solvent technique using silicon as the molten zone and the carbon derived from the carbon pedestal. In the present technique the layer on the top surface grew because the evaporation of silicon from the liquid silicon layer saturated with carbon caused supersaturation and precipitation of carbon in the presence of silicon to epitax silicon carbide on the surface.

Plate 6 is a 100X magnification of a surface as deposited by this technique.

Plate 7 is an electron diffraction pattern showing that the deposit is single crystal (6H) alpha silicon carbide.

Plate 8 shows a fractured edge of the substrate and film. In this case the deposited film is 25 micrometers thick.

In order to obtain further information on the nature of this process, a few qualitative experiments were made on the following:

(a) The wettability of SiC by molten silicon versus temperature.

(b) The evolution of the process as a function of time at a given temperature.

(c) The effects of pressure in retarding the evaporation of the silicon charge.

A liquid epitaxy process is strongly influenced by surface preparation of the substrate and the "wettability" of the surface by the solvent. To gain insight into the mechanism of how molten silicon wetted silicon carbide vs. temperature, the following series of experiments was conducted. The crucible was prepared by processing 200 mg of silicon in a normal cycle, thus coating the walls and pedestal with SiC. A substrate was then placed on the pedestal and a cube (10^{-3} inches on a side) of silicon was placed on the substrate. The furnace was then heated at temperature for 1 minute and allowed to cool. This experiment was repeated for 1500°C, 1600°C, 1700°C, 1800°C, 1900°C, 2000°C and 2100°C. The top portion of Plate 9 shows the results. It can be seen that with increasing temperature, the silicon melt became less viscous and wetted the surface adjacent to the silicon charge. As the silicon cooled it retracted to a hemispherical ball. As the temperature is raised the surface wetting becomes greater but is not too uniform. By 2100°C the entire surface is wetted by the silicon and the excess silicon has evaporated before it could cool sufficiently to form a hemisphere. To improve the surface conditions a layer of 1000°A of silicon was sputtering onto the SiC surfaces and the experiments were repeated (Plate 9 - lower portion). A rather dramatic improvement in uniformity of wetting by the silicon which acts as a surface etch can be seen. On cooling, the silicon coalesced into several balls as the liquid layer contracted due to increased surface tension of the silicon at the lower temperatures. These silicon hemispheres are very reminiscent of similar phenomena that sometimes occur on vapor grown crystals and strongly indicate a layer of molten silicon in certain crystal growth processes. These experiments also indicate that a proper surface cleaning followed by a sputter etch and a sputter deposition of a thin silicon layer would be quite advantageous. This thin layer would act as a protective layer and a surface etch in much the same way as the nickel intermediate layers acted in the C.V.D. process described by Berman in the 1968 Conference[8].

Another series of experiments was performed to observe the evolution of the epitaxial process as a function of time. It would be advantageous if one could stop the process after 1 minute, look at the results and then continue the process a second minute and so on. This cannot be done on the same crucible because the molten silicon seals the crucible until the cycle is essentially complete, i.e., all silicon is lost or converted to SiC. Hence, five crucibles were prepared and processed identically as far as possible except for time at temperature of 2250°C which was 1, 2, 4 and 5 minutes respectively. The caps were sawed off the crucibles for examination. The results are shown on Plate 10. After one minute the silicon had wetted the sides of the crucible, the pedestal, and had covered the substrate. There was still excess silicon available in the trough around the pedestal. After two minutes most of the silicon had evaporated from the surface of the substrate leaving the newly formed silicon carbide film clearly visible. However, the caps were still sealed to the crucible as was the substrate to the pedestal. After five minutes at temperature the cap and substrate could be removed in the normal manner.

These results suggested that it would be desirable to maintain or increase the temperature to obtain improved carbon solubility and at the same time to slow down the vaporization rate so that the crystal growth rate would be somewhat lower and the control of layer thickness would be better. In order to do so the normal film experiments were repeated in a pressure furnace[9] which had a 20 atm pressure capability. While some changes in geometry and shielding were necessary, the experiments and results at one atmosphere were consistent with the previous experiments. Experiments at 20, 10, 6 and 3 atmospheres pressure at 2250°C showed that 3 atmospheres of helium was sufficient to suppress vaporization at 2250°C so that the time for taking an experiment to completion was 20 minutes as opposed to 6 minutes at 1 atm. Thus an over pressure of three atmospheres was sufficient to drastically influence the experiments as shown on Plate 11. The crucible on the left has been heated to 2250°C for 1 minute at 1 atm. The center crucible has been heated to 2250°C for 6 minutes at 3 atm. The right hand crucible has been heated to 2250°C at 3 atm for 20 minutes. We hope to extend these measurements to other temperatures and pressures to obtain greater reproducibility of the film quality and thickness.

In order to make p type deposits boron and/or aluminum was added to the silicon in the amount of about 1 gram dopant in 100 grams silicon.

These films of alpha silicon carbide prepared on alpha silicon carbide substrates were used as the alpha layer onto which beta silicon carbide was sputtered and then thermally annealed into a single crystal beta film. The procedure closely followed that presented by I. Berman[9] on the first day of this conference and will not be repeated here.

Plate 12 is an electron diffraction pattern of one of these beta surfaces before and after annealing.

Plate 13 is a cross section of an unannealed beta layer on alpha silicon carbide.   After annealing the boundary between α and β SiC layers can no longer be identified by polishing or etching techniques.

Conclusions

It is concluded that a high temperature liquid epitaxy process involving silicon as the solvent and the evaporation of silicon form a liquid layer of molten silicon saturated with carbon has been demonstrated.

It appears that control of the variables by better surface treatments such as sputtering a deposited silicon layer of the order of $1000A^{o}$ preceded by sputter etching can improve the epitaxed beta layers.

It also appears that further investigations of temperature-pressure schedules may lead to more precise control of thickness and quality of layers.

It has been shown that the liquid epitaxy process is compatible with the sputter deposition process described by Berman et al for making thin films of single crystal beta silicon carbide upon liquid dpitaxed films of alpha silicon carbide.

Acknowledgement:

The interest and support of Mr. R. M. Barrett, Director of the Solid State Sciences Laboratory of AFCRL is gratefully acknowledged. Also, acknowledgement is due to J. Comer for the electron diffraction patterns and to J. Hawley for assistance in conducting the experiments.

REFERENCES

1.   H. Nelson, RCA Review, Vol. 24, p. 603, 1963.

2.   J. M. Blum and Kwang Shih, "The Liquid Epitaxy of $Al_xGa_{1-x}As$ for
     Monolithic Planar Structures", PIRE, Vol. 59, No. 10, Oct 1971.

3.   H. T. Mindon, "A Comparison of Liquid Phase Epitaxy and Chemical Vapor
     Epitaxy of III-V Semiconductors", Solid State Technology, pp 31-38,
     Jan 1973.

4.   J. R. O'Connor and J. Smiltens, ed., "Silicon Carbide", Pergamon Press,
     pp 24-30, 1960.

5.   G. A. Wolff et al., "Principles of Solution and Traveling Solvent Growth
     of Silicon Carbide", Silicon Carbide - 1968, pp S67-71, Pergamon Press.

6.   P. W. Pellagrini and J. M. Feldman, "Solution Growth of SiC using Transi-
     tion Metal-Silicon Solvents", International Conference on Silicon
     Carbide - 17-20 Sep 73, Miami Beach, Florida.

7.   G. Sanjiv Kamath, "Silicon Carbide Junction Diode", U. S. Patent No.
     3,565,703, 23 Feb 71.

8.   I. Berman and J. J. Comer, "Heteroepitaxy of Beta SiC Employing Liquid
     Metals", International Conf. on Silicon Carbide - 1968; publ. in
     Materials Rsch. Bulletin, Vol. 4, pp S107-S118, 1969.

9.   R. C. Marshall, "High Pressure, High Temperature Crystal Growth System",
     AFCRL Report No. 67-0656 (1967).

10. I. Berman, R. C. Marshall and C. E. Ryan, "Annealing of Sputtered β-SiC",
     International Conference on Silicon Carbide - 17-20 Sep 73, Miami
     Beach, Florida.

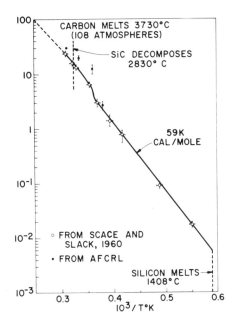

Plate 1. The solubility of Silicon in Carbon as a function of Reciprocal Temperature (after Scace and Slack).

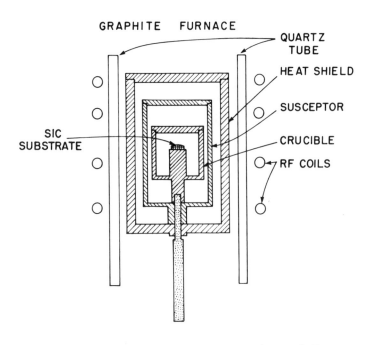

Plate 2. Schematic of Construction of Furnace.

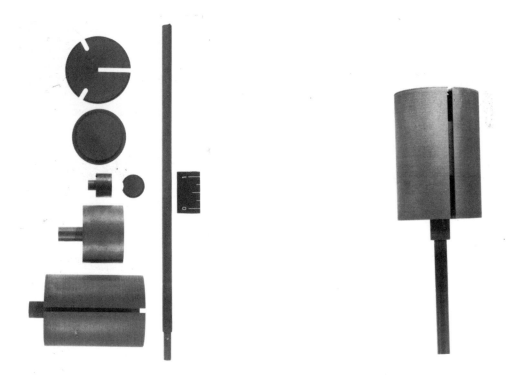

Plate 3. Individual Graphite Components of the Furance.

Plate 4. The Fully Assembled Furnace.

Plate 5.   The Furnace in Operation.

Plate 6.   100X Magnification of the
Surface of a Layer of Alpha SiC.

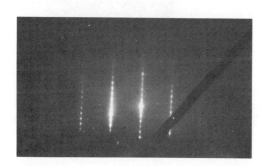

Plate 7.   Electron Diffraction
Pattern (6H-Alpha Silicon Carbide
Deposited Layer).

Plate 8.   Beveled Edge of Substrate and
Deposited Layer.

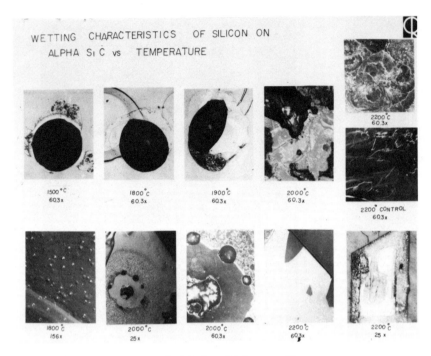

Plate 9. Wetting of Alpha SiC vs. Temperature

(a) on etched SiC surface - top row
(b) on 1000Å Silicon sputtered surface - bottom row.

Plate 12.
    Electron Diffraction Pattern of Sputter Deposited Beta SiC Layer

Plate 12a.
    before annealing

Plate 12b.
    after annealing

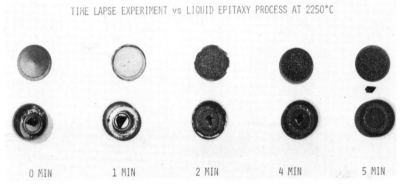

Plate 10. Liquid Epitaxy Process versus Time at Temperature of 2250°C.

Plate 11. Effects of Pressure on Suppression of Vaporization of Silicon

Plate 13. Cross Section of Unannealed Beta Layer on Alpha Substrate.

Annealing Behavior of Neutron Irradiated β-SiC

Hiroshige Suzuki and Takayoshi Iseki

1. Introduction

N.F. Pravdyuk et al.[1] reported the lattice expansion studies on irradiated α-SiC and diamond in the temperature range 120–420°C, and suggested that the materials might be used as an indicator of neutron fluence at fluence below the saturation level and of irradiation temperature. Thereafter many researches on the SiC irradiation temperature monitor were reported[2-4], since SiC monitors have several advantages of small size, low residual radioactivity and so on. The lattice constants of SiC increased with the fluence of fast neutron, and a saturation of the change took place after the irradiation of 3×10^{20} n/cm²[2]. In the post-irradiation annealing, the lattice constants were unaffected until the annealing temperature exceed the irradiation temperature, above which the dimension decreased linearly with increasing temperature. The irradiation temperature was estimated as the temperature at which the recovery of the radiation-induced dimensional change began. The recovery of this change can be also determined by the measurement of macroscopic length and the linear thermal expansion. Recently an attempt of the estimation of the irradiation temperature by the annealing behavior of electrical resistanse of α-SiC was done by R.J. Price[4].

The present investigation was undertaken to see if a practical method could be found for the estimation of the irradiation temperature. Polycrystalline β-SiC was irradiated to 5.0×10^{19} n/cm² (E>0.18MeV) at temperatures of 290–500°C.

The thermal expansion measurement by high temperature X-ray diffraction and usual dilatometry, and the specific electric resistivity measurement were carried out in order to know their annealing behavior, in addition to the isochronal annealing measurement of lattice constant as a typical method of the estimation of the irradiation temperature.

2. Materials

Powder and sintered rod of β-SiC were used as specimens for irradiation. The mixture of semiconductor-grade silicon metal and spectroscopically pure graphite was heated at 1500°C for 3 hours in vacuum, then the mixture converted into β-silicon carbide powder. The powder was heated at about 800°C in air to remove unreacted graphite, and treated with hot HNO_3-HF solution for 1 hour to remove silica and unreacted silicon metal.

In the next place, the graphite and β-SiC powder were pressed into rods.

603

The reaction-sintered β-SiC was fabricated by the conversion of the rod into β-SiC by means of a siliconizing atmosphere at temperatures below 1550°C. The rod was then machined to correct size and shape, 3 mm in diameter and 15 mm long, with diamond tools. Unreacted graphite and silicon were removed similarly by the method which was used above, and further heating at 1800°C in vacuum was added in order to anneal any stress accumulated in the rod.

The rod specimen used in the present study gave X-ray diffraction pattern typical of cubic β-polytype SiC, and a small amount of graphite or silicon metal was found in some cases by microscopic observation, which was used without further purification. Apparent density of these rods was 90-95% of the theoretical, and typical impurity were as follows; Fe: 100-900 ppm, Mg, Al, Ti, Mn, Cu and Ag: less than 10 ppm respectively, and Co and Sb: less than 1 ppm respectively.

3. Irradiation
   The first irradiation was carried out in hole VT-1 of JRR-2. Fast neutron fluences were 2.4×10^{19} and 5.0×10^{19} n/cm^2. The powder which was encapsuled in vitreous silica tube and the rods were sleeved in graphite, and sealed in air in aluminum capsule thermally insulated from the reactor coolant. The specimens were settled between two thermocouples, one indicated 430±10°C and another 520 ±10°C. According to the arrangement of the specimens and thermocouples, the specimens were irradiated at 500±20°C. The recording of the temperature showed that the specimens were cooled down below 100°C within 20 minutes after shut down of the reactor.

   The second irradiation was carried out in hole VT-5 in a fast neutron fluence of 1.5×10^{17} to 7.0×10^{17} n/cm^2. The specimens were sealed in helium. The irradiation temperature was estimated at 300±30°C from earlier experiment.

   The third irradiation was carried out in the O-5 position of the JMTR (Japanese Material Testing Reactor) in a fast neutron fluence of 1.5×10^{19} n/cm^2 (E>0.18MeV). A thermocouple and specimen were spaced by NaK at a distance of 10 cm from each other, and the thermocouple indicated the temperature of 290 ±10°C.

4. Experiments

4.1. Lattice constant change in isochronal annealing
   The d-spacing of (422) reflection of β-SiC specimen was determined by the powder X-ray diffraction as usual. The (531) reflection of silicon metal was used for an internal standard of diffraction angle. The measurement of the d-spacing was carried out at 15°C between each annealing of the temperature interval of 100°C. The heat treatment was carried out in vacuum of 10^{-4} torr for 1 hour at each temperature. The X-ray measurement was repeated three times at each temperature, and the precision was ±0.00003Å or ±0.003% in Δd/d.

4.2. Thermal expansion measurement by high temperature X-ray diffraction
   The apparatus used in the experiment was made by Rigaku Denki Co., Ltd. Tungsten plate was used both as a main heater and as a sample holder. The temperature was measured by the W.Re5-26 thermocouple which was welded to the back surface of the sample holder. The measurement was done 10 minutes after the temperature became constant. The d-spacing of (311) of β-SiC was employed and (400) of silicon metal was used as an internal standard. It took 20 minutes to scan the both diffraction peaks once. The thermal expansion of silicon metal itself was corrected by the data in the literature[5];
$$\Delta\ell/\ell (\%) = 2.417 \times 10^{-4}(t-20) + 1.837 \times 10^{-7}(t-20)^2 - 5.651 \times 10^{-11}(t-20)^3$$

where t is temperature in °C. The condition of X-ray diffraction was virtually identical with the usual method. The X-ray measurement was repeated three times at every temperature, and its precision was ±0.0001Å or ±0.008% in Δd/d.

4.3. Dilatometry

The linear thermal expansion of β-SiC rod was measured from room temperature to 900°C with a precision vitreous silica dilatometer. The measurement cycled twice without taking away the specimen from vitreous silica support. After the measurement the recovery was calculated by the curves obtained in first and second runs. The measurement was carried out by heat rate of 5°C/min. and in vacuum.

4.4. Specific electric resistivity change in isochronal annealing

The electric resistivity of the specimen was measured using a four-contact circuit. Current terminals were attached by silver paint to both ends of the rod. The voltage drop between a pair of voltage probes, which was also attached by silber paint, was measured with a potentiometer. The contact between SiC and silver was ohmic in the experimental condition. The voltage drops were about 0.2-50 mV/cm. The value was practically the same in the case of reversal current direction. The isochronal annealing was done in the same manner as the lattice constant measurement by means of X-ray diffraction.

5. Results and Discussion

X-ray studies showed that the specimen was in β-polytype with no detectable α-polytype phase, and that no phase change occurred during the irradiation. After irradiation, d-spacing changed as is shown in Table 1. The change was not apparent in a fluence of less than 7.0×10^{17} n/cm^2.

Table 1. d-spacing of (422) reflection of irradiated β-SiC specimen

Reactor	Fast neutron fluence (n/cm^2)	Irradiation temperature (°C)	d-spacing*(Å) ±0.00003Å	Δd/d (%)
	Non-irradiated		0.88988	
JRR-2, VT-5	7.0×10^{17}	300**	0.88988	0.0
VT-1	2.4×10^{19}	500	0.89192	0.23
VT-1	5.0×10^{19}	500	0.89313	0.37
JMTR, 0-5	1.5×10^{19}	290	0.89119	0.15

   * measured at 15°C
  ** estimated

Recovery of radiation-induced lattice expansion of the specimen irradiated to 5.0×10^{19} n/cm^2 at 500°C by post-irradiation annealing is shown in Fig. 1. The recovery of the change began at higher temperature than the irradiation temperature, and was tending to saturate in some level by about one hour annealing.

Non-irradiated β-SiC was *p*-type semiconductor, after irradiation it changed to *n*-type. By annealing, changed to *p*-type again.

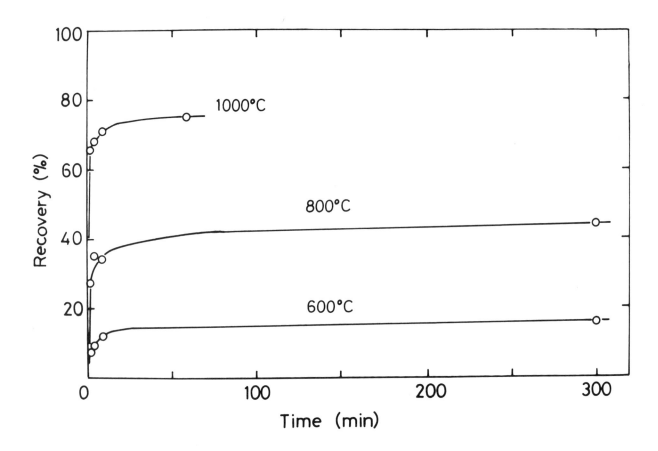

Fig. 1   Recovery of radiation-induced lattice expansion of
the specimen by post-irradiation annealing

5.1. Lattice constant change in isochronal annealing

The annealing behavior of the change is given in Fig. 2. It was found the
point of deflection of the curve indicated the irradiation temperature within
precision in the order of $\pm20°C$. The accuracy and precision can be improved by
the decreasing interval of the temperature in isochronal annealing. In the
experiment, the irradiation temperature could not be estimated unless more than
0.05% change in d-spacing or lattice constant caused. This amount of change
caused in the irradiation of more than 5×10^{18} n/cm^2 at 500°C. The results
obtained here were consistent with R.P. Thorne 's data[2].

5.2. Thermal expansion measured by high temperature X-ray diffraction

The results of the thermal expansion measured by high temperature X-ray
diffraction appear in Fig. 3. The d-spacing change of (311) of β-SiC was
written in the form of $\Delta d/d_o$, where d_o was the value of the non-irradiated
specimen at room temperature. The d-spacing of the non-irradiated specimen
increased monotonously with temperature, but on the other hand the d-spacing of
the irradiated specimen showed a crook at about the irradiation temperature.
And the change above the temperature was composed of the recovery of both
radiation damage and thermal expansion. The curve of irradiated specimen
approached gradually that of non-irradiated specimen above the irradiation
temperature. On the basis of these data, in spite of limited experiments, it

seemed reasonable to assume that the irradiation temperature could be estimated from the crook in the case of irradiation to more than 2×10^{19} n/cm^2 at 500°C.

Usually lattice parameter determination with a high degree of accuracy was done by using a high-order reflection, but in the present experiment a low-order

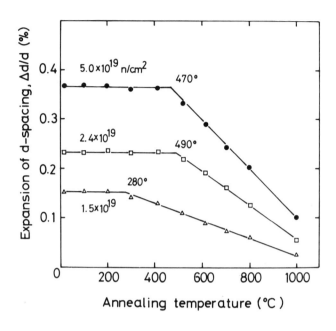

Fig. 2 Lattice constant change of irradiated specimen in isochronal annealing

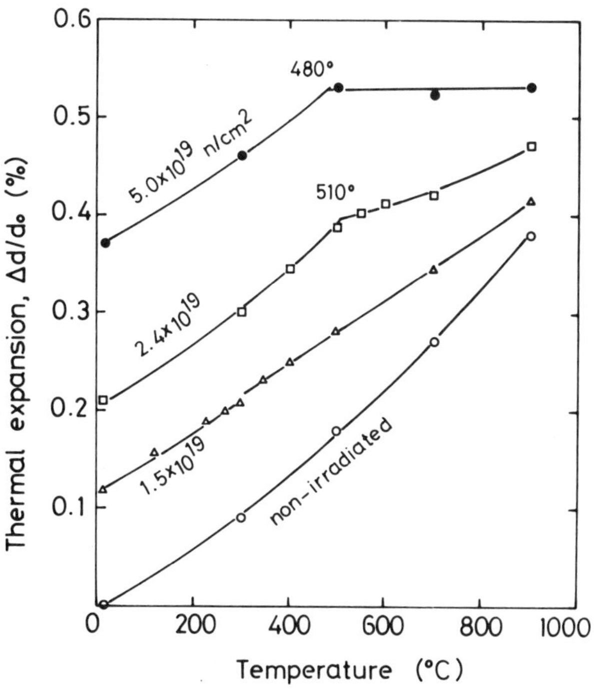

Fig. 3 Thermal expansion measured by high temperature X-ray diffraction

reflection, its diffraction angle about 70°2θ was used. That is because the diffraction intensity was small in high-order reflections, and the superposition of desired diffraction and tungsten's diffraction was occurred in some cases. Furthermore it took much time to scan the goniometer from Si's peak to SiC's peak in higher angle diffraction. As the specimen is held almost horizontally and it does not fall off after heating at 900°C, the application of high power X-ray generator, such as rotating anode type, enables to measure at higher angle and to improve an accuracy.

X-ray method has the advantage of requirement of a small amount of specimen, for example, β-SiC powder of about 10mg. Comparing with the isochronal annealing measurement, thermal expansion measurement by high temperature X-ray diffraction could reduce the time required. It is necessary for the former measurement to take the specimen in and out at every heating and cooling and to set the specimen on the sample holder of X-ray diffractometer. As for the latter experiment, it is difficult to obtain the absolute quantity of linear thermal expansion of the specimen, because the d-spacing of the internal standard changes at the same time.

5.3. Linear thermal expansion measured by means of dilatometer

The result is shown in Fig. 4. It shows the data which was calculated from the difference between first run and second run. The recovery of the specimen which was irradiated to higher fluence occurred more steeply, in the case of the specimen irradiated at the same temperature. And the crook of the change indicated overestimated temperature by about 20-80°C. J.I. Bramman et al.[3] pointed out, would be expected that the annealing behavior might be affected by the rate of temperature rise in the dilatometer and the onset of annealing was ill-defined. In the present experiment, similar conclusion was obtained. But to cope with large numbers of monitors, a more rapid data extraction method than isochronal annealing must be established. Judging from this point of view, a continuous differential dilatometry should be reconsidered. The sensitivity of the dilatometer should be improved and the heating rate of the experiment should be decreased. Furthermore a step-wise heating schedule may be accepted.

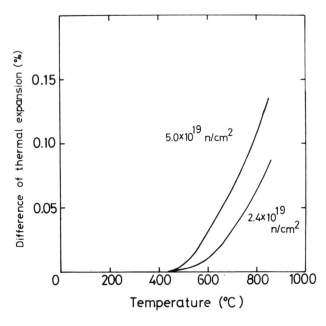

Fig. 4 Difference of linear thermal expansion in conventional dilatometry

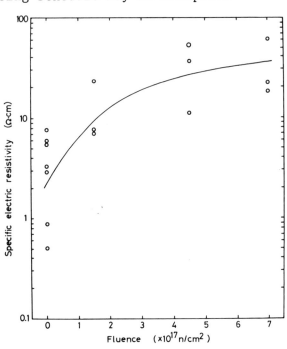

Fig. 5 Specific electric resistivity of β-SiC with neutron irradiation

5.4. Specific electric resistivity change in isochronal annealing

The specific electric resistivity of the specimen was about 1 Ωcm, and increased with the neutron fluence. Fig. 5 shows the specific electric resistivity change of the specimen irradiated to the fluence of 7.0×10^{17} n/cm². The increased value with the irradiation returned to the value of the non-irradiated specimen above the irradiation temperature in isochronal annealing, as is shown in Fig. 6. The crook of the annealing curve occurred at about the irradiation temperature. In the case of irradiation to 2.4×10^{19} n/cm² in VT-1 of JRR-2, the crook of the electric resistivity recovery shows 480±20°C. On the other hand from the isochronal annealing of lattice constant the irradiation was estimated as 490±20°C, and from thermal expansion measurement by high temperature X-ray diffraction it was 510±20°C, while the thermocouple indicated at 500±20°C. Therefore the specific electric resistivity measurement with isochronal annealing could be one of the estimation method of the irradiation temperature. By the method the irradiation temperature could be estimated of the specimen irradiated to lower fluence, such as 1.5 x 10^{17} n/cm^2, which could not be estimated by X-ray methods.

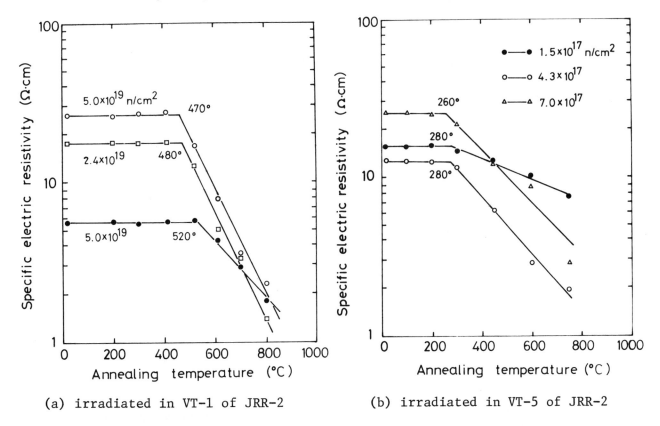

(a) irradiated in VT-1 of JRR-2   (b) irradiated in VT-5 of JRR-2

Fig. 6 Specific electric resistivity change of irradiated specimen in isochronal annealing

Recently R.J. Price[4] reported the same experiment by using hot-pressed α-SiC contained aluminum. According to his result, in the isochronal annealing, measurable recovery started a little below the irradiation temperature and the electrical resistance increased above the temperature with increasing temperature. In the present study, the recovery caused the decrease of the specific resistivity, and the behavior was contrary to the J.R. Price's result. The contrary would cause partly because the differences of polytype of SiC specimen and of the neutron fluence. Partly because SiC was an semiconductor, and the electric resistivity was influenced by impurity and microstructure. R.P. Thorne et al.[2] found the specific electric resistivity of β-SiC increased by 10-10000 fold with the neutron irradiation. It suggests that the electric resistivity of the irradiated specimen decreased above the irradiation temperature.

The carrier concentrations of a number of semiconducting materials are decreased during irradiation with fast neutrons. Fast neutron irradiation decreased the carrier concentration in β-SiC[6], and the predominant defect produced by neutron irradiation was an effective doner state[7].

6. <u>Conclusion</u>

In order to find a practical method for the estimation of the irradiation temperature, the recovery of the irradiated polycrystalline β-SiC was measured by the determination of the lattice constant and the specific electric resistivity in isochronal annealing, by the measurement of lattice expansion with high temperature X-ray diffractometer and by the conventional dilatometry. The estimated temperatures from lattice constant and specific electric resistivity change in isochronal annealing, and from thermal expansion measurement by high temperature X-ray diffraction agreed well with the thermocouple recording. In the present experiment the conventional dilatometry resulted in an overestimate of the irradiation temperature by about 20-80°C, and further improvements should be required in the experimental condition and equipment.

Acknowledgement

The research was supported in part by a grant from Power Reactor and Nuclear Fuel Development Corporation. The authors wish to thank the members of Research Reactor Utilization Section, Japan Atomic Energy Research Institute for their assistance in the irradiation and Dr. Y. Inomata, National Institute for Researches in Inorganic Materials, for the preparation of β-SiC powder.

References

1. N.F. Pravdyuk, V.A. Nikolaenko, V.I. Karpuchin and V.N. Kuznetsov, Properties of Reactor Materials and the Effects of Radiation Damage, D.J. Littler (ed.), (Butterworths, 1962) p.57

2. R.P. Thorne, V.C. Howard and B. Hope, Proc. Brit. Ceram. Soc., (7) (1967) p.449

3. J.I. Bramman, A.S. Fraser and W.H. Martin, J. Nucl. Energy <u>25</u> (1971) 223

4. R.J. Price, Nucl. Technol. <u>16</u> (1972) 536

5. Y.S. Touloukian (ed.), Thermophsical Properties of High Temperature Solid Materials, Vol. 1 (Macmillan, 1967) p.892

6. L.W. Aukerman, H.C. Gorton, R.K. Willardson and M.V.E. Bryson, Silicon Carbide, J.R. O'Connor and J. Smiltens (ed.), (Pergamon, 1960) p.388

7. P. Nagels and M. Denayer, 7th Inter. Conf. on the Physics of Semiconductors, Vol. 3, (Academic, 1964) p.225

Silicon Carbide Junction Thermistor

Robert B. Campbell

A thermistor consisting of a SiC p-n junction has been developed for high temperature use. The response of the thermistor is essentially logarithmic with temperature. The response curve can be varied by changing the junction structure, and thermistors have been fabricated with resistances at $1000^{\circ}C$ varying from 5 ohms to 10^5 ohms. When packaged in stainless steel housings, the thermistors have been tested at $1050^{\circ}C$ for over 2500 hours with no measurable change in calibration.

INTRODUCTION

The silicon carbide (SiC) junction thermistor was developed as a sensitive, reliable temperature sensor for elevated temperatures.

The thermistor consists of a small SiC chip containing a p-n junction. This chip is alloyed between two tungsten disks to provide mechanical and ohmic contact, and this tungsten-silicon carbide assembly is packaged in a suitable sheath to protect the thermistor element from the ambient.

In operation for temperature measurement, a small voltage (nominally 1 volt) is applied across the junction in either the forward or reverse direction of the diode and the current measured. This current varies logarithmically with the temperature at the junction, providing a measure of the ambient temperature.

FABRICATION

The crystals used for the thermistors were grown by the sublimation technique. To grow the p-n junction, the initial crystal growth is carried out in a small vapor pressure of aluminum to form p-type material. During the run the aluminum is gradually replaced with nitrogen to form n-type material, and a n-type skin on the p-type core. Thus a double p-n junction is formed in each crystal during growth.

These crystals are cut into small dice (about 1 mm on a side) with diamond impregnated wheels. These dice are the individual thermistor elements. The tungsten contacts are alloyed directly to the SiC die (without a braze material) at 1750°C. At this temperature the tungsten reacts with the silicon on the surface of the SiC and the bond is a mixed phase region of tungsten silicides and carbides. The use of tungsten for contacts to the SiC is dictated by the temperature extremes the thermistor is expected to see. The device has been tested from -200° to 1100°C, and tungsten is the only material whose coefficient of thermal expansion matches that of SiC over a wide temperature range. In addition, the direct tungsten bonding gives a mechanically strong and ohmic contact.

Although SiC is quite inert to oxidizing and reducing ambients below 1400°C, the tungsten contacts and leads from the thermistor must be protected from the ambient atmosphere. In addition to protecting the tungsten and other components, the package must have a low thermal mass so that rapid changes in the ambient temperature can be monitored. These are two nearly opposing requirements. For high temperature operation we have chosen a sheath of stainless steel 304 or Inconel 600. These materials are resistant to oxidation and can be used with relatively thin wall sections to reduce thermal mass. In the present design, one side of the thermistor is in direct contact with the base of the sheath and thus the sheath forms one contact. The other contact is brought through the center of the tube (insulated with alumina beads) and through a modified semiconductor header. After assembly, the entire unit is evacuated and welded. Completed units of various designs are shown in Plate 1. Contact to the thermistor is made between the sheath and the center post of the header.

TEST RESULTS

Figure 1 shows a typical response curve for the thermistor. The response is logarithmic from -200°C to 700°C. Above 700°C the rate of change of resistance is somewhat slower. We believe this is due to the bulk resistance of the crystal becoming comparable to the junction resistance. As alluded to earlier, each thermistor will have two calibration curves, depending on the direction of current during the measurement, i.e., forward or reverse direction. These two curves are generally nearly parallel, with the curve taken in the forward direction being displaced toward a lower resistance at a given temperature.

These thermistors have been cycled between 1050°C and 30°C (occasionally to -200°C) with no noticable change in the calibration curve.

Devices have been made with resistance values at 1000°C from 10^5 ohms to 10 ohms. The different values are due to the junction structure grown during the growth process.

The temperature coefficient, β , is defined as:[1, 2]

$$R = RoA \exp \beta \left(\frac{1}{T} - \frac{1}{To} \right) \text{ where}$$

Ro = resistance at ambient temperature To,

R = resistance at temperature T and

A = constant, nearly independent of temperature and taken as unity here.

Figure 2 shows values of β as a function of T for two thermistors of quite different charac-teristics. AX-134 had a forward resistance of 30 ohms at 1000°C while AX-135 had a forward resistance of 10^5 ohms at 1000°C. These values of β, ranging from 3000° at 200°C to more than 8500° at 1000°C are larger than those reported for most bulk type thermistors; and extend to much higher temperatures than the bulk thermistors.

The temperature variation of a (the small increment temperature coefficient, or sensitivity) is shown in Figure 3. These numbers are given in percent change in resistance per degree centigrade and vary from 0.5%/°C to 2.0%/°C.

The time constant (the time in seconds for the thermistor to fall 63% from temperature T to To) is determined almost entirely by the thermal mass of the package used. The thermistor element itself has a time constant of a few seconds, Figure 4 shows a plot of log (T-To) versus time for a thermistor in a stainless steel sheath. In this test, the thermistor was soaked at temperature T ($\approx$ 900°C) and then removed to temperature To = 30°C. The decrease in temperature is shown in the plot. This curve follows the form:

$$T = To \exp t/\tau .$$

The time constant τ in this case is several hundred seconds. By using thinner walled sheath material and a thinner bottom plug, we feel this can be reduced by more than an order of magnitude. However, if the wall is too thin, it could possibly oxidize through in a few hundred hours and ruin the device.

Devices of this type have been run for over 2500 hours at 1050°C with frequent thermal cycling to 30°C with no measurable change in calibration. They have also been run over 5000 hours at 700°C in an automobile exhaust system running programmed stops and starts with no noticable change.

In summary, the SiC junction thermistor is a reliable, reproducible sensitive device for measuring temperatures from -200°C to 1100°C.

REFERENCES

1.   C. Bossun, F. Guttman, and L. M. Simmons, J. Appl. Phys. 21, 1267, 1950.

2.   John N. Shive; Semiconductor Devices, D. Van Nostrand Co., Inc., Princeton, N.J., 1959.

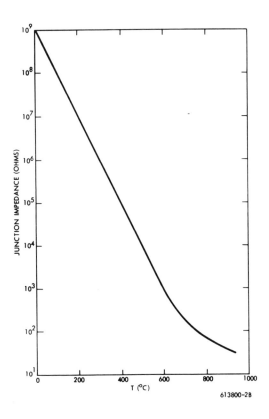

613800-2B

Figure 1. SiC Thermistor, Typical Response Curve

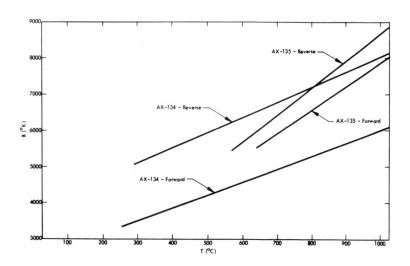

Figure 2. Temperature Coefficient versus Temperature – SiC Thermistor

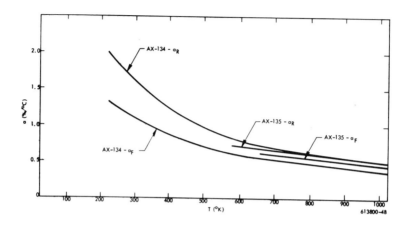

Figure 3.   Sensitivity versus Temperature

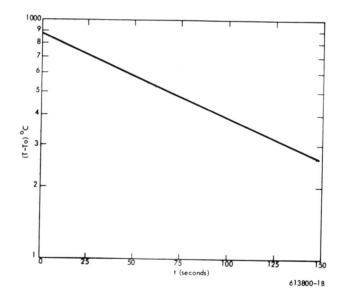

Figure 4.   Cooling Curve – SiC Thermistor

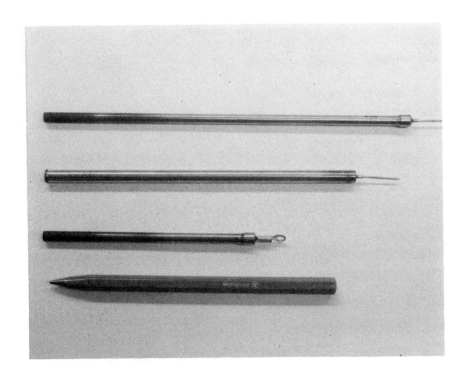

Plate 1.   Various Designs of SiC Thermistor

Photoluminescence of 4H-SiC Single Crystals Grown from Si Melt

Hiroyuki Matsunami, Akira Suzuki and Tetsuro Tanaka

Single crystals of 4H-SiC were prepared from Si melt in a
graphite crucible. Photoluminescence of the crystals was
studied in the temperature range from 90K to 400K. Green
luminescence at room temperature and violet luminescence
at lower temperatures are attributed to the recombination
through boron-center and aluminum-center, respectively.

Introduction

Since silicon carbide (SiC) has a wide energy gap, this material
seems to be useful for optoelectronics. A great deal of researches
have been done on electrical and optical properties of SiC (1,2).
Although there are various kinds of polytypes in SiC, two of them
(3C-type and 6H-type) have been mainly investigated because
preparation of single crystals of other polytypes is very difficult.
There are few reports on the optical properties of 4H-SiC (3,4).
Choyke studied the absorption spectra of various polytypes and
obtained 3.26eV as the exciton energy gap of 4H-SiC at 0K (3). Potter
measured photoluminescence of 4H-SiC, but the precise data were not
presented (4).

In this report the method for preparation of 4H-SiC single
crystals from Si melt is described. Photoluminescence in the visible
light region of 4H-SiC is studied in the temperature range between
90K and 400K. Green luminescence at room temperature and violet
luminescence at lower temperatures thus obtained are analysed.

Crystal Growth

The method of solution growth from Si melt in a graphite crucible
has been adopted by several researchers for preparation of SiC single
crystals (5,6). They, however, studied the growth condition for obtaining only 3C-SiC, and did not apply this method to praparation of α-SiC.

Single crystals of 4H-SiC can be obtained by this solution method
if the following conditions are satisfied: 1) The temperature of Si
melt should be kept about $1800^{\circ}C$; 2) The temperature gradient of the
crucible is maintained as small as possible (less than $5^{\circ}C$ /cm). In the
experiments, the temperature of the upper part of the crucible was
lower than that of the bottom to realize the condition(2). Then 4H-SiC
grew at the upper part of the crucible. (In the case of 3C-SiC, the
temperature gradient is rather large and crystals grow at the bottom
(6).)

The crucibles were made of graphite (density: 1.60, ash:0.02%)
and the typical shape of them was a cylinder of 25mm(inner diameter)
×40mm(depth). The crucibles were baked for 2∿3 hours at about $1850^{\circ}C$
under Ar gas flow before crystal growth. Undoped polycrystals of Si
were etched with a mixture of hydrofluoric acid and nitric acid.
Boron(B)-doped(2×10^{-3}at%) Si was used as a raw material in order to get
B-doped 4H-SiC. After putting Si in the crucible, pure Ar gas was
flowed (about 5ℓ/h) and the crucible was maintained at about $1300^{\circ}C$
for 20 minutes. After that the temperature was increased to about
$1800^{\circ}C$ in a few seconds to avoid cracking of the crucible, and Si melt
was kept at that temperature for about 10 hours. The color of single
crystals thus obtained was transparent greenish-blue and the typical
size was about 0.8×0.8×0.2 mm^3.

The polytype of 4H-SiC was identified by the oscillation method
of X-ray diffraction. The lattice constant of the c-axis of 4H-SiC is
$10.05\overset{\circ}{A}$ (2), and our experimental results were 10.35, 10.16 and $10.22\overset{\circ}{A}$
in several samples. Spectroscopic analysis showed that traces of
several elements (Aℓ, Fe, Mg, Cu and Ca) existed in undoped 4H-SiC.
These impurities may have been introduced from the graphite crucible.

Experimental Procedures

The method to obtain photoluminescence in the visible light region was an ordinary one. The excitation light was UV light from a mercury lamp, and the wavelength of 365nm(3.40eV) was selected by using an interference filter and a filter of a solution of 10% copper sulfate. The excitation light was chopped with a frequency of 360Hz, and the luminescence from the sample was gathered with a condenser lens and introduced to a spectrometer.The spectrometer is Czerny-Turner type using a grating, the blaze wavelength of which is 750nm. The signal was detected by a photomultiplier(HTV R592) and the Lock-in amplifier.

Experimental Results

Figure 1 shows the spectra of photoluminescence of undoped 4H-SiC. Violet luminescence and green luminescence were observed at 95K and 300K, respectively. This means the color of luminescence changes with temperature. At 300K, the peak of the spectrum locates at 2.37eV, and its half width is about 365meV. At 95K, the spectrum has a number of emission peaks, among which the highest photon energy locates at 3.07eV. The green luminescence generally diminishes at lower temperatures as shown in Fig.1, but in a certain sample it remains even at 95K.

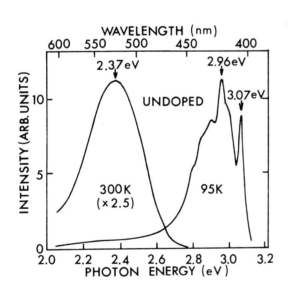

Fig. 1

Photoluminescence spectra of undoped 4H-SiC.

In Fig.2 the spectra of violet luminescence band at various temperatures are shown. Though 3.07eV peak is observed at higher energy side at 101K, a new peak of 3.10eV

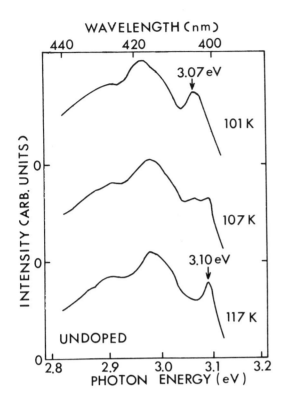

Fig. 2

Spectral change of violet
band with temperature
(undoped 4H-SiC).

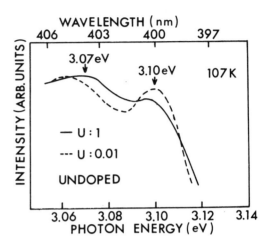

Fig. 3

Spectral change of violet
band with excitation intensity
(undoped 4H-SiC).
U: excitation intensity.

emerges at higher temperature as shown in the figure. And this becomes
more dominant than the former peak with temperature increase. The half
width of this peak is narrower than that of the former.

The change of the spectra of these two peaks with the excitation
intensity at 107K is shown in Fig.3. Increasing the excitation intensity
brings a displacement of the 3.07eV peak towards higher energy side,
whereas the 3.10eV peak remains at the same energy.

Figure 4 shows how the emission intensity of the violet and the
green luminescence changes with temperature. The violet luminescence
which is dominant below 100K abruptly decreases with temperature rise.
The activation energy (E_a) of the decrease is about 184∿186meV. Because
of the decrease, the color of the luminescence suddenly changes from
violet to green. This green luminescence gradually increases with

temperature, and reaches to the maximum intensity at around room temperature. Its intensity at room temperature is 2∿3 times as large as at low temperature. Above room temperature it decreases with an activation energy of 510∿524meV.

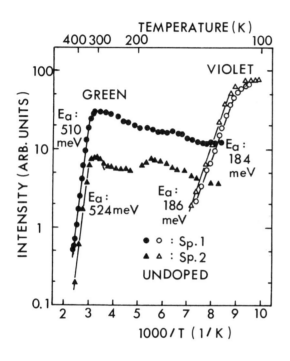

Fig. 4

Temperature dependence of emission intensity of violet and green luminescence (undoped 4H-SiC). E_a: activation energy.

Figure 5 shows the spectral change of B-doped 4H-SiC at 300K and 90K. Very strong green luminescence was obtained at room temperature. The spectrum is the same as that of undoped sample (see Fig. 1), but the intensity is 5∿10 times as strong as that of undoped one. At lower temperatures the intensity decreases a little and the color changes to yellowish-green. Fine structure appears in the spectrum at 90K and the peak of the highest energy can be observed at 2.60eV as a shoulder. The spectrum shifts to lower energy side and the maximum peak shifts from 2.37eV to 2.31eV. The half width increases from 365meV to 408meV and this causes the color change from green to yellowish-green. The intensity of this luminescence decreases with an activation energy of 536meV with temperature rise above room temperature.

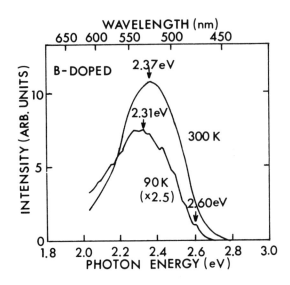

Fig. 5

Photoluminescence spectra
of B-doped 4H-SiC.

Discussion

The violet luminescence of undoped 4H-SiC may be due to an
aluminum(Aℓ) acceptor by the following reasons.
1) Spectroscopic analysis shows the presence of an aluminum impurity
in undoped samples.
2) The photoluminescence spectrum of the Aℓ-doped sample is very
similar to that of the violet band of undoped sample, whereas the
former intensity is stronger.
3) The characteristics of the spectrum shown in Figs.2 and 3 are
resemble to those of orange luminescence in Aℓ-doped 3C-SiC reported
by Zanmarchi (7) and Choyke and Patrick (8).
4) The activation energy (184∿186meV) with which the violet luminescence
decreases is nearly equal to that of orange luminescence in Aℓ-doped
3C-SiC (about 180meV) (9).

From the above results, the 3.07eV peak is attributed to the
recombination of a hole bound to an Aℓ-acceptor with an electron bound
to an unknown donor (probably nitrogen(N)). (donor-acceptor pair
recombination) The 3.10eV peak may be due to the recombination of
such a bound hole with a free electron. The energy difference between
3.10eV peak and the energy gap is $0.16eV+E_x$, where E_x is the exciton
binding energy of the order of 0.01eV (10). Assuming the above process

this corresponds to an ionization energy of the Aℓ-acceptor, and fairly agrees with 0.18eV+E_x reported by Choyke and Patrick in 3C-SiC (8). The other peaks in the violet band seem to be phonon replicas of the above two peaks.

The green luminescence may be due to boron(B)-center, comparing the characteristics of the spectrum with those reported before (11,12). This was verified by observing photoluminescence of B-doped samples. Although boron is not detected in undoped samples by spectroscopic analysis the luminescence efficiency of B-center is so high that even the traces of boron may cause the green luminescence. The following features should be noticed: 1) The emission intensity at room temperature is higher than at low temperature; 2) The maximum peak of the spectrum tends to shift to the higher energy side with temperature rise. This tendency is opposite to the temperature dependence of the energy gap.

Mechanism of light emission through the B-center in other polytypes was scarecely studied (11,12). The energy level of boron was obtained electrically as 0.39eV from the valence band (13). Considering that the energy difference between the energy gap and the maximum peak is about 0.86eV, there is a possibility of the transition from an unknown donor ($E_d \approx 0.50$eV from the conduction band) to the B-center. The no-phonon peak, however, is very weak and its phonon replicas are dominant in the luminescence spectrum through a deep level. Based on such general tendency, each peak at 90K in Fig.5 is considered to be a phonon replica of a certain no-phonon peak, which probably locates in the range of 2.70∿2.80eV estimated from the zero intensity at higher energy side of the spectrum. This energy is 0.45∿0.55eV smaller than the energy gap. This value seems to be related to the activation energy of the intensity decrease above room temperature. A certain deep level (about 0.50eV) seems to take part in the green luminescence, but the unusual features above described suggest the existence of rather complicated mechanism.

Conclusion

The following conclusions were obtained in this study.

1) Single crystals of 4H-SiC were reproducibly obtained by solution growth from Si melt in the graphite crucible under some peculiar conditions.

2) The violet luminescence of 4H-SiC is attributed to the recombination through Aℓ-center.

3) The green luminescence of 4H-SiC may be due to B-center, whereas the experimental results suggest rather complicated mechanism.

Acknowledgements

The authors wish to express their thanks to Mr. A. Komuro for crystal preparation. They also want to thank to Mr. T. Sonoda for spectroscopic analysis. This work was partially supported by Grant-in= Aid for Scientific Research from the Ministry of Education.

References

1. J.R.O'connor and J.Smiltens, Ed., Silicon Carbide- A High Temperature Semiconductor, Pergamon Press, (1960).

2. H.K.Henisch and R.Roy, Ed., Silicon Carbide- 1968, Pergamon Press, (1969).

3. W.J.Choyke, Silicon Carbide- 1968, p.S 141, Pergamon Press, (1969).

4. R.M.Potter, Silicon Carbide- 1968, p.S 223, Pergamon Press, (1969).

5. F.A.Halden, Silicon Carbide- A High Temperature Semiconductor, p.115, Pergamon Press, (1960).

6. W.E.Nelson, F.A.Halden and A.Rosengreen, J. Appl. Phys., 37, 333 (1966).

7. G.Zanmarchi, J. Phys. Chem. Solids, 29, 1727 (1968).

8. W.J.Choyke and L.Patrick, Phys. Rev. B, 2, 4959 (1970).

9. A.Suzuki, unpublished.

10. D.S.Nedzvetskii, B.V.Novikov, N.K.Prokf'eva and M.B.Reifman, Soviet Physics- Semiconductor, 2, 914 (1969).

11. A.Adamiano, R.M.Potter and V.Ozarow, J. Electrochem. Soc., 110, 517 (1963).

12. V.I.Pavlichenko and I.V.Ryzhikov, Soviet Physics- Solid State, 10, 2977 (1969).

13. H.J.van Daal, W.F.Knippenberg and J.D.Wasscher, J. Phys. Chem. Solids, 24, 109 (1963).

*Subsurface Silicon Carbide Formation in Silicon after High Energy C⁺ Implantation**

G. H. Schwuttke and K. Brack

High energy ion-implantation of $Si:C^+$ is used to produce sub-surface silicon carbide in single crystal silicon. Formation of cubic SiC is achieved after annealing of C^+ implanted Si in the range from 900° to 1200°C. The SiC distribution in the implanted silicon is determined after the 1200°C anneal. The maximum concentration of SiC is found 1.5μm below the silicon surface. The SiC crystals are embedded in a poly-crystalline silicon matrix.

INTRODUCTION

Recently, it was shown that subsurface silicon nitride layers can be formed in silicon wafers through high energy implantation of N^+ into silicon followed by proper annealing.[1] This paper describes experiments conducted to study the formation of subsurface silicon carbide films in silicon after high energy C^+ implantation.

EXPERIMENTAL

The investigations are conducted with 1 MeV and 2 MeV single charged carbon ions. To implant the ions the AFCRL van de Graaff generator is used. Accordingly, carbon dioxide gas is passed through a thermo-mechanical leak into a radio frequency activated source. The positive carbon ions are driven into the van de Graaff with the help of a variable voltage probe. Emerging from the accelerator the ions drift into a magnetic analysing system and here the carbon ions are bent 90 degrees into the exit port.

* Sponsored in part by Air Force Contract Number F19628-72-C-0274.

The ion beam leaving the analyser is about 30% defocused by drifting down a 4 ft. long tube to hit the silicon target. At this position the 1μamp beam has a circular cross-section of approximately 20mm. Due to the partial defocusing the ion beam has a hot spot in the center.

The implantation target is a silicon wafer, 37mm in diameter, 1mm thick and of <100> or <111> orientation. The slice is mechanically-chemically polished. The silicon has 2 ohm-cm resistivity. The target slice is kept at room temperature during bombardment. The silicon is bombarded to a fluence of $5 \times 10^{16} C^+/cm^2$.

POST BOMBARDMENT MEASUREMENTS

No visible marks of surface damage due to ion bombardment appear at the surface of the C^+ implanted wafer. Electron transmission micrographs and selected area diffraction patterns of the surface indicate no damage to the crystal surface. Measured down from the surface the silicon produces well defined Laue spots and Kikuchi lines. However, some radiation damage is present in the top layer as evidenced in electron transmission micrographs. A fairly sharp boundary line separates this surface layer from a highly damaged layer which extends further down into the silicon. The position of this damage layer is determined through the ion energy. Position and extent of such a damage zone can be measured optically on a bevel. An example of a 1 MeV C^+ bombarded and subsequently bevelled wafer is shown in Plate 1. By measuring the bevel precisely the position of the center of the damage zone layer is calculated to be 1.5μm below the surface. Systematic electron diffraction studies of the damage zone as shown in Plate 1 indicate that the disturbed layer consists of a continuous lens-shaped sheet of amorphous silicon. The sheet reaches a maximum thickness of ∿3000Å in the area of most intense bombardment which is approximately $10^{17} C^+ cm^{-2}$ for the sample under discussion.

POST ANNEALING MEASUREMENTS

To induce silicon carbide formation in the damage layer a series
of annealing experiments was performed. The samples were annealed
to 1200°C most of the time in intervals of 100°C. Each annealing
cycle lasted for one hour.

WIDTH OF DAMAGE ZONE

The width of the damage zone was measured after every annealing
cycle on a 1 or 4 degree bevel. Representative and important
examples are shown in Plate 2. It is interesting to note that
up to 500°C the damage zone width stays fairly constant. The
damage zone obtained after the 500°C 1 hour anneal is shown in
Plate 2b. At 600°C the width of the damage zone starts to
decrease. This is shown in Plate 2c. In general we find about
half the thickness of the amorphous layer after the 600°C anneal.
Crystallization of the amorphous layer starts from top and bottom
interface simultaneously. The auto-epitaxial crystallization
of the amorphous zone is completed after a 700°C anneal. After
annealing at 700°C the defect zone is not visible anymore on
the bevel. This is shown in Plate 2d which represents the amorphous
layer after the 700°C anneal. The original position of the
amorphous zone is indicated on the photomicrograph of Plate 2d.
After 1200°C anneal a subsurface layer is again visible on the
bevel. This is shown in Plate 2e.

ELECTRON TRANSMISSION MICROSCOPY

This section discusses detailed electron microscopy results
obtained on the C^+ implanted samples after the different annealing
steps up to 700°C and at higher temperatures.

ANNEALING UP TO 700°C

Annealing up to 700°C is primarily characterized by the crystallization of the amorphous sub-surface amorphous layer into quasi-
crystalline silicon. Electron transmission microscopy investigations show that small changes in the damage structure are already
apparent after the 500°C. Such changes are stronger after the
600°C anneal and are the result of commencing crystallization
of the amorphous silicon phase. A typical example of the
interface between damage layer and surface layer is shown in
Plate 3. The diffraction pattern shown in Plate 3 provides
insight into the start of the annealing mechanism. The presence
of two different phases are clearly identified in the diffraction
pattern. The diffuse ring pattern corresponds to amorphous
silicon and superimposed one recognizes a single crystal
diffraction pattern of the [100] single crystal silicon zone.
The different silicon spots show additional spots and streaking
typical for the presence of heavy twinning and stacking faults
in the diffracting single crystal silicon.

Based on such electron diffraction data and optical measurements
as shown in Plate 2, we interpret the first annealing phase of
the sub-surface amorphous layer as being due to auto-epitaxial
crystallization starting at the original interfaces formed by
the amorphous layer within the single crystal silicon above or
below the damage zone. As a result of the crystallization of the
amorphous silicon one observes a volume reduction in the damage
zone. Consequently, the original sub-surface layer shrinks in
width. This is readily observed in Plates 2.

Electron microscopy of the damage zone after the 700°C anneal
shows only quasi-crystalline single crystal silicon. No evidence
of poly-crystalline silicon or silicon carbide precipitation is
obtained after the 700°C anneal. Note, that the damage layer is
not visible anymore on the bevel after 700°C anneal (Plate 2d).

ANNEAL ABOVE 700°C

Detailed transmission electron microscopy studies of the damage
structure were made of samples after a four hour anneal at
1200°C. After such an anneal the damage structure is again
visible at a bevel as shown in Plate 2e. The following results
relate to the high dose area indicated in Plate 1. The original
thickness of the amorphous layer in this area was 3000Å. Electron
transmission micrographs and selected area diffraction patterns
of the upper and lower half of the damage layer corresponding to
Plate 1 after the 1200°C anneal are shown in Plate 4a and 4b.

The diffraction pattern shown in Plate 4a shows quasi-crystalline
silicon and poly-crystalline silicon carbide (weak). The
diffraction pattern of the lower half is shown in Plate 4b. It
shows poly-crystalline Si and SiC (strong). The complete damage
profile after the 1200°C anneal is obtained by recording the
diffraction patterns at different positions across a bevel
as shown for instance in Plate 1. The profile obtained is
shown in Plate 5. Starting with the top layer the crystal
is single crystal silicon down to the interface layer crystal/
damage zone. The interface is single crystal silicon but
crystallization of the amorphous phase has lead to twinning.
The upper and lower half of the crystallized damage layer were
discussed in connection with Plates 4a and 4b. These diffraction
patterns were obtained at positions C and D respectively. At the
lower interface (position E) the lattice is heavily twinned and
heavy SiC precipitation is also present. At position F (bulk)
the silicon has the perfection of the original bulk lattice.
However, silicon-carbide precipitation is found down to a depth
of 2 to 3 times the average depth of the original damage layer
(1.5µm). In the sample under discussion we found SiC precipitation
as deep as 3µm under the surface.

DISCUSSION AND SUMMARY

Experiments were conducted to produce sub-surface silicon-carbide
films through high energy ion implantation. It was found that
silicon carbide formation can be achieved after large dose
implantation and four hour annealing at 1200°C. A continuous
silicon carbide film was not produced in these investigations. SiC
crystals formed embedded into a poly-crystalline silicon matrix.
We conclude that the maximum achievable dosis used in these
investigations ($\sim 10^{17}$ C^+/cm^2) was too low to allow the formation
of a continuous SiC film. The density of carbon atoms in a
SiC-crystal is 4.8×10^{22} atoms/cm^3. To achieve the proper
stoichiometric relation between silicon and carbon atoms
2.5×10^{18} ions/cm^2 would have to be implanted as a minimum
dose. This would lead to excessively long implantations of
250 hours based on the ion-current available to us.

The formation of SiC on silicon surfaces was previously
investigated by I. H. Khan and R. N. Summergrad[2] for the
temperature range 800°C - 1000°C, by J. Graul and E. Wagner[3] for
the range 1180°C - 1260°C and also by E. Biedermann and
K. Brack[4] at 1100°C. All these investigators report the cubic
modification for the SiC formed. In our investigations the
SiC formation occurred under the silicon surface. It is
noteworthy that we observed SiC formation between 900°C - 1200°C
and that the cubic modification of SiC is typically for this
temperature range.

REFERENCES

1. G. H. Schwuttke and K. Brack. Transaction of the Met. Soc. of
       AIME, Vol. 245, 475, (1969).

2. I. H. Khan and R. V. Summergrad, Appl. Phys. Letters, Vol. 11,
       12, (1967).

3. J. Grant and E. Wagner, Appl. Phys. Letters, Vol. 31, 67, 1972.

4. E. Biedermann and K. Brack, J. Appl. Phys. 37, 4288, (1966).

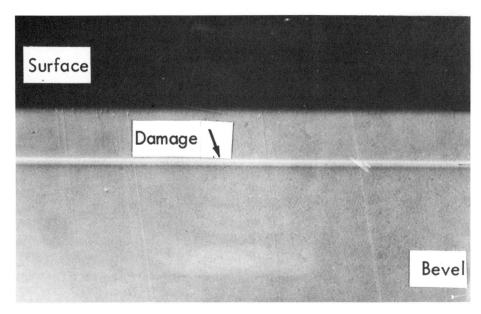

Plate 1. Photomicrograph of a 1 degree bevel showing the subsurface layer in
silicon after 1 MeV C^+ implantation to a fluence of 3×10^{16} C^+/cm^2 at a depth of
1.5μm.

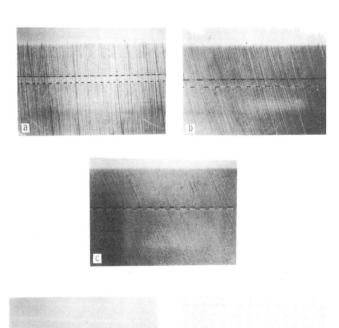

Plate 2. Photomicrographs showing
damage zone on a bevel after 1 MeV C^+
bombardment. Bevel angles are about
45 minutes and are rough polished.
The depth of the layer is 1.5μm below
the surface.

   (a) No anneal
   (b) After 500°C anneal
   (c) After 600°C anneal
   (d) After 700°C anneal
   (e) After 1200°C anneal and
        fine polish

Plate 3. Electron diffraction pattern of the interface of the subsurface amorphous layer after 650°C anneal.

Plate 4a. Electron diffraction pattern from the upper half of the damage layer showing quasi single-crystalline Si and weak poly-crystalline SiC rings (arrow) after 1200°C anneal.

Plate 4b. Electron diffraction pattern from the lower half of the damage layer showing poly-crystalline silicon and SiC (arrow) after 1200°C anneal.

Plate 5. Schematic of damage profile in C^+ bombarded Si after 1200°C anneal. Note SiC distribution. Diffraction patterns shown in Plates 4a,b correspond to position C and D. Diffraction patterns (not shown in this paper) were also obtained at position A, B, E and F.

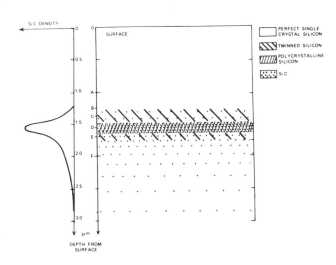

Production of Silicon Carbide from Rice Hulls

J. V. Milewski, J. L. Sandstrom and W. S. Brown

INTRODUCTION

In the conventional process for producing silicon carbide, silica in the form of sand and carbon in the form of coke are reacted at 2,400°C in an electric furnace. The silicon carbide produced is in relatively large grains which are subsequently ground to the desired size. A new process for producing a fine grained high purity silicon carbide product has been developed which utilizes rice hulls as the raw material. Rice hulls have a high silica content (15 to 20% by weight), and the carbon atoms are intimately mixed with the silica as a result of the natural plant growth. The intimate mixture of silica and carbon permits the reaction to occur at a lower temperature than in the conventional process and it also results in a material of submersion particle size.

PROCESS DESCRIPTION

The rice plant extracts silica from soil as it grows and deposits much of the silica in the hull. The hulls, which are normally discarded in the rice milling operation, contain cellulose and other organic substances as well as silica and traces of inorganic compounds.

The process for converting rice hulls into silicon carbide or silicon nitride has two basic steps. Step one is a coking process in which the raw rice hulls are heated in the absence of oxygen at about 700°C for a time sufficient to drive the volatile components out of the rice hulls. In step two the coked hulls are heated to a much higher temperature in an inert or reducing atmosphere to cause a solid state reaction to take place between the carbon and silica. This will produce silicon carbide by the following reaction:

$$3C + SiO_2 \rightarrow SiC + 2CO$$

If nitrogen is introduced into the atmosphere, silicon nitride is also formed. The amount of nitrogen, the temperature and the furnace geometry determine the ratio of silicon nitride to silicon carbide in the product. Since there is an excess of carbon in the mixture before calcining, the product contains some free carbon.

PREVIOUS WORK

Cutler and his collaborators (1 - 3) first demonstrated the feasibility of this process with small quantities of rice hulls in a high temperature furnace. They measured weight loss as a function of time as the conversion process proceeded for a variety of operating conditions. Experiments were performed using hulls which had been ground and pre-coked as well as hulls in their natural state. They demonstrated that the addition of iron as a catalyst improved the reaction kinetics. A variety of sweep gases were introduced into their furnace to reduce the CO partial pressure. Nitrogen, argon and ammonia were all successfully used to improve the kinetics of the reaction.

PROCESS DEVELOPMENT

Further development work has been directed to providing data which would be useful in developing a commercial process. From thermogravimetric studies conducted at Rutgers University, the initial and final reaction temperatures were determined as well as an estimate of the time necessary to have the reaction go to near completion. Results are shown in Table I.

TABLE I. Thermogravimetric Analysis Results for Rice Hull Conversion to Silicon Carbide under Various Conditions

Sample Number	Atmosphere	Hulls Used	Initiation Temp. °C	Temp.@ Max. dw/dt*	Est. Completion Temp.°C	Est. Max Time to Completion, min.
4	N_2	Coked	1505	1600	1600[+]	60[+]
5	NH_3	Coked	1425	1540	1580[+]	30
3	H_2	Coked	1400	1480	1550	25
7	NH_3	Fe Catalyzer Coked	1300	1440	1480	20

*$\frac{dw}{dt}$ = rate of weight loss

In production larger charges will be used resulting in longer times and higher temperatures due to the insulation effect of the product formed and to the inhomogeneous heat distributions inherent in the equipment.

The processing is presently done in two different types of furnaces. The coking operation is performed in a cylindrical stainless steel retort which has a volume of about one cubic foot. The raw hulls are weighed and placed in the retort, which is then lowered into a preheated circular furnace. During the subsequent 3 to 4 hours of coking time, all the volatiles are driven off and the resulting "coke" reaches a temperature of approximately 700°C. The coked hulls are then allowed to cool to 250°C before exposing them to air. Next, the pre-coked hulls are weighed and charged into a high temperature verticle 6" diameter 50" long cylindrical tube, gas fired furnace (Figures 1 and 2). A plug at the bottom of the central tube contains a passage which admits sweep gas to the furnace for atmosphere control. Ports are provided at several elevations along the furnace wall which permit observing the outside temperature of the silicon carbide tube with an optical pyrometer. The temperature variation from the top to the bottom of the furnace is approximately 40°C.

The furnace is preheated to a temperature in the range of 1,500°C to 1,600°C. To produce silicon carbide, ammonia gas is fed into the bottom of the furnace tube. In two to four hours the reaction is completed -- depending upon the temperature and other operational parameters. At the end of the run an insulation plug is removed from the bottom of the furnace tube and the product is allowed to fall into a receptacle. About one pound of product is formed per cycle with the present equipment.

MATERIAL CHARACTERIZATION

Approximately one-tenth of the rice hulls by weight convert to silicon carbide. Silicon nitride is also produced in some regions of the furnace where conditions are favorable. The material produced in a normal batch contains a significant quantity of free carbon which is removed by heating in air at 800°C. The unreacted silica is removed by washing with either hydrofluoric acid or an aqueous sodium hydroxide solution. Results are reported here for material produced by using pre-coked hulls as the charge material with three liters per minute of ammonia flowing into the bottom of the furnace. The furnace outside wall temperature was 1,600°C ± 20°C and the batch time was three hours.

The crystal morphology of the silicon carbide has been determined by X-ray analysis to be approximately 25% beta and 75% alpha. It is possible to vary this ratio by modifying the process variables. A typical chemical analysis for the material after burning off the carbon is shown in Table II.

TABLE II.   Chemical Analysis

	Prior to HF Cleaning	After HF Cleaning
Free Carbon	0.1 %	--
Silicon	64.21	67.00
Oxygen	5.90	1.76
Nitrogen	3.60	3.39
Total Carbon	21.39	22.92
	95.2	95.07

The results of an emission spectrograph analysis are shown in Table III; and
Table IV shows additional mass spectrograph analysis results.

TABLE III.   Emission Spectrograph Semi-Quantitative Analysis

Al	0.020
Ca	0.300
Cr	0.001
Cu	0.001
Fe	0.040
Mg	0.090
Ti	0.001
Si	High
Total	0.453 Not Counting Silicon

TABLE IV.   Mass Spectrograph Analysis of Elements not
Covered or Found by Emission Spectrograph

Found in Traces Only	Not Found
Cl	B
F	P
Ba	S
Sr	
Some alkali earth	
OH	
H_2O	

The HF acid treatment indicates a major reduction on the SiO_2 as would be
expected, but both analyses fail to account for about four percent of the
material.   Further analytical work is being considered to clear this anomaly.
One possible explanation is that the nitrogen content may be low by as much as
a factor of two because of the difficult analytical procedure.   Mass spectrographic analysis did not indicate the presence in any significant quantity of
any element other than those already analyzed.   The analysis indicates that by
relatively simple processing to remove the unreacted silica and the excess
carbon a product can be produced containing less than one-half percent impurity.

The particle size was determined by two methods, the first being scanning

electron microscope photographs. These photographs reveal fibers ranging from
0.1 to 0.5 microns in diameter with the fiber length to diameter ratio being
in excess of 100 which are thought to be the beta phase. The bulk of the material is in the form of particles which appear to have a diameter of 0.5 microns
or less. The sizes are in general agreement with results obtained from a Mine
Safety Apparatus (MSA) particle size analysis test, which is based on application of Stokes Law to settling of particles in a fluid. MSA tests indicate that
the product contains no particles greater than 40 microns in diameter, 80% of the
product is less than three microns in diameter, and about 50% is less than one
micron in diameter.

Further evidence of the generally small particle size was obtained from
surface area measurements using a nitrogen absorption technique. These measurements indicated that the surface area of the processed material prior to washing with HF was approximately ten square meters per gram, while the post-HF
treated material has an area of 15 to 18 square meters per gram. (As a reference, a one micron diameter fiber has a surface area measurement of about one
square meter per gram.)

SUMMARY

In summary, this process produces submicron size particles of silicon
carbide with a minimum of impurities other than silica, carbon and silicon
nitride. The silicon nitride content can be varied by changes in process parameters from 0 to ten percent and the residual carbon and silica can be easily
extracted. The material produced is a mixture of submicron powder and submicron
fibers (whiskers). The fiber content can also be varied by modifying the
process variables.

The high fiber content material is being considered for use in reinforced
plastics while the electrically conductive submicron powder is being considered
for high temperature applications such as heating elements, emission control
catalyst supports, combustion catalyst supports, heat exchangers, and turbine
blades.

REFERENCES

1. "Production of Silicon Carbide from Rice Hulls," I. B. Cutler, U. S. Patent
        No. 3,754,076 assigned to University of Utah, Salt Lake City, Utah
        (Filed October 30, 1970).

2. "Production of SiC and Si_3N_4 from Rice Hulls," I. B. Cutler and W. R. Rao,
        Presented at 75th Annual American Ceramic Society Conference,

Symposium No. 2, Cincinatti, Ohio, April 30, 1973.

3. "Kinetics of Silicon Carbide Formation from Rice Hulls," June Gunn Lee,
    M. S. Thesis, University of Utah, Salt Lake City, Utah, June 1973.

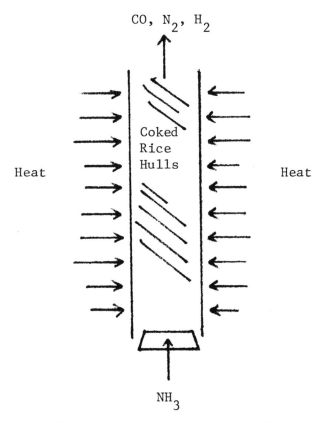

Fig. 1. Schematic of Calcining Furnace

Fig. 2. Photograph of Calcining Furnace During Operation

Damage and Penetration
of Implanted Ions in SiC

D. Eirug Davies and J. J. Comer

The visual effects arising from reflectivity changes on implanting silicon are also seen when silicon carbide is implanted. Such changes in appearance can be rendered visible on bevelled sections of implanted silicon carbide and can be used to estimate the depth of penetration of structural damage and as an indicator of the junction depths that can be subsequently expected with appropriate annealing.

The advantages of ion implantation as a general semiconductor doping process are well known and diode formation in SiC was accomplished as early as 1967.[1] These particular devices were fabricated in aluminum doped hexagonal SiC with multi-energy phosphorus implants up to 400 KeV. The SiC surface was covered with sputtered quartz and an evaporated aluminum mask was used to give a planar diode array. The sputtered gold contacts used were shown to be non-rectifying by forming dual contacts onto single implanted regions. Though the units were only annealed to 850°C, which is rather marginal in view of more recent damage annealing studies, some indication that irradiation damage was not dominant and responsible for rectification was obtained by lightly re-implanting the diodes. The additional unannealed damage resulted in rectification deterioration and the initial characteristics were only restored on re-annealing to 800°C and above.

The following is a continuation of this work in an attempt to characterize further the implantation parameters. In particular, little is known of the junction depths achieved. A delineation effect has been found on cross-sectioning implanted layers in SiC that is readily visible without any etching, plating or staining. Here, this effect is examined to see how it can be used to estimate either the depth of penetration of the ions or of the associated irradiation damage.

An example of the delineation on a shallow angle section is shown in Fig. 1. In this particular instance, 5×10^{15} phosphorus ions cm^{-2} were implanted at 200 and 400 KeV into hexagonal p-SiC. A mask covered part of the surface area during implantation and is responsible for the contrast seen on the bevel with the delineation appearing on the implanted region alone. A modified bevelling technique is employed that has the sectioning jig inverted from the normal method of use with the SiC facing upwards and the glass optical flat on top. Moving the optical flat provides the lapping action and the film of abrasive is sufficiently thin for the bevelling portion to be visible in a low powered microscope while actually lapping. In this manner, extremely small samples can be used, less rounding occurs at the edge, and any surface scratching can be terminated at its initiation. While the above example is of a 3° bevel, 1° sections with the added magnification are used for most measurements.

The observed delineation is considered to be caused by damage rather than dopant action from the ions. It is visible before annealing and tends to be less pronounced after annealing to~ 1000°C. Recrystallization of SiC implanted layers takes place between 900°C and 1000°C. This can be seen from the reflection electron diffraction patterns in Fig. 2 for a part of a sample that was implanted with phosphorus at 100 KeV. The layer became amorphous and remained so after annealing to 900°C. By 1000°C it has re-ordered and appears similar to an unimplanted part that is also shown. Similar annealing has also been found necessary in the case of 10^{15} cm^{-2} nitrogen implants. This is in agreement with the Rutherford backscattering studies of Hart, et al[2] who find re-ordering at 750°C and above and little disorder remaining above 1200°C.

That the delineation does not disappear completely on annealing beyond what is required for recrystallization may be due to some residual defects or to some coloration change arising from doping effects. Carroll,[3] for example, has used color observations as an indication of whether diffusion is taking place in samples heated to various temperatures in different impurity environments.

Another reason for believing that the delineation is primarily due to damage is that it is also obtained on implanting the inert ion argon. It is presumed to be similar in nature to the so called "milky white" appearance characteristic of implanted regions on silicon surfaces and attributed to Raleigh scattering[4] or changes in the average dielectric properties.[5] No delineation is normally visible below the surface on sectioned silicon and the

failure to see it may be attributable to the work damage introduced by the
bevelling itself and which would tend to mask the effect. The polishing of SiC
is thought not to introduce a plastically deformed layer at the surface.[6]

The depth dependence of the delineation on energy has been investigated
for energies ranging from 100 KeV to 1 MeV. The resulting penetrations for
boron, nitrogen and phosphorus are shown in Fig. 3. The implanted dose at each
energy was 10^{15} cm^{-2} and most of the implants with the same ion were into the
same substrate in closely spaced adjacent strips. The calculated projected
range of the three ions under consideration is also included in Fig. 3 (broken
curves). If the delineation is taken as being indicative of the width of the
amorphous layer, it is seen that it follows the projected range in the case of
nitrogen while being slightly deeper for boron and phosphorus. While the
maximum in the damage distribution should be shallower than the ion range,[7]
it is evident that the amorphous layer will be distributed about and extend
beyond the damage maximum so that the experimental data is not at variance with
the calculated projected range.

The delineation depth is not as sensitive to other parameters as it is to
energy. Of the limited number of samples examined, the depth was found not to
be particularly dependent on either the dose implanted or the initial substrate
material. With boron, adjacent regions on one sample were implanted to 10^{15}
cm^{-2} and 6×10^{15} cm^{-2} at 300 KeV. The common section for the two implants showed
no difference other than the 6×10^{15} cm^{-2} delineation being far more pronounced.
As to the substrate material, the nitrogen implants, other than the 250 KeV
implant, were into a heavily colored substrate. The 250 KeV implant was into a
colorless Al doped substrate and as can be seen from the depth dependence on
energy in Fig. 3, there is no apparent shift in the measured depth. The delineation itself is generally most marked towards its deeper extreme but does not
disappear completely even at the surface. The gradation, however, allows the
deeper limits of multiple energy implants (into a common area) to be rendered
visible. Thus, the 100 KeV B implant measured in Fig. 3 was into a region that
was also implanted to 300 KeV.

From the device viewpoint, it is of interest to relate the damage delineation to junction depths. Junctions can be rendered visible by preferential
electrolytic etching in hydrofluoric acid/methanol[8] and an example is shown
in Fig. 4. Two phosphorus implanted planar diodes that were annealed to 900°C

prior to etching are seen. The interference pattern that is used in the depth
determination serves to emphasize where the converted n-type material on the
bevel has not been etched. In two instances where phosphorus junctions were
used and both the damage and etch delineation measured, the damage was found to
extend to only 65% and 70% of the junction depth. Bearing in mind the
relatively high background doping normally obtained in SiC and that doses of
device interest will be $\sim 10^{15}$-10^{16} cm^{-2} as used here, an assumption of the
damage extending to two thirds of the junction depth may be used as a rough
guide for predicting junction depths from the damage delineation.

The use of range-energy data computed by S. Roosild is gratefully
acknowledged. Part of the work reported was initiated while one of the authors
(D. E. Davies) was employed at Ion Physics.

1. F. A. Leith, W. J. King, P. McNally, D. E. Davies, C. M. Kellett, "High
    Energy Ion Implantation of Materials", Final Report, Contract No. AF19(628)-
    4970 (Jan 1967).

2. R. R. Hart, H. L. Dunlap and O. J. Marsh, Radiation Effects 9, 261 (1971).

3. P. Carroll, Silicon Carbide, p. 341, Pergamon Press (1960).

4. R. S. Nelson and D. J. Mazey, Can Journ Phys. 46, 689 (1968).

5. S. Kurtin, G. A. Shifrin, T. C. McGill, Appl Phys Lett 14, 223 (1969).

6. J. W. Faust, Silicon Carbide, p. 403, Pergamon Press (1960).

7. P. Sigmund and J. B. Sanders, Proc Int Conf on Applications of Ion Beams to
    Semiconductor Technology, p. 215, Ophrys (1967).

8. H. C. Chang, C. LeMay and F. Wallace, Silicon Carbide, p. 496, Pergamon
    Press (1960).

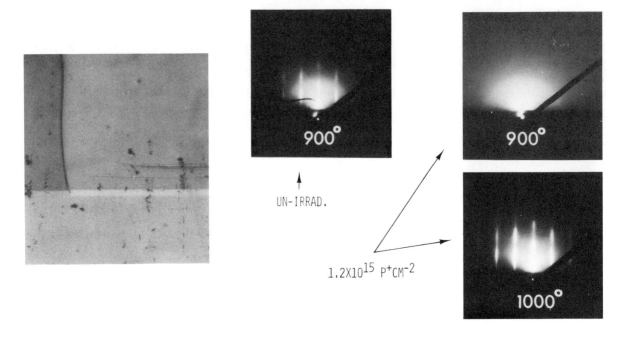

Figure 1

Figure 2

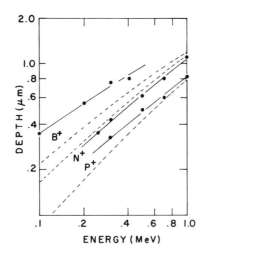

Figure 3

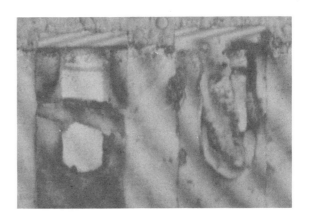

Figure 4

Au-SiC Schottky Barrier Diodes

S. Y. Wu and R. B. Campbell

INTRODUCTION

SiC Schottky barrier diodes, unlike those in Si or GaAs, have never been extensively studied. Mead and Spitzer,[1] in a study of the position of Fermi levels at metal-semiconductor interfaces, measured barrier heights of SiC Schottky diodes using differential capacitance-voltage techniques. They obtained barrier heights on Au and Al deposited on chemically etched n-type α-SiC of 1.95 and 2.0 V, respectively. Hagen[2] studied barrier heights on cleaved and etched 6H and 15R polytype SiC single crystals. The barrier height on n-type samples, as deduced from photoresponse and differential capacitance-voltage techniques, was 1.45 ($\pm$ 0.10) V, independent of the work function of the metals (Au, Ag, Al). Berman and Campbell[3] also used both photoresponse and capacitance methods and found for Au on n-type SiC a barrier height of 2.6 V.

These widely differing and ambiguous values of the barrier height prompted us to study the SiC Schottky barrier again. It is the purpose of this work to re-evaluate the barrier height on SiC Schottky barrier diodes and, furthermore, to investigate in detail the properties of Schottky barriers on SiC which result in barrier heights exceeding 1 eV.

In this paper, we describe the experimental methods and the results on the evaluation of the height of the Schottky barrier which is formed by vacuum-evaporating gold on chemically etched n-type 6H polytype SiC single crystal surfaces. Three independent techniques were used: differential capacitance versus voltage, photoresponse, and forward current versus voltage methods. The data will be compared with the barrier height values obtained previously by other workers. The results on the study of the current transport mechanism of the metal (Au) n-type SiC system will be presented. This is done with the forward current-voltage characteristics of the diodes prepared.

SAMPLE PREPARATION

The silicon carbide single crystals studied were of n-type hexagonal (6H) polytype which were grown by the sublimation technique[4] in our laboratories. They were doped with nitrogen to a concentration level of about $10^{17}/cm^3$. To eliminate or reduce the residual stress and surface damage introduced during the crystal growth process, as-grown crystal platelets of about 10-15 mils thickness were first treated with repeated oxidation and etching processes. They were then mounted and bonded to tungsten disks on one basal plane in vacuum at $1900^{\circ}C$ to form back ohmic contacts. The other face was cleaned with concentrated HF and etched in molten sodium peroxide for a few seconds. After the samples were cooled, they were rinsed thoroughly in diluted HCl and deionized water, and dipped finally in methanol and dried with a nitrogen jet.

The samples were next mounted on a metal evaporation holder covered with a tungsten mask and loaded in a vacuum system. The system, equipped with a liquid nitrogen cold trap, was pumped down to a pressure of less than 10^{-6} Torr. The gold metal was then evaporated from a tungsten-crucible boat to a thickness ranging from 500 to 1000 $\overset{o}{A}$. Two sizes of circular gold contacts were used: 8 and 50 mils diameter. All samples for testing were mounted on transistor headers. Contacts on the front gold metal were made with 1 mil gold wires bonded with gold paste.

EXPERIMENTAL METHODS AND RESULTS

Differential Capacitance Versus Voltage Method

The first method used to determine barrier heights was made by measuring small signal differential capacitances at various applied voltages. The measurements were carried out on a Boonton capacitance bridge, model 75C, at 10 kHz. A typical result is presented in Figure 1, which shows $1/C^2$ versus the applied voltage. The curve is linear with the extrapolated intercept on the voltage axis V_i of 1.34 volts.

The barrier height is related to the intercept voltage, V_i, by the equation

$$\phi_{Bn} = V_i + \zeta - \Delta\phi + \frac{kT}{q} \qquad (1)$$

where ζ is the distance from the bottom of the conduction band to the Fermi level, $\Delta\phi$ the image force lowering of the barrier height at the interface,[5] and kT/q due to the contribution of the mobile carriers in the reserve layer. The value ζ can be calculated from the doping concentration

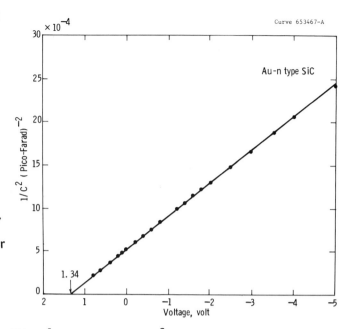

Fig. 1   Measured C^{-2} versus applied voltage.

obtained from Hall and resistivity measurements, or from the slope of the $1/C^2$ vs voltage curve shown in Figure 1 and is found to be 0.138 V at 300°C for $N_D = 3.3 \times 10^{17}/cm^3$.

The barrier lowering by the image force can be calculated from[5]

$$\Delta\phi = \left[\frac{q^3 N_D \left[V_{bi} - V - \frac{kT}{q} \right]}{8\pi^2 \epsilon_d^2 \epsilon_s^2 \epsilon_o^3} \right]^{1/4} \qquad (2)$$

where V_{bi} is the built-in potential, ϵ_o the permittivity of free space, ϵ_d the dielectric constant at optical frequencies (about 6.7 for SiC[6]), and ϵ_s the low-frequency dielectric constant. The value of $\Delta\phi$ calculated from Eq. (2) at zero bias is equal to 0.0917 volt. Substituting into Eq. (1) all the values on the righthand side, we obtain the barrier height ϕ_{Bn} of about 1.41 V.

Photoresponse Method

A value of the barrier height was also determined by measuring the short circuit photocurrent. A monochromatic light, obtained from a tungsten projection lamp with a Jarell-Ash 1/4 meter grating monochromator, was incident on the gold metal contact of the diode. The gold was about 1000 Å thick and 50 mils in diameter. The incident light beam was focused on the gold layer with a lens after passing through a slit. The short circuit current was measured with a Keithley 610BR electrometer. The measurements were made for photon energy from 1.3 eV to 2.6 eV, which is lower than the SiC band gap energy of 3.0 eV.[7] The photon density of the incident beam at different energies was first calibrated with a silicon PIN photodiode.

A typical result of the measurements is given in Figure 2, which shows the square root of the photocurrent per incident photon as a function of photon energy. The experimental result is almost linear in the energy range from 1.8 eV to 2.4 eV. At higher energies, the experimental points tend to fall below the linear curve. This is partly due to the fact that the thickness of the gold layer is greater than the electron attenuation length L in gold of 740 Å[8] as described by Spitzer and Mead.[9] The absorption coefficient of gold in the range of photon energy measured was assumed constant.

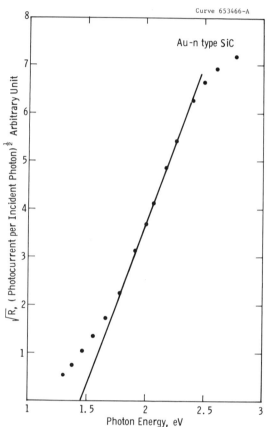

Curve 653466-A

Fig. 2 Square root of short-circuit photocurrent per incident photon as a function of photon energy at 300°K.

According to simple Fowler theory,[10] the photo-current per absorbed photon, R, can be obtained from the equation

$$\sqrt{R} = K(h\nu - q\phi) \text{ for } h\nu - q\phi > 3kT \qquad (3)$$

where K is a constant. Therefore, the extrapolated value of the linear curve on the energy axis will give the barrier height. The barrier height obtained from the photoresponse method is 1.45 eV.

Current Versus Voltage Method

The third method used to evaluate the barrier height was the forward current-voltage technique. The barrier height is determined from the saturation current density which is obtained by extrapolating the linear part of the current-voltage characteristic, in a semi-logarithmic plot, to the current axis.

The current-voltage characteristic of a Schottky barrier diode under thermionic emission is given by[11]

$$J = A^* T^2 e^{-\frac{q}{KT}(\phi_{Bo} - \Delta\phi)} \left[e^{\frac{qV}{kT}} - 1 \right] \qquad (4)$$

where J is the current density, A^* the effective Richardson constant, ϕ_{Bo} the zero-field asymototic barrier height, $\Delta\phi$ the image force Schottky barrier lowering, V the applied voltage, k the Boltzmann constant, and T the absolute temperature. Under forward bias conditions for $V > 3kT/q$, the equation can be represented by $J = J_S \exp(qV/nkT)$, where J_S is the saturation current density. The factor n which takes into account the lowering of barrier height by the applied voltage is given by[5]

$$n = 1 + \frac{1}{4} \left[\frac{q^3 N_D}{8\pi^2 \epsilon_d^2 \epsilon_s \epsilon_o^3 (\phi_{Bo} - \zeta - V - kT/q)^3} \right]^{1/4} \qquad (5)$$

Figure 3 shows, in a semi-logarithmic plot, a typical curve of forward current versus voltage measured at room temperature. For low forward voltages $V < 0.35$ volt, the thermionic emission current over the high barrier is small. The current is almost dominated by leakage currents. In the range of voltages from 0.35 to 0.85 volt the thermionic emission becomes important. The curve is linear, extending over seven orders of magnitude in current. The slope of the linear part gives the factor n of about 1.07. This value is in good agreement with that calculated from Eq. (5) for voltages from 0.35 to 0.85 volts, using the value of ϕ_{Bo} obtained below, and N_D and ζ calculated before. The calculated n value is about 1.05 ± 0.02. For voltages higher than 0.9 volt, the sub-

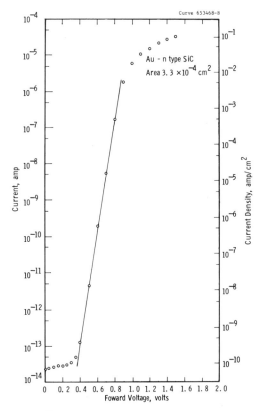

Fig. 3   Typical forward current-voltage characteristic at 300°K.

strate resistance comes into effect and the current starts to fall off from the linear curve.

The saturation current density J_S obtained by extrapolating the linear part of the curve to the current axis is about 1.5×10^{-16} A/cm^2. The barrier height is related to the saturation current density by

$$\phi_B = \frac{kT}{q} \ln \left[\frac{A^* T^2}{J_S} \right] \tag{6}$$

Using $A^* = (m^*/m_o)A$, where $m^* = 0.6 m_o$, [11] and $A = 120$ amp/cm^2/$^\circ$K^2, the Richardson constant for thermionic emission into a vacuum, we obtain the barrier height of 1.35 V.

SUMMARY

In conclusion, we have studied the SiC Schottky barrier diodes made by vacuum-evaporating gold on chemically etched n-type hexagonal (6H) SiC. The Schottky barrier height was determined from three independent techniques to be 1.40 ± 0.05 V.

ACKNOWLEDGMENT

The authors wish to acknowledge the support of the Air Force Cambridge Research Laboratories under Contract F-19628-73-C-0072.

REFERENCES

1. C. A. Mead and W. A. Spitzer, Phys. Rev. 134, A713 (1964).

2. S. H. Hagen, J. Appl. Phys. 39, 1458 (1968).

3. H. S. Berman and R. B. Campbell, Report N00014-71-C-0405 (1972).

4. D. R. Hamilton, Proc. Conf. on SiC, p. 43, Pergamon Press, New York (1960).

5. S. M. Sze, C. R. Crowell and D. Kahng, J. Appl. Phys. 35, 2534 (1964).

6. W. G. Spitzer, D. Kleinman, and D. Walsh, Phys. Rev. 113, 127 (1959).

7. W. J. Choyke and L. A. Patrick, Phys. Rev. 127, 1868 (1962).

8. C. R. Crowell, W. G. Spitzer, L. E. Howarth and E. E. Labate, Phys. Rev. 127, 2006 (1962).

9. W. G. Spitzer and C. A. Mead, J. Appl. Phys. 34, 3061 (1963).

10. R. H. Fowler, Phys. Rev. 38, 45 (1931).

11. S. M. Sze, Physics of Semiconductor Devices, New York: Wiley, 1969, p. 20.

International Conference on SiC—1973: Closing Remarks

C. E. Ryan

When I was asked to make a few closing remarks at this Conference, I was rather pleased because I have also had the honor of making the opening remarks at the First Conference on Silicon Carbide at Boston in 1959. Again at the Second Conference at Penn State I had the opportunity of participating in the start of the Conference. Since I suspect this will be the last International Conference on Silicon Carbide under AFCRL aegis, I will have had, in one sense, the last word as well as the first word.

I do not need to give any new perspective on Silicon Carbide. Mr. R. C. Marshall did that very well at the start of this Conference. Nor would I presume to review the Conference on a paper by paper, session by session, or even a day by day basis. We all have our special interests and orientations and all of us have been subjected to a mass of new and refined information in at least seventy-five technical papers. Certainly the comprehensive invited papers will require very careful study just as they required careful preparation by critical experts. Many of the other papers present new ideas, new processes and quite significantly, a new degree of sophistication that clearly indicates silicon carbide is coming of age.

I will eagerly await the Proceedings of this Conference. I believe they will be even more important than the Proceedings of the two previous Conferences which are the worlds most frequently cited references on Silicon Carbide.

Let me put this series of Conferences into context with each other and with the Air Force Cambridge Research Laboratories. Why did we sponsor these Conferences in the first place? Primarily because we recognized the importance and the significance of silicon carbide as a semiconductor material with unique properties, particularly those which render it suitable for use in difficult environments and in situations where the utmost in reliability is required. We at AFCRL feel that we have done our job with a minimum of expense to the United States Air Force. We have encouraged the development of this extremely stable

651

and superbly reliable semiconductor to a point where it now has a technology
ready for exploitation whenever the need arises. Moreover this technology is
fully documented in the Proceedings of the three International Conferences and
the references to the many papers contained therein.

Where, then, do we go from here? I visualize two directions which should be
pursued in parallel. The first is the exploitation of the available silicon
carbide technology for specific devices such as shown in Table I. If the need
is there, the basic technology is available. I will not dwell on this direction.
The direction is to extend this technology of high temperature refractory
material to its neighbors. Dr. Glenn Slack discussed this possibility rather
well in his paper. Let me be clear about this. We are not looking for a substitute for silicon carbide as a refractory high temperature wide band gap semiconductor. We are suggesting that the physics and chemistry of this material
have much in common with those of related highly refractory materials which are
largely centered in the small portions of the periodic table covered by B, C, N,
Al, Si, P and O as shown in Table II. This is an extremely important group of
elements for our modern scientific technology. It includes the important single
crystals, Diamond, Silicon, Quartz, Sapphire, Boron Nitride, Aluminum Nitride,
Boron Phosphide as well as Silicon Carbide. In the polycrystalline and amor-
phorous areas our leading high temperature conductor is graphite. The best
insulators are Al_2O_3 and Boron Nitride. Our best liquid encapsulant is B_2O_3.
Quartz is our essential laboratory ware as well as our surface mask for silicon
and silicon carbide. SiO and Si_3N_4 are very important to the semiconductor and
high temperature refractory technologies.

Diamond, Silicon Carbide and Aluminum Oxide are the leading abrasive materials and Boron Carbide is very important in the machine tool industry. There
are many other important real and potential applications for materials composed
of these elements.

This small area of the periodic table (Table II) has many difficult and un-
solved problems. If there is enough interest, AFCRL will seek to sponsor a new
conference in due time which will include these refractory materials as well as
silicon carbide. Your comments and opinions are sincerely solicited.

On behalf of Air Force Cambridge Research Laboratories and all of our Con-
ferees, I wish to thank all of you for attending and contributing by your papers
and by your discussion. This includes the people who are here for the first time

as well as those of you who are veterans of the 1968 Conference and of the 1959 Conference. I sincerely hope that each of you has been stimulated as I have by the new ideas, concepts, techniques and information which has been shared in the last few days. I suspect I am not the only one here who has been reinventing devices that are no longer too advanced for the state of the material technology.

I also wish to thank the Speakers, Invited Speakers, Session Chairmen, and Day Organizers. A particular word of thanks and praise is due to Professor J. W. Faust, Jr., of the University of South Carolina, who worked very hard and succeeded in outdoing the excellent job he did in directing the 1968 Conference.

One final word. As it has been in the past, our door is always open to discuss any aspect of silicon carbide with you. We sincerely hope to continue to act as an information exchange center or "focal point" on all aspects of silicon carbide as an electronic and/or optical material.

Thank you.

TABLE I

SUMMARY OF SILICON CARBIDE DEVICE POTENTIAL

1. High Temperature Devices (above 500°C)
2. High Power Devices (EB-PN Devices)
3. High Frequency Devices (Impatt Diodes)
4. High Radiation Resistant Devices
5. High Reliability Devices
6. Cold Cathode Devices
7. LED Devices
8. Schottky Diode Devices
9. Specialized Devices:

      (a) UV Detectors
      (b) Radiation Detectors
      (c) High Temperature Photocells
      (d) Heterojunction Devices

TABLE II

III-A	IV-A	V-A	VI-A
B	C	N	O
Al	Si	P	

PART VI
APPENDIXES

Appendix I

Tables of Etchants for SiC

J. W. Faust, Jr., and H. M. Liaw

The subjects of surface preparation and of various processes for etching silicon carbide were reviewed by Faust (27). This review included a table of etchants and covered the literature up to 1959. The literature on surface preparation and etching, including tables of etchants, in the next ten years was reviewed by Jennings (37). Since Jennings review in 1969, more information on etching of silicon carbide has been published. All of the data reported as etchants have been compiled into the tables in this appendix. Materials that were reported as having no attack on SiC or that leave a reaction product on the surface are thus excluded. Thus the work on oxidation and the action of phosphoric acid, as two examples, that were reviewed in references 27 and 37 are not included. Also excluded are reports that merely say that SiC is attacked by a material.

Under the column "Etchant", ratios of etchants are given by weight for solids or by volume for liquids. Under "Remarks", temperature, times, and etch rates are those given in the articles and should be considered as approximate. Words or phrases in quotes are those used by the individual authors. The silicon face is designated as (0001) for α-SiC and (111) for β-SiC while the carbon face is designated as (000$\bar{1}$) for α-SiC and ($\bar{1}\bar{1}\bar{1}$) for β-SiC. Where ever possible, the polytype or polytypes used in the study are given in the Ramsdell notation. In the very old literature, the terms type I, II etc. were used. The Ramsdell notation for these old terms are: type I, 15R; type II, 6H; type III, 4H; type IV, 21R; type V, 51R; and type VI, 33R. It must be remembered that many of the crystals used in the early work are probably not pure polytypes. The term "distorted" hexagon, used in the "Remarks", refers to a hexagon with three long sides alternating with three short sides. Four references were left out of the table for lack of information.

657

One paper referred to differentiating between hexagonal and rhombohedral polytypes by etch figures (63), but the etchant could not be identified. The other three were patents (64), (65), and (66). To the best of our knowledge, these tables give all of the data on etchants through 1972.

Molten Salt Etching

Etchant	Temp °C	Remarks	Ref.
KOH:KNO₃ (1:4)	650-700	Etch for 5 min to clean surface before alloying. Also cleans p-n junctions.	(1) (23)
KOH:KNO₃ (1:1)	600	Etch rate 1μ/min on (000$\bar{1}$) and 0.1μ/min on (0001). Both sides smooth and clean.	(21)
KOH	600-800	Etch rate 1μ/min on (000$\bar{1}$) and 7μ/min on (000$\bar{1}$). Report they "deduced" the identity of the faces.	(22) (46)
K₂CO₃:KNO₃ (2:1)	900	Type II crystals. Regular hexagonal pits on one base, other base quite dull. Pyramids and prisms had modified triangles.	(19)
K₂CO₃	400-900	Reveals etch pits.	(46)
KNO₃:NaOH (2:1)	500-700	Reveals dislocations.	(13)
K₂CO₃:Na₂CO₃ (3:1)	1000	Reveals dislocations.	(8) (14)
K₂CO₃:Na₂CO₃ (1:1)	950	Reveals etch pits.	(46)
K₂CO₃:Na₂CO₃ (1:3)	920	Etch for 1.5 min. Reveals dislocations.	(26)
K₂CO₃Na₂CO₃ (1:3)	900	Etch rate 0.5 mg/cm²/min. No reaction at 500°C.	(27)
KNO₃:K₂SO₄:KOH (2:1:1)	700	p-n junction treatment.	(1)
KNO₃:Na₂CO₃:KOH (1:1:1)	700	p-n junction treatment.	(1)
KNO₃:K₂SO₄:NaOH (2:1:1)	500-700		(13)
KOH:NaNO₃:Na₂CO₃	------	"Distorted" hexagonal pits on one face of Type I crystals. Regular hexagons on one base of Type II crystals.	(7)
KClO₃:KNO₃:K₂CO₃ (1:1:1)	500-700		(13)
Na₂CO₃	900	Reveals dislocations on (0001). Etch rate 0.1 mg/cm²/min; 0.3μ/min.	(2) (27) (46)
Na₂B₄O₇ (Borax)	1000	Reveals dislocations on (0001). Etch rate 0.3μ/min. Type I, IV, and VI give	(2) (19) (10) (11)

Molten Salt Etching

Etchant	Temp °C	Remarks	Ref.
$Na_2B_4O_7$ (Borax)	1000	distorted hexagons. Type II regular hexagonal pits (19) says on both faces and bisymmetrical rectangular pits on prism faces	(36)
$NaNO_3:Na_2CO_3$	-----	Type II crystals contained equilateral triangle or two equilaterial triangles one turned 60° to the other.	(5)
		One face always bright with "distorted" hexagons. Other face etched faster and had no etch figures.	(6)
$NaNO_3:Na_2O_2$ (1:1)	400-600	"Selective" and "Symmetrical".	(15)
$NaOH:Na_2O_2$ (3:1)	700	Reveals dislocations on (111) but not $(\bar{1}\bar{1}\bar{1})$ on β-SiC. Etch rate 7μ/min.	(3) (4) (16)
$NaOH:Na_2O_2$ (3:1)	400-600	Etch for 10 min. Reveals dislocations on β-SiC.	(17)
$NaNO_2:Na_2O_2$ (9:1)	500	Etch for 1 min. 6H smooth surface on Si face; rough surface on C face by x-ray determination.	(18)
$NaNO_2:Na_2O_2$ (1:1)	550	Reveals dislocations. Etch rate at 500°C ~0.1 mgm/cm²/min; at 900°C ~1 mgm/cm²/min.	(25) (15) (18) (20)
Na_2O_2	350-900	Etch for 15 to 30 secs. Fine details of surface.	(27) (58)
NaOH	600-800	Etch for 1 to 3 min. Reveals dislocations.	(24) (46) (26)
NaOH	800	Etch rate 1.5μ/min.	(25)
NaOH	900		(27)
$NaF:Na_2SO_4$ (1:1)	950	Melt stirred by stream of gas. Well defined pits on both basal planes. Etch rate 5μ/min.	(9)
PbO	800	Etch rate 100μ/hr.	(12)
$PbO:PbF_2$	600		(12)

Molten Salt Etching

Etchant	Temp °C	Remarks	Ref.
KCl:LiCl:NaCl (ternary eutectic with 1-10% NaOH or Na$_2$CO$_3$)	450-500	Etch 10 to 15 mins. Fine details of surface.	(24)

Gas-Phase Etching

Etchant	Temp °C	Remarks	Ref.
H$_2$	1530-1743	On β-SiC (111) etches faster than ($\bar{1}\bar{1}\bar{1}$). Etch rates for solngrown crystals 0.3-0.5μ/min at 900°C, 1.0μ/min. at 1000°C. Less than 0.02μ/min for undoped epitaxial grown layers.	(28)
	1600-1700	Flow rate 8.5 cm/sec. Got smooth surfaces (non-preferential) etch rate 0.25-4.0μ/min.	(29)
		1550°C rate affected by susceptor. Temperature, flow rate, and partial pressure studied.	(30) (31)
Cl$_2$ (6%), O$_2$ (26%) in Ar.	850-900	Etch pits on (111) but not on ($\bar{1}\bar{1}\bar{1}$) of β-SiC.	(3) (39)
Cl$_2$:O$_2$ gas mixture	1100	Shows polarity-matt surface is (000$\bar{1}$) (carbon face); pits on (0001) (silicon face). Determined by x-rays.	(32)
Ar:Cl$_2$:O$_2$ (68:26:6)	900	Flow rate 175 cc/min. non-preferential etch. Etch rate 0.25μ/min.	(33)
Cl$_2$:O$_2$ (2:1)	1000	Etch rate 1μ/min. rate faster on (000$\bar{1}$).	(34)
Cl$_2$	1000	Flowrate 10 cc/min. to thin sample for TEM. Etch rate 0.5μ/min. Layer of carbon left on surface.	(12) (35) (34) (36) (27) (37)
CO	1300	Attacks SiC leaving a carbon layer.	(38) (57)
F$_2$	300	Preferential etch at low gas flow.	(40) (41)

Gas-Phase Etching

Etchant	Temp °C	Remarks	Ref.
ClF_3	200–400	Etch at 400°C for 2 to 3 min. Etch pits on Si face only. Also pits on {111}.	(42) (43)

Thermal Etching

Etchant	Temp °C	Remarks	Ref.
Ar	1900–2100	SiC decomposes. Activation energy 111.4 kcal/mole rate not dependent on impurities.	(44)
vacuum	------	Hexagonal pits on (0001) at dislocations. Say better than chemical etching.	(45)
vacuum	1500	Surface decomposition.	(52)

Molten Metal Etchants

Etchant	Temp °C	Remarks	Ref.
Pt	------	Pt reacts at high temperature. Reaction product removed by aqua regia.	(48)
Cr – Si alloy	1600–1700	Etch for one hr. Etch pits on both {0001} faces. Extract from alloy by ClF_3.	(43)
Mo, W, Au–W	------	Similar to Pt above.	(53) (54) (37)
Ta, Fe, Co, Ni			(55) (56)
Pd			

Electrolytic Etching

Electrolyte	Voltage	Current or c.d.	Time	Remarks	Ref.
HF, HNO₃ in alc. or glycol				Improves electrical characteristics.	(1)
H₃PO₄, 17% alc. soln.				Improves electrical characteristics.	(1)
Molten KOH	0-10	5 A/cm²	few min.	Electrolytic polish for 6H and 15R. Use a stainless steel or carbon crucible as the cathode.	(47)
20% aq. KOH	3-5	1A	4-5	Stainless steel cathode.	(49)
10% chromic acid (CrO₃:H₂SO₄:H₂O)	3-5	1A	4-5	Stainless steel cathode.	(49)
HClO₄:HAc:H₂O (1:3:6)	3-5	1A	4-5	Stainless steel cathode.	(49)
K₂CO₃:10% aq. NaNO₃	3-5	1A	4-5	Stainless steel cathode.	(49)
10% aq. Oxalic acid	3-5	1A	4-5	Stainless steel cathode.	(49)
HAc:HF:H₂O (1:1:1)	3-5	1A	4-5	Stainless steel cathode.	(49)
10% H₂C₂O₄ (Oxalic acid)	5-10	1A	1/2	Anodizing β-SiC.	(50)
HF Solution				For preparing both surfaces smooth and clean. Use only for polytype.	(21)
HF:HNO₃:glycol				For cleaning diodes.	(23)
20% aq. KOH	6	1A	1/3	Revealing microstructure.	(51)
400 ml HAc, 120 gm CrO₃, 380 ml H₂O, 30 ml H₂SO₄, 20 ml C₂H₅OH	6	.4A	1/10	"Beautiful" coloration.	(51)
HF:H₂O (1:50)		20.50 mA		Reveals p-n junction in p-type material.	(33)

Other Etching Techniques

Etchant	Remarks	Ref.
Sat'd aqueous soln. of $K_3Fe(CN)_6$:NaOH (1:1)	Etch at 110°C. Pits form on (0001) face.	(59)
Ar ions	SiC crystal made cathode in a discharge at 2kV in argon at 3×10^{-2} torr. C.d. 1 mA/cm^2. Etch rate 3 to 4 μ/hr.	(60) (61)
----------	Electron Bombardment	(62)

REFERENCES

(1) Afanaseva, G. M., Ryzhikov, I. V., Kmita, T. G., and Pavlichenko, V. I., Karbid Kremniya Dokl. Vses. Konf., Kiev, 265 (1964).

(2) Amelinckx, S., Strumane, G., and Webb, W. W., J. Appl. Phys. $\underline{31}$, 1359 (1960).

(3) Bartlett, R. W., and Barlow, M., J. Electrochem. Soc. $\underline{117}$, 1436 (1970).

(4) Barlett, R. W., and Martin, G. W., J. Appl. Phys. $\underline{39}$, 2324 (1968).

(5) Baumhauer, H., Zeits. Krist. $\underline{55}$, 249 (1915).

(6) Becke, F., Zeits. Krist. $\underline{24}$, 537 (1895).

(7) Espig. H., Abhand, Math - Phys. kl. Sach. Akad. Wiess. Leipzig $\underline{38}$, 53 (1921).

(8) Faust, J. W. Jr., in "Methods of Experimental Physics" vol. 6, page 147, Academic Press, New York 1959.

(9) Gabor, T., and Jennings, V. J., Electrochem. Techn. $\underline{3}$, 31 (1965).

(10) Gevers, R., Amelinckx, S., and Dekeyser, W., Naturwiss. $\underline{39}$, 448 (1952).

(11) Gevers, R., Nature $\underline{171}$, 171 (1953). also J. Chem. Phys. $\underline{50}$, 321 (1953).

(12) Gabor, T., and Stickler, R., Nature $\underline{199}$, 1054 (1963).

(13) Ellis, R. C. Jr., in "Silicon Carbide" (eds. O'Connor and Smiltens), Pergamon Press, New York 1960, page 420.

(14) Horn, F. H., Phil. Mag. $\underline{43}$, 1210 (1952).

(15) IBM Deutschland Internationale Buero-Machinen GmbH, Brit. Patent 1,023,749 March 23 (1966).

(16) Jennings, V. J., Sommer, A., and Chang, H. C., J. Electrochem. Soc. $\underline{113}$, 728 (1966).

(17) Faust, J. W. Jr., Electrochemical Society Meeting, Ottawa, Canada, September (1958).

(18) Liebmann, W. K., J. Electrochem. Soc. $\underline{111}$, 885 (1964).

(19) Weigel, O., Nach. Ges. Wiss. Gott. Math - Phys. Kl, 264 (1916).

(20) Brack, K., J. Appl. Phys. $\underline{36}$, 3560 (1965).

(21) Brander, R. W., and Boughey, A. L., Brit. J. Appl. Phys. $\underline{18}$, 905 (1967).

(22) Chang-Lin Kuo, and Shih-Hsin, Tang, Wu Li Hsueh Pao $\underline{22}$, 831 (1966).

(23) Duisenbaev, M., Soviet Physics, Semiconductors $\underline{4}$, 1163 (1971).

(24) Shaffer, P. T. B., J. Appl. Phys. <u>39</u>, 5332 (1968).

(25) Brander, R. W., and Sutton, R. P., Brit. J. Appl. Phys. 2, 309 (1969).

(26) Patel, A. R., and Mathai, K. J., J. Phys. Chem. Solids <u>30</u>, 2482 (1969).
      also Indian J. Pure Appl. Phys. <u>7</u>, 486 (1969).

(27) Faust, J. W. Jr., in "Silicon Carbide", (eds. O'Connor and Smiltens),
      Pergamon Press, New York 1960, page 403.

(28) Bartlett, R. W., and Mueller, R. A., Mat. Res. Bull. <u>4</u>, S-341 (1969).

(29) Chu, T. L., and Campbell, R. B., J. Electrochem. Soc. <u>112</u>, 955 (1965).

(30) Harris, J. M., Gatos, H. C., and Witt, A. F., J. Electrochem. Soc. <u>116</u>,
      381 (1969).

(31) Kumagawa, M., Kuwabarg, H., and Yamada, S., Jap. J. Appl. Phys. <u>8</u>, 421 (1969).

(32) Wallace, C. A., J. Appl. Cryst. <u>3</u>, 328 (1970).

(33) Campbell, R. B., and Berman, H, Mat. Res. Bull. <u>4</u>, S-199 (1969).

(34) Smith, R. C., Electrochem. Soc., New York meeting, abst. 123 (1963).

(35) Lea, A. C., Trans. Brit. Ceram. Soc. <u>40</u>, 93 (1941).

(36) Thibault, N. W., Am. Min. <u>29</u>, 249 (1944).

(37) Jennings, V. J., Mat. Res. Bull. <u>4</u>, S-199 (1969).

(38) Ervin, G. Jr., J. Amer. Ceram. Soc. <u>41</u>, 347 (1958).

(39) Haga, L. J., and Tucker, T. N., U. S. Patent 3,398,033 August 20, 1968.

(40) Knippenberg, W. F., Philips Res. Repts. <u>18</u>, 270 (1965).

(41) Purdy, R. C., J. Am. Ceram. Soc. <u>17</u>, 39 (1934).

(42) Cooks, F. H., Das, B. N., and Wolff, G. A., J. Mat. Sci. <u>2</u>, 470 (1967).

(43) Wolff, G. A., Das, B. N., Lampert, C. B., and Mlavsky, A. I., Mat. Res. Bull.
      <u>4</u>, 567 (1969).

(44) Ghoshtagore, R. N., Solid State Electronics <u>9</u>, 178 (1966).

(45) Savitskii, K. V., Ilynshchenkov, M. A., Burrakov, K. K., and Bykonya, A. F.,
      Kristallographiya <u>11</u>, 341 (1966).

(46) Yasuda, Sadao, and Nakamura, Takayuki, Tokai Denkyoku Giho <u>23</u>, 16 (1963).

(47) Barrett, C. S., Barrett, M. A., Mueller, R. M., and White, W., J. Appl. Phys.
      <u>14</u>, 2727 (1970).

(48) Biedermann, E., French Patent 1,498,862, (Cl, Hol1) 2° Oct. (1967).

(49) Walker, D. E. Y., J. Mat. Sci. $\underline{2}$, 197 (1967).

(50) Walker, D. E. Y., Prakt. Metalographie $\underline{5}$, 376 (1968).

(51) Robinson, G. W. Jr., and Gardner, R. E., J. Am. Ceram. Soc. $\underline{47}$, 201 (1964).

(52) Marsh, O. J., and Dunlap, H., Radiation Effects $\underline{6}$, 301 (1970).

(53) Marsh, O. J., private communication.

(54) Harmon, C. G., and Mixer, W. G. Jr., "A Review of Silicon Carbide",
      Battelle Memorial Institute No. 748, June (1952).

(55) Hall, R. N., J. Appl. Phys. $\underline{29}$, 914 (1958).

(56) Campbell, R. B., and Chu, T. L., J. Electrochem. Soc. $\underline{113}$, 825 (1966).

(57) Ruff, O., Trans. Electrochem. Soc. $\underline{68}$, 87 (1935).

(58) Duval, C., in "Inorganic Thermogravimetric Analysis", 2nd Edition, Elsevier,
      New York (1963).

(59) Harris, J. M., Gatos, H. C., and Witt, A. F., J. Electrochem. Soc. $\underline{116}$,
      672 (1969).

(60) Drum, C. M., Phys. Status Solidi $\underline{9}$, 635 (1965).

(61) Dillon, J. A., in "Silicon Carbide" (eds., O'Connor and Smiltens), Pergamon
      Press, New York, 1960, page 235.

(62) Wright, M. A., J. Electrochem. Soc. $\underline{112}$, 1114 (1965).

(63) Padurav, N. N., Neves Jahrb. Mineral., Geol., Monatsh, $\underline{1945-48A}$, 144 (1948).

(64) Ebert, E., and Spielman, W., U. S. Patent 3,421,956.

(65) Chong, H. C., U. S. Patent 3,078,219 (1963).

(66) Int. Bus. Mach. Corp., British Patent 1,141,513 (1969).

Appendix II

Tables of Data on Silicon Carbide

"Electronic Properties Information Center (EPIC) Data Sheets on Silicon Carbide" were published in June 1965 (DS-145 by M. Neuberger AD465-161). In August 1968, Neuberger issued a supplement to DS-145 as "Interim Report No. 62" for the attendees at the Second International Conference on Silicon Carbide held on 21-23 October 1968 at Penn State University. The tables in this report were included as an appendix (minus the references) in the conference proceedings.

The Executive Committee and the Program Committee of the International Conference on Silicon Carbide-1973 felt it highly desireable to revise these tables and bring them up to date for inclusion in this conference proceedings. The Conference Director, Professor J. W. Faust, Jr., undertook this task with valuable assistance from the International Committee on Silicon Carbide; in particular from Drs. R. B. Campbell, W. J. Choyke, and Lyle Patrick.

PHYSICAL AND ELECTRONIC PROPERTIES

Physical Properties	Type	Value	Unit	Temp (^{o}C)	Reference
Formula		SiC			
Molecular Weight		40.1			
Density	2H	3.214	g/cm^3	20	16
	6H	3.211			
	β	3.210			
	β	3.166	g/cm^3	20	25
	β	3.125		1000	
	β	3.075		2000	

Color				
Pure α colorless	N-doped 6H green			24
Al-doped α blue, black	N-doped 8H yellow-orange			
B-doped α brown, black	N-doped 15R orange-yellow			
Pure 3C yellow	N-doped 27R red			
N-doped 3C yellow-green	N-doped 24R purplish-grey			
N-doped 4H yellow-orange				

	Type	Value	Unit	Temp (^{o}C)	Reference
Melting point (35 atm.) (decomposes)	β	2830 ± 40	^{o}C		34

Lattice Symmetry	Type	Value			Reference
cubic (fcc)	β	$T_d^2 - f\bar{4}3m$			45
hexagonal	α	$C_6^4 v5 - C6mc$			
rhombohedral	α	$C_{3v}^5 - R3m$			

Lattice Parameter		a_o	c_o	Unit	Reference
cubic	β	4.3596		$\overset{o}{A}$	41
hexagonal	2H	3.0763	5.0480		1
	4H	3.076	10.046		20
	6H	3.08065	15.11738		41
	8H	3.079	20.147		20
	15R	3.073	37.30		
	21R	3.073	52.78		
	33R		82.5		51
	66H		165.88		
	126R		371.4		

PHYSICAL AND ELECTRONIC PROPERTIES

Physical Properties	Type	a_o	α	Z	Reference
Lattice Parameter (continued)					
hexagonal cell	15R	12.691Å	13° 54' 5"	5	20
rhombohedral cell	21R	17.683Å	9° 58'	7	

	Type	Value	Unit	Scale	Temp.	Reference
Hardness	β	9.2–9.3		Mohs	20°C	25
||c-axis	α	2130	kg/mm^2	Knoop (100g)	20°C	36
⊥ c-axis	α	2755				
Polished,||(100)	β	2670				
Polished, ⊥(111)	β	2815				
	β	3100–3475	kg/mm^2	Knoop (100g)	20°C	25

	Type	Value	Unit	Temp. (°C)	Reference
Compressibility (to 500 atm)		0.21	10^{-6} kg/cm^2	20	22
Thermal Expansion	α	5.12	10^{-6} cm/°C	25–1000	26
		5.48		25–1500	
		5.77		25–2000	
		5.94		25–2500	
	β	3.8	10^{-6} cm/°C	200	25
		4.3		400	
		4.8		600	
		5.2		800	
		5.8		1000	
		5.5		1400–1800	
Coefficient of Thermal Expansion					
(1/a)(da/dT)	6H	4.2	°C^{-1}	700°K	41
(1/c)(dc/dT)		4.68			
Specific Heat	β	0.17	cal/gm°C	20	25
		0.22		200	
		0.28		1000	
		0.30		1400–2000	
	α	0.165	cal/gm°C	27	7
		0.27	cal/gm°C	700	37
		0.35	cal/gm°C	1550	

PHYSICAL AND ELECTRONIC PROPERTIES

Physical Properties	Type	Value	Unit	Temp. (°C)	Reference
Thermal Conductivity	α	0.410 0.335 0.255 0.213	W/cm°C	20 600 800 1000	37
$\perp$(111)	β	0.255 0.155 0.121 0.125	W/cm°C	200 1000 1400 2000	25
$\parallel$(111)		0.226 0.155 0.138	W/cm°C	200 1000 1400–2000	
Shear Modulus	α	27	10^6 psi	20	30
	α	27.85	10^6 psi	20	9
Bulk Modulus	α	14.01	10^6 psi	20	37
Youngs Modulus	α α	69 64	10^6 psi	20 1500	30
	α	65	10^6 psi	20	9
Elastic Constants c_{11} c_{12} c_{33} c_{44} $c_{66} = (c_{11}-c_{12})/2$	6H	5.00 0.92 5.64 1.68 2.04	10^{12} dyne/cm^2	70–300°K	2
c_{11} c_{12} c_{44}	β	2.89 2.34 0.554	10^{12} dyne/cm^2	300°K	22

Electronic Properties	Symbol	Type	Value	Unit	Temp. (°K)	Ref.
Energy Gap	E_g	2H	3.30	eV	2–8	31
		6H	2.86	eV	300	32
		β	2.2, 2.6	eV		17
		β	2.2	eV	300	32
		β	2.6	eV	300	47
Exciton Energy Gap	E_{gx}	β	2.390	eV	4.2	10
		6H	2.023		4.2	12
		8H	2.80		4.2	10
		33R	3.002		4.2	12
		15R	2.986		4.2	12
		21R	2.853		4.2	22a
		4H	3.265		4.2	13
		24R	2.728		8	46

PHYSICAL AND ELECTRONIC PROPERTIES

Electronic Properties	Symbol	Type	Value	Unit	Temp. ($^\circ$K)	Ref.
Energy Gap Shift with Temperature	dEg/dT	β	-5.8 ± 0.3	10^{-4}eV/$^\circ$K	295–700	14
		6H	-3.3	10^{-4}eV/$^\circ$K	300–700	11
Deformation Potential for Conduction Band	Ξ	α	11.5	eV	300–1300	43
Work Function	ϕ	α	4.6	eV	300	18
Effective Mass	m_n^*	β	0.41 ± 0.04	m_o	300	33
	m_n^*	6H	0.25 ± 0.02	m_o	300	21
		15R	0.28 ± 0.02			
Lifetime	τ_n	α	10^{-8}–10^{-9}	sec.	300	40
Dielectric Constant	ε_s	β	9.72		300	48
	$\varepsilon_s \perp$	6H	9.66		300	
	$\varepsilon_s \parallel$	6H	10.03		300	
	$\varepsilon_\infty \perp$	6H	6.52		300	48
	$\varepsilon_\infty \parallel$	6H	6.70		300	
	ε_∞	β	6.52		300	
Refractive Index λ_{Na}	n	β	2.48		300	42
	ε	4H	2.712		300	49
λ_{Na}	ω	4H	2.659			49
	ε	6H	2.69			42
	ω	6H	2.647			42
	ε	15R	2.697			49
	ω	15R	2.650			49
Piezoelectric Stress Constant	ε_{33}	6H	0.2	Coul/m^2	300	44
	ε_{15}	6H	0.08			
Debye Temperature	θ	β	1430	$^\circ$K		33
		α	1200		300	39
Thermal EMF	Q		-70	μV/$^\circ$C	293	6
			-110		1273	
Phonon Energy	LO(x)	β	102.8	meV	300	23
	LA(x)		79.4			
	TO(x)		94.4			
	TA(x)		46.3			

PHYSICAL AND ELECTRONIC PROPERTIES

Electronic Properties	Symbol	Type	Value	Unit	Temp.	Ref.
Phonon Energy	LO(Γ)	β	120.5		300	50
	TO(Γ)		98.7			
	LO(Γ)	All α types	103.9			
	TO(Γ)		95.0			
	LA(Γ)		75.6			
	TA(Γ)		33.0			
Magnetic Suscepti-bility	χ	6H	-10.6×10^{-6}	$(\text{g-mole})^{-1}$	300	15
Spectral Emissivity	$\varepsilon(\lambda)$		0.94 ($\lambda=9\mu$)		1800	19
Mobility	μ_n	β	1000	$\text{cm}^2/\text{V-sec}$	300	38
Electrical Resis-tivity	ρ	β	$10^{-2}-10^3$	Ω cm	300	28
		β	$>10^6$			29

MOBILITY AND RESISTIVITY OF ALPHA SiC

Type	μ (cm^2/V sec.)			ρ (Ω cm)			N_O (cm^{-3})		
	-196°C	25°C	600°C	-196°C	25°C	600°C	-196°C	25°C	600°C
6H n	710	462	34	725	0.9	8.2	10^{13}	10^{16}	2×10^{16}
6H n*		171	21		95	32		4×10^{14}	9×10^{15}
6H p		16			5.2			7×10^{16}	
6H n**	936	607	248	0.01	0.002	0.001	7×10^{17}	5×10^{18}	2×10^{19}
6H n†	41	37	21	.67	.31	.23	2×10^{17}	5×10^{17}	10^{18}
15R n		382						$< 10^{17}$	

 * compensated

** epi-layer (P)

† epi-layer (As)

REFERENCES: (3), (4), and (8)

REFERENCES

(1) Adamsky, R. F. and Merz, K. M., Z. f. Krist., 111, 350 (1959).

(2) Arlt, G. and Schodder, G. R., J. Acoust. Soc. of Amer., 37, 384 (1965).

(3) Barrett, D. L., J. Electrochem. Soc., 113, 1215 (1966).

(4) Barrett, D. L. and Campbell, R. B., J. Appl. Phys., 38, 53 (1967).

(5) Belle, M. L., et al., Soviet Phys. Semiconductors, 1, 315 (1967).

(6) Union Carbide Corp., Parma Res. Lab., Thermoelectric Materials, by
        Breckenridge, R. G., Bi-Monthly PR no. 8, Mar. 28 - May 28, 1960.
        Contract no. Nobs-77066. June 15, 1960. AD-245 092.

(7) Royal Aircraft Establishment, Silicon Carbide - A Review, by Brown, A.R.G.
        Tech. Note no. Met/Phy 325. AD-249 685.

(8) Campbell, R. B., Westinghouse Astronuclear Labs., private communication.

(9) Carnahan, R. D., J. Amer. Ceram. Soc., 51, 223 (1968).

(10) Choyke, W. J., et al., Phys. Rev., 133, A1163 (1964).

(11) Choyke, W. J. and Patrick, L., "Silicon Carbide", (ed. by O'Connor and
        Smiltens) Pergamon Press, New York, 1960, 306.

(12) Choyke, W. J., et al., Phys. Rev., 139, A1262 (1965).

(13) Choyke, W. J., et al., In Int. Conf. on Semiconductor Phys., Proc., 7th,
        Paris, 1964., v. 1 (ed. by Hulin, M.) N. Y., Acad. Press, 1964, 751.

(14) Dalven, R., J. of Phys. and Chem. of Solids, 26, 439 (1965).

(15) Das, D., Indian J. of Phys., 40, 684 (1966).

(16) De Mesquita, A. H. G., Acta Cryst., 23, 610 (1967).

(17) Picatinny Arsenal, Dover, N. J., Feltman Res. Labs., The Band-Gap Photocon-
        ductivity of High-Purity Single Crystals of Cubic (beta) Silicon Carbide,
        Tech. memo., by Detrio, J. A., Feb. 1968. 31 p. AD 666 210.

(18) Dillon, J. A., Jr., et al., J. Appl. Phys., 30, 675 (1959).

(19) Martin Co., Martin-Marietta Corp., Orlando, Fla. Infrared Signature Char-
        acteristics by, Durand, J. L. and Houston, C. K. ATL-TR-66-8. Con-
        tract no. AF 08 635 5087. Jan. 1966. 174 p. AD 478 597

(20) Donnay, J. D. H., Crystal Data, Determinative Tables. 2nd Ed., American
        Crystallography Assoc., 1963.

(21) Ellis, B. and Moss, T. S., Royal Soc. of London, Proc., $\underline{299A}$, 383 (1967).

(22) Gmelins Handbuch der Anorganischen Chemie; 8th edition, Silicium. Part B.
       Weinheim, Verlag Chemie, GmbH, 1959.

(22a) Hamilton, D. R., et al., Phys. Rev., $\underline{138}$, A1472 (1965).

(23) Choyke, W. J., Hamilton, D. R., and Patrick, L., Phys. Rev., $\underline{133}$, A1163 (1964).

(24) Golightly, J. P., Can. Mineral., $\underline{10}$, 105 (1969).

(25) Kern, E. L., Hamill, D. W., Deem, H. W., and Sheets, H. D., Mat. Res. Bull.,
       $\underline{4}$, S25 (1969).

(26) California. Univ., Livermore. Lawrence Radiation Lab. Thermal Expansion
       of High Temperature Materials, by Krikorian, O. H. Contract W-7405-
       eng-48. Sept. 6, 1960. 7 p.

(27) Lely, J. A. and Kroeger, F. A., In: Semiconductors and Phosphors. Halbleiter
       und Phosphore. Proc. of Internat. Colloquium-Partenkirchen. (Ed. by
       Schoen, M. and Welker, H.) N. Y., Intersci. Pub., Inc., 1958. p. 525.

(28) Stanford Res. Inst., Beta-Silicon Carbide and its Potential for Devices.
       FR no. 21. By Nelson, W. E., et al. Contract no. NObrs-87235.
       Dec. 31, 1963. AD-428 005.

(29) Stanford Res. Inst., Menlo Park, Calif. Growth and Characterization of
       Beta-Silicon Carbide Single Crystals. Final Rept., May 16, 1964-
       Jan. 15, 1967. By Nelson, W. E., et al. Contract no. AF 19-628-4190.
       Feb. 1967. 80 p.

(30) Shaffer, P. T. B. and Jun, C. K., Mat. Res. Bull., $\underline{7}$, 63 (1972).

(31) Patrick, L., et al., Phys. Rev., $\underline{143}$, 526 (1966).

(32) Philipp, H. R., and Taft, E. A. In Silicon Carbide, A High Temperature
       Semiconductor, (ed. by O'Connor, J. R. and Smiltens, J.) N. Y.,
       Pergamon Press, 1960. p. 366.

(33) Engineering Sciences Lab., Dover, N. J. Research on Optical Properties
       of Single Crystals of Beta Phase Silicon Carbide. Summary Tech.
       Rept. no. OR8394, May 5, 1965-June 5, 1966. By Pickar, P. B., et al.
       Contract no. DA-28-017-AMC-2002, A, Oct. 1966. 85 p. AD-641 198.

(34) Scace, R. I. and Slack, G. A. In Silicon Carbide, A High Temperature Semi-
       conductor, (ed. by O'Connor and Smiltens) N. Y., Pergamon Press, 1960.
       p. 24.

(35) Shaffer, P. T. B., J. Amer. Ceram. Soc., $\underline{47}$, 466 (1964).

(36) Shaffer, P. T. B., J. Amer. Ceram. Soc., $\underline{48}$, 601 (1965).

(37) Plenum Press Handbook of High Temperature Materials - No. 1. Materials
          Index by Shaffer, P. T. B., Plenum Press, 1964. p. 107.

(38) Stanford Res. Inst., Menlo Park, Calif. Development of Manufacturing Methods
          for Growing Large beta-Silicon Carbide Single Crystals. IR-8-222 VI,
          by Silva, W. J., et al. Contract F336 15-62-C-1328. Dec. 1967. p. 52.

(39) Slack, G. A. J. of Appl. Phys., 35, 3460 (1964).

(40) International Committee on Silicon Carbide, private communication.

(41) Taylor, A. and Jones, R. M. Silicon Carbide, A High Temperature Semicon-
          ductor, (ed. by O'Connor, J. R. and Smiltens, J.) N. Y., Pergamon
          Press, 1960. p. 147.

(42) Thibault, N. W. Amer. Miner., 29, 327 (1944).

(43) Van Daal, H. J., et al. J. Phys. Chem. Solids, 24, 109 (1963).

(44) Philips Res. Repts. Mobility of Charge Carriers in Silicon Carbide. Rept.
          Suppl. no. 3. By van Daal, H. J. 1965. N65-30608.

(45) Wyckoff, R. W. G. "Crystal Structures". Wiley & Sons, N. Y., v. 1, p. 114.

(46) Zanmarchi, G. Int. Conf. on Semiconductor Phys., Proc., 7th, Paris, 1964.
          v. 1. (ed. by Hulin, M.) N. Y., Acad, Press, 1964. p. 57.

(47) Ziomek, J. S. and Pickar, P. B. Phys. Status Solidi, 21, 271 (1967).

(48) Patrick, L., and Choyke, W. J., Phys. Rev., B2, 2255 (1970).

(49) Shaffer, P. T. B., The Microscope, 18, 179 (1970).

(50) Feldman, D. W., Parker, J. H. Jr., Choyke, W. J., and Patrick, L., Phys. Rev.
          173, 787 (1968).

(51) Verma, A. R., and Krishna, P., "Polymorphism and Polytypism in Crystals",
          John Wiley & Sons, New York, 1966.

List of Participants

ADDAMIANO, Arrigo, 4222 Robertson Boulevard, Alexandria, Virginia   22309

ALLIEGRO, R. A., Norton Company, Worcester, Massachusetts

BATHA, H. D., The Carborundum Company, Niagara Falls, New York

BERMAN, H. S., Westinghouse Astronuclear Laboratory, Pittsburgh, Pennsylvania

BERMAN, I., Air Force Cambridge Research Laboratories (AFSC), Solid State Sciences
    Laboratory, L. G. Hanscom Field, Bedford, Massachusetts   01730

BLANK, J. M., General Electric Company, Nela Park, Cleveland, Ohio   44112

BRACK, K., IBM East Fishkill Laboratories, Hopewell Junction, New York   12533

BRADT, R. C., The Pennsylvania State University, University Park, Pennsylvania

BRANDER, R. W., Post Office Research Department, London NW 2, United Kingdom

BROWN, W. S., College of Engineering, University of Utah, Salt Lake City, Utah

BRUCE, J. A., Solid State Sciences Laboratory, Air Force Cambridge Research
    Laboratories (AFSC), L. G. Hanscom Field, Bedford, Massachusetts   01730

BUCHNER, E., Cesiwid Elektrowarme GmbH, 852 Erlangen, Neumuhle 4, West Germany

CALL, R. L., College of Electrical Engineering, University of Arizona, Tucson,
    Arizona

CAMPBELL, A. B., Department of Engineering Physics, McMaster University, Hamilton,
    Ontario, Canada

CAMPBELL, R. B., Westinghouse Astronuclear Laboratory, Pittsburgh, Pennsylvania

CHOYKE, W. J., Westinghouse Research Laboratories, Pittsburgh, Pennsylvania

COMER, J. J., Air Force Cambridge Research Laboratories, Air Force Systems Command,
    Bedford, Massachusetts   01730

COPPOLA, J. A., Carborundum Company, Niagara Falls, New York

CRANE, R. L., Metals and Ceramics Division, U. S. Air Force Materials Laboratory,
    Wright-Patterson Air Force Base, Ohio   45433

CUOMO, J. J., IBM Thomas J. Watson Research Center, Yorktown Heights, New York

DAVIES, D. Eirug, Air Force Cambridge Research Laboratories, Air Force Systems
    Command, Bedford, Massachusetts 01730

DAVIES, J. A., Atomic Energy of Canada Limited, Chalk River Nuclear Laboratories,
    Chalk River, Ontario, Canada

DeBOLT, Harold E., Avco Systems Division, Lowell, Massachusetts   01851

DIEFENDORF, R. J., Rensselaer Polytechnic Institute, Materials Division, Troy,
    New York   12181

DUBEY, M., Department of Physics, Banaras Hindu University, Varanasi, India

DUBROVSKII, G. B., A. F. Ioffe Physico-technical Institute, Academy of Sciences of
    the USSR, Leningrad, USSR

EBI, R., Institut fur Chemische Technik der Universitat Karlsruhe, 75 Karlsruhe,
    West Germany

FAGEN, E. A., Energy Converison Devices, Incorporated, Troy, Michigan   48084

FAUST, Jr., J. W., College of Engineering, University of South Carolina, Columbia,
    South Carolina   29208

FELDMAN, J. M., Northeastern University, Boston, Massachusetts   02115

FERGUSON, I. F., Materials Science Group, Reactor Fuel Element Laboratories,
    UKAEA, Springfields, Salwick, Preston, Lancashire, United Kingdom

FITZER, E., Institut fur Chemische Technik, der Universitat Karlsruhe, 75 Karlsruhe,
    Kaiserstrasse 12, 7500 West Germany

FONG, C. Y., Department of Physics, University of California, Davis, California

GATOS, H. C., Department of Metallurgy and Materials Science, Massachusetts
    Institute of Technology, Cambridge, Massachusetts   02139

GEICZY, I. I., Institute of Semiconductor Physics, Siberian Branch of the Academy
    of Sciences, Novosibirsk, USSR

GILLESSEN, K., Institut A fuer Werkstoffkunde, Technische Universitaet, Hannover,
    West Germany

GRAUL, J., Siemens AG, Balanstrasse, Munich, West Germany

HARDY, L. H., The Carborundum Comapny, Niagara Falls, New York

HARRIS, R. C. A., Hughes Aircraft Company, Tucson, Arizona

HARTLINE, S. D., The Pennsylvania State University, University Park, Pennsylvania

HEMSTREET, L. A., Naval Ordnance Laboratory, Silver Spring, Maryland   20910

HENG, T. M., Westinghouse Research and Development Center, Pittsburgh, Pennsylvania

HILBORN, Jr., R. B., University of South Carolina, College of Engineering,
    Columbia, South Carolina   29208

HOLLENBERG, G. W., Metals and Ceramics Division, U. S. Air Force Materials
    Laboratory, Wright-Patterson Air Force Base, Ohio   45433

HUBATSCHEK, R. M., Institut fur chemische Technologie anorganischer Stoffe,
    Technical University, Vienna, Austria

INOMATA, Y., National Institute for Researches in Inorganic Materials, Kurakake,
    Sakura-mura, Niihari-gun, Ibaraki-ken, 300-31, Japan

INOUE, Z., National Institute for Researches in Inorganic Materials, Kurakake,
    Sakura-mura, Niihari-gun, Ibaraki-ken, 300-31, Japan

ISEKI, Takayoshi, Tokyo Institute of Technology, Research Laboratory of Nuclear
    Reactors, Ookayama, Meguro-ku, Tokyo 152, Japan

ISHIWATA, M., Department of Physics, Saitama University, Urawa, Japan

KANG, H., University of South Carolina, College of Engineering, Columbia, South
    Carolina   29208

KASPRZYK, M. R., The Carborundum Company, P. O. Box 337, Niagara Falls, New York

KEHR, D., Institut fur Chemische Technik, 12 Kaiserstrasse, Karlsruhe, 7500 West
    Germany

KEMENADE, v., A. W. C., Philips Research Laboratories, Eindhoven, The Netherlands

KENNEDY, P., Ceramics Group, Reactor Fuel Element Laboratories, UKAEA, Salwick,
    Preston, England

KEYES, Robert W., IBM Thomas J. Watson Research Center, Yorktown Heights, New York

KHOLUYANOV, G. F., A. F. Ioffe Physico-Technical Institute of the Academy of
    Sciences of the USSR, Leningrad, USSR

KIEFFER, A. R., Institut fur chemische Technologie anorganischer Stoffe, Technical
    University, Vienna, Austria

KNIPPENBERG, W. F., Philips Research Laboratories, Eindhoven, The Netherlands

KOMATSU, Hiroshi, National Institute for Researches in Inorganic Materials, Kurakake,
    Sakura-mura, Niihari-gun, Ibarake-ken, 300-31, Japan

KRISHNA, P., Physics Department, Banaras Hindu University, Varanasi, India

KRUKONIS, Val J., Avco Systems Division, Lowell, Massachusetts   01851

KUWABARA, Hiroshi, Research Institute of Electronics, Shizuoka University, Johoku
    3-5-1, Hamamatsu 432, Japan

LEAMY, H. J., Bell Laboratories, Murray Hill, New Jersey   07974

LIAW, H. M., Motorola Incorporated, Semiconductor Products Division, 5005
    E. McDowell Road, Phoenix, Arizona   85005

LITTLER, J., Air Force Cambridge Research Laboratories (AFSC), Solid State Sciences
    Laboratory, L. G. Hanscom Field, Bedford, Massachusetts   01730

LOMAKINA, G. A., A. F. Ioffe Physico-Technical Institute of the Academy of Sciences
    of the USSR, Leningrad, USSR

MARSH, O. J., Hughes Research Laboratories, Malibu, California   90265

MARSHALL, R. C., Air Force Cambridge Research Laboratories (AFSC), Solid State
    Sciences Laboratory, L. G. Hanscom Field, Bedford, Massachusetts   01730

MATSUNAMI, Hiroyuki, Department of Electronics, Faculty of Engineering, Kyoto
    University, Kyoto, Japan

McMURTRY, C. H., The Carborundum Company, P. O. Box 337, Niagara Falls, New York

MILEWSKI, J. V., Esso Research and Engineering Company, P. O. Box 8, Linden,
    New Jersey   07036

MITCHELL, J. B., Department of Engineering Physics, McMaster University, Hamilton,
    Ontario, Canada

MOGAB, C. J., Bell Laboratories, Murray Hill, New Jersey   07974

MOKHOV, E. N., A. F. Ioffe Physico-Technical Institute of the Academy of Sciences
    of the USSR, Leningrad, USSR

MUENCH, v. W., Institut A fuer Werkstoffkunde, Technische Universitaet, Hannover,
    West Germany

NATHENSON, H. C., Westinghouse Research and Development Center, Pittsburgh,
    Pennsylvania   15235

NAUM, R. G., The Carborundum Company, P. O. Box 337, Niagara Falls, New York   14302

NESTEROV, A. A., Institute of Semiconductor Physics, Siberian Branch of the Academy
    of Sciences, Novosibirsk, USSR

NOAKES, J. E., Scientific Research Staff, Ford Motor Company, Dearborn, Michigan

OLSON, B. A., Norton Company, Worcester, Massachusetts

OTA, M., National Institute for Researches in Inorganic Materials, Kurakake,
    Sakura-mura, Niihari-gun, Ibaraki-ken, 300-31, Japan

PANDEY, Dhananjai, Physics Department, Banaras Hindu University, Varanasi, India

PATRICK, Lyle, Westinghouse Research Laboratories, Pittsburgh, Pennsylvania 15235

PELLEGRINI, P. W., Air Force Cambridge Research Laboratories (AFSC), L. G. Hanscom
    Field, Bedford, Massachusetts   01730

POSEN, H., Solid State Sciences Laboratory, Air Force Cambridge Research Labora-
    tories (AFSC), L. G. Hanscom Field, Bedford, Massachusetts   01730

PROCHAZKA, Svante, General Electric Research and Development Center, P. O. Box 8,
    Schenectady, New York   12301

RAM, U. S., Department of Physics, Banaras Hindu University, Varanasi, India

RANDON, J. L., Centre des Materiaux, Ecole des Mines, B.P, 114, 91102 Corbeil,
    France

RUBISCH, O., Sigri Elektrographit GmbH, 8901 Meitingen, West Germany

RUTZ, R. F., IBM Thomas J. Watson Research Center, Yorktown Heights, New York

RYAN, C. E., Air Force Cambridge Research Laboratories (AFSC), Solid State Sciences
    Laboratory, L. G. Hanscom Field, Bedford, Massachusetts   01730

SAHEBKAR, M., Institute fur Chemische Technik der Universitat Karlsruhe,
    West Germany

SANDSTROM, J. L., Hulco Incorporated, Salt Lake City, Utah

SATO, H., Scientific Research Staff, Ford Motor Company, Dearborn, Michigan 38121

SCHWUTTKE, G. H., IBM East Fishkill Laboratories, Hopewell Junction, New York

SHAFFER, Peter T. B., The Carborundum Company, Niagara Falls, New York

SHENNAN, J. V., Ceramics Group, Reactor Fuel Element Laboratories, UKAEA, Salwick,
    Preston, England

SHEWCHUN, J., Department of Engineering Physics, McMaster University, Hamilton,
    Ontario, Canada

SHINOZAKI, S., Scientific Research Staff, Ford Motor Company, Dearborn, Michigan

SINGH, G., Department of Physics, Banaras Hindu University, Varanasi, India

SLACK, Glen, General Electric Research and Development Center, Schenectady,
    New York

SLAMA, G., Centre des Materiaux, Ecole des Miens, B.P. 114, 91102 Corbeil, France

SUZUKI, Akira, Department of Electronics, Faculty of Engineering, Kyoto University,
    Kyoto, Japan

SUZUKI, Hiroshige, Tokyo Institute of Technology, Research Laboratory of Nuclear
    Reactors, Ookayama, Meguro-ku, Tokyo 152, Japan

TAIROV, Yu. M., V. F. Ulyanov (Lenin) Leningrad Electrical Engineering Institute,
    Prof. Popov Street 5, Leningrad, USSR

TANAKA, H., National Institute for Researches in Inorganic Materials, Kurakake,
    Sakura-mura, Niihari-gun, Ibaraki-ken, 300-31, Japan

TANAKA, Tetsuro, Department of Electronics, Faculty of Engineering, Kyoto Univer-
    sity, Kyoto, Japan

THOMPSON, D. A., Department of Engineering Physics, McMaster University, Hamilton,
    Ontario, Canada

TOMITA, T., Department of Physics, Saitama University, Urawa, Japan

TSVETKOV, V. F., V. F. Ulyanov (Lenin) Leningrad Electrical Engineering Institute,
    Prof. Popov Street 5, Leningrad, USSR

TUNG, Y, University of South Carolina, College of Engineering, Columbia,
    South Carolina   29208

VELDKAMP, J. D. B., Philips Research Laboratories, Eindhoven, The Netherlands

VENDL, A. F., Institut fur chemische Technologie anorganischer Stoffe, Technical
    University, Vienna, Austria

VERSPUI, G., Philips Research Laboratories, Eindhoven, The Netherlands

VIGNES, A., Centre des Materiaux, Ecole des Mines, B.P. 114, 91102 Corbeil, France

VIOLIN, E. E., V. F. Ulyanov (Lenin) Leningrad Electrical Engineering Institute,
    Prof. Popov Street 5, Leningrad, USSR

VODAKOV, Yu. A., A. F. Ioffe Physico-Technical Institute of the Academy of Sciences
    of the USSR, Leningrad, USSR

WAGNER, E., Institut fur Technische Elektronik, Technische Universitat Munchen,
    Munich, West Germany

WALKER, D. E. Yeoman, Materials Science Group, RFL, UKAEA, Springfields, Preston,
    Lancs, United Kingdom

WAWNER, Jr., Franklin E., Department of Materials Science, University of Virginia,
    Charlottesville, Virginia   22901

WEAVER, G. Q., Norton Company, Worcester, Massachusetts

WEISS, J. R., Rensselaer Polytechnic Institute, Materials Division, Troy, New York

WESSELS, B., General Electric Corporate Research and Development Center, Schenectady
    New York   12345

WITT, A. F., Department of Metallurgy and Materials Science, Massachusetts Institute of Technology, Cambridge, Massachusetts   02139

WRUSS, W. H., Institute fur chemische Technologie anorganischer Stoffe, Technical University, Vienna, Austria

WU, S. Y., Westinghouse Research Laboratories, Pittsburgh, Pennsylvania   15235

YAMADA, Shoji, Research Institute of Electronics, Shizuoka University, Johoku 3-5-1, Hamamatsu 432, Japan

YESSIK, M., Scientific Research Staff, Ford Motor Company, Dearborn, Michigan

PART VII

INDEXES

Author Index

Subject Index